Comprehensive Treatise of Electrochemistry

Volume 10
Bioelectrochemistry

COMPREHENSIVE TREATISE OF ELECTROCHEMISTRY

Comprehensive Treatise of Electrochemistry

Volume 10
Bioelectrochemistry

Edited by

Supramaniam Srinivasan

Los Alamos National Laboratory
Los Alamos, New Mexico

Yu. A. Chizmadzhev

Institute of Electrochemistry of the
  Academy of Sciences of the USSR
Moscow, USSR

J. O'M. Bockris

Texas A&M University
College Station, Texas

Brian E. Conway

University of Ottawa
Ottawa, Ontario, Canada

Ernest Yeager

Case Western Reserve University
Cleveland, Ohio

PLENUM PRESS ▪ NEW YORK AND LONDON

Library of Congress Cataloging in Publication Data

Main entry under title:

Bioelectrochemistry.

(Comprehensive treatise of electrochemistry; v. 10)
Includes bibliographies and index.
1. Bioelectrochemistry. I. Srinivasan, S. (Supramaniam), 1932–   . II. Series.
QD552.C64 vol. 10   541.3′7 s [574.19′283]   84-24781
[QP517.B53]

ISBN-13: 978-1-4612-9444-3   e-ISBN-13: 978-1-4613-2359-4
DOI: 10.1007/978-1-4613-2359-4

Contributors

H. Berg • Academie der Wissenschaften der DDR, Zentral Institut fur Mikrobiologie und Experimentelle Therapie DDR. 69 Iena Beutenbergstrasse 11, German Democratic Republic

Michael N. Berry • Department of Clinical Biochemistry, Flinders Medical Center, Bedford Park, South Australia 5042, Australia

Henry N. Blount • Center for Catalytic Science and Technology, and Brown Chemical Laboratory, The University of Delaware, Newark, DE 19711

Edmond F. Bowden • Dept. of Chemistry, Virginia Commonwealth University, Richmond, VA 23284

Yu. A. Chizmadzhev • Institute of Electrochemistry, Academy of Sciences of the USSR, 117071 Moscow, V-71, Leninsky Prospekt 31, U.S.S.R.

James C. Conti • Ethicon Inc. (Div. J & J), Rt. 22, Somerville, NJ 08876

Glenn Dryhurst • Department of Chemistry, University of Oklahoma, Norman, OK 73019

Eugene Findl • Department of Energy and Environment, Building 801, Brookhaven National Laboratory, Upton, NY 11973

Anthony R. Grivell • Dept. of Clinical Biochemistry, School of Medicine, Flinders University of South Australia, Bedford Park, 5042, South Australia, Australia

Fred M. Hawkridge • Department of Chemistry, College of Humanities and Sciences, 1101 West Main Street, Richmond, VA 23284

Shinpei Okhi • Department of Biophysical Sciences, SUNY at Buffalo, 118 Cary Hall, Buffalo, NY 14214

V. F. Pastushenko • Academy of Sciences of the U.S.S.R., Institute of Electrochemistry, 31 Leninsky Prospekt, Moscow V-71, U.S.S.R.

Arthur A. Pilla • Department of Orthopedics, Mount Sinai Hospital, Gustav Levy Place, Room 1702, Annenberg Building, New York, NY 10029

Supramaniam Srinivasan • Los Alamos National Laboratory, P.O. Box 1663, MS D429, Los Alamos, NM 87545

Elaine R. Strope • Personal Products Co. (Div. J & J), Van Liew Ave., Milltown, NJ 08850

M. R. Tarasevich • Institute of Electrochemistry, Academy of Sciences of the U.S.S.R., 117071 Moscow, V-71, Leninsky Prospekt, U.S.S.R

Patricia G. Wallace • Dept. of Clinical Biochemistry, School of Medicine, Flinders University of South Australia, Bedford Park, 5042, South Australia, Australia

Preface to Comprehensive Treatise of Electrochemistry

Electrochemistry is one of the oldest defined areas in physical science, and there was a time, less than 50 years ago, when one saw "Institute of Electrochemistry and Physical Chemistry" in the chemistry buildings of European universities. But after early brilliant developments in electrode processes at the beginning of the twentieth century and in solution chemistry during the 1930s, electrochemistry fell into a period of decline which lasted for several decades. Electrochemical systems were too complex for the theoretical concepts of the quantum theory. They were too little understood at a phenomenological level to allow the ubiquity in application in so many fields to be comprehended.

However, a new growth began faintly in the late 1940s, and clearly in the 1950s. This growth was exemplified by the formation in 1949 of what is now called The International Society for Electrochemistry. The usefulness of electrochemistry as a basis for understanding conservation was the focal point in the founding of this Society. Another very important event was the choice by NASA in 1958 of fuel cells to provide the auxiliary power for space vehicles.

With the new era of diminishing usefulness of the fossil fuels upon us, the role of electrochemical technology is widened (energy storage, conversion, enhanced attention to conservation, direct use of electricity from nuclear–solar plants, finding materials which interface well with hydrogen). This strong new interest is not only in the technological applications of electrochemistry. Quantum chemists have taken an interest in redox processes. Organic chemists are interested in situations where the energy of electrons is as easily controlled as it is at electrodes. Some biological processes are now seen in electrodic terms, with electron transfer to and from materials which would earlier have been considered to be insulators.

It is now time for a comprehensive treatise to look at the whole field of electrochemistry.

The present treatise was conceived in 1974, and the earliest invitations to authors for contributions were made in 1975. The completion of the early volumes has been delayed by various factors.

There has been no attempt to make each article emphasize the most recent situation at the expense of an overall statement of the modern view. This treatise is not a collection of articles from *Recent Advances in Electrochemistry* or *Modern Aspects of Electrochemistry*. It is an attempt at making a mature statement about the present position in the vast area of what is best looked at as a new interdisciplinary field.

Texas A&M University	John O'M. Bockris
University of Ottawa	Brian E. Conway
Case Western Reserve University	Ernest B. Yeager
Texas A&M University	Ralph E. White

Preface to Volume 10

The Comprehensive Treatise of Electrochemistry is a presentation of the various frontiers of electrochemical knowledge; however, there is no doubt that of all the areas of electrochemistry, the electrochemical application to biology is the one in which the frontier seems to be the longest and most exciting.

The seminal contribution to this area was the suggestion by Albert Szent-Gyorgyi in 1949 that electronic conductivity could occur in proteins. Prior to this time the idea that in biological ("non-conducting") substances electrons could travel, transfer, and even form pd's at interfaces would have been inconceivable. The linking of the high efficiency of biological energy conversion with a fuel cell concept of biological cells, and the experimental establishment of electronic conductivity in proteins in 1969 have opened areas of great significance in terms of molecular biology. Indeed, at present (1985) some of these considerations seem to reach out toward aspects of the cancer problem.

There is much else in bioelectrochemistry apart from these modern developments; in particular, there is the vast area associated with the names of Hodgkin and Huxley and the subject of electrophysiology, seen as an application of the Nernst–Planck equation.

The application of electrical concepts to biological phenomena is surely one of the more exciting in all science and the tenth volume of the Comprehensive Treatise of Electrochemistry is the most detailed description yet made of that field.

Institute of Hydrogen Systems   S. Srinivasan
Academy of Sciences of the U.S.S.R.   Yu. A. Chizmadzhev
Texas A & M University   J. O'M. Bockris
University of Ottawa   B. E. Conway
Case Western Reserve University   E. Yeager

Contents

3. Electrochemistry of Biopolymers

H. Berg

4. Bioelectrocatalysis

M. R. Tarasevich

5. Electrochemical Aspects of Bioenergetics

Edmond F. Bowden, Fred M. Hawkridge, and Henry N. Blount, III

6. Electrochemical Aspects of Metabolism

Michael N. Berry, Anthony R. Grivell, and Patricia G. Wallace

7. Electrochemistry of the Nervous Impulse

Yu. A. Chizmadzhev and V. F. Pastushenko

8. Electrochemical Approach for the Solution of Cardiovascular Problems

Supramaniam Srinivasan

9. Electrochemical Techniques in the Biological Sciences

Eugene Findl, Elaine R. Strope, and James C. Conti

1

The Origin of Electrical Potential in Biological Systems

S. OHKI

Electrical potential of biological systems orginates from various sources such as the existence of free ions in the biological systems, ionized molecular groups or electrical polarization of biomolecules, and various forms of assembly of biomolecules possessing such ionized molecular groups or electrically polarized or polarizable molecular groups. Another source of electrical potential may come from the occurrence of electron-transfer reactions in certain biochemical reaction systems. In certain biological systems, chemoelectrical, mechanoelectrical, and thermoelectrical coupling processes may produce a special type of electrical potential. Nonpolar molecules may exert electrical forces on each other by way of instantaneously fluctuating electrical polarization within the molecules.

Here, we would like to review systematically the electrical potentials which originate from, first, individual molecular events, then, molecular assemblies as biointerfaces, followed by discussion of the potential differences across biological membranes, and lastly, interaction forces acting between two cells.

S. OHKI • Department of Biophysical Sciences, School of Medicine, State University of New York at Buffalo, Buffalo, New York 14214.

1. Electrical Potential of Biomolecules

1.1. Electrostatic Potential

1.1.1. Electrostatic Potential Due to Electric Charges

Let us consider an electrical potential at a point in a space where electric charges are distributed as point charges in a uniform dielectric medium. The electrical potential at a point in the space is expressed by:

$$\psi = \sum_{j=1} \frac{q_j}{\varepsilon r_j} \quad \text{(C.G.S. unit)} \tag{1}$$

where q_j is the jth point charge, r_j the distance between the jth point charge and the point being considered, and ε the dielectric constant of the medium.

In an aqueous solution containing positive and negative ions, however, the electroneutrality condition must prevail and the average of the total charge over any large region of such an electrolyte solution is zero:

$$\rho = \sum_j q_j = \sum_{-j} e Z_j n_j = 0 \tag{2}$$

where e is the electronic unit charge, and Z_j and n_j are the valency and the number density of the jth ionic species, respectively. Therefore, the average electric potential over any large region of such an electrolyte solution is constant.

In a local region, however, taken around a particular ion as center, there is an unequal distribution of the surrounding ions, owing to electrostatic forces. A simple electrostatic example is the situation where the space charge, ρ, is distributed with spherical symmetry about a center, so that ρ is a function of r (distance from the center ion) only.

Now, consider a region where a spherical ion of radius b and charge Ze is at its center, and this charge Ze is considered to be distributed uniformly over the surface of the ion. The electrical potential function of the central ion can be designated ψ. It may be expressed as:

$$\psi = \psi_K + \psi_i \tag{3}$$

where ψ_K is the potential due directly to the central ion K under consideration and ψ_i is the potential due to the ion atmosphere.[1] Then, the concentration of the jth species near the central ion is given by the Boltzmann distribution equation:

$$n_j = n_j^0 \, e^{-Z_j e \psi / kT} \tag{4}$$

and the total charge per unit volume will be:

$$\rho = \sum_{j=1} n_j Z_j e = e \sum_{j=1} Z_j n_j^0 \, e^{-Z_j e \psi / kT} \tag{5}$$

where n_j^0 is the number density of the jth species at an infinite distance from the central ion K.

Now, the Poisson equation relates the potential function ψ to the charge density:

$$\nabla^2 \psi = -\frac{4\pi}{\varepsilon}\rho \tag{6}$$

where the symbol ∇^2 is equivalent to the Laplacian operator, Δ.

From Eqs. (5) and (6), we obtain:

$$\nabla^2 \psi = -\frac{4\pi e}{\varepsilon} \sum_{j=1} Z_j n_j^0 \, e^{-Z_j e\psi/kT} \tag{7}$$

This equation is known as the Poisson–Boltzmann equation. By solving this equation, we may know the potential distribution and the ion distribution around the ion concerned. The Debye–Hückel solution is based on the approximation of the exponential in Eq. (7) by taking the leading terms (the first two terms) in the series expansion and applying the electroneutrality condition for bulk ionic solution ($\sum_{j=1} Z_j n_j^0 = 0$):

$$\nabla^2 \psi = \varkappa^2 \psi \tag{8}$$

where $\varkappa^2 = (4\pi e^2/\varepsilon kT) \sum_{j=1} Z_j^2 n_j^0$. Thus, the Debye–Hückel equation [Eq. (8)] is a good approximation for the Poisson–Boltzmann equation [Eq. (7)] when $e\psi/kT \ll 1$. Since the surface potentials of many biological membranes have magnitudes comparable to $kT/e = 25.6\,\text{mV}$ (25°C), the Debye–Hückel approximation may not necessarily be a good expression for the analysis of the surface electrical phenomena in such systems. The quantity $\varkappa$ is called the Debye constant and its inverse, $1/\varkappa$, corresponds to the effective thickness of the ionic atmosphere around the ion concerned.

In order to determine ψ_i, the potential at the surface of an ion due to the presence of the surrounding ion atmosphere, we consider the following model of the system. We treat the ions as conducting spheres, and assume that the distance of closest approach of any two ions is a (see Figure 1). For distance

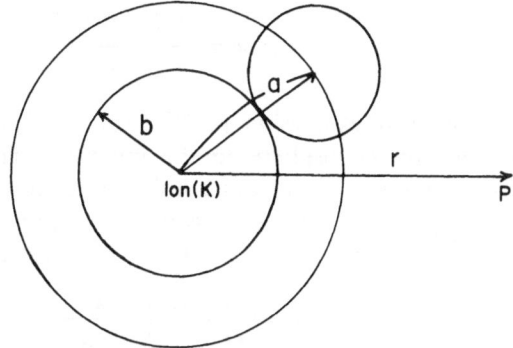

Figure 1. The central ion K of radius b and the collision diameter a which represents the distance of closest approach between the centers of two ions.

greater than a, we assume that the forces between ions are purely electrostatic. On the average the neighboring ions will be distributed around the central ion K with spherical symmetry. It is assumed that the charge of each surrounding ion, as if it were a point charge, is located at the center of each ion. Then, in the region between $r = b$ and $r = a$, there is no charge. In the region of $r > a$, we treat the medium as continuous; a space charge at any point is expressed as that in its corresponding volume element of the body having a continuous space charge determined by the concentrations of the various ions, and the dielectric constant ε in the region $r > a$ is constant.

In the region between $r = b$ and $r = a$, since there is no charge, ψ must be a solution of Laplace's equation:

$$\nabla^2 \psi = 0 \tag{9}$$

After solution of the equation with the aid of Gauss's law, ψ is expressed by:

$$\psi = \frac{Ze}{\varepsilon r} + A \qquad (b < r < a) \tag{10}$$

where A is a constant.

In the region $(r > a)$ where Poisson's equation holds [Eq. (7)], the Debye–Hückel approximation [Eq. (8)] yields:

$$\psi = \frac{Be^{-\kappa r}}{r} \qquad (r > a) \tag{11}$$

Since the electric intensity, $\partial\psi/\partial r$, and the potential ψ are continuous functions of r throughout the region $(r > b)$, at the surface $r = a$ we can equate the values of ψ and $\partial\psi/\partial r$ given by Eqs. (10) and (11).

$$\frac{Ze}{\varepsilon a} + A = \frac{Be^{-\kappa a}}{a}$$
$$B = \frac{Ze\, e^{\kappa a}}{\varepsilon(1 + \kappa a)} \tag{12}$$

In the outer region $(r > a)$, the electrical potential is

$$\psi = \frac{Ze}{\varepsilon r}\left[\frac{e^{-\kappa(r-a)}}{1 + a\kappa}\right] \qquad (r > a) \tag{13}$$

In the above expression, $[e^{-\kappa(r-a)}/(1 + a\kappa)]$ corresponds to the contribution of the ion atmosphere to the potential. This can be compared with the expression $\psi = Ze/\varepsilon r$ which holds in the absence of the ion atmosphere. The actual potential is always less than the latter.

In the inner region $b < r < a$, the electrical potential is given by:

$$\psi = \frac{Ze}{\varepsilon r} - \frac{Ze}{\varepsilon}\left(\frac{\varkappa}{1 + \kappa a}\right) \qquad (b < r < a) \tag{14}$$

The first term on the right is due to the potential of the central ion ψ_K, and the second term corresponds to ψ_i in Eq. (3), which is the potential due to the surrounding ion atmosphere.

1.1.2. Free Energy of a Charged Sphere

In the previous section, we discussed the electrical potential of a central ion which has a spherical shape. The free energy of such a charged sphere due to electrical potential, W_K, is expressed by the work of charging the sphere from zero to q, which is the final charge on the sphere of the central ion, while all the other ions have their full charges; so κ is a constant during the charging process,[2]

$$W_K = \int_{\lambda=0}^{\lambda=1} \lambda \psi \, d(\lambda q) = \psi_K q_K / 2 \tag{15}$$

Thus, the electrical free energy of each sphere is equal to half its charge multiplied by its potential.

The free energy associated with the central ionic charge of Ze spread uniformly over the sphere surface b is given by:

$$W_K = \tfrac{1}{2} q \psi |_{r=b} = \tfrac{1}{2} \frac{(Ze)}{\varepsilon} \frac{(Ze)}{b} \tag{16}$$

since ψ at $r = b$ is $\lambda q / \varepsilon b$. This energy is called the "self energy of the sphere."

The electrical energy of the central ion due to the presence of the surrounding ions can be given by:

$$W_i = -\int_0^1 \frac{\lambda Ze}{\varepsilon} \frac{\varkappa}{1 + \kappa a} (Ze d\lambda) = -\frac{Z^2 e^2}{2\varepsilon} \left(\frac{\varkappa}{1 + \kappa a} \right) \tag{17}$$

Therefore, the total electrostatic free energy of a central ion of radius b is given from Eqs. (16) and (17):

$$W_{\text{tot}} = W_K + W_i = \frac{Z^2 e^2}{2\varepsilon} \left(\frac{1}{b} - \frac{\varkappa}{1 + \kappa a} \right) \tag{18}$$

The work of the electrical energy of the central ion K due to the presence of the surrounding ions, W_i, which corresponds to the change in free energy per ion due to a finite concentration, is related to the activity coefficient, f_i, of the electrolyte under consideration:

$$W_i = kT \ln f_i \tag{19}$$

The activity, a, of the jth ionic species is expressed as the product of the activity coefficient f_j and concentration C_j.

$$a_j = f_j C_j \tag{20}$$

For ion–ion interaction in the solution, the logarithm of activity coefficient, log f, is proportional to the square root of the ionic strength $\omega \equiv \frac{1}{2}\sum C_i Z_i^2 = 1000/2N \sum_j n_j Z_j^2$ (N: Avogadro's number). This is clear from Eqs. (8), (17) and (19).

1.1.3. Ionization of Biomolecules

Biological molecules include some inorganic and organic molecules, lipids and proteins.

Most biological molecules in aqueous solution have dissociable groups which either release H^+ or other positive or negative ions, and these dissociated groups themselves become ionized forms depending on the H^+ concentrations

Table 1
Chemical Structure of Various Amino Acids and Lipids

$$\text{Amino acids} \quad R \!\!-\!\! \underset{\overset{|}{\underset{+}{NH_3}}}{\overset{\overset{|}{H}}{C}} \!\!-\!\! COO^-$$

Amino acids with polar R groups	Amino acids with nonpolar R groups

Table 1 (continued)

Amino acids with polar R groups	Amino acids with nonpolar R groups

Arginine

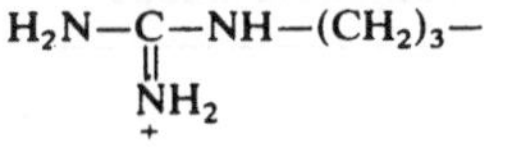

$$H_2N-\underset{\underset{+}{\overset{\|}{NH_2}}}{C}-NH-(CH_2)_3-$$

Histidine

$$HC=\!\!=C-CH_2-$$
$$HN\underset{+}{\overset{}{=}}\underset{\underset{H}{\overset{|}{C}}}{\overset{}{C}}\;NH$$

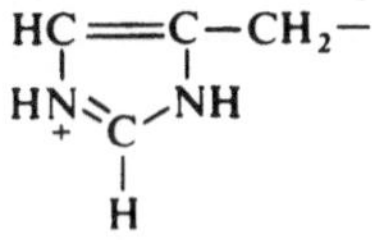

Lipids Phospholipids
Glycolipids
Steroids

Phospholipids

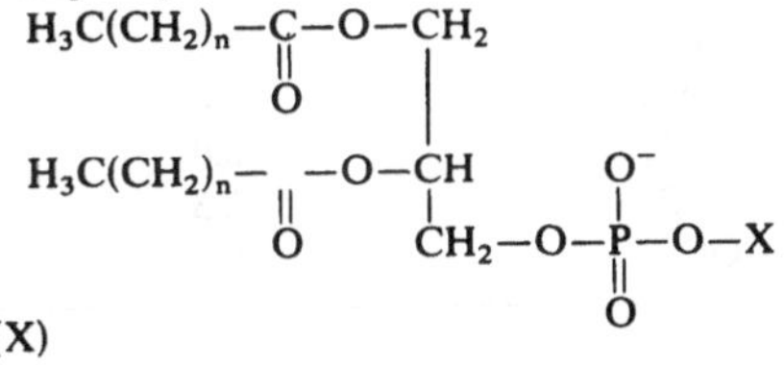

$$H_3C(CH_2)_n-\underset{\overset{\|}{O}}{C}-O-CH_2$$

$$H_3C(CH_2)_n-\underset{\overset{\|}{O}}{}-O-CH \qquad \overset{O^-}{\underset{\overset{\|}{O}}{}}$$
$$CH_2-O-\overset{}{P}-O-X$$

(X)   (at neutral pH)
$CH_2CH_2NH_3^+$ (phosphatidylethanolamine)   Zwitterion
$(CH_2)_2-N^+(CH_3)_2$ (phosphatidylcholine)   Zwitterion
$CH_2-CHNH_3^+COO^-$ (phosphatidylserine)   One net negative charge

Glycolipids

$$H_2C(CH_2)_n-\underset{\overset{\|}{O}}{C}-O-CH_2$$ (at neutral pH)

$$H_3C(CH_2)_n-\underset{\overset{\|}{O}}{C}-O-CH$$
$$CH_2-O$$

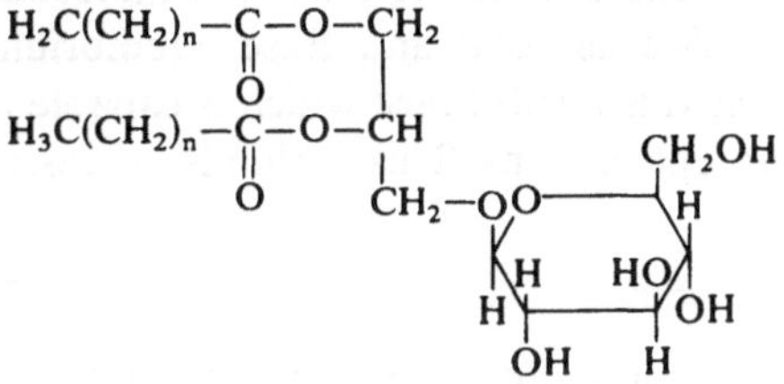

Monogalactosyl
Diacylglycerol
(neutral polar group)

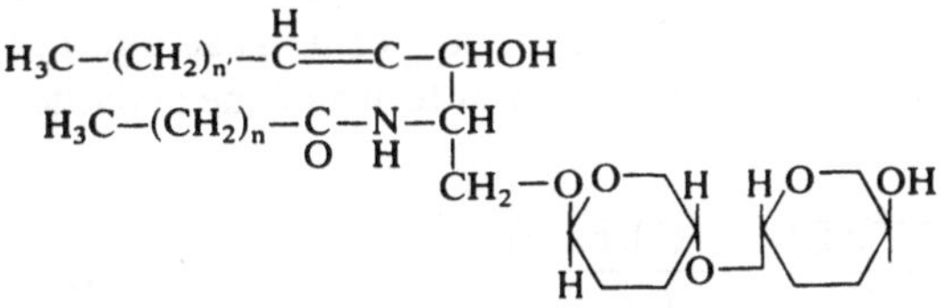

$$H_3C-(CH_2)_{n'}-\overset{\overset{H}{|}}{C}=C-CHOH$$
$$H_3C-(CH_2)_n-\underset{\overset{\|}{O}}{C}-\underset{\overset{|}{H}}{N}-CH$$
$$CH_2-O$$

glycosphingolipids (ceramides)
Gangliosides = ceramides containing sialic acid residues
(dissociable charged group)

N-Acetylneuraminic acid

Sterols
Cholesterol

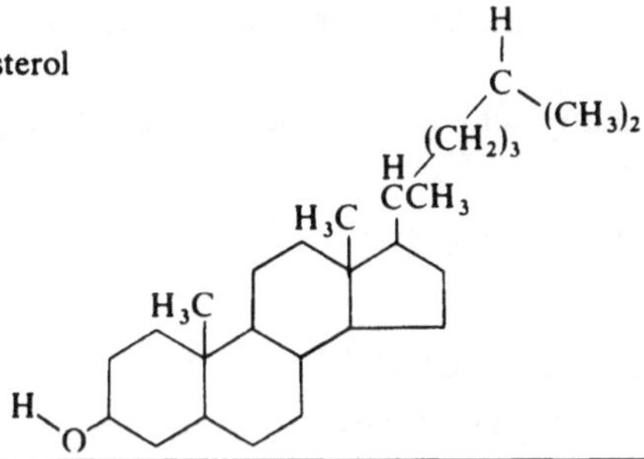

(at neutral pH)
Uncharged polar group

Table 2
Some Acids and Their Conjugate Bases[a]

Acid	Conjugate base	$pK_a \equiv -\log K_a$ in water at 25°[b]
H_3PO_4	$H_2PO_4^-$	2
$H_2PO_4^-$	HPO_4^{--}	7
HPO_4^{--}	PO_4^{---}	12
H_3O^+	H_2O	—
H_2O	OH^-	15.7
CH_3COOH	CH_3COO^-	4.75
NH_4^+	NH_3	9.3
$CH_3NH_3^+$	CH_3NH_2	10.7
$HOOC-CH_2-COOH$	$HOOC-CH_2-COO^-$	2.8
$HOOC-CH_2-COO^-$	$^-OOC-CH_2-COO^-$	5.7
$^+H_3N-(CH_2)_2-NH_3^+$	$^+H_3N-(CH_2)_2-NH_2$	7.0
$^+H_3N-(CH_2)_2-NH_2$	$H_2N-(CH_2)_2-NH_2$	10.0
$^+H_3N-CH_2-COOH$	$^+H_3N-CH_2-COO^-$	2.3
$^+H_3N-CH_2-COO^-$	$H_2N-CH_2-COO^-$	9.7

[a] Reference 1.
[b] The pK_a values shown are an index of the relative strength of the acids shown, in water at 25°; the smaller the pK_a value, the stronger the acid.

as well as on the concentrations of the ions dissociable from the molecular groups (Table 1). This problem is studied as Acid and Base equilibrium. According to Brønsted and Lowry, an acid is a substance which is capable of yielding protons and a base is able to accept protons. This notion is expressed as:

$$A^+ \rightleftarrows H^+ + B \tag{21}$$

where A^+ is an acid and B is its conjugate base. Some examples of acids and their conjugate bases are listed in Table 2. From the law of mass action, the equilibrium constant K_a of this reaction is expressed in terms of the activities of H^+, acid, and base:

$$K_a = \frac{(H^+)(B)}{(A^+)} = \frac{(H^+)(\text{base})}{(\text{acid})} \tag{22}$$

where K_a is the acid dissociation constant, and ( ) represents activity for the corresponding species.

In the case of a base in water, many acid–base equilibria may be expressed in terms of a "base dissociation constant", K_b.

$$B + H_2O \rightleftarrows BH^+ + OH^- \tag{23}$$

K_b is expressed by:

$$K_b = \frac{(\text{conjugate acid})(OH^-)}{(\text{total conjugate base})} \tag{24}$$

On the other hand, this acid–base equilibrium can be expressed in terms of an "acid dissociation constant", K_a:

$$BH^+ \rightleftarrows B + H^+ \tag{25}$$

$$K_a = \frac{(B)(H^+)}{(BH^+)} = \frac{(\text{total conjugate base}) (H^+)}{(\text{conjugate acid})} \tag{25}$$

However, K_b and K_a are related by the equation:

$$K_b = \frac{K_w}{K_a} \quad \text{or} \quad pK_b = pK_w - pK_A \tag{26}$$

where $pK = -\log K$ and $pK_w = -\log K_w$. Here, K_w is the ion product constant for water $[K_w = (H_3O^+)(OH^-)]$ which is 1.0×10^{-14} at 25°C. Therefore, we can formulate all acid–base equilibriums in terms of K_a values.

With the aid of a pH meter, it is easy to titrate an acid of known concentration with a strong base of known concentration and vice versa to find out the pK_a value. According to Eq. (22) pK_a is:

$$\log K_a = \log H^+ + \log \frac{(\text{base})}{(\text{acid})}$$

or $\hspace{11cm}$ (27)

$$pH = pK_a + \log \frac{(\text{base})}{(\text{acid})}$$

where $pH = -\log (H^+)$. In most cases, the ion concentration is used instead of its activity as an approximation for dilute electrolyte solutions. The more accurate methods to determine acid ionization constants are (1) from conductance measurements and (2) from electromotive force cells.[3,4]

The ionization constants of some biologically important weak acids are given in Table 3. The acid–base titration curves of some acids are shown in Figure 2.

1.1.3.1. Dipolar Ions

Many biological molecules such as amino acids, proteins, phospholipids, etc., are zwitterionic forms in the appropriate acidic solution in which the molecule contains an equal quantity of negatively and positively charged groups (Table 4). These charges have a discrete separation resulting in a high dipole moment of the molecule, although the dipoles in a protein tend to cancel each other resulting in a moderate overall dipole moment of these larger biomolecules. The ionization scheme for glycine is shown in Figure 3.

The zwitterionic molecules (dipolar ions) can be positively or negatively charged as a whole or neutral depending on the pH of the solution.

Table 3
pK_a Values and Related Thermodynamic Functions for Certain
Carboxylic Acids, Phosphoric Acids, Ammonium Ions, Amino
Acids, and Carbonic Acid (All Data for 25°)[a,b]

Substance	pK_a	ΔF^0	ΔH^0	ΔS^0	ΔC_p^0
Water (K_w)	13.997	19,089	13,519	−18.7	−47
Carboxylic acids					
Formic acid	3.752	5,117	−23	−17.6	−42
Acetic acid	4.756	6,486	−92	−22.1	−37
Propionic acid	4.874	6,647	−163	−22.8	−38
Chloroacetic acid	2.861	3,901	−1,158	−17.0	−40
Glycolic acid	3.831	5,225	175	−16.9	−39
Lactic acid	3.860	5,267	−99	−18.0	−41
Succinic acid, pK_1	4.207	5,740	762	−16.7	−32
Succinic acid, pK_2	5.636	7,693	−108	−26.1	−52
Benzoic acid	4.202	5,732	104	−18.9	−48
Phosphoric acid and derivatives					
Phosphoric acid, pK_1	2.148	2,930	−1828	−16.0	−37
Phosphoric acid, pK_2	7.198	9,823	987	−29.6	−54
Glycerol 2-phosphoric acid, pK_1	1.335	1,820	−2,893	−15.8	−78
Glycerol 2-phosphoric acid, pK_2	6.650	9,069	−412	−31.8	−54
Glucose 1-phosphoric acid, pK_2	6.504	8.870	−431	−31.2	−47
Ammonium ion and derivatives					
Ammonium ion	9.245	12,614	12,480	−0.4	0
Methylammonium ion	10.615	14,484	13,088	−4.7	8
Dimethylammonium ion	10.765	14,687	11,859	−9.5	23
Trimethylammonium ion	9.791	13,358	8,815	−15.2	44
Ethanolammonium ion	9.498	12,958	12,080	−0.6	−1
Tris(hydroxymethyl) aminomethane	8.076	11,018	10,900	−0.3	—
Amino acids					
Glycine, pK_1	2.350	3,205	1,156	−6.9	−32
Glycine, pK_2	9.780	13,340	10,550	−9.4	−12
α-Alanine, pK_1	2.348	3,203	773	−8.2	−37
α-Alanine, pK_2	9.867	13,458	10,980	−8.3	−17
α-Amino-n-butyric acid, pK_1	2.284	3,123	298	−9.5	−34
α-Amino-n-butyric acid, pK_2	9.831	13,408	10,695	−9.1	−17
α-Aminoisobutyric acid, pK_1	2.357	3,215	492	−9.1	−38
α-Aminoisobutyric acid, pK_2	10.206	13,919	11,531	−8.0	−15
Serine, pK_1	2.186	2,980	1,366	−5.4	−32
Serine, pK_2	9.205	12,550	10,405	−7.2	−2
Threonine, pK_1	2.096	2,859	1,180	−5.6	−32
Threonine, pK_2	9.100	12,410	9,960	−8.2	−15

Table 3 (continued)

Substance	pK_a	ΔF^0	ΔH^0	ΔS^0	ΔC_p^0
Hydroxyproline, pK_1	1.815	2,476	918	−5.2	−34
Hydroxyproline, pK_2	9.660	13,160	9,835	−12.7	−18
Proline, pK_1	1.970	2,663	342	−7.8	−38
Proline, pK_2	10.640	14,510	10,310	−14.1	−12
β-Alanine, pK_1	3.551	4,845	1,179	−12.3	−30
β-Alanine, pK_2	10.235	13,963	12,570	−4.7	−3
γ-Aminobutyric acid, pK_1	4.031	5,500	405	−17.1	−34
γ-Aminobutyric acid, pK_2	10.556	14,400	12,070	−7.8	−5
ε-Aminocaproic acid, pK_1	4.373	5,965	−8	−20.0	−40
ε-Aminocaproic acid, pK_2	10.804	14,740	13,560	−4.0	8
Glycylglycine, pK_1	3.148	4,140	862	−12.9	−40
Glycylglycine, pK_2	8.252	11,260	10,600	−2.0	−10
Aspartic acid, pK_1	1.995	2,720	1,783	−3.1	−29
Aspartic acid, pK_2	3.910	5,335	1,110	−14.2	−33
Aspartic acid, pK_3	10.006	13,650	9,025	−15.5	−21
Carbonic acid, pK_1	6.352	8,666	2,240	−21.6	−90
Carbonic acid, pK_2	10.329	14,092	3,603	−35.2	−65

[a] Reference 1.
[b] All values are given on the molarity scale (concentrations as moles per kilogram of water), but, at 25^0, values on the molarity scale (moles per liter of solution) differ by only about 0.001 in pK_a. ΔF^0 and ΔH^0 in cal mol^{-1}; ΔS^0 and ΔC_p^0 in cal deg^{-1} mol^{-1}

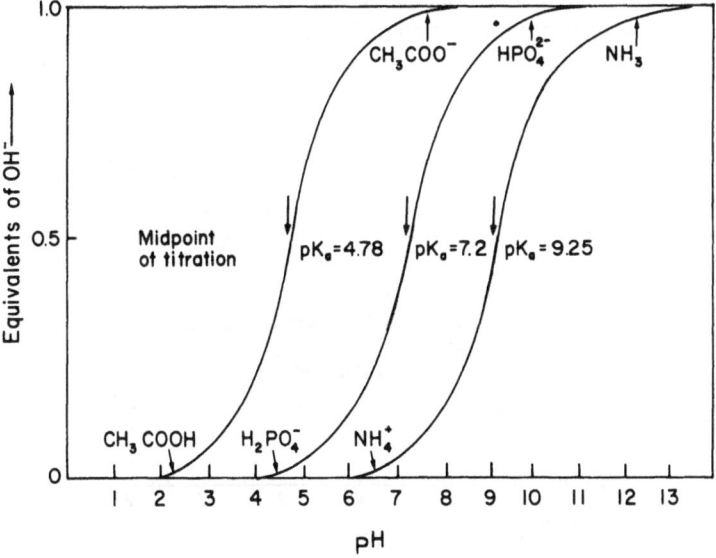

Figure 2. Acid–base titration curves of some acids.

Table 4
Effective Dipole Moments of Some Amino Acids
and Proteins in Aqueous Solution
at 25°C

Molecule	$\mu_{\text{eff}}(D)$
Glycine	15.7
Glycine betame	14.1
α-Alanine	15.9
β-Alanine	19.5
α-Aminobutyric acid	16.0
β-Aminobutyric acid	19.3
γ-Aminobutyric acid	23.6
ε-Aminocaproic acid	28.6
Glycine ethyl ester	2.11
α-Aminobutyric ethyl ester	2.09
β-Alanine ethyl ester	2.14
Myoglobin in water	170
Carboxyhemoglobin in water	480

Also, another type of amino acids, including lysine, arginine, histidine, glutamic acid, aspartic acid, tyrosine, and cysteine, which have three groups capable of ionization, have 3 pK_a values. Examples are listed in Table 5.

1.1.3.2. Effect of Molecular Groups and Temperature on pK_a

The pK_a value of a dissociable group of a molecule depends upon its molecular structure, namely neighboring atomic or molecular groups, and environmental temperature.

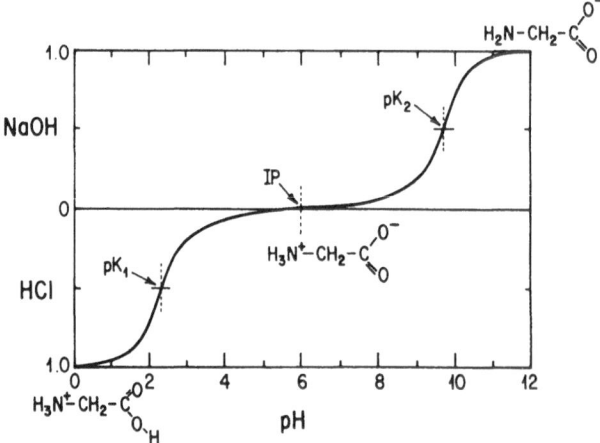

Figure 3. Acid–base titration curve of glycine. IP: Isoelectric point.

Table 5
Summary of pK Values of Amino Acids

Amino acid	pK_1	pK_2	pK_3
Aspartic acid	1.995	3.910	10.006
Glutamic acid	2.19	4.25	9.67
Histidine	1.82	6.00	9.17
Lysine	2.16	9.18	10.79
Arginine	1.81	9.01	12.5
Tyrosine	2.20	9.1(NH_2)	10.95(OH)
Cysteine	1.8	8.3	10.8

From temperature dependence data and other environmental factors, it is possible to evaluate the standard free energy change, ΔF^0, the molal changes in heat content (ΔH^0), entropy (ΔS^0) and heat capacity at constant pressure (ΔC_p^0) which occur when the dissociation reaction occurs under standard conditions.

These thermodynamic quantities are related as follows:

$$\Delta F^0 = -RT \ln K_a = 2.3026 \, RT \, pK_a$$

$$\Delta H^0 = RT^2 \frac{\partial \ln K_a}{\partial T} = 2.3026 \, R \frac{\partial pK_a}{\partial(1/T)}$$

$$\Delta S^0 = -\frac{\partial(\Delta F^0)}{\partial T} = \frac{\Delta H^0 - \Delta F^0}{T}$$

$$\Delta C_p^0 = \frac{\partial(\Delta H^0)}{\partial T}$$

(28)

pK_a values and related thermodynamic functions for biologically important dissociable groups are listed in Table 3. It is seen from Table 3 that neighboring molecular groups affect the dissociation constant. Polar but uncharged substances adjoining the ionizing groups have a marked effect on the pK values. Within carboxylic acids, chloroacetic acid is nearly one hundred times as strong an acid as acetic acid. Glycolic and lactic acids, with a hydroxyl group in the α position, are nearly ten times as strong. On the other hand, the pK_a values for any homologous series, such as the fatty acids, are in general independent of the length of an alkyl group attached to the acidic group (Table 6).

The values of ΔH^0 show a close relation to structure. For all the carboxylic acids, these values are relatively small (generally less than 2 Kcal/mole). The same considerations apply to the phosphoric acid groups. On the other hand, ΔH^0 values for the ammonium ions are very large, on the order of 10 to 12 Kcal/mole.

Table 6
Influence of Substituents (R) in Different Positions
on the Dissociation (pK') of the Carboxyl
Group in R(CH₂)ₙCOOH at 25°ᵃ

Substituent (R)	Dipole moment of CH₃R (Debye units)	pK' for position of substituent			
		$\alpha(n = 1)$	$\beta(n = 2)$	$\gamma(n = 3)$	$\delta(n = 4)$
CH_3	0	4.87	4.83	4.80	4.85
$CH_2{=}CH$	0.34	4.42			
C_6H_5	0.39	4.26			
HO	1.65	3.82			
HS	1.39	3.60			
COOR'	1.7	3.34	4.52		4.60
I	1.6	3.15	4.05	4.64	4.77
Br	1.8	2.86	4.01	4.58	4.72
Cl	1.8–1.9	2.81	4.07	4.52	4.69
COOH	1.7	2.92	4.24	4.36	4.42
O_2N	3.0–3.8		3.79		
$N{\equiv}C$	3.1–3.5	2.44			
NH_3^+		2.35	3.55	4.03	4.21

ᵃ Reference 1.

Certain structural regularities are more closely associated with entropy effects for ionization (Table 2). Some types of acid–base equilibria, where an uncharged acid reacts with a neutral molecule, such as water, to produce an anion and a cation, are associated with a large negative value of ΔS^0.

1.1.3.3. Effects of Electric Charges and Dipoles on pKₐ

The presence of a neighbouring substituent molecular group carrying a net charge, or of a dipole, may affect greatly the characteristic pK_a value of the molecule. Comparison of the ionization of the carboxyl group in glycine ($H_3N^+CH_2COOH$) and in propionic acid ($H_3C{-}CH_2COOH$), which are identical in electronic structure, reveals the influence of the adjoining positive charge in increasing the acid strength of the carboxyl group from 4.87 (propionic) to 2.35 (glycine) in pK or 3240 calories in ΔF^0. Similarly, as the distance between the carboxyl and amino groups is increased—compare the values for β-alanine ($H_3N^+(CH_2)_2COOH$)(pK_a 3.55), γ-aminobutyric ($H_3N^+(CH_2)_3COOH$)(4.03), and ε-aminocaproic ($H_3N^+(CH_2)_5COOH$)(4.37) acids—the influence of the positive charge progressively decreases. Some of these examples are shown in Tables 3, 6, and 7. Bjerrum[5] gave the first quantitative formulation for the electrostatic effect on pK_a. According to him, the contribution of the charged group to the electrical potential at the acidic group is:

$$\psi = \frac{Ze}{\varepsilon r} \tag{29}$$

Table 7
Influence of Substituents (R) in Different Positions
on the Acidity (pK') of the Ammonium
Group in $R(CH_2)_n NH_3^+$ at 20° or 25°[a]

Substituent (R)	Dipole moment of CH_3R (Debye units)	$\alpha(n=1)$	$\beta(n=2)$	$\gamma(n=3)$	$\delta(n=4)$
		pK' for position of substituent			
CH_3-	0	10.66	10.59	10.68	10.70
$CH_2=CH-$	0.34	9.76			
C_6H_5-	0.39	9.38			
$HO-$	1.65		9.48		
H_2N-	1.23		9.98	10.62	10.86
$-COOR'$	1.7	7.75	9.13	9.71	10.15
$-COO^-$		9.72	10.19	10.40	10.69
$-NH_3^+$			6.98	8.58	9.32

[a] Reference 1.

where r is the distance between the electric charge (Ze) and a proton from the dissociation group. Then, the electrical work done under the influence of the potential ψ in removing a proton from the acidic group at r to infinity is, per mole,

$$\Delta W = Ne\psi = \frac{NZe^2}{\varepsilon r} \quad (N: \text{Avogadro's number}) \tag{30}$$

This value is approximately equivalent to the difference ΔF^0 in free energy of ionization of an acid containing a charged substituent group and a similar acid containing no charged group.

$$\Delta W = \Delta F^0 = 2.303 RTpK_a \tag{31}$$

When the substituent group is further removed from the acidic group, then Bjerrum's formula gives a good approximation. Kirkwood and Westheimer[6] and Westheimer and Shookhoff[7] made further improvements on Bjerrum's hypothesis.

The effect of a dipole of moment μ in a homogeneous medium is also given similarly by:

$$pK_a = \frac{Ne\mu \cos \theta}{2.303 RT\varepsilon r^2} \tag{32}$$

where θ is the angle between the dipole direction and the line joining its center to the ionizable proton.

1.1.3.4. Electrostatic Effect on Ionization in Polybasic Acids

In the ionization of a proton from a large spherical central ion (such as a positively charged protein ion) having n_0 identical and independent sites available of which n sites are occupied by protons, the reaction is:

$$P^{+n} \rightleftharpoons P^{+(n-1)} + H^+ \tag{33}$$

The hydrogen ion concentration (H^+) is equal to $Kn/(n_0 - n)$ where K is the dissociation constant. That is,

$$pH = pK + \log\frac{n_0 - n}{n} \tag{34}$$

The free energy change for the ionization of the proton can be expressed by two terms:

$$\Delta F = \Delta F_i - \frac{\partial W(\bar{Z})}{\partial \bar{Z}} \tag{35}$$

where ΔF_i is the intrinsic free energy charge due to specific interacting factors and $W(\bar{Z})$ is the electrostatic energy due to the central ion and its ion atmosphere, and $\bar{Z}$ is the valency (the average net charge) of the large central ion. The free energy change of ionization per mole is equal to $2.3026RTpK_a$ [from Eq. (28)]:

$$pK = pK_i - \frac{N}{2.3RT}\frac{\partial W(\bar{Z})}{\partial \bar{Z}} \tag{36}$$

The potential of the large spherical central ion is given by Eq. (14):

$$pK \cong pK_i - \frac{\partial\left[\dfrac{\bar{Z}^2 e^2 N}{2.3RT\varepsilon}\left(\dfrac{1}{b} - \dfrac{\kappa}{1+\kappa a}\right)\right]}{\partial \bar{Z}} \tag{37}$$

By setting:

$$\omega = \frac{e^2 N}{Z\varepsilon RT}\left(\frac{1}{b} - \frac{\kappa}{1+\kappa a}\right) \tag{38}$$

Eq. (37) is rewritten as:

$$pK = pK_i - \frac{2\omega\bar{Z}}{2.3} = pK_i - 0.868\omega\bar{Z}$$

$$= pH - \log\frac{n_0 - n}{n} \tag{39}$$

The magnitude of ω can be obtained from Eq. (39). The value of ω depends on the size and shape of the large central ion as well as on the ionic strength. Since the effect of asymmetry would be to increase the surface area of the particle, the increase in asymmetry of the large polyvalent ion tends to decrease the value ω. Increase in the ionic strength also decreases ω [see Eq. (38)]. The calculation of ω and the use of Eq. (39) also gives various information on the large polyvalent ion. This classical treatment for hydrogen ion titration curves of proteins regards the molecule as an impenetrable sphere on which amino acid residues are grouped into classes of intrinsically identical sites with their charges uniformly distributed over the surface. This treatment, however, is offset by its inability to yield electrostatic information about individual groups, their specific roles, and their interactions.

In the more realistic discrete-charge electrostatic theory (Tanford and Kirkwood[8]), the amino acid groups are point charges positioned at fixed sites on the surface of the protein or are buried at a short distance within the interior of the molecule which is assumed to be a continuous medium of low dielectric constant. The theory was successfully tested on a variety of model compounds. However, this calculation was also limited to the mutual effect of two groups only, such as the iron atom and the amino acid in a hemoglobin molecule.

From experiments on hydrogen ion titration of proteins, Orttung[9,10] found that all charges should be to a certain extent buried at the surface. Later, Shire et al.[11] introduced a modification into the previous discrete-charge model whereby, for each individual group, the magnitude of electrostatic intramolecular interaction was reduced in direct proportion to the extent of the group's exposure to the solvent, "solvent accessibility." With this treatment, it was shown that the discrete-charge model can be fruitfully employed to study several aspects of electrostatic effects in myoglobin[12] and hemoglobin.[13]

1.1.3.5. Effect of Surface Charge and Surface Dielectric Constant on pK_a

The apparent pK'_a of a dissociable group at the biosurface differs from the intrinsic pK_a of the group in bulk solution, due to the electrostatic enhancement of the hydrogen ion concentration (activity) at the interface, caused by the surface charge, and also due to the shift in the acid–base equilibrium arising from the different local polarity at the interface.[14]

$$pK'_a = pK_a + \Delta pK_a^{el} + |\Delta pK_a^{p}| \tag{40}$$

where ΔpK_a^{el} is the electrostatic-induced shift and ΔpK_a^{p} is the polarity-induced shift. The sign of the polarity-induced shift depends on the change in the total number of charges on protonation. For the dissociation of a molecular acid, the pK is increased as a result of the lower dielectric constant at the interface than in the bulk solution, whereas, for the dissociation of a cationic acid the situation is reversed. The electrostatic shift in pK_a is

$$\Delta pK_a^{el} = -e\psi(0)/2.3kT \tag{41}$$

where $\psi(0)$ is the surface potential at the surface calculated from the diffused double layer theory. The apparent pK'_a is therefore

$$pK'_a = pK_a - \frac{e\psi(0)}{2.3kT} + |\Delta pK_a^{p}| \tag{42}$$

Since $\psi(0)$ is a function not only of the surface charge, but also of ionic concentrations in the bulk solution (see Eq. (70) or (72)), the apparent pK_a' depends also on the ionic concentrations of the bulk solution. Also, if ions other than H^+ can bind to the dissociable groups at the biosurface, an additional term, $\Delta pK_a^{binding}$, accounting for this factor should be included in the expression for the apparent dissociation constant:

$$pK_a' = pK_a - \frac{e\psi(0)}{2.3kT} + |\Delta pK_a^p| + \Delta pK_a^{binding} \tag{43}$$

This will be discussed in the following section.

1.1.3.6. Effect of Ion Bindings of pK_a

The determination of pK_a values for biopolymers generally is carried out for a particular situation where these molecules are in a given ionic environment. Since there may be different binding affinities for different ions with a given biopolymer, the observed pK_a value must correspond to an "apparent" dissociation constant, not the intrinsic dissociation constant. Such an "apparent" dissociation constant of H^+ would therefore vary with differences in the ionic environment, such as ionic species and ionic strength, as well as the nature of the surface charge of the biomolecules.

Recently, Ohki and Kurland[15] have demonstrated this experimentally as well as theoretically for lipid monolayer systems where H^+ and other cations can bind to the charged sites of the membrane. An example of this is shown in Figure 4. Suppose that an acidic phospholipid molecule has an acidic group with an H^+ association constant of 10,000 M^{-1}, and its monomolecular film, which has an area of 65 Å^2 per molecule, is formed on different 0.1 M monovalent salt solutions; the cations of the salts also have different binding constants to the phospholipid molecules. Then, the variation of surface potential (Gouy–Chapman double layer potential) of such a monolayer with respect to the pH of the subphase can be calculated, provided that the binding constants are known. Figure 4 shows the fraction (f) of a charge per molecule (or the fraction of charge per charged group of a molecule) in such cases. Curve A corresponds to the case where a monovalent cation does not bind to the lipid charged group, B, the case where a monovalent cation binds with a binding constant of 0.08 M^{-1}, and C the case of a monovalent cation having a binding constant of 0.6 M^{-1}. The pH values corresponding to the mid-points of these curves are taken to be the apparent pK_a for H^+ (5.65 for A, 5 for B, and 4.3 for C) which are indicated by ▲ in Figure 4. Although the given intrinsic $pK_a = 4$ is the same in each case, the apparent pK_a values obtained from a fraction of charge versus pH plot are different for different ionic conditions.

It is clear that the "apparent" dissociation constant of H^+ depends on the binding affinities of other ions at the H^+ binding site and the ionic strength, as well as the nature of surface charge (surface charge and surface potential) which, in turn, pertains to the enhancement of ionic concentration at the

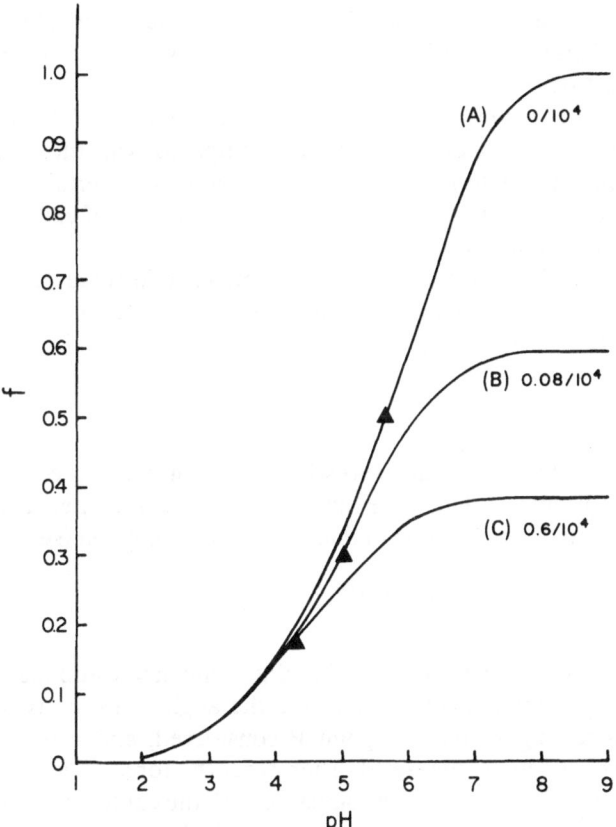

Figure 4. Theoretical values of the fraction of charge density ($f = \sigma/\sigma^{\text{int}}$) vs. pH of the subphase solution for a monolayer having initially one net negative charge site per 65 Å². The monolayer has been assumed to form on a 0.1 M monovalent salt solution and the cation binding constants have been taken as: (A) 0 M^{-1}; (B) 0.08 M^{-1}; and (C) 0.6 M^{-1}, respectively. The binding constant for H^+ is assumed to be 1×10^4 M^{-1}. σ is the surface charge density and σ^{int} the surface charge density with no ion binding (Reference 15).

membrane surface. These situations have been recognized for polyelectrolyte solutions.[16] The determination of the correct ion binding constants, including that for H^+ binding, is significant not only in the study of the mechanism of membrane adhesion and fusion, but also in studies of many other biochemical processes involving charged membranes.

1.2 Electrical Potential Due to Molecular Polarization

Many biological molecules exhibit electrical polarization moments due to asymmetrical distribution of nuclei and their associated electronic charges within the molecule. Also, an externally applied electrical field or a local electric field caused by the surrounding molecular groups can induce a displacement of the charge distribution within a molecule, which is called the induced electric polarization.

The electrical potential energy and its force due to these electrical polarizations are of rather a short-range nature compared with those arising from the coulombic force with respect to distance. However, certain assemblies of these polarized molecules could result in the properties of rather long-range electrical potential and force.

Here, we shall review the electrical potential and its force due to individual polarized or polarizable molecules. Those for the molecular assembly will be discusssed later (Section 2).

1.2.1. Dipole Potential

Let us consider a molecule consisting of two charges $\pm q$ of opposite signs separated by distance l. Then, the dipole moment μ is defined as $\mu = lq$. The electrical potential at a given point due to the dipole is given (see Figure 5) by:

$$\psi \cong \frac{\mu \cos \theta}{\varepsilon r^2} \qquad (r \gg l) \qquad (44)$$

where r is the distance between the midpoint of the dipole and the point where the electrical potential is considered, θ is the angle between two vectors, $\vec{\mu}$ and $\vec{r}$, $\vec{\mu}$ connecting μ and the point P considered, and ε is the dielectric constant of the medium (Figure 5). Therefore, the force $(-\partial\psi/\partial r)$ is proportional to $1/r^3$ which is short-range compared with the coulombic force $(\sim 1/r^2)$.

As mentioned above, the majority of biological molecules may be expected to have some permanent dipole moment. The limiting case of nonpolar molecules is represented by molecules having a high degree of symmetry such as benzene and many of the hydrocarbons. The opposite extreme is represented by a class of molecules known as dipolar ions or dipolar molecules. The dipolar ions are molecules, such as the amino acids in neutral solution, which contain two or more oppositely charged groups, resulting from ionization (such as $-COO^-$, $-NH_3^+$, $-PO_3^-$, etc.). They form the dipolar ions or multipolar ions

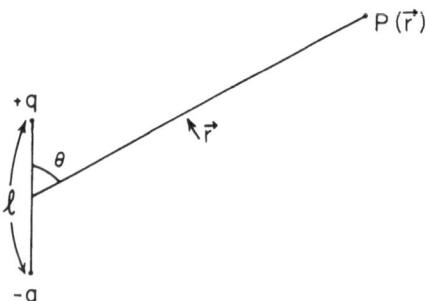

Figure 5. A schematic diagram of a charge dipole and a point subject to the dipole potential.

discussed in the previous section. On the other hand, dipolar molecules are not required to have ionized forms since the difference in electronegativity of each atom results in electronic charge displacement within the molecule and causes the shift of the electronic cloud from that in the symmetrical atomic state. Examples for these are H_2O, O_2, various alcohols, some amines, etc. Some of the biologically important dipole ions and dipolar molecules are listed in Tables 1, 4–8.

1.2.1.1. Solubility

Since polar molecules associate preferentially with water, another polar molecule, the degree of solubility of a molecule in the aqueous phase depends on the degree of polarity of the molecule. This is also true for the case of nonpolar molecules in nonpolar solvents.

Dipolar ions, like ionic salts, are generally far more soluble in water than in media of lower dielectric constant. Some examples are given in Table 8. A dipolar ion [such as glycine ($H_3^+NCH_2COO^-$)] is far more soluble in water ($\varepsilon_{H_2O} = 78.5$ at 25°C), relative to ethanol ($\varepsilon_{ethOH} = 24$ at 25°C), than the uncharged isomer [such as glycolamide ($HOCH_2CONH_2$)]. Also, in a homologous series such as glycine, α-alanine ($CH_3—CH(NH_3)—COO^-$), α-amino-n-butyric acid ($CH_3—CH_2—CH(NH_3)—COO^-$), α-amino-n-valeric

Table 8
Solubilities of Some Dipolar Ions and Uncharged Isomers in Water and Ethanol at 25°[a]

Substance	Solubility (moles/liter)		Solubility ($\log N$)[c]		$\Delta \log N$[b]
	In water	In ethanol	In water	In ethanol	
Glycine	2.886	0.00038	−1.247	−4.638	3.391
DL-α-Alanine	1.656	0.00076	−1.491	−4.347	2.856
DL-α-Amino-n-butyric acid	1.800	0.00260	−1.440	−3.818	2.378
DL-α-Amino-n-caproic acid	0.0866	0.00104	−2.801	−4.215	1.414
L-Leucine	0.171	0.00128	−2.503	−4.125	1.622
L-Asparagine	0.186	0.000023	−2.468	−5.870	3.402
β-Alanine	6.123	0.00189	−0.816	−3.955	3.139
ε-Aminocaproic acid	3.848	0.00194	−0.975	−3.947	2.972
Diglycine	1.512	2.22×10^{-5}	−1.522	−5.889	4.367
Triglycine	1.0229	1.06×10^{-5}	−2.241	−7.206	4.965
Glycolamide (uncharged)	5.509	0.342	−0.900	−1.699	0.799
Lactamide (uncharged)	8.779	2.847	−0.506	−0.759	0.253

[a] These data are taken from Cohn, E. J. and Edsall, J. T. (1943) "Proteins, Amino Acids, and Peptides," Reinhold Publishing Corp., New York (Chapter 9), where many other data are also given.
[b] The value of $\Delta \log N$ denotes the solubility increment in passing from ethanol to water.
[c] N: the molar fraction of solute.

acid $(CH_3-(CH_2)_2-CH(NH_3)-COO^-)$, and α-amino-n-caproic acid $(CH_3-(CH_2)_3-CH(NH_3)-COO^-)$, the value of $\Delta \log N_j$ (N_j: the mole fraction of solute) decreases approximately 0.50 for each additional CH_2 group inserted in the molecule.

The solubility of dipolar ions or dipolar molecules also depends on the ionic strength of salt and salt species in the medium; the addition of salt increases the solubility of dipolar ions. The lower the dielectric constant of the medium, the greater is the rate of increase in solubility for a given increment in ionic strength. This phenomenon is called the "salting-in effect." However, in some cases, a further increase in the ionic strength in the medium decreases the solubility of dipolar molecules which is called the "salting-out effect." Such effects are due to ion–dipole interaction and image charge interaction, respectively. The latter type of interaction is relevant to the calculation of electrostatic potentials near the boundary between two different dielectric media.

1.2.1.2. Activity Coefficient and Solubility

The determination of relative solubility in different solvents provides data for calculation of the free energy of transfer of the solute from one medium to another. The chemical potential of the solute in different solvents is the same when the two solution systems are in equilibrium with each other:

$$\mu_i^{(\alpha)} \equiv \mu_i^{0(\alpha)} + RT \ln a_i^{(\alpha)} = \mu_i^{(\beta)} \equiv \mu_i^{0(\beta)} + RT \ln a_i^{(\beta)} \tag{45}$$

$$\text{or} \quad \mu_i^{0(\alpha)} - \mu_i^{0(\beta)} = RT \ln \frac{a_i^{(\beta)}}{a_i^{(\alpha)}}$$

where μ_i, μ_i^0 and a_i are the chemical potential, the standard chemical potential, and the activity for the ith species, respectively. α and β refer to different solvent phases.

Denoting the mole fraction of a solute in the aqueous phase by $N^{(W)}$ and in the other solvent by $N^{(A)}$, the activities are expressed by

$$a^{(W)} = f^{(W)}N^{(W)} \quad \text{and} \quad a^{(A)} = f^{(A)}N^{(A)}. \tag{46}$$

If the standard state is taken as being the same for both phases, we have the following relation:

$$a_i = f^{(W)}N^{(W)} = f^{(A)}N^{(A)} \tag{47}$$

If $f^{(A)^0}$ is the activity coefficient of a solute in a given solvent at zero ionic strength, and $f^{(A)}$ is the activity coefficient of the solute in the same medium but at finite ionic strength (ω), the relative activity coefficeint γ is expressed by:

$$\log \gamma = \log (f^{(A)}/f^{(A)^0}) = \log (N^{(A)^0}/N^{(A)}) \tag{48}$$

1.2.2. Ion–Dipole Interaction

When we consider the interaction energy between a dipolar ion of fixed orientation and an ion j which are both in the same aqueous solution (ε), the most simple approximate form for the interaction is expressed as:

$$W_j(\cos\theta) = Z_j e \psi_{\text{dipole}} = \frac{Z_j e \mu \cos\theta}{\varepsilon r^2} \tag{49}$$

where $Z_j e$ is the charge of a species j ion. The total energy due to the dipole and the surrounding ions should be the summation over all ionic species as well as the average over all possible orientations of the dipolar ions. If the dipolar ion is considered as a point dipole of moment μ embedded in a continuous medium of dielectric constant ε and ions can approach up to the distance b from the point dipole, then the final expression[1] is:

$$W = kT \ln f = -kT \left(\frac{\mu e}{\varepsilon kT}\right)^2 \left(\frac{2\pi N}{2303}\right)\frac{2\omega}{3b} \tag{50}$$

where $\omega = 1000/2N\sum n_i Z_i^2$ (the ionic strength). In this case, the logarithm of activity coefficient f is proportional to the first power of the ionic strength ω [contrast to the case of ion–ion interaction where $\log f \propto \omega^{1/2}$, see Eq. (19)], to the square of the dipole moment, and to the inverse square of the dielectric constant.

This very simple treatment has ignored the microscopic dielectric properties of the dipolar ion. Kirkwood[17] treated the above problem by introducing the dipolar molecule as a cavity of low dielectric constant in the medium of relatively high dielectric constant. The limiting law valid for infinite dilution is expressed as:

$$\log f = -K\omega \tag{51}$$

where ω is the ionic strength and K is:

$$K = \frac{2\pi N e^2}{2303\varepsilon kT}\left[\frac{3\mu^2}{2\varepsilon akT} - \frac{b^3\alpha(\rho)}{a}\dagger\right] = -\left(\frac{\partial\log f}{\partial\omega}\right)_{\omega=0} \tag{52}$$

Here ρ is the ratio b/a and $\alpha(\rho)$ is a function of ρ, the value of which is 1 when $\rho = 0$ and 1.96 when $\rho = 0.9$.

Equation (52) shows that the activity coefficient of the dipolar ion again depends on the first power of the ionic strength, ω. Another point is that the overall constant K consists of two parts; the first part is proportional to $(\mu/\varepsilon T)^2$ which results from the electrostatic interaction of the dipole and the surrounding ions, and corresponds to the decrease in activity coefficient due to the ion atomsphere (Table 9). This term is related to the "salting-in effect". The second part of the equation, which is proportional to $(\varepsilon T)^{-1}$, does not depend on the moment of the dipolar ion, but only on its geometrical properties

† a: the distance of the closest approach of the dipolar ion and a surrounding ion (see Fig. 1).

Table 9
Salting-in Constants, K'_R, and Dipole Moments
of Certain Amino Acids and Peptides[a]

Substance	Salting-in constant, K'_R	Dipole moment, μ	μ/K'_R
Glycine	0.33	15.5	47
α-Aminocaproic acid	0.33	$(15.5)^b$	$(47)^b$
Diglycine	0.58	27.6	48
ε-Aminocaproic acid	0.7	29	41
Triglycine	0.8	35	44
Lysylglutamic acid	1.2	61	51

[a] Reference 2.
[b] The dipole moment of α-aminocaproic acid has been assumed to be the same as that of glycine, since both are α-amino acids.

(a, b) as a cavity of low dielectric constant in which image charges are produced by the surrounding ions. It corresponds to a "salting-out" effect, which may be thought of in terms of a repulsion of the ions and their image charges.

1.2.2.1. Image Charge

One of the elementary methods used to solve Laplace's equation under certain boundary conditions (when electrical charges are situated in an inhomogeneous dielectric media) is that of the image method. The electrostatic construction known as image charges are useful in problems of electrostatic interaction relating to molecules in solution. For example, they are used in the theory of the salting-out effect, to calculate the interaction between ions and a surrounding dielectric medium.

Let us consider the case where a point charge is situated at the center of a spherical region of dielectric constant ε_1 and of radius a, and on the outside of this sphere $(r > a)$ the dielectric constant is everywhere ε_2 out to infinity (see Figure 6). Then, the potential at any point r inside the region is expressed

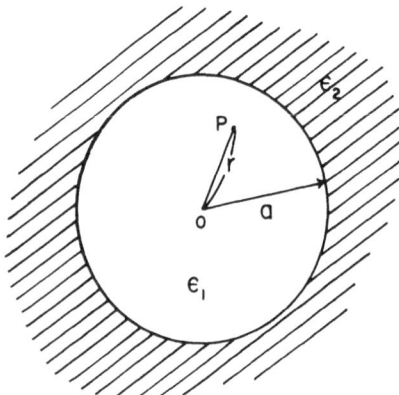

Figure 6. Two-dimensional diagram of a spherical molecule (its radius: a) having dielectric constant ε_1 in a medium dielectric constant of ε_2.

by two parts:

$$W = -\int_{\infty}^{a} \frac{e\,dr}{\varepsilon_2 r^2} - \int_{a}^{r} \frac{e}{\varepsilon_1 r^2}\,dr$$

$$= \frac{e}{\varepsilon_1 r} + \frac{e}{a}\left(\frac{1}{\varepsilon_2} - \frac{1}{\varepsilon_1}\right) \tag{53}$$

The second term is the effect of the surrounding medium on the inside region. This corresponds to the effect of an additional total charge $e[(1/\varepsilon_2) - (1/\varepsilon_1)]\varepsilon_1$ uniformly distributed over the spherical surface of the sphere $r = a$ assuming the dielectric constant is everywhere the same as the inside region ε_1 instead of ε_2; depending on the relative magnitudes of ε_1 and ε_2 (whether ε_1 is less or greater than ε_2), this fictitious (image) charge is either negative or positive. It may be considered as an image of the central charge, e, resulting from the presence of the surrounding different dielectric medium.

As seen from Eq. (53), this fictitious charge effect depends on the size of the cavity domain with dielectric constant ε_1, and the magnitude of the dielectric constant of the solvent medium, ε_2.

1.2.2.2. Salting-out and Salting-in

Gases and slightly soluble organic molecules are usually rendered less soluble in water by the addition of salts. Since $\log f = \log (S_0/S)$† the activity coefficients of most uncharged molecules are increased by the addition of neutral salts. Empirically, a plot of $\log (S_0/S)$ is found to vary linearly with the molal concentration of added salt, $\log f = -Kn_3$ (n: number of moles; component 1 is solvent, component 2 is the uncharged molecule, and component 3 is the salt), as shown in Figure 7. The salting-out constant K [see Eq. (52)] is inversely proportional to the difference between the reciprocal dielectric constants of solution and water, $1/\varepsilon - 1/\varepsilon_0$, so that acetone, which lowers the dielectric constant of water, is salted out (K is negative), while HCN, which raises the dielectric constant of water, is salted in. These results may be accounted for by preferential solvation of the ions by the higher-dielectric constant component. Theoretical considerations also indicate that K is dependent upon the valence type of the salt and is inversely proportional to the mean ionic radius of the salt. Thus the highly solvated lithium ion has less effect than sodium or potassium ions. Glycine increases the dielectric constant of aqueous solutions and hence is salted in, like HCN, on addition of salts.

1.2.2.3. Effects of Salts on Proteins

Concentrated solutions of $(NH_4)_2SO_4$, Na_2SO_4, or phosphate buffers have long been used to salt proteins out of solution.

---

† If n_2 and n_3 are both very small compared to n_1, the ratio of the two solubilities of component S_{02}/S_2 in two different media expressed in moles per liter, is virtually identical to the ratio of the two solubilities, N_{02}/N_2.

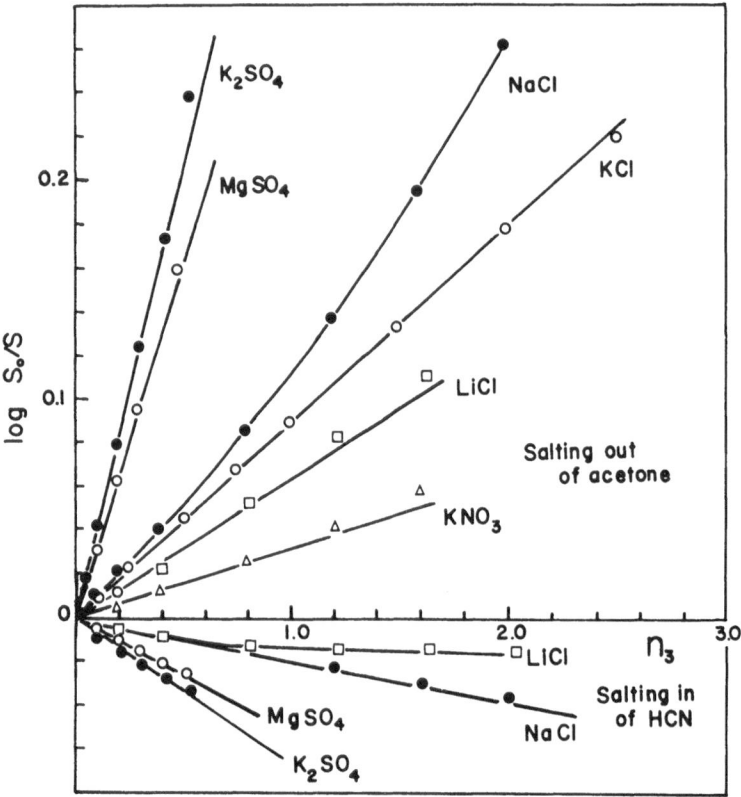

Figure 7. Log (S_0/S) for acetone (upper curves) and HCN (lower curves) versus molality of added salt in aqueous solutions. (After Reference 1, p. 271).

Figure 8 illustrates the effects of several salts on the solubility of horse carboxyhemoglobin. Before the linear salting-out region at high ionic strength is reached, proteins exhibit an initial salting-in region at low ionic strengths. For this reason the extrapolation of the straight-line salting-out portion of the curve will not intersect the ordinate at log $(S/S_0) = 0$, but at some greater value. The salting-out constant K is then defined by the equation log $(S'/S_0) = -K\omega$, where S'/S_0 is the intercept value of the extrapolation. Though the solubility is dependent on temperature and pH, the nature of the salt and the protein size and shape are the main factors determining K, and, in general, the larger a protein molecule is, the greater is K. Amino acids with no net charge behave similarly to proteins, but the effects of added salts are quantitatively much smaller.

Kirkwood[7] has given a theoretical treatment for ion–dipole interactions for various models, the simplest of which is a sphere with a point dipole at its center. The result has been described in the previous section. In all the models

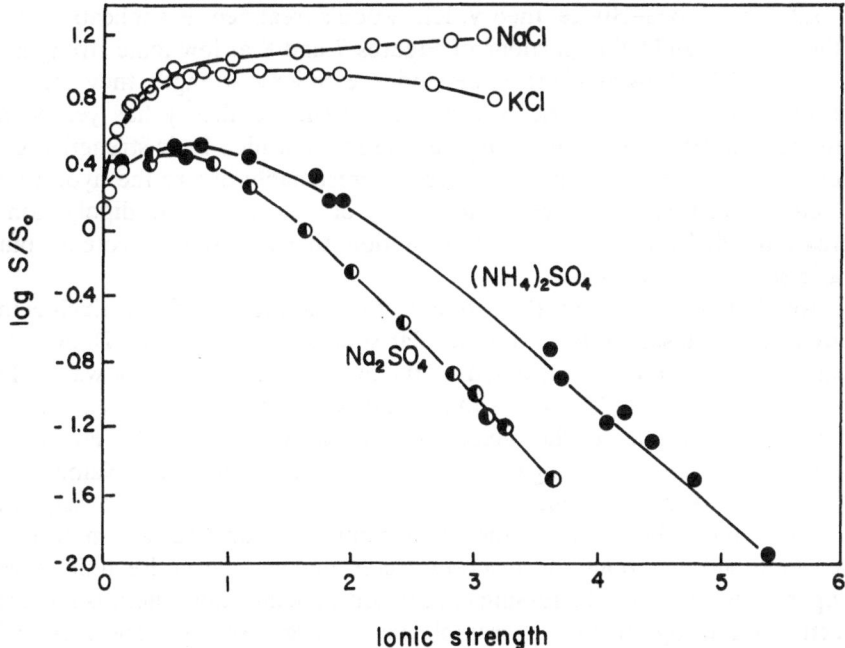

Figure 8. Logarithm of the relative solubility of horse carboxyhemoglobin in salt solutions at varying ionic strengths. The section of the curves at ionic strengths less than 0.1 is linear. (After Reference 1. p. 318).

he investigated, for the dipolar molecule, $-(\log f)$ was equal to $K\omega$, exhibiting a first-power dependence on ionic strength in contrast to the square-root dependence for ion–ion interactions. The value of K is dependent upon the assumed model. For the spherical model, given in Eq. (52), K is separable into two parts, $K = K_i - K_o$, where the salting-in term is K_i and the salting-out term is K_o. The salting-in term arises from electrostatic interactions between the dipole and surrounding ions and is proportional to $(\mu/kT)^2$. No dependence upon the dipole moment appears in the salting-out term, which is proportional to $1/\varepsilon T$ and is dependent upon the shape of the dipolar molecule considered as a cavity of low dielectric constant in a medium of high dielectric constant. In Eq. (52), the low dielectric constant of the cavity was lost. This salting-out term arises only because of the explicit consideration of the dipolar molecule as a cavity with a microscopic dielectric constant much less than the bulk dielectric constant of the solvent (image potential). Kirkwood has also derived the salting-in term for a prolate ellipsoid of revolution with charges at the foci. In this model the salting-in term is proportional to $\mu/(\varepsilon T)^2$. The first-power dependence of the dipole moment agrees better with experiment than the square dependence of the spherical model (see Figure 7) (Table 9). No salting-out term was calculated for the ellipsoidal model.

Like the Debye–Hückel theory, Kirkwood's treatment is applicable only to the initial straight-line portions of Figures 7 and 8 at low ionic strengths. Since K_i and K_o show an identical dependence on ionic strength, they are not resolved unless the dielectric constant is varied. No theory has yet been presented, incorporating the salting-out observed at high ionic strengths K', where the increase in activity coefficient is presumably due to the hydration of ions and consequent decrease in activity of water. Because dipolar ions increase the dielectric constant of the solution, they also salt-in proteins, but to a lesser extent than salts.

An alternative theoretical treatment of the salting-out effect, which is in many respects closer to the real molecular situation, was given by Debye.[18] Qualitatively, his treatment is based on the fact that in a mixture of water (1) and an organic solute (2), the introduction of salts (ions) causes a rearrangement of the molecules in the mixed solvent, the water molecules crowding around the ions and squeezing out the molecules of component 2. Namely, as salt is added to the solution, its concentration increases, with a consequent decrease in the Debye length of the counterionic cloud around each molecule. The cloud shrinks about the macromolecule making the entire unit more compact. Also, with the increasing counterion concentration, there is a more effective screening of the macromolecular charge, so that the identical macromolecules in solution can now approach each other more closely. When the macromolecules approach closely enough, van der Waals force becomes large enough to hold them together creating a larger mass which in turn attracts even more macromolecules. When the mass of aggregates becomes large enough, the entire complex will fall out of solution forming a precipitate.

It is known that various anions have different capabilities of salting-out proteins and other large molecules (lyotropic series). The lyotropic numbers of the anions are determined by the relative concentrations required to cause the coagulation of protein solutions[19,20] (see also Table 10). Figure 9 shows the relation between the threshold concentration and the lyotropic number of anions. The various species of anions with a common cation fall on a straight line for the respective cation when the log C_{th} are plotted against the lyotropic numbers of the anions, and the slope of the straight lines depend on the species of cations in the following order:

$$H > Li > K > Na > Rb > Cs > NH_4.$$

1.2.3. Dipole–Dipole Interaction

1.2.3.1. Interaction Energy

For two fixed dipoles (placed in the same plane), the interaction energy is expressed by:

$$W = -(\mu_1\mu_2/\varepsilon r^3)\{2\cos\theta_1\cos\theta_2 - \sin\theta_1\sin\theta_2\} \qquad (54)$$

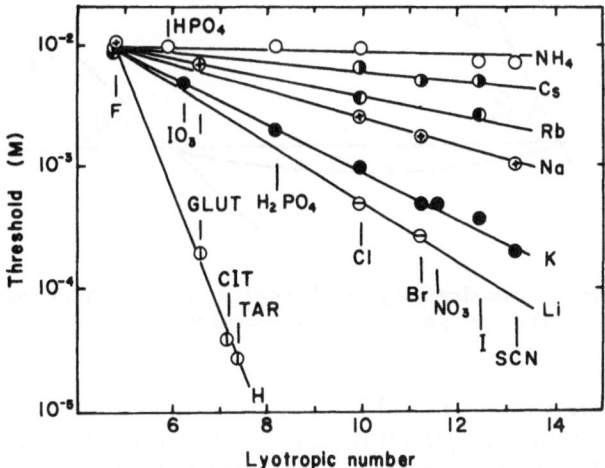

Figure 9. Relation between the threshold concentration and the lyotropic number of anions. Cation species are indicated in the figure at the right of the respective lines (ref. 19, 20).

where μ_1 and μ_2 are the dipole moments of dipole 1 and dipole 2, θ_1 and θ_2 are the angles of the dipoles with respect to the line connecting the two centers of dipoles 1 and 2, and r is the distance between the two dipoles (see Figure 10).

In general, the dipole–dipole interaction is expressed by:

$$W = \frac{\mu_1 \cdot \mu_2}{r^3} - \frac{3(\mu_1 \cdot r)(\mu_2 \cdot r)}{r^5} \tag{55}$$

Table 10
Lyotropic Numbers[a]

(Anion)	(Number)
F	4.8
Phosphate	5.9
Glutamate	6.6
Aspartate	6.6
Citrate	7.2
Tartrate	7.4
Propionate	9.8
SO$_4$	6.8
Acetate	8.7
Cl	10.0
NO$_3$	10.3
Br	10.8
I	11.2
SCN	12.3

[a] Reference 154.

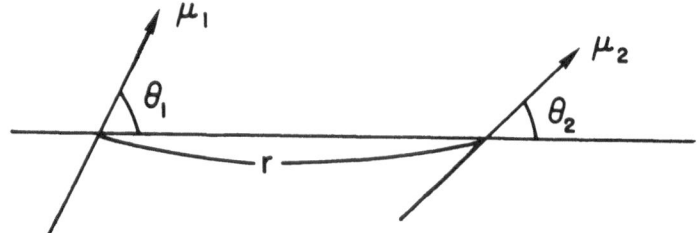

Figure 10. Interaction of two electric dipoles. The angles θ_1 and θ_2 are the orientation of the two dipoles in the same plane with respect to a line that passes through the dipole centers.

Five special cases of Eq. (54) are shown in Figure 11. Considering the five cases shown in the figure, it is seen that the most stable arrangement of the two dipoles occurs when they are collinear and are aligned head to tail. In this case, the energy of interaction is negative and has maximum magnitude:

$$W = -\frac{2\mu_1\mu_2}{\varepsilon r^3} \quad \text{(collinear, head to tail alignment)} \tag{56}$$

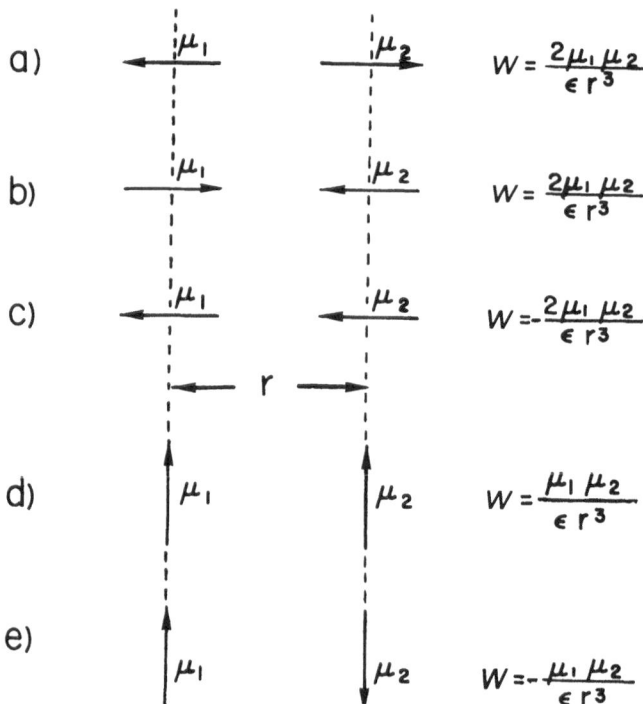

Figure 11. Five different ways in which two dipoles may orient and interact with one another.

If one or both dipoles are not fixed, we have another expression for the interaction energy between two dipoles.

Let us consider the case where dipole 1 is fixed and dipole 2 is free to rotate in the field of dipole 1. The potential energy of a dipole in an electric field (E) is given by:

$$W = -\boldsymbol{E} \cdot \boldsymbol{\mu} = -E\mu \cos\theta \tag{57}$$

The average component of dipole moment μ_2 in the electric field due to dipole 1 can be written as:

$$\overline{\mu_2} = \frac{\mu_2^2 E_1}{2kT} = \frac{\mu_2^2 \mu_1 (1 + 3\cos^2\theta)^{1/2}}{3kT\varepsilon r^3} \tag{58}$$

Therefore, the energy of interaction is:

$$W = -\frac{\mu_1^2 \mu_2^2}{3kT\varepsilon^2 r^6}(1 + 3\cos^2\theta) \tag{59}$$

In the event that dipole 1 is also free to rotate, the energy of interaction is found by averaging Eq. (59) over all possible values of θ. Since the average value of $\cos^2\theta$ is $1/3$, the energy of interaction between the two freely rotating dipoles is given by:

$$W = -2\mu_1^2\mu_2^2/3kT\varepsilon^2 r^6 \tag{60}$$

This equation would probably be the most appropriate to use for molecules in solution because individual molecules in solution generally have freedom to move (or to vibrate or to rotate).

If we express dipole moments in Debye units ($D = 10^{-18}$ esu $\cdot$ cm) and r (distance) in angstroms, these interaction energies [Eqs. (56), (59), and (60)] are expressed as:

$$W = \begin{cases} -2 \times 14.3 \, \mu_1\mu_2/\varepsilon r^3 \text{ Kcal/mole,} & \text{(fixed dipole–dipole)} \\[2mm] -\dfrac{3.5 \times 10^4 \, \mu_1^2\mu_2^2(1 + 3\cos^2\theta)}{\varepsilon^2 T r^6} \text{ Kcal/mole,} & \\[2mm] & \text{(fixed dipole–rotating dipole)} \\[2mm] -\dfrac{7.0 \times 10^4 \, \mu_1^2\mu_2^2}{\varepsilon^2 T r^6} \text{ Kcal/mole} & \text{(rotating dipole–rotating dipole)} \end{cases} \tag{61}$$

respectively.

One would expect that the dipole–dipole interactions would be relatively weak over relatively large distances of separation because of the $1/r^3$–$1/r^6$ distance dependence. However, over a short distance of separation, dipole–dipole interactions can be quite important, as for example in hydrogen bonding.

1.2.3.2. The Hydrogen Bond[21]

The hydrogen bond can be represented schematically as shown in Figure 12, where the hydrogen atom is covalently bonded to electronegative atom X. R is a general chemical group and Y is also an electronegative atom. All elements have been assigned an electronegativity number that indicates the relative strength of this property. Fluorine has the highest value of 4.0. Biologically important atoms and their electronegativities are: F(4.0), 0(3.45), N(2.98), Br(2.75), C(2.55), S(2.53), I(2.45), H(2.13), and P(2.10). Only the first three in this list are generally important atoms in hydrogen bonding in biological systems. Because atoms X and Y are more electronegative with respect to hydrogen and the group R, fractional charges result at each atomic site; and hence small dipoles are formed which interact with one another in the head–tail arrangement.

According to this model for the hydrogen bond, the greater the electronegativity of X and Y, the stronger the hydrogen bond. This has been supported experimentally. Although the model of hydrogen bond in terms of dipole–dipole charge interaction is successful overall[22] it does not necessarily explain all the experimental data.[23] Other theoretical explanations have also been proposed for hydrogen bonding.[24,25,26] The disagreement over the fundamental nature of the hydrogen bond has not been settled.

Hydrogen bonds range in strength between 2 and 10 Kcal/mole and as such are relatively weak. The energies for some hydrogen bonds are given in Table 11. Individually, they can be easily broken by thermal agitation, but in large numbers they can give great stability to a molecular assembly. Hydrogen bonds can be formed between chemical groups in different molecules or between different groups in the same molecule. It has been found that hydrogen bonds are directional like covalent bonds, and that individual hydrogen bonds have characteristic bond lengths. Generally the strongest bonds are formed when the atoms X, Y, and H are collinear, and the bond strength usually lessens as the bond is bent or distorted. Because of the electrostatic nature of the bond, the bond strength is affected by the pH and the amount of free ions in solution.

The importance of hydrogen bonds in maintaining biological structure is represented with three well-known examples: the α-helix and β structure of polypeptides, and the double helix of DNA (Figure 13).[27,28]

Other types of electrical interactions among biomolecules will be discussed in the forthcoming section (Section 4) about molecular interaction.

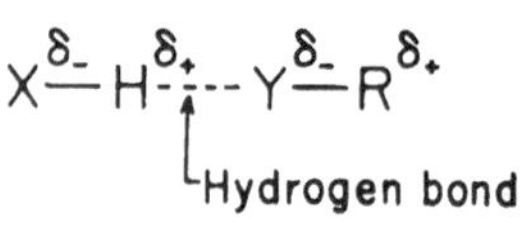

$$X \overset{\delta_-}{-} H \overset{\delta_+}{\underset{\uparrow}{\text{---}}} Y \overset{\delta_-}{-} R^{\delta_+}$$

└Hydrogen bond

Figure 12. An illustration of the hydrogen bond between the molecule X—H and the molecule Y—R. The symbols X and Y represent general electronegative atoms. R is an arbitrary chemical group. The symbols δ_- and δ_+ indicate a slight negative and positive charge, respectively, and — — — represents the hydrogen bond interaction.

Table 11
ΔH (Enthalpy) Values for Different Types of Hydrogen Bonds[a]

Type Substance	$-\Delta H$ (kcal/mole)
$O-H\cdots O$	
Water (H_2O)	Gas: 4.4–5.0, liquid: 2.8
Methanol (CH_3OH)	Gas: 3.2–7.3, liquid: 4.7
Ethanol (C_2H_5OH)	Gas and liquid: 4.0
$N-H\cdots N$	
Ammonia (NH_3)	Gas: 3.7–4.4
Amines (CH_3NH_2, $C_2H_5NH_2$)	Gas: 3.1–3.6
$F-H\cdots F$	
Hydrogen fluoride (HF)	Gas: 6.7–7.0
$C-H\cdots N$	
Hydrogen cyanide (HCN)	Gas: 3.3, liquid: 4.6

[a] Reference 27.

So far, we have discussed the electrical potential due to ionized molecules or the electrical polarization within a molecule. It is known that there is another type of potential due to electron transfer reactions, a part of which can be measured as an electrical potential by conventional electrodes.

1.3. Potential Due to Electron Transfer Reaction

1.3.1. Redox Potential of Biomolecules[29]

Oxidation–reduction reactions (also called redox reactions) are those in which there is a transfer of electrons from an electron donor (the reducing agent) to an electron acceptor (the oxidizing agent).

Oxidizing and reducing agents function as conjugate redox pairs, corresponding to the conjugate nature of acid–base pairs:

Acid–base reaction:

$$\text{proton donor} \rightleftharpoons H^+ + \text{proton acceptor} \qquad (62)$$
$$\text{(acid)} \qquad\qquad\qquad \text{(base)}$$

Oxidation–reduction reaction:

$$\text{electron donor}^{Z-n} \rightleftharpoons n e^- + \text{electron acceptor}^{Z} \qquad (63)$$
$$\text{(reductant)} \qquad\qquad\qquad \text{(oxidant)}$$

where Z is the valence, and n is the number of electrons transferred per molecule. In the oxidation–reduction reaction, valence changes are accompanied by electron transfer. Just as an equilibrium constant may be used to express the tendency of an acid to dissociate a proton, it may also be used to express the tendency of a reducing agent to lose electrons.

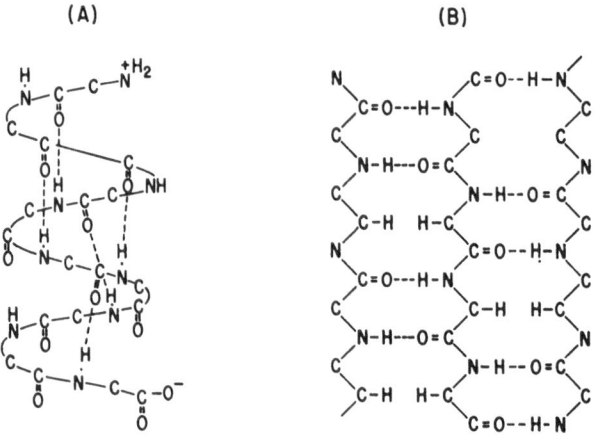

(A)   (B)

(C)

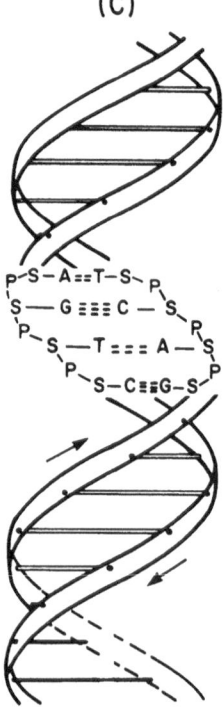

Figure 13. (A) The α-helix structure of a polypeptide chain and the hydrogen bonding within the chain. (B) The β-sheet structure of polypeptide chains and the hydrogen bonding between the chains. (C) Schematic diagram of the DNA double helix and hydrogen bondings. Hydrogen bonds are indicated with dashed lines (———).

For biological systems this tendency is expressed by the standard reduction potential, E_0', defined as the electromotive force (emf) in volts given by a half-cell in which the reductant and oxidant species are both present at 1.0 M concentration unit activity, at 25°C and pH 7.0 in equilibrium with an electrode which can reversibly accept electrons from the reductant species, according to the equation:

$$\text{Oxidant} + ne^- \rightleftarrows \text{Reductant} \qquad (64)$$

where n is the number of electrons transferred. The standard reduction potential is a measure of the electron pressure which a given reductant–oxidant pair generates at equilibrium under the specified conditions. The ultimate standard of reference is the reduction potential of the reaction:

$$2H^+ + 2e^- \rightleftarrows H_2 \qquad (65)$$

which is by convention set at $E^0 = 0.0$ volts (reference voltage) under conditions in which the pressure of H_2 gas is 1.0 atm (fugacity = 1), activity of H^+ is 1.0, that is pH = 0.0, and the temperature is 25°C. When this value is corrected to pH 7.0 ($[a_{H+}] = 1 \times 10^{-7} M$), the reference pH assumed in all biochemical calculations, the standard reduction potential of the hydrogen–hydrogen-ion system becomes $E_0' = -0.42$ volt.

The standard reduction potentials of a number of biologically important redox couples are given in Table 12. Systems having a more negative standard reduction potential than the H_2–$2H^+$ couple have a greater tendency to lose electrons than hydrogen; those with a more positive potential have a lesser tendency to lose them.

Table 12
Standard Reduction Potentials of Some Conjugate Redox Couples[a,b]

Reductant	Oxidant	E_0', volts
Acetaldehyde	Acetate	−0.60
H_2	2H+	−0.42
Isocitrate	α-Ketoglutarate + CO_2	−0.38
NADH + H^+	NAD^+	−0.32
NADPH + H^+	$NADP^+$	−0.32
Lactate	Pyruvate	−0.19
NADH dehydrogenase (reduced)	(oxidized)	−0.11
Cytochrome b		
[Fe(II)]	[Fe(III)]	0.00
Cytochrome c		
[Fe(II)]	[Fe(III)]	+0.26
H_2O	$\frac{1}{2}O_2$	+0.82

[a] Reference 29.
[b] The data are calculated on the basis of two-electron transfers at pH $\cong$ 7.0 and T = 25 to 37°C.

In soluble oxidation–reduction systems, the electron pressure can be measured by an electromotive cell as an electrical potential. The hydrogen ion concentration of most organic redox systems must usually be considered, since in these systems the reduced form can exist as an anion which can accept hydrogen ions and so become inoperative in terms of contributing to the potential; thus, the potential of such systems is greatly influenced by the hydrogen ion concentration. For example, an oxidation–reduction system equilibrates with a metallic electrode, usually platinum, used to sense the electron pressure at any given ratio of reductant to oxidant species. The potential generated by such an electrode is measured by comparing it with that of another half-cell of known potential, for example, the calomel reference half-cell.

The common properties of these systems are the ability to convert chemical energy (or photon energy) into electrical energy.

The free energy change for an oxidation–reduction reaction is:

$$\Delta F = \Delta F^0 - RT \ln \frac{Ox}{Red} \tag{66}$$

and if the reaction can be conducted in such a manner as to produce a reversible potential, then Eq. (66) can be changed to:

$$E = E^0 + \frac{RT}{nF} \ln \frac{Ox}{Red} = E^0 + \frac{2.303}{nF} \log \frac{(Ox)}{(Red)} \tag{67}$$

where E^0 is the standard redox potential (all concentrations at $1\,M$ and temperature 25°C), and E is the observed electrode potential. At 25°C, the term $2.303\,RT/nF$ has the value 0.059 when $n = 1$ and 0.03 when $n = 2$. Since it is customary to calculate equilibria of biological redox couples in terms of two-electrons, the Nernst equation [Eq. (67)] simplifies to:

$$E' = E'_0 + 0.03 \log \frac{(Ox)}{(Red)} \tag{68}$$

where E'_0 is the standard redox potential (pH 7.0, $T = 25$°C, all concentrations at 1.0 M). This expresses mathematically the shape of a titration curve of a reductant with an oxidant (Figure 14), just as the Henderson–Hasselbach equation [Eq. (27)] does the same for titration of an acid with a base. When the ratio of oxidizing agent to reducing agent is unity, the Nernst equation becomes:

$$E' = E'_0 \tag{69}$$

The standard reduction potential is therefore the emf in volts at the precise midpoint of the titration curve of a given reductant at pH 7.0, 25°C and 1.0 atm, just as the pK_a of an acid equals the pH at the midpoint. Therefore, E'_0 is often called the *midpoint potential* (see Figure 14).

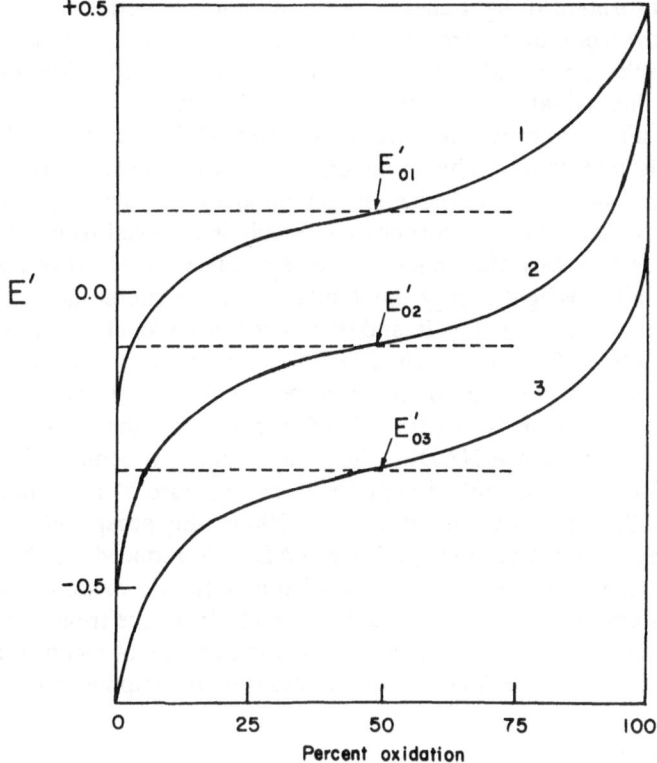

Figure 14. Oxidation–reduction titration curves of three reducing agents, showing midpoint potential E'_0.

The standard reduction potentials of various biological oxidation–reduction systems allow us to predict the direction of flow of electrons from one redox couple to another under standard conditions. Furthermore, we can calculate the final equilibrium resulting when electrons flow from one redox system of known standard potential to another of known potential, as well as the free-energy changes during such reactions which will be discussed later.

1.3.2. Photochemical Potential of Biomolecules

Photosynthetic cells contain specific pigments which are excited by absorbing a special wave of light. For example, a chlorophyll molecule in green algae has characteristically a light absorption maximum at 700 nm, which undergoes bleaching (P700)† when the cell is illuminated. Since this type of change in the pigment can also be produced by the oxidizing agent ferricyanide, the bleaching is believed to correspond to loss of an electron from this pigment. Pigments other than chlorophyll, which act as electron transfer agents, are yellow carotenoids, phycobilins, etc.

† P700: a photoexcitable pigment.

The absorption of light energy by these photosynthetic cells causes electrons to flow from an electron donor to an electron acceptor in a direction opposite to that predicted from the standard oxidation–reduction potential of the interacting oxidation–reduction systems. Namely, on illumination, green plant cells cause electrons to flow from water to $NADP^+$ (nicotinamide adenine dinucleotide phosphate). This is understood in the following way (also see Figure 15): The standard reduction potential of the ground state of the photoreactive center of a chlorophyll assembly is believed to be about $+0.4$ volt. It therefore has little tendency to lose an electron. However, when it is excited and absorbs light energy, the standard reduction potential of the excited state (P700) is about -0.60 volt, and therefore it loses an electron much more readily. Electrons from the excited (P700) molecule can thus easily reduce $NADP^+$, which has a standard reduction potential of -0.32 volt.

When an electron is lost from P700 and reduces some electronegative electron acceptor, such as $NADP^+$, the electron hole left behind in the ground state of P700 must be refilled with electrons. Because of its rather positive potential ($+0.4$ volt), the ground state of p700 readily accepts electrons from relatively electropositive donors. Once $NADP^+$ is reduced by this electron transfer from the pigment in photosynthetic cells, the oxidation–reduction enzyme systems may be so arranged that the electrons are transported across a chain of electron carrier systems and then the energy of electron flow may be utilized to synthesize ATP and other energy-rich compounds to store light

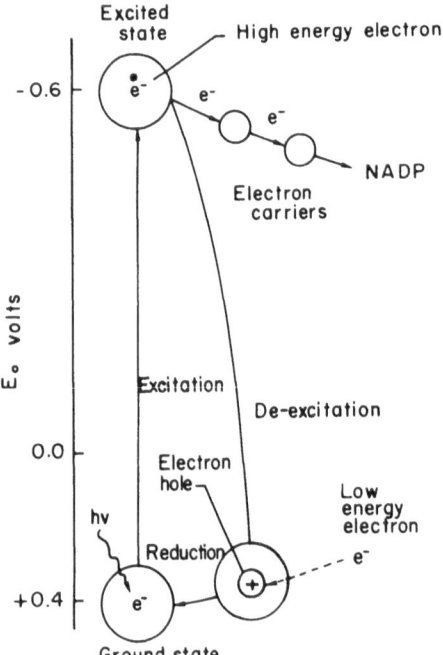

Figure 15. Boosting the energy level of transferable electrons of the pigment assembly. The high-energy electrons leaving the excited pigment are used to reduce electronegative acceptors such as $NADP^+$. Low-energy electrons coming from a more electropositive donor can fill the electron holes (Reference 29).

energy as chemical energy. This latter process is similar to various electron transport processes in respiratory systems (e.g., in mitochondria).[29]

2. Electrical Potential at Biomolecular Interfaces

Biochemical and biophysical activity often occurs at the interfaces of biomolecules and various types of assemblies of such molecules.[30] Several biomolecular interfaces considered are listed in Table 13.

One of the typical examples of such biomolecular interfaces is a "membrane", a two-dimensional molecular array which forms the boundary of biological cells. These membranes are composed of lipids and proteins. Of these, the lipid molecules which are amphipathic (a molecule consisting of two parts, each of which has an affinity for a different phase), are considered to be responsible for maintaining the two-dimensional molecular structure; the protein molecules, which perform a variety of biochemical functions, are often associated with (in/on) these lipid bilayer membranes (Figure 16).

In order to elucidate the physicochemical properties of such a biological membrane interface, several model membrane systems (lipid monolayer, lipid bilayers, and protein-incorporated lipid model membrane systems) which mimic biological membrane interfaces have also been studied.[30–36] In particular, many properties at the membrane surface are intimately related to the electrical potential originating from the fixed charge or electrical polarization of the membrane constitutents.

Table 13
Biological Interfaces[a]

Type of interface	Examples
Biological macromolecule–biological macromolecule	Deoxyribonucleic acid (DNA)–histone
	Antibody–antigen interactions
Biological macromolecule–small molecule or ion	Membrane protein–phospholipid
	Membrane protein–cholesterol
	Adenosine triphosphatase (ATPase)–adenosine triphosphate (ATP)
Biological macromolecule–solution	Ribonucleic acid (RNA)–cytoplasm
	Any protein–solution interface
Cell organelle–cytoplasm	Nuclear membrane–cytoplasm
	Endoplasmic reticulum–cytoplasm
Cell–solution	Blood cell membranes–blood serum
Cell–cell	Any cellular tissue
Tissue–gas	Skin, lung alveoli
Tissue–solid	Surgical implants, e.g., synthetic heart valves, dental adhesives

[a] Reference 30.

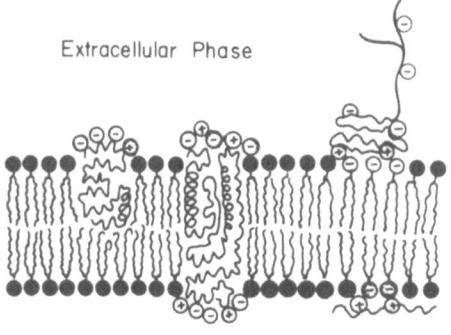

Extracellular Phase

Intracellular Phase

Figure 16. A schematic diagram of bio-membrane structure (a cross section normal to the membrane surface), showing the charged hydrophilic amino acid side groups projecting into the aqueous phase and the uncharged hydrophobic groups in contact with the lipid phase of the bilayer.

We shall consider the electrical potential and its related phenomena occurring at the monomolecular membrane surface.[30–33] The monomolecular membrane may be considered to correspond to half of the biological membranes. Although such a membrane interface in zero approximation is a two-dimensional array of discrete molecules, as depicted in Figure 17, we idealize it as a smooth, continuous two-dimensional interface and consider the cases of two simplified model surfaces (a flat plane and a spherical surface (Figure 18A, B).

2.1. Equilibrium Interfacial Electrical Potentials

2.1.1. Interfacial Potential Due to Surface Charge and Electrolyte

2.1.1.1. Double Layer Potential for a Charged Flat Surface
Let us consider an infinite flat surface, which has a surface charge density σ; it is also assumed that phases separated by the surface are a nonaqueous phase having a dielectric constant ε', and an electrolyte phase with dielectric constant ε. In this situation, the electrical potential at a distance x in the electrolyte solution from the surface is given by the Poisson–Boltzman equation [Eq. (7)].

The electrical potential at such a membrane surface, with respect to the potential at infinite distance as zero, can be obtained under the following ideal conditions: (1) the surface charge is uniformly distributed, (2) ions in the aqueous solution are considered as point charges, (3) the dielectric constant of the aqueous phase is ε everywhere up to the boundary plane.

$$\sigma = -\left(\frac{\varepsilon kT}{2\pi}\right)^{1/2}\left[\sum_j n_j^0(\infty)(e^{-Z_j e\psi(0)/kT} - 1)\right]^{1/2}$$

$$= -\left(\frac{\varepsilon kTN}{2000\pi}\right)^{1/2}\left[\sum_j C_j(\infty)(e^{-Z_j e\psi(0)/kT} - 1)\right]^{1/2} \qquad (70)$$

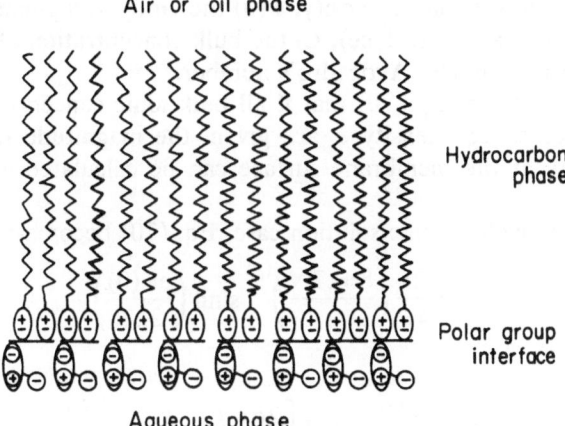

Figure 17. A schematic diagram of a phospholipid monolayer at the air/water or oil/water interface.

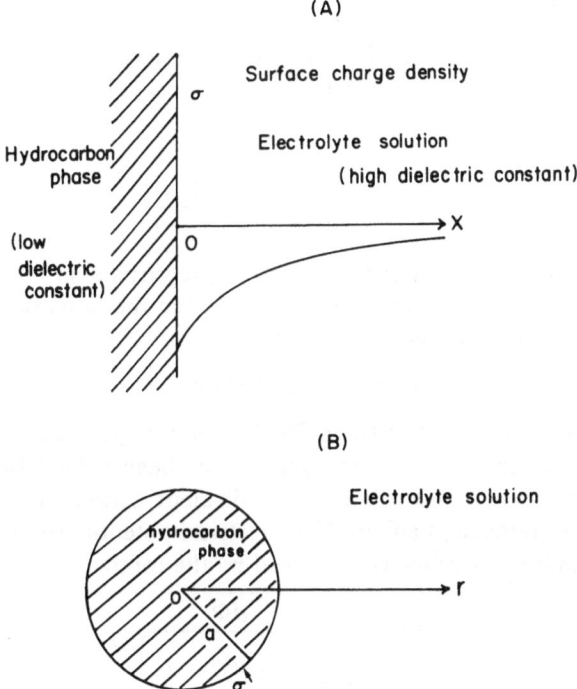

Figure 18. Idealized interfaces with planar (A) and spherical (B) boundaries. The shaded area and the nonshaded area refer to the hydrocarbon phase and the aqueous phase, respectively.

where σ is the surface charge density, $\psi(0)$ the Gouy–Chapman double layer potential at $x = 0$ (at the surface), C_j the bulk concentration of the jth ionic species (mole/liter) and N Avogadro's number.

It is obvious from Eq. (70) that if all bulk ionic concentrations, $C_j(\infty)$, and the surface charge density σ are given, the magnitude of the surface potential, $\psi(0)$, at the membrane surface can be calculated analytically or numerically.

For a uni-univalent ionic solution case, Eq. (70) becomes:

$$\sigma = \frac{2\varepsilon}{4\pi} \left(\frac{8\pi kTN}{\varepsilon} \right)^{1/2} \sinh \left(\frac{e\psi(0)}{2kT} \right) \tag{71}$$

or

$$\psi(0) = \frac{2kT}{e} \sinh^{-1} \left[\frac{2\sigma\pi}{\varepsilon} \left(\frac{\varepsilon}{8\pi kTN} \right)^{1/2} \right] \tag{72}$$

The double layer potential at a finite point away from the membrane is difficult to obtain except for a few special cases, such as uni-univalent, and uni-univalent and one divalent electrolyte cases.

2.1.1.1.1. Uni-univalent Electrolyte Case. The potential ψ at a point x is expressed[80] by:

$$\psi(x) = \frac{2kT}{e} \ln \left[\frac{1 + \alpha \exp(-\kappa x)}{1 - \alpha \exp(-\kappa x)} \right] \tag{73}$$

and

$$\alpha = \frac{\exp[e\psi(0)/(2kT)] - 1}{\exp[e\psi(0)/(2kT)] + 1} \tag{74}$$

where ψ is the potential at a distance x from the surface, in a diffuse double layer and κ is the Debye constant [Eq. (8)]. When the potential is relatively low ($kT \gg e\psi$), this equation is approximated as:

$$\psi = \psi(0) \exp(-\kappa x) \tag{75}$$

2.1.1.1.2. Uni-univalent and One Divalent Electrolyte Case. The electrical potential of a smeared charged membrane has been solved by Abraham–Shrauner[37] for 2–1–1 electrolytes (one divalent cation, one monovalent cation, and one monovalent anion). The electrical potential $\psi(x)$ at a distance x from the membrane surface is expressed by the following equation:

$$X + D = \int_x^\infty \frac{d\Phi}{\left[\sum\limits_{j=1}^{n} \chi_j (e^{-Z_j\Phi} - 1) \right]^{1/2}} \tag{76}$$

where $\Phi = e\psi/kT$, $\chi_j = n_j/n$, $X = x/\lambda_D$, D is an integral constant and $\lambda_D^2 = kT\varepsilon/2q^2n$ [λ_D is proportional to the Debye length ($1/\kappa$)]. Here, n_j is the number density of the jth species, n is the number density of the monovalent cation, q the unit positive charge and ε is the permittivity of the solution (MKS Unit).

2.1.1.2. Double Layer Potential for a Charged Spherical Surface

When the interface is spherical, the electrostatic potential is solved explicitly only for a linearized case (Debye–Hückel approximation [Eq. (8)], although there have been a number of attempts to obtain the exact solution.[38–42,82]

We consider a hard sphere of radius a which is in an electrolyte solution, and the spherical surface possesses a uniform charge density. We also assume the system has spherical symmetry. Then, Eq. (11) is correct when the distance from the center of the sphere, r, is greater than a (more exactly $r > a + r'$, where r' is the radius of the surrounding ion):

$$\psi(r) = \frac{Ze}{\varepsilon r} \frac{e^{\kappa(a-r)}}{1 + \kappa a} \qquad (r > a) \qquad (77)$$

This equation is equivalent to Eq. (14). The screening parameter κ is seen to regulate the rate of decay of the electrical potential. For large particles, for which the double layer thickness $(1/\kappa)$ is small compared with the particle radius, the equation must approach the corresponding equation [Eq. (75)] for small potentials for the flat double layer.

For small particles, however, the factor $1/r$ [in Eq. (77)] causes the electrical potential to fall off more rapidly than the purely exponential expression. If indeed the particle is small in comprison to the thickness of the double layer, the charge in the surrounding ionic atmosphere going from the particle surface to the bulk of the solution, must be distributed among spherical shells of increasing volume. This explains the characteristic difference between spherical and flat double layers, and shows that the equations for the spherical double layer are especially important for $\kappa a \ll 1$. Müller[43] calculated numerically the electrical potential around spherical particles for large values of the potential. Comparison of the results of Müller with those corresponding to the Debye–Hückel equations shows that the differences are not very large, though it should be considered that Müller's results do not go to very high values of the potential where the deviations should be greater (Table 14).

2.1.1.2.1. Ion Binding and Adsorption: Binding Site Model. When the surface charged sites interact with ions in aqueous solution and bind to them (similar to an ionic bond), the net fixed surface charges will be altered. In order to obtain the free net charge, several models can be used to describe such ion binding reactions coupled with the surface electrical potential:

1. One charged site to one ion binding model
2. Two charged sites to one ion binding model
3. Complex binding model

Table 14
Illustration of the Difference Between the Approximation
of Debye and Hückel and the
Theory of Müller for Spherical Particles

κr	$\varkappa(r - a)$	$\dfrac{Ze\psi}{kT}$	
		Müller[43]	Debye–Hückel
0.2	0.0	2.83	2.78
0.5	0.3	0.82	0.82
1.0	0.8	0.25	0.25

For model 1 we assume that one ion binds to one charged site. Then such binding reactions can be described as:

$$[A^-] + [M^+] \rightleftarrows [AM] \tag{78}$$

where $[M^+]$ is the concentration of monovalent cation, $[A^-]$ is the surface concentration of surface fixed charge sites, and K_1 is the association constant for the above binding reaction:

$$K_1 = \frac{[AM]}{[A^-][M^+]} \tag{79}$$

Therefore, the free net charge density σ is given by:

$$\sigma = \frac{\sigma^{int}}{1 + K_1[M^+]^s} = \frac{\sigma^{int}}{1 + K_1[M^+]_0\, e^{-e\psi(0)/kT}} \tag{80}$$

where $[M^+]^s$ and $[M^+]_0$ are the concentrations of monovalent cation, M^+, at the membrane surface and in the bulk solution, respectively, $\psi(0)$ is the electrical potential at the membrane surface, and σ^{int} is the surface charge density for the case of no ion-binding. With use of Eqs. (70) and (80), one can compute for one of the unknown quantities in the equations.

For model 2 the free net charge density σ is expressed by:

$$\sigma = \frac{\sigma^{int}}{1 + K_2[M^{2+}]^s} = \frac{\sigma^{int}}{1 + K_2[M^{2+}]_0 \exp\left(-\dfrac{2e\psi(0)}{kT}\right)} \tag{81}$$

where K_2 is the association constant for the following binding reaction:

$$[A^{--}] + [M^{++}] \overset{K_2}{\rightleftarrows} [AM] \quad \text{and} \quad K_2 = \frac{[AM]}{[A^{--}][M^{2+}]} \tag{82}$$

where $[A^{--}] \equiv [A -]/2$. This corresponds to the case where one divalent cation binds two nearest neighboring negative charge sites for which the concentration is considered as half of the total concentration of negative charge sites. K has the dimension of M^{-1}.

By use of the above treatment, several investigations on ion binding to phospholipid membranes have been carried out[15,44–50,249,276,277] (Table 15).

For model 3 the binding scheme has been considered by Cohen and Cohen[51] in a more rigorous manner, especially for divalent cation binding to the membrane sites. They consider one site for one divalent cation in addition to two fixed charge sites for one divalent cation. By use of a statistical average method, various possibilities for nearest neighbor sites of divalent cations were taken into account. Although a more realistic situation for ion binding schemes was introduced and gave a new binding constant at high divalent ion concentrations, all methods (the previously mentioned simple ion binding reaction and the latter more complete ion binding reaction scheme) generated approximately the same results at low concentrations of the divalent ion (in the range less than mM). The similar studies to the above have been done also by others (ref. 305, 306).

2.1.1.1.2. Ion Binding and Adsorption: Ion Condensation Model. In the ion condensation model the solution of the Poisson–Boltzmann equation for a charged infinitely long cylinder (cylindrical polymer ion) in an electrolyte solution leads to the following important result[52–54]: the counterion con-

Table 15

Intrinsic Ion Binding Constants (M^{-1}) to Various Phospholipid Membranes

Ion-Membrane System	Reference

Phosphatidylcholine membranes. 1:1 binding scheme

Mn^{2+}	Mg^{2+}	Ca^{2+}	Ni^{2+}	Sr^{2+}	Ba^{2+}								
0.3	1.0	1.0	1.2	2.8	3.6								(277)

Phosphatidylserine membranes. 2:1 binding scheme

La^{3+}	Mn^{2+}	Ba^{2+}	Ca^{2+}	Sr^{2+}	Mg^{2+}	Li^+	Na^+	K^+	Reference
			35		40		0.8		(45)
100									(49)
	49	37	30	25	10	0.4	0.6	0.2	(249)
			30		10		0.6		(15)
							0.4–12		(276)

Phosphatidylserine membranes. 1:1 binding scheme

Ni^{2+}	Co^{2+}	Mn^{2+}	Ba^{2+}	Sr^{2+}	Ca^{2+}	Mg^{2+}	Li^+	Na^+	NH_4^+	K^+	Cs^+	TEA	Reference
40	28	25	20	14	12	8							(50)
							0.8	0.6	0.17	0.15	0.05	0.03	(47)

Phosphatidylglycerol membranes. 1:1 binding scheme

Mn^{2+}	Ca^{2+}	Ni^{2+}	Co^{2+}	Mg^{2+}	Ba^{2+}	Sr^{2+}	Reference
11.5	8.5	7.5	6.5	6.0	5.5	5.0	(48)

centration at the cylinder surface (i.e., the concentration within the "condensation" layer of the order of one or two molecular diameters from the surface plane) is independent of bulk counterion concentration. The counterions are assumed to bind to the polymer ion to reduce the effective charge density on the polymer ion surface to a constant value (σ_{crit}). For a given charge density (σ) on the polymer ion ($\sigma > \sigma_{crit}$), once the number of bound ions is large enough to give a net charge density of the critical value, no more ions are bound. Here, a critical value of the effective charge density is one unit charge per $eZ/\varepsilon kT$ of length,[52,53] where Z is the valence of the counterions.

The solution of the Poisson–Boltzmann equation for the case of two parallel charged plates with an intervening solution containing only counterions also gave the ion condensation.[54,55] However, in this case the critical parameter is determined by the product of the surface charge density on the plate (σ) and the distance (d) between the plates:

$$(2|\sigma|Zd)_{crit} = 1.31\,\varepsilon kT/e \tag{83}$$

In considering ion binding to a membrane surface, the above mentioned condensation layer partially neutralizes the initial surface charge density, so that the quantity $f = 1 - \sigma/\sigma_0$ is independent of the concentration of counterions in bulk solution: here σ_0 is the surface charge density when no counterions are present in the condensation layer, σ is the apparent charge density when neutralization of the surface by the counterions in the condensation layer is taken into account, and f represents the fraction of surface charge neutralized by condensed ions. Specificity should be taken into account in the ion condensation model, as it might be manifested in competition between two ionic species i and j for the condensation layer, by introducing a relative distribution parameter, K_{ij}:

$$[C_i]^s/[C_i]^s = K_{ij}[C_j]/[C_i] \tag{84}$$

where $[C_i]^s$ and $[C_j]^s$ are the respective concentrations of counterionic species i and j in the condensation layer and $[C_i]$ and $[C_j]$ are the respective concentrations of ions i and j in bulk solution. If the parameters f_i and f_j are defined as the fractions of surface charge neutralized, respectively, by ion species i and j in the condensation layer (with $f_i + f_j = f$), then one obtains the following relation:

$$f_i = f/(1 + K_{ij}[C_j]/[C_i]) \tag{85}$$

where f, the total fraction of surface charge neutralized, is, as before, independent of the concentration of ions in bulk solution.

On the other hand, the binding site model mentioned in the preceding section assumes that counterions of type i bind independently to single sites at the membrane surface with an intrinsic binding constant K_i. Then, the fraction of surface charge neutralized by ion species i is:

$$f_i = Y_{i0}K_i[C_i] / \left(1 + \sum_j Y_{j0}K_j[C_j]\right) \tag{86}$$

where $Y_{j0} = \exp\left(-eZ_j\psi(0)/kT\right)$.

These two models have in common the following features: (1) the discrete nature of the surface and the solution at the interface is ignored (i.e., ignored in terms of a detailed molecular structure picture); thus the negatively charged membrane surface is represented by a uniform surface charge distribution, with density σ at the interfacial plane, and (2) the solvent is treated as a continuous medium, with uniform dielectric constant.

One obvious distinction between these two theoretical models is that the ion condensation theory gives a fractional neutralization (cation concentration in the condensation layer) which is independent of cation concentration in the bulk solution, while the binding site model does not explicitly predict such a result.

By comparing the above equations for the two ion adsorption models and investigating adsorption of counterions on membrane surfaces, we may argue which theoretical adsorption scheme is a preferable model for ion binding to lipid bilayer membranes as well as biological membranes. A recent study of monovalent cation adsorption on a phosphatidylserine vesicle membrane suggests that the ion binding site model is more appropriate for ion binding to negatively charged phospholipid membrane surfaces.[56]

2.1.1.3. Surface Potential

When amphipathic molecules are spread on either an aqueous surface or an aqueous/oil interface, they usually form a thin film composed of a molecular layer at the interface, and it is possible to measure a change in potential across this interface, which is a function of the electrolyte concentration in the aqueous phase.

This potential is caused by a partial separation of positive and negative ions of the electrolyte brought about by the interface dipole, dielectric constant, van der Waals forces, and fixed electric charges of the thin-layer surface. "Surface potential" is the usual name for this potential which is a static electrical polarization potential. This potential was first measured at the air–aqueous interface.[57] An accurate theoretical prediction of the measured surface potential in the above system is obviously difficult because of the uncertainty involved in assessing the surface electric dipole moment or polarization and an accurate dielectric constant of the electrolyte solution near the interface and of the film. Also a possible breakdown in the Poisson–Boltzmann equation causes some difficulties since the ionic strengths needed to produce accurately measurable potential changes are large. However, the surface potential changes ΔV are interpreted approximately by means of the equation (also see Figure 19):

$$\Delta V = \psi_D + \psi \tag{87}$$

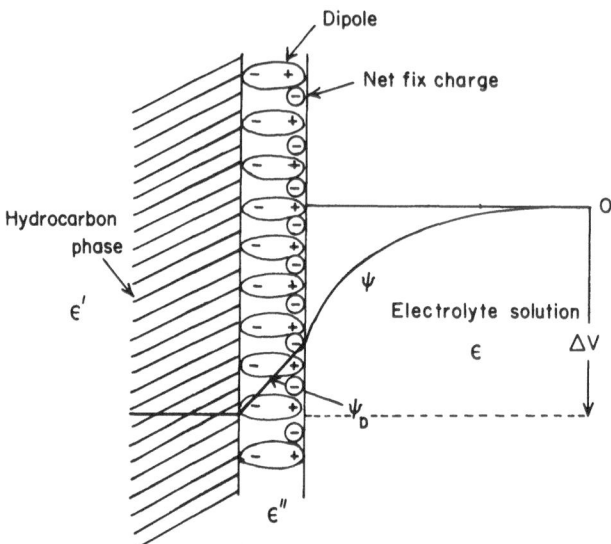

Figure 19. A schematic diagram of the surface potential profile. Surface potential: $\Delta V = \psi_D + \psi$, where ψ_D is the polarization potential of the membrane interface and ψ is the Gouy–Chapman diffuse layer potential. Here, it is assumed that either dipoles, μ_1 and μ_2, are zero.

where ψ_D is the effective dipole change normal to the interface created by the addition of one film and counterions, and ψ is the double layer potential in the interface relative to that far away in the bulk. ψ can be derived by solving the Poisson–Boltzmann equation for a certain ideal case, which has been discussed in the previous section.

The potential change due to the effective dipole normal to the interface can be divided into three parts (Figure 20):

$$\psi_D = 4\pi n \mu_1 + 4\pi n \mu_2 + 4\pi n \mu_3 \tag{88}$$

where μ_1 is the dipole moment or polarization of the film (monolayer) molecule at the air–film interface, μ_2 is the dipole moment or polarization of the bulk component of the film molecule, and μ_3 the dipole moment (polarization) of the hydrophilic polar group/aqueous molecular (H_2O + ions) complex at the film–water interface and n is the number of molecules per unit area. The value of μ_1 is the quantity which can be measured (Table 16); the value of μ_2 is usually negligibly small or small but constant since the membrane interior is composed of hydrocarbon chain substances and as a whole does not possess a permanent dipole, or if any, its magnitude is small. The value of μ_3 can be evaluated by subtracting the contribution of μ_1, μ_2, and ψ from the total surface potential change ΔV, or by summing up the normal component of each molecular bond dipole of the molecular complex. However, the theoretically detailed molecular orientation of the polar groups, H_2O and ions, is difficult to determine. The orientation of each polar segment (bond polarization) depends upon the structure of the molecular complex and the molecular

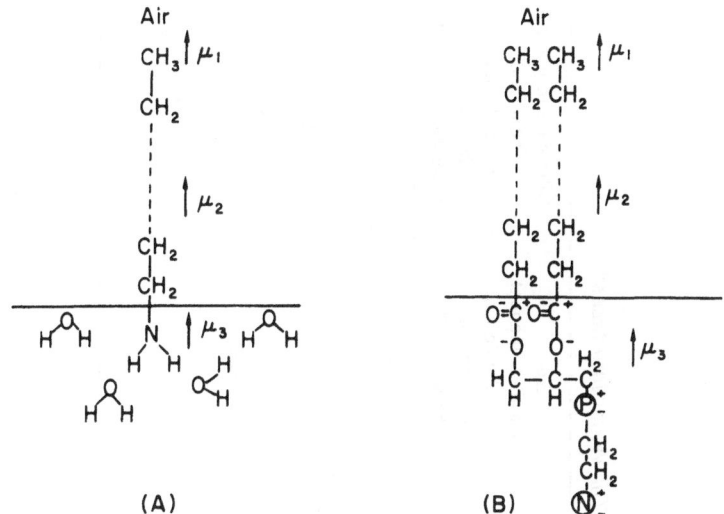

Figure 20. Components μ_1, μ_2, and μ_3 of the surface potential ΔV for electrically neutral (A) and charged (B) monolayers, respectively.

packing, and also depends upon the type of interaction with H_2O molecules and ions from the electrolyte solution. Spectroscopic studies—Raman, infrared and magnetic resonance (ESR and NMR)—may provide useful information about detailed molecular structure and packing. Some of the dipole moments of biologically important molecular bonds or molecular groups are listed in Table 17.

The surface potential due to a film formed at an oil–aqueous interface can also be analyzed in a manner similar to that above.

As for the calculation of the double layer potential, ψ, there are several assumptions[32,75] used in calculating ψ theoretically which do not correspond to the real systems, as briefly mentioned in the above. They are:

a) *Point charge assumption*: It is well known that if high but possible values of the potential ψ and bulk ion concentrations are substituted into the Boltzmann equation [Eq. (4)], impossibly high surface concentrations of ions are often predicted. This problem arises from the assumption of point charges instead of ions of finite volume. A number of authors have attempted to introduce the ionic volume correction into double layer theory by rewriting the Boltzmann equation with some form of space restriction factor.[58,59]

b) *Uniform dielectric constant assumption*: The Poisson equation contains a term for the continuous dielectric constant of the medium in which the space charge is located. This simple assumption is usually used in calculation of the diffuse double layer potential: the dielectric constant is the same as that in the aqueous solution. However, it is expected that the dielectric constant would vary from a high value in the bulk aqueous solution to a lower value in the vicinity of the membrane surface. The accumulation of ions in the diffuse double layer should influence the dielectric constant of that medium and also

Table 16

Variations in μ_1 for Various Films at the Air–Water and Oil–Water Interfaces[a]

Film	ΔV (mV)	$\Delta(\Delta V)$ (mV)	$\Delta\mu_1$ (vertical component of dipole differences in ω-bonds) (mD)	Difference in μ for ω-bond and C—H bond, from bulk measurements (mD)
Myristic acid with carboxyl group ionized (25 Å²). Air–water	−50	−900	−600	−1800
Perfluorodecanoic acid with carboxyl group ionized (25 Å²). Air–water	−950			
Stearic acid (pH 8.2). Air–water	0	−1190	−800	−1800
ω-Trifluorostearic acid (pH 8.2). Air–water	−1190			
Hexadecanoic acid on 1 N NaOH (at 66 Å²). Air–water	−28	0	0	−1900
ω-Bromohexadecanoic acid on 1 N NaOH (at 66 Å²). Air–water	−28	−132	−230	−1900
ω-Bromohexadecanoic acid on 1 N NaOH (at 66 Å²). Oil–water	−160			
Hexadecanoic acid (pH 4). Air–water (20 Å²)	+390	−1260	−660	−1900
ω-Bromohexadecanoic acid (pH 4). Air–water (20 Å²)	−870			

[a] Reference 31.

Table 17
Dipole Moments of Molecular Bonds
and Molecular Groups

Dipole Moments of Bonds $(mD)^a$

Group Moments in Debye Units[b,c]

Group	Aliphatic compounds	Aromatic compounds
CH_3-C	0 (0°)	0.4 (0°)
$C-Cl$	2.1 (0°)	1.6 (0°)
$C-Br$	2.0 (0°)	1.5 (0°)
NH_3-C	0.8 (100°)	1.5 (142°)
$C-CN$	3.8 (0°)	3.9 (0°)
$C-NO_2$	3.5 (0°)	3.9 (0°)
$C-OH$	1.7 (60°)	1.6 (60°)
$C-OCH_3$	1.3 (70°)	1.3 (70°)
$C-COOH$	1.7 (74°)	1.6 (74°)

[a] Reference 31.
[b] Reference 278.
[c] The angle in brackets denotes the deviation (in degrees) of the group moment from the bond axis.

the molecular adsorption, including water molecules as well as chemically bound ions at the membrane surface which contribute to change of the dielectric constant.[60,61] There are several semiempirical formulas used to calculate the dielectric constant near the membrane surface. The complete theoretical analysis of the behavior of the dielectric constant near the membrane surface is not yet well established, although there have been a number of attempts to elucidate this problem.[60-62,81]

c) *Non Ion polarization assumption*: When an ion is brought from the bulk phase to the diffuse double layer by an electric field, the ion becomes polarized. This polarization contributes an extra energy term to the Boltzmann equation.[58,62]

d) *The self-atmosphere effect of the counterions*: In the double layer as in the bulk phase, there may be a finite free energy of interaction between neighboring ions. Loeb[63] and Williams[64] have both considered this correction to the Poisson–Boltzmann equation. By comparison with the uncorrected Poisson–Boltzmann result, the self-atmosphere effect reduces the potential and the degree of the reduction becomes less as κx increases.

e) *Smeared charge assumption for surface charges*: The surface charges of the biomembrane surface are obviously discrete rather than uniformly distributed. The effect of discrete membrane surface charges has been studied by a number of investigators.[65-73].

f) *Other assumptions*: Levine[74] pointed out that the *electrostriction* term should be corrected for the Poisson–Boltzmann equation. However, the magnitude of this term depends on ψ^2 and is still negligibly small even at the highest potentials normally encountered in diffuse double layers. Also, as was pointed out above, when the dielectric medium is not uniform, the existence of charge in the medium gives rise to a complex electrical potential treated as an "image potential" effect. This would affect the ion distribution near the membrane which in turn affects the diffuse double layer potential. Other noncoulombic interaction factors between the counterions and the membrane surface should be considered as a correction to the Poisson–Boltzmann equation.

Nevertheless, there have been many attempts to measure surface potential in order to correlate the surface charge of cells and their function. These methods include:

1) measurements of surface potential of amphipathic monomolecular films formed at either the air/water interface or the oil/water interface[75-79,46,135] as mentioned before. Since the measurable quantity as the surface potential is $\Delta V = \psi_D + \psi$ [Eq. (87)], if such quantities are measured at two different subphase electrolyte solutions, the difference between the two quantities will be equal to the difference in the corresponding double layer potentials, provided that the polarization potential term ψ_D is constant.

$$\Delta(\Delta V) = \Delta V(1) - \Delta V(2) = \psi_D(1) + \psi(1) - \psi_D(2) - \psi(2) \cong \psi(1) - \psi(2). \tag{89}$$

2) Measurements of membrane conductance caused by small amounts of ions[47,83-87] and zwitterion surfactants.[88] The conductance G may be written as

$$G = \sum_j (Z_j F)^2 U_j(C_j)/h = \sum_j (Z_j F)^2 (U_j/h) K_j[C_j] \exp\left[-Z_j F(\psi_0 - \psi_m)/RT\right], \tag{90}$$

where (c_j) and $[C_j]$ are the concentration of jth ionic species in the membrane and in the bulk phases, respectively, U_j its mobility, K_j its partition coefficient between the membrane/aqueous phases, h the thickness of the membrane, and ψ_0 and ψ_m are the electrical potential in the bulk and at the membrane surface, respectively; $\psi_0 - \psi_m$ corresponds to the surface potential of the membrane. Therefore, by measuring membrane conductances under different ionic conditions, we can relate the difference in conductances to the difference in surface potential of the membrane under the two conditions.

3) The distribution of paramagnetic amphiphiles between membrane and bulk phases measured by electron spin resonance spectroscopy.[89]

4) The measurements of the electrophoretic mobility of liposomes, cells, and organelles.[90,91] By this method, one can observe the ζ-potential of the cell surface. It will be seen later (electrokinetic potential) that this potential (ζ) can be related to the surface potential.

5) Another method to measure cell surface potential is the use of fluorescence spectroscopy. The fluorescent chromophores attached to a long hydrocarbon chain can be introduced to the membrane surface as a probe and, utilizing the variation of fluorescence intensity of these probes as a function of H^+ concentration at the membrane surface, the magnitude of the surface potential can be determined.[17,92,93]

2.1.2. Interfacial Potential Due to Molecular Adsorption and Polarization

2.1.2.1. Molecular Adsorption

One basic difficulty with the Gouy–Chapman Theory in systems involving an impenetrable flat surface or electrode is that, since the ions have finite size, the distance of closest approach of their centers to the surface is finite. Thus, the potential which appears in the Gouy–Chapman equation is not equal to the surface potential $\psi(0)$, but is the potential in the plane of closest approach of the counterions to the surface.

This problem was recognized and a nondiffuse region of the double layer was introduced by Stern.[94] Stern divided the liquid phase into two parts. One part is represented in the theory by a surface charge concentrated in a plane at a small distance δ from the surface charge on the surface. This distance δ is assumed to be of the order of magnitude of several Å. The second part of the liquid charge is then taken to be a diffused space charge extending from the plane $x = \delta$ to infinity (Figure 21). Between $x = 0$ (surface) and $x = \delta$ (the adsorbed layer), the electrical potential changes linearly.

According to Stern, the ratio, f_j, of the number of sites occupied by the jth (monovalent cation only) ionic species, N_j, to the total number of ion adsorbable sites, N_o, in the adsorption layer, is expressed by

$$f_j \equiv \frac{N_j}{N_o} = \frac{\dfrac{n_j}{n_o} \exp\left[-\dfrac{eZ_j\psi_{\delta j}}{kT}\right] \exp\left[-\dfrac{\phi_j}{kT}\right]}{1 + \sum_j \dfrac{n_j}{n_o} \exp\left[-\dfrac{eZ_j\psi_{\delta j}}{kT}\right] \exp\left[-\dfrac{\phi_j}{kT}\right]} \qquad (91)$$

where n_j and n_o are the number densities of the jth ionic species and the solvent, respectively, ψ_δ is the potential at the boundary between the adsorption (nonspecific) and diffused layers, and ϕ_j allows for any additional specific adsorption potential. There, ψ_β is the potential energy for specific ion adsorption. For nonspecific adsorption, $\psi_\beta = \psi_\delta$, there are only electrostatic forces to cause adsorption of ions. The energies involved in the specific adsorption are considered to be chemical in nature. For ions, the term $1/n_o \exp\left[-(\phi_j/kT)\right]$

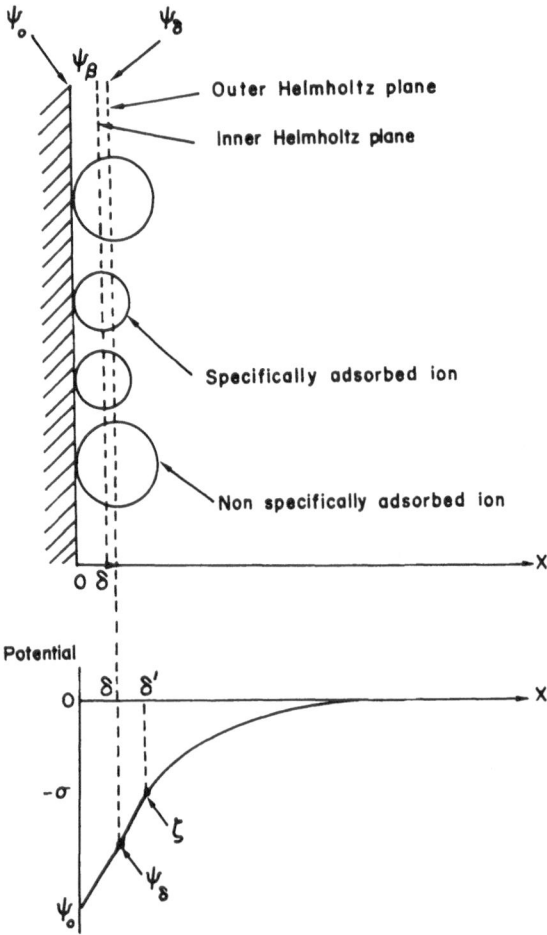

Figure 21. A schematic diagram of the Stern adsorption layer (top) and the average potential profile of the Stern layer and Gouy–Chapman diffuse double layer.

is considered as the binding constant to the adsorption site. The potential ψ_δ is related to $\psi(0)$ in the following equation:

$$\sigma = \frac{\varepsilon'}{4\pi\delta}(\psi(0) - \psi_\delta) \tag{92}$$

where σ is the surface charge density and ε' the dielectric constant of the Stern layer region.

The above-mentioned adsorption is similar to that for a Langmuir type monomolecular adsorption where the adsorbed molecules are not mobile. The modification required to take into account the mobility of adsorbed molecules has been discussed by several authors.[75]

Nonspecific adsorption phenomena have often been discussed in terms of discrete charge effects. In order to describe the Stern type of adsorption of ions, the adsorption of a particular ion is considered to involve the rearrangement of the existing ions so as to form a hole for the ion to be adsorbed and then the adsorption of this ion at the hole, where the potential would be smaller than that calculated on the basis of a smeared surface charge. It has been shown that the rearrangement of ions to form a hole requires only a little energy[95] so that the important problem is the calculation of the potential in the hole, known as the micropotential (local potential).

2.1.2.1a. Local Potential (Discrete Charge Potential). The surface charges on the biomembrane surface are obviously discrete rather than uniformly distributed. Several authors[69-72] have studied the electrical potential produced by an arbitrary arrangement of discrete surface charges in membrane systems. In most cases, a mathematical procedure based on a linearized treatment of the discrete charge potential problem was used which would lead to an underestimated value of the micropotential in systems involving localized structure. The so-called "discreteness of charge effect" in metal electrode–electrolyte solution or ionized monolayer–electrolyte solution systems has also been studied by a number of authors.[65-68] These theoretical treatments include a change in dielectric constant near the metal–water interface and calculations of the electrostatic energy of ion adsorption, in addition to the discrete surface charge distribution.

Recently, Sauve and Ohki[73] have studied the discrete charge potential of a membrane–electrolyte system having three dielectric constant regions (nonpolar, fixed charge, and aqueous phases) (Figure 22). The mathematical description of the local potential produced by an arbitary arrangement of polar groups in contact with an electrolyte solution was derived. The main conclusions derived from the theory, are that (1) it is accurate to describe membrane phenomena which depend upon the space average value of the surface charge potential in terms of a smeared charge theory, (2) in systems involving localized structure such as specific ion channels, a uniform charge (smeared charge) approach will lead to an underestimate of the membrane surface charge density, and (3) the microenvironment of a localized structure is an extremely important factor in the general behavior of the local potential acting at that point. This effect results from the mutual interaction of the surface charges with each other's ionic atmosphere, and cannot be taken into account with a discrete charge theory based on a linearized form of the Poisson–Boltzmann equation.[71] The discrete charge approach should be required in cases of (1) discussion of the molecular mechanism of ion passage through ionic channels which have localized structures and (2) elucidation of specific ion binding or adsorption at the membrane binding sites which have also localized structures.

The specific nature of ion binding onto the charged membrane surface may be analyzed in terms of this micropotential, image potential, the size of

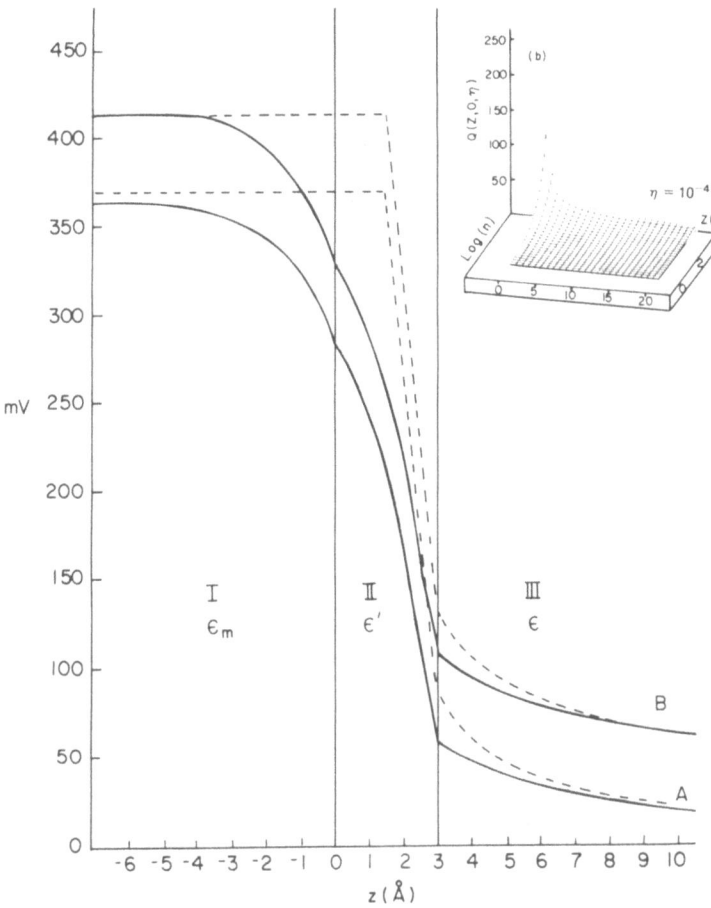

Figure 22. Comparison between spatially averaged discrete charge (continuous line) and smeared charge (dashed line) potentials, where (A) $[M^{2+}M_2^-] = 35$ mM, (B) $[M^{2+}M_2^-] = 1$ mM and $[M \cdot M^-] = 100$ mM. A membrane–electrolyte system having three dielectric constant regions [nonpolar ($\varepsilon_m = 2$), fixed charge ($\varepsilon' = 20$), and aqueous ($\varepsilon = 80$) phases] is represented, and the surface charge density was set to $e/48$ Å^2. The fixed charges were placed at a distance of 1.5 Å from the surface of the membrane on the hexagonal lattice. Q is a quantity relating to the electrical potential at a distance from the membrane surfaces at certain electrolyte solutions. n: the number of univalent negative ions in the bulk aqueous phase per cm^3, $\eta = 2C^{2+}/n$ where C^{2+} is the bulk concentration of divalent cations in number of ions/cm^3.

an ion, and its degree of hydration. Also, the structured water at the membrane surface seems to be an important factor for ion binding onto the membrane surface.

2.1.1.2. Molecular Polarization at Biosurfaces

As has been mentioned before, most biomembrane surfaces are exposed to aqueous phases where the polar groups of membrane molecules are situated.

They may form complex arrays of polar groups at the membrane surface (see Figure 19).

We may consider that such structures are two-dimensional arrays of dipoles on the membrane surface, and that the net dipole moment is directed perpendicular to, or at some angle with respect to, the plane of the membrane surface.

Here, let us consider a simple case where the molecular dipoles, μ, are all oriented perpendicular to the surface of the membrane (Figure 23A), and consider the electrical potential at a point situated at a distance z from the plane of the molecular dipole array. The electrical potential at this point due to a single molecular dipole (x, φ, z) is expressed in a way similar to that in

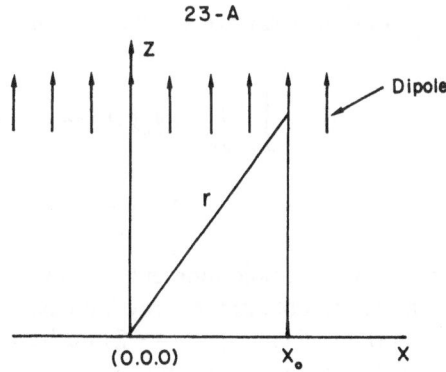

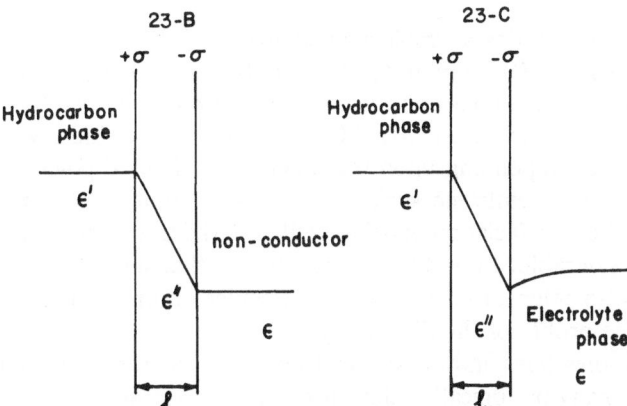

Figure 23. (A) A schematic representation of a cross section of a two-dimensional electric dipole array. (B) Electrical potential profile across a two-dimensional dipole array of infinite dimension, where the dipole array is at the hydrocarbon/vacuum interface. (C) The same dipole array as in (B) except that the dipole array is at the hydrocarbon/aqueous interface in this case.

Eq. (49) as

$$\psi_d = \frac{\mu \cos(\theta + \pi)}{\varepsilon r^2} = -\frac{\mu \cos \theta}{\varepsilon r^2} = -\frac{\mu}{\varepsilon} \frac{z}{(x^2 + z^2)^{3/2}} \tag{93}$$

Here, the cylindrical coordinate (x, φ, z) is used, the z-axis is parallel to the normal to the membrane surface, and the point concerned is chosen as the origin $(0, 0, 0)$ of the coordinate system.

The potential energy at the origin due to a disc (of radius x) of the molecular dipole array $\bar{\psi}_d$ is given as:

$$\bar{\psi}_d = \sum_{\text{disc}} \psi_d \tag{94}$$

The summation in Eq. (94) may be replaced by an integration for the case where the dipoles in the array are small and compact enough. Then, Eq. (94) becomes

$$\bar{\psi}_d \cong \int_0^{x_0} \int_0^{2\pi} n\psi_d x \, dx \, d\varphi$$

$$= -\frac{2\pi n\mu}{\varepsilon} \left[1 - \frac{z}{(z^2 + x_0^2)^{1/2}} \right] \tag{95}$$

where n is the density of the dipole moment per unit area. When x_0 is zero, $\bar{\psi}_d = 0$. As x_0 increases, the potential energy increases and, for a dipole disc of infinite radius, the potential energy $\bar{\psi}_d = -2\pi n\mu/\varepsilon$, which does not depend on the distance from the dipole array. This is shown schematically in Figure 23B. However, when such a dipole array faces an aqueous solution, the potential profile may be similar to that of the Stern–Gouy–Chapman potential (Figure 23C).

In this case, besides a "molecular condenser" due to the dipole array, $4\pi\sigma l = \varepsilon''(\psi(0) - \psi(l))$, similar to that of the Stern layer, a diffuse layer potential due to electrolytes and the charges of dipoles may be formed in the aqueous solution which is slightly different from the double layer potential discussed above. Depending upon the magnitude of the dipoles μ and their orientation at the membrane surface, the contribution of such polarization potentials to the interfacial potential as well as to the transmembrane potential could be considerable. In addition, it is possible that molecular (nonspecific or specific) adsorption of ions or water molecules occurs. This would further complicate the profile of the diffuse layer potential.

Many studies have indicated that there are structured water molecules adsorbed on many biologically relevant membrane surfaces[96] although such molecular adsorption is not only due to the coulombic and dipolar interactions but also nonelectrostatic molecular interactions. The structured water on the membrane surface seems to play an important role in the stability of the membrane surface, or in membrane–membrane interactions.[97–98] Recently, a rigorous calculation of the free energy of the dipole–dipole interaction between a structured water dipole and an ion dipole was made by Cevc and co-workers[99,100] suggesting that this interaction energy has a decisive role in the specificity of adsorption of ions to the membrane surface.

2.2. Nonequilibrium Interfacial Potential

2.2.1. Electrokinetic Potential

If an electric field is applied tangentially along a charged surface, a force is exerted on both parts of the electric double layer. The charged surface (plus attached material) tends to move in an appropriate direction, while the ions in the mobile part of the double layer show a net migration in the opposite direction, carrying solvent along with them, thus causing its flow. Conversely, an electric field is created if the charged surface and the diffuse part of the double layer are made to move tangentially relative to each other,

There are four electrokinetic phenomena (electrophoresis, electro-osmosis, streaming potential, and sedimentation potential), all of which involve both the theory of the electric double layer and that of liquid flow. Among them electrophoresis has the greatest practical applicability to the study of biomolecules and biocell surface porperties. In this section, the relation between electrophoretic mobility and its related electrokinetic potential ζ will be discussed.

When a particle surrounded by an electric double layer is subjected to an electric field, the Stern layer and a part of the diffuse double layer move with the particle. The electrical potential at the plane of shear between the bound and free parts of the double layer is called the zeta potential (ζ). It is considered that the shear plane is usually located at a small distance further out from the surface than the Stern plane and that ζ is generally marginally smaller in magnitude than the Stern potential (ψ_δ) (see Figure 21).

If we consider a spherical particle of radius a and electric charge Q, moving with a uniform velocity u in a field of unit electrical potential gradient, the electrical force on the particle, $QE = Q$, will be balanced by the viscous resistance which is assumed to be expressed by Stokes' law;

$$Q = 6\pi a\eta u \qquad (96)$$

where η is the viscosity of the medium.

If it is assumed that the particle radius a includes that part of the double layer which moves with the particle, then the zeta potential ζ can be expressed as the work done in bringing a unit positive charge from infinity to a distance a from the center of the particle. The force, F, on the unit positive charge at any point r from the center of the particle is:

$$F = \frac{Q}{\varepsilon r^2} \qquad (97)$$

Therefore,

$$\zeta = -\int_{\infty}^{a} F\, dr = -\frac{Q}{\varepsilon}\int_{\infty}^{a} \frac{dr}{r^2} = \frac{Q}{\varepsilon a} \tag{98}$$

Eliminating a between Eqs. (96) and (98), the mobility u of the particle is expressed in terms of ζ, ε and η:

$$u = \frac{\varepsilon\zeta}{6\pi\eta} \tag{99}$$

An alternative approach to this problem is to regard the double layer as a parallel plate condenser in which one plate is the particle surface and the other plate is a plane of counterions at a potential ζ located a distance from the surface and moving with a velocity u relative to the particle surface. If the surface charge density is σ, the electrical force per unit area of the particle plate in a field of unit potential gradient will be σ and this force will be balanced by the viscous resistance, which for an assumed Newtonian flow, leads to the equation:

$$\sigma = \eta\frac{u}{a} \tag{100}$$

The capacitance (C) per unit area of the double layer will be given by:

$$C = \frac{\varepsilon}{4\pi a} = \frac{\sigma}{\zeta} \tag{101}$$

Eliminating σ and a from Eqs. (100) and (101), the mobility is:

$$u = \frac{\varepsilon\zeta}{4\pi\eta} \tag{102}$$

Equation (99) differs by a factor of 2/3 from Eq. (102). The latter equation was first derived by Smoluchowski[101] and the former by Hückel.[102] The latter equation is used for values of ka greater than 10^3, and the former for values of ka smaller than 1.0.[103] Both equations are limiting forms of a more general expression which was derived by Henry[104]:

$$u = \frac{\varepsilon\zeta}{6\pi\eta} f(ka) \tag{103}$$

The equations discussed above are all first-order approximations. Actually, the movement of particles causes a deformation of the diffuse double layer (relaxation effect), which alters the above equations.[105,313]

However, the above simplified equations, and Eqs. (73) and (74) for the surface potential of a plane surface, with use of a certain value for a slipping layer distance (δ'), have been used in a number of studies on biological cell surfaces in regard to ζ potentials and their surface charge densities (Table 18).[106–108] However, in view of the complicated molecular structures of biomembrane surfaces, the use of these simplified equations and the interpreta-

Table 18
Electrophoretic Data for Various Cells[a]

Cell type	Electrophoretic mobility $(\mu\text{m sec}^{-1}\text{ V}^{-1}\text{ cm})$ (25°C)	σ electronic charge $\text{cm}^{-2} \times 10^{-13}$
Human erythrocytes	-1.16^{b}	0.73
Dog erythrocytes	−1.28	1.1
Toad erythrocytes	−0.81	0.5
Pig erythrocytes	−0.88	0.75
Chicken erythrocytes	−0.82	0.69
Epithelium (Toad)	−1.13	—
Kidney tumour (Hamster)	−1.15	1.0
SV-3T3	−1.72	2.2
Escherichia coli	−1.12	1.31
Hela cells	−0.98	—
Ehrich ascites	−1.07	1.31
Rat liver cells	−1.01	—

[a] References 107 and 108.
[b] Standard saline, pH 7.2, 25°C.

tion of the results from such analysis should be carried out with great care. Further studies of theoretical analysis of cell electrophoresis data have been made (ref. 309–313).

2.2.2. Electrical Potential Due to Ion or Electron Movement across an Oil–Water Interface

2.2.2.1. Ion Movement across an Oil–Water Interface

Let us suppose that the water and oil phases are in contact with each other and form an interface, and that each phase contains electrolytes A_wB_w and A_oB_o, respectively, where A is a cation and B an anion.

When the two phases are in equilibrium, the electrochemical potentials of the jth ionic species in the two phases are equal:

$$\mu_{jw} = \mu_{jo} \tag{104}$$

where $\mu_{j\beta} = \mu_{j\beta}^0 + RT \ln a_{j\beta} + Z_j F \phi_\beta$ (β = water or oil) and $a_{j\beta}$ is the activity of the jth ionic species in phase β.

The potential difference between the two phases at equilibrium is

$$\Delta\phi = \phi_w - \phi_o$$

$$= \Delta\phi_j^0 + \frac{RT}{Z_j F} \ln \frac{a_{jo}}{a_{jw}} \tag{105}$$

where

$$\Delta\phi_j^0 = \frac{-\Delta^0 G_j^{o \to w}}{Z_j F} = \frac{-1}{Z_j F} (\mu_{jw}^0 - \mu_{jo}^0) \tag{106}$$

Here, $\Delta\phi_j^0$ is called the standard ion transfer potential at the water/oil interface and is specific for each ionic species.

Let us suppose that $\Delta^0 G_{A0}^{o \to w}$ and $\Delta^0 G_{Bw}^{o \to w}$ have large positive values and $\Delta^0 G_{Bo}^{o \to w}$ and $\Delta^0 G_{Aw}^{o \to w}$ have large negative values, respectively. Then, from Eq. (106), $\Delta\phi_{Ao}^0$ and $\Delta\phi_{Bw}^0$ have large negative values and $\Delta\phi_{Bo}^0$ and $\Delta\phi_{Aw}^0$ have large positive values. When an external electrical potential is applied through the interface by way of two reversible electrodes, and the potential difference between the two phases, $\Delta\phi = \phi_w - \phi_o$, is changed so as to increase the magnitude of $\Delta\phi$ in the direction of the negative sign, a new equilibrium state will be attained at a certain external electric potential. For a cation in the oil phase, when $(\Delta\phi - \Delta\phi_{Ao}^0)(ZF/RT) \gg 1$, a_{Ao} is much greater than a_{Aw}. Therefore, there will not be any appreciable current due to cation movements from the oil phase to the water phase. However, as $\Delta\phi$ increases further in the same negative direction and becomes close to the $\Delta\phi_{Ao}^0$ value, then a_{Aw} should be a comparable amount with a_{Ao} at equilibrium. In other words, appreciable cation movements across the interface from the oil to water phases should occur. For anions in the water phase the situation is the same. The less negative value of these two standard potentials, $\Delta\phi_{Ao}^0(A:o \to w)$ and $\Delta\phi_{Bw}^0(B:w \to o)$, is defined as the cathode equilibrium potential. Then, if $|\Delta\phi|$ is greater than the $|\Delta\phi_{cath}|$, an observable current across the interface results. When the potential difference is increased in the positive direction, a similar phenomenon occurs with respect to the current and the applied potential.

As long as the potential difference between the two phases, $\Delta\phi$, lies between the values of $\Delta\phi_{cath}$ and $\Delta\phi_{anod}$ (either $\Delta\phi_{Bo}^0(B:o \to w)$ or $\Delta\phi_{Aw}^0(A:w \to o)$, whichever has smaller magnitude), no appreciable ionic current will be observed across the interface (such an interface is called an "unpolarized interface"), but for the values of $\Delta\phi$ outside the range, $\Delta\phi_{cath}$ to $\Delta\phi_{anod}$, there will be an appreciable ionic current flow observed, as shown in Figure 24.

When another ionic species C is in the system in addition to an assumed relatively high concentration of the above mentioned ionic species AB, the following relation holds for the ionic species C at the equilibrium state:

$$\Delta\phi = \Delta\phi_C^0 + \frac{RT}{Z_j F} \ln \frac{a_{Co}}{a_{Cw}} \tag{107}$$

Depending on the value of $\Delta\phi$ and $\Delta\phi_C^0$, there will be a current flow of the ionic species C only (see Figure 24). In this case, the AB electrolytes serve as supporting electrolytes for the species C.

The above principles have been applied to a number of studies on ionic transfer behavior across such interfaces, by use of voltammetry and polarographic techniques.

Although such ionic transfer processes surely occur at biological interfaces, not much work has been done for systems of biological cells.

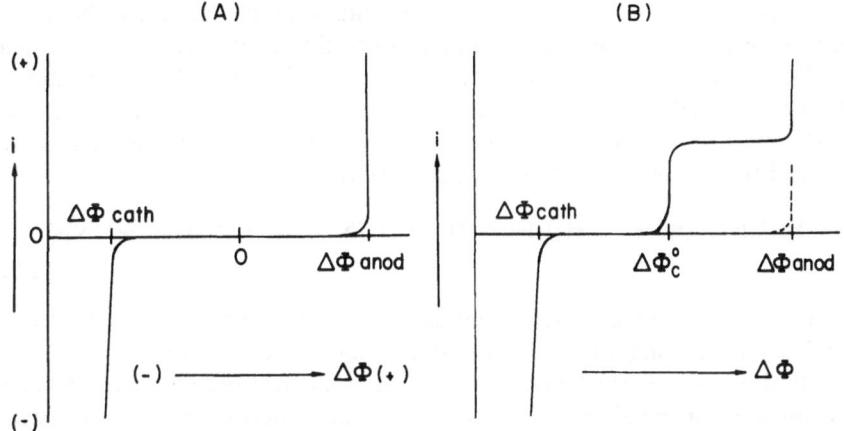

Figure 24. (A) Polarization of the oil–water interface and ionic movement (A^+B^- electrolyte) due to applied electric voltage. (B) Current–voltage relation due to ion movement (C ionic species) in A^+B^- supporting electrolyte solution across the oil/water interface.

2.2.2.2. Electron Movement across an Oil–Water Interface

Consider an electron exchange reaction occurring at the oil/water interface where the aqueous phase contains a redox agent pair Ox(w)/Red(w) and the oil phase contains another redox agent pair Ox(o)/Red(o). The electron exchange reaction at the interface can be written as follows:

$$Ox(w) + Red(o) = Red(w) + Ox(o) \qquad (108)$$

and for an n electron transfer, we have the following equilibrium:

$$\Delta\phi = \phi^0_{Ox(o)/Red(o)} - \phi^0_{Ox(w)/Red(w)} + \frac{RT}{nF}\ln\frac{a^o_{Ox}a^w_{Red}}{a^o_{Red}a^w_{Ox}} \qquad (109)$$

where

$$\phi^0_{Ox(\beta)/Red(\beta)} = -\frac{1}{nF}(\mu^0_{Red(\beta)} - \mu^0_{Ox(\beta)}) \quad (\beta = o \text{ or } w)$$

The first two terms in Eq. (109) are denoted by:

$$\phi^0_{Ox(o)/Red(o)} - \phi^0_{Ox(w)/Red(w)} \equiv \Delta\phi^{0o \to w}_{Ox/Red} \qquad (110)$$

which is the difference in standard redox potential of each redox agent pair in each solvent phase referenced to that of the standard redox potential. When the electron transfer reaction across the oil–water interface occurs within the range of the ideal "unpolarized interface" with respect to the other electrolytes in the system, the current due to an applied external potential represents the net current due to the electron exchange reaction of the redox agents at the interface.

Recently, Samec et al.[109,110] have studied at the interface the redox reaction system composed of an aqueous solution containing $Fe(CN)_6^{3-}$ and $Fe(CN)_6^{4-}$ and a nitrobenzene solution containing $(C_6H_5)_2Fe$. When an external electric potential difference was applied through such an interface, a characteristic electric current was observed, the behavior of which was analyzed in terms of the following electron transfer reaction:

$$Fe(CN)_6^{3-}(w) + (C_6H_5)_2Fe(NB) = Fe(CN)_6^{4-}(w) + (C_6H_5)_2Fe^+(NB)$$
(111)

where w and NB stand for the aqueous and nitrobenzene phases, respectively. Similar studies using another redox system have also been reported.[111]

Redox systems which resemble those in biological systems have also been investigated for the above charge transfer process across the water/oil interface. Boguslarvsky et al.[112] found that the electron transfer reaction between NADH(w) and Vitamin K_3 (decane) was catalyzed by the presence of chlorophyll molecules adsorbed at the interface

$$(NADH)^w + (A)^o \xrightarrow{\;^s(Chl)\;} (NAD)^w + (A^-)^o + (H^+)^w \qquad (112)$$

where $^s(\ )$ refers to the interface concentration. In this reaction system, the electron transfer reaction is coupled with proton release into the water phase. This resembles the situation in a chloroplast membrane system, where the electron and proton transfer reactions are coupled, which means that at least one of the membrane pumps contains molecules for which redox reactions involve the participation of protons.

Also, the same authors observed the interfacial potential change due to the proton movement (from the water to oil phases) across the octane/water interface in the presence of ATPase adsorbed at the interface, in conjunction with the hydrolysis of an ATP molecule in the water phase.[113] Anderson et al.[114] have reported an electron transfer phenomenon across membranes using reducing agent (w) in aqueous phase|Vitamin K_1 or Coenzyme Q_{10}(hexane)|reducible substance (w) systems [for example, methyl viologen (reducing agent) (w)|Vitamin K_1 (hexane)|$FMNH_2$ (reducible substrate) (w)], where coenzyme Q_{10} or Vitamin K_1 is present as a carrier molecule across the membrane and mediates the redox reaction at each interface (Figure 25). The neutral lipid coenzyme Q_{10} is an essential link in the electron transfer chain in mitochondria, and Vitamin K_1 may have a similar function in chloroplasts and certain bacteria.

Recently, a number of studies have been made on photoelectric redox potentials across thin lipid membranes.[115] These studies may give information about the molecular mechanism by which electrons (or photo-induced electrons) are transported and about the related energy synthesis phenomena in some biological systems.

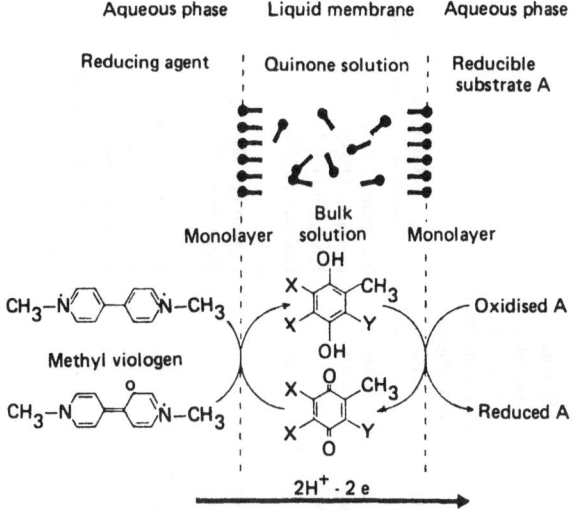

Figure 25. Schematic representation of an electron transport membrane model (after Ref. 114).

3. Transmembrane Potential across Cell Membranes

For many years, it has been known that many individual biological cells maintain different distributions in ionic concentrations and an electrical potential difference between their intracellular and extracellular phases at the resting state (Table 19). In some cells, upon application of an appropriate stimulus (electrical depolarization or chemical stimulus), the cells exhibit a time-dependent response via a potential difference across the cell membranes which does not necessarily follow Ohm's law. The former potential is called a "resting membrane potential" and the latter an "excitation potential." We would like to review the origins of these electrical potential differences across the cell membrane.

3.1. Resting Membrane Potential

3.1.1. Potential due to Ionic Distribution or Movement across Membranes

In order to explain this transmembrane potential, a number of authors have proposed membrane potential theories. Historically, the first membrane potential theory for biological systems was the use of the concept of the Donnan membrane equilibrium.

3.1.1.1. Equilibrium Potential: Donnan Membrane Equilibrium[116]

When ionic molecules are in an aqueous solution, the ion has electrochemical potential

$$\mu_j = \mu_j^0(p \cdot T) + RT \ln a_j + Z_j F \psi \qquad (113)$$

Table 19
Ion Distributions, Relative Ionic Permeabilities, and Resting Potential of Various Cells

Specimen	Extracellular (intracellular) ion concentration (mM)			Relative permeability	Resting potential $E = \psi_i - \psi_o$ (mV) at normal physiological pH	Reference[a]
	$K^+(K^+)$	$Na^+(Na^+)$	$Cl^-(Cl^-)$			
Loligo nerve axon	10 (400)	440 (50)	560 (40–150)	K > Cl > Na	−60	Hodgkin (1964)
Frog sartorius muscle	2 (125)	110 (15)	77 (1.2)	K > Cl > Na	−95	Hodgkin (1951)
Dog skeletal muscle	4 (140)	150 (12)	120 (−)	K > Cl > Na		Hodgkin (1951)
Visceral smooth muscle					−50–54	Holman (1958)
Red cells						
man	5.0 (.36)	155 (19)	112 (78)	Cl > K ≅ Na	−15	Lassen (1972)
rabbit	5.5 (142)	150 (22)	110 (80)	Cl > K ≅ Na		Bernstein (1954)
dog	4.8 (10)	153 (135)	112 (87)	Cl ≥ K		
L-cell	5.6 (171)	136 (8.6)	147 (70)	Cl > K ≅ Na	−15	Lamb and MacKinnon (1971)
Mouse fibroblasts L-cell					−10	Potapova et al. (1972)
Cortical cell (guinea pig fetus)[c]					0–7.6	Sedlacek (1969)
Fibroblast line (hamster kidney)	6.3 (−)	148.7 (−)		K > Na	−55	Sachs and McDonald (1972)
HeLa cell					−17	Borle and Loveday (1968)

[a] Cited in Reference 147.
[b] Potential value from Lassen (1972); pH 7.2.
[c] 150 days of gestation.

where μ_j^0 is the standard chemical potential of the jth ionic species and a_j its activity.

If the membrane is semipermeable to a certain ion, because of the electroneutrality as well as the equality ($\mu_{jo} = \mu_{ji}$) of the electrochemical potential for the permeable ionic species across the membrane at equilibrium state, the electrical potential difference between the two phases could be maintained at equilibrium:

$$\psi_i - \psi_o = \frac{RT}{F} \ln \frac{(a^+)_o}{(a^+)_i} = \frac{RT}{F} \ln \frac{(a^-)_i}{(a^-)_o}, \tag{114}$$

where (a^+), (a^-) are the bulk activities of monovalent cations and anions, respectively, and suffixes o and i refer to the outside and inside of the membrane.

We denote the ratio by:

$$\frac{(a^+)_o}{(a^+)_i} = \frac{(a^-)_i}{(a^-)_o} \equiv \gamma_D \tag{115}$$

which is called the Donnan ratio for membrane equilibrium.

When there are multivalent ions in the system, the Donnan ratio is generally written as:

$$\frac{(a_o^{Z+})^{1/Z}}{(a_i^{Z+})^{1/Z}} = \frac{(a_i^{Z-})^{1/Z}}{(a_o^{Z-})^{1/Z}} = \frac{(a^+)_o}{(a^+)_i} = \gamma_D \tag{116}$$

where Z refers to the valency of an ionic species. Here, the activity of a molecular species may often be replaced by its concentration when the concentration is sufficiently dilute.

The electrical potential difference between the two phases is expressed by:

$$\psi_i - \psi_o \equiv E = \frac{RT}{F} \ln \gamma_D \tag{117}$$

Formerly, the membrane potential in biological cells was thought to be due to this Donnan equilibrium potential. Bernstein[117] suggested that the resting membrane potential was determined by the ratio of potassium ion concentration inside and outside the cell. The relative impermeability of Na^+ across the cell membrane was observed by Boyle and Conway.[118] The validity of the formula of the membrane potential for biological cells

$$E = \frac{RT}{F} \ln \frac{(K^+)_o}{(K^+)_i} \tag{118}$$

was suggested by analogy with the case of muscle tissue, where the Nernst potential for potassium ions was in good agreement with experimental results over a wide range of external potassium concentrations.[119,120] However, the assumption of a sodium impermeable membrane had to be abandoned as soon as more accurate measurements of membrane potential by the use of the micropipet electrode[121] were carried out and tracer measurements were applied to muscle and nerve tissues.[122,123] These experiments revealed that the Donnan equilibrium potential theory is not sufficient to explain the observed membrane potentials (Table 20).

3.1.1.2. Nonequilibrium Potential: Diffusion Potential Theory

Near the end of the last century, Nernst[124] proposed the basic concept of ionic movement in an electrolyte system. In an electrolyte system, ions are acted upon by two kinds of forces relevant to our discussion: the coulombic force and the osmotic force, which is essentially equal to the force of diffusion due to the ionic concentration difference. According to the Nernst theory, the liquid junction potential exhibited at the interface of two electrolyte solution phases with different ionic concentrations arises essentially from the difference in ionic mobilities across the liquid junction.

A simple example of the above case is that of two solutions of a uni-univalent electrolyte of different concentrations, in contact. In this case, the liquid junction potential (Nernst diffusion equation) is expressed by:

$$E = \psi_i - \psi_o = \frac{RT}{F} \frac{U^+ - V^-}{U^+ + V^-} \ln \frac{C_o}{C_i} \tag{119}$$

where E is the difference between the electrical potentials ψ_i and ψ_o of the two bulk phases, where subscripts i and o refer to "inside" and "outside" phases, respectively; U^+ and V^- are the mobilities of the positive and negative ions; and C_o and C_i are the ionic concentrations of the outside and inside phases. As seen from Eq. (119), the potential difference E will be set up at the liquid junction when there is a difference in mobilities between mobile ions. Nernst's work was largely used for the development of membrane potential theory. When one of the mobilities of the ionic species is much

Table 20
Equilibrium Potentials

Frog muscle	Squid axon[a]
$E_K = \dfrac{RT}{F} \ln \dfrac{[K]_o}{[K]_i} \cong -103 \text{ mV}$	$E_K = \dfrac{RT}{F} \ln \dfrac{[K]_o}{[K]_i} = -93 \text{ mV}$
$E_{Na} = \dfrac{RT}{F} \ln \dfrac{[Na]_o}{[Na]_i} \cong +50 \text{ mV}$	$E_{Na} = \dfrac{RT}{F} \ln \dfrac{[Na]_o}{[Na]_i} = +55 \text{ mV}$
$E_{Cl} = \dfrac{RT}{F} \ln \dfrac{[Cl]_o}{[Cl]_i} = -104 \text{ mV}$	$E_{Cl} = \dfrac{RT}{F} \ln \dfrac{[Cl]_o}{[Cl]_i} = -65(-33) \text{ mV}$
$E_{rest} = -95 \text{ mV} \neq E_K$	$E_{rest} = -60 \text{ mV} \neq E_K$

[a] Ionic concentrations are shown in Table 19; $T = 24°C$.

greater than the other, this equation approximately equals the Donnan equilibrium potential.

Planck[125] derived a more general theory concerning ionic movement in solution by using the so-called Nernst–Planck diffusion equation. The following is the most simple expression of ionic fluxes for the one-dimensional case,

$$J_j = -C_j U_j \frac{\partial \mu_j}{\partial x} \qquad (120)$$

With Eq. (113), Eq. (120) can be written as:

$$J_j = -U_j C_j \left(RT \frac{\partial \ln C_j}{\partial x} + F Z_j \frac{\partial \psi}{\partial x} \right) \qquad (121)$$

where subscript j refers to the jth ionic species, J_j its flux, U_j its mobility,† Z_j its valency, C_j its concentration, and ψ the electrical potential at a position x. Planck succeeded in integrating Eq. (121) under fairly unrestricted conditions. When the net current is zero the diffusion potential E (because of the potential caused by the diffusion of ions) is expressed by:

$$E = \psi_i - \psi_o = \frac{RT}{F} \ln \xi \qquad (122)$$

where ξ satisfies the following relation (for a detailed derivation, see MacInnes[126]):

$$\frac{\xi U_i - U_o}{V_i - \xi V_o} = \frac{\ln (C_i/C_o) - \ln \xi}{\ln (C_i/C_0) + \ln \xi} \cdot \frac{\xi C_i - C_o}{C_i - \xi C_o}, \qquad (123)$$

where C_i and C_o are the total ionic concentrations of the inside and outside of the solution ($C = \sum_i C_j^+ = \sum_j C_j^-$) and $U = \sum_j U_j^+ C_j^+$, $V = \sum_j U_j^- C_j^-$.

In this expression, there are no jumps in values of electrical potential and ionic concentration at the phase boundary. Although Eqs. (122) and (123) have a quite complex form, this equation can be reduced easily to the Nernst diffusion equation and the constant field equation (Goldman[127]), by assuming proper physical conditions. The former, the Nernst diffusion equation [Eq. (119)], can be obtained from Eqs. (122) and (123) for the case of solutions containing only a single salt (such as NaCl) and the latter, the constant field

---

† U is the mobility defined as the velocity acquired due to a unit force. The dimension of mobility U is $cm^2 sec^{-1} joule^{-1}$ mole. In electrochemicstry, however, the gradient of electrical potential in volts/cm is usually considered as the driving force and hence the dimension of mobility u is $cm^2 sec^{-1} volt^{-1}$. The relation between two mobilities U and u is:

$$u = ZFU$$

Mobility U has been called the "diffusion mobility" and mobility u the "electrical mobility" by Spiegler and Wyllie.[268]

equation, is obtained from Eq. (122) under the following assumptions:

(a) the electrical potential gradient is constant in the membrane;
(b) each ion flux is at a steady state and the total net current is zero.

In the above case the flux of the jth ionic species can be expressed by:

$$J_j = \text{const} = Z_j F U_j \frac{E}{h} \frac{(C_j)_i - (C_j)_o\, e^{-Z_j FE/RT}}{e^{-Z_j FE/RT} - 1} \tag{124}$$

where h is the thickness of the membrane. Then, the membrane potential $E = \psi_i - \psi_o$ is expressed in the following form:

$$E = \frac{RT}{F} \ln \left[\frac{\sum_j U_j^+(C_j^+)_o + \sum_j U_j^-(C_j^-)_i}{\sum_j U_j^+(C_j^+)_i + \sum_j U_j^-(C_j^-)_o} \right] \quad \text{(constant field equation)} \tag{125}$$

where $U_j^\pm$ is the mobility of the cation $^{(+)}$ or anion $^{(-)}$ of the jth species, and $(C_j)_o$ and $(C_j)_i$ are the concentrations of jth ionic species at the outside and inside surfaces in the membrane, respectively.

Hodgkin and Katz[128] modified the above constant field equation, introducing partition coefficients for ions between the aqueous and membrane phases. The expression is as follows:

$$E = \frac{RT}{F} \ln \frac{\sum P_j[C_j^+]_o + \sum P_j[C_j^-]_i}{\sum P_j[C_j^+]_i + \sum P_j[C_j^-]_o} \tag{126}$$

where

$$P_j \equiv \frac{RT U_j b_j}{h} \tag{127}$$

is the membrane permeability constant for the jth ion, and b_j is the partition coefficient of the jth ionic species between the bulk ($[C_j]$) and membrane ((C_j)) phase concentration [$b_j = (C_j)/[C_j]$] (Figure 26A).

The membrane potential of many biological cells has been analyzed quantitatively by using the above equation [Eq. (126)] with known bulk ion concentrations and ion permeabilities across the membrane (Figure 27). The latter (permeabilities) were determined from isotope flux measurements across membranes together with the use of the flux equation [Eq. (124)]. However, it should be remembered that the flux equation [Eq. (124)] used to obtain the permeability was derived under the conditions of the constant field assumption. Therefore, if this membrane potential theory is not a correct expression for biological membranes, the permeability values obtained using Eqs. (124) and (126) will lose their validity.

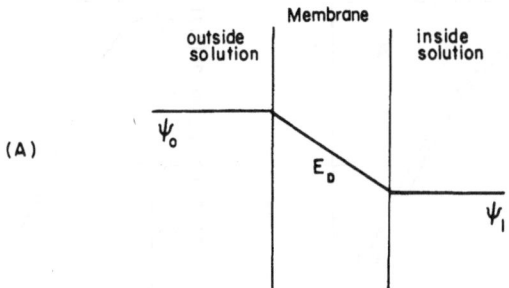

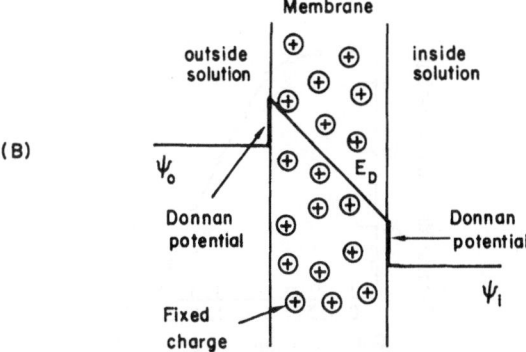

Figure 26. Membrane potential profiles: (A) Goldman–Hodgkin–Katz constant field model (diffusion potential-dominated); (B) fixed-charge membrane model with positive fixed charges.

The thermodynamic expression (under isothermal and isobaric conditions) for the diffusion potential is

$$dE = -\frac{dG}{F} = -\frac{RT}{F}\sum_j \frac{t_j}{Z_j} d(\ln a_j) \tag{128}$$

where E is the potential, G the Gibbs free energy, t_j the transference number (the relative quantity of the total current carried by the jth ion, $t_j = u_j/\sum_j u_j$, and $\sum_j t_j = 1$), and a_j is the activity of the jth ion. The total diffusion potential is

$$E = -\frac{RT}{F}\int_{a_o}^{a_i}\sum_j \frac{t_j}{Z_j} d(\ln a_j) \tag{129}$$

This equation is also derived from Eq. (122) with the conditions of electroneutrality, zero electrical current, and quasi-stationary state for each ion.

Another simplified solution for Eq. (122) is obtained by assuming a constant concentration gradient,[129] namely, the concentration in the mem-

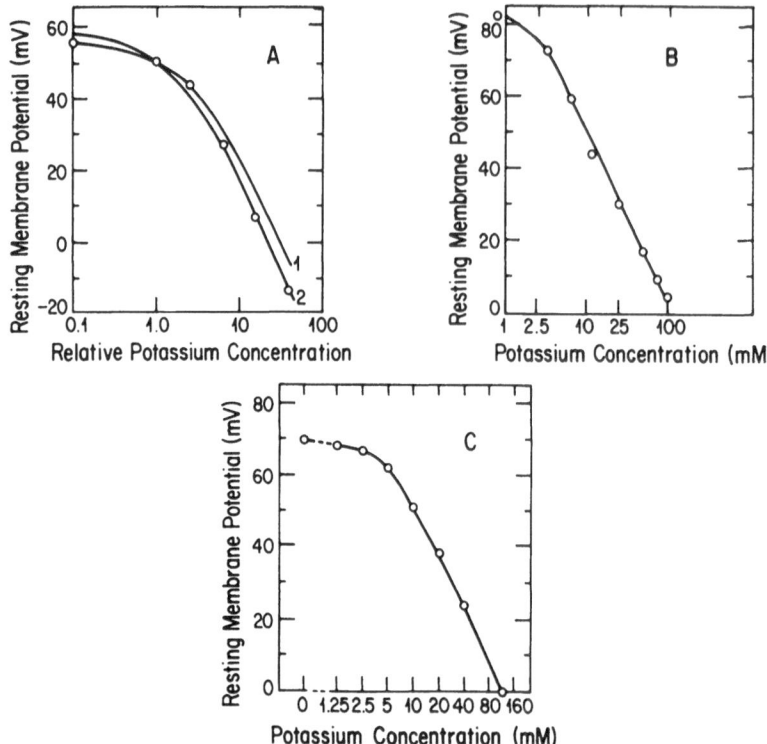

Figure 27. Relation between potassium concentration in external solution and potential difference across resting membrane. (A) Squid giant axon. (Data of Curtis and Cole[272]). (B) Frog sartorius muscle. (Data of Ling and Gerard[121]). (C) Frog myelinated nerve. (Data of Huxley and Stümpfli[273]). Abscissa: potassium concentration on logarithmic scale, (B) and (C) in mM; (A) as multiple of concentration in standard solution (13 mM). Ordinate: potential difference across resting membrane (outside potential minus inside potential). The curves in (A) give the calculated changes in resting potential, drawn according to Eq. (126) with $P_K : P_{Na} : P_{Cl} = 1 : 0.04 : 0.45$ (curve 1) and $P_K : P_{Na} : P_{Cl} = 1 : 0.025 : 0.3$ (curve 2). (For details, see Reference 128). The membrane potentials in (A) and (B) are the observed values and have not been corrected for the junction potentials between axoplasm or myoplasm and the electric field. In (C) the potentials are given with reference to a node which had been depolarized with isotonic potassium chloride (117 mM) (Reference 194).

brane at the distance x from the outside surface being

$$C(x) = C_o + \frac{C_i - C_o}{h} x \qquad (0 \le x \le 1) \qquad (130)$$

where h is the thickness of the membrane. Then, Eq. (129) is expressed by the following for the case of all ions being univalent:

$$E = \frac{RT}{F} \frac{(U_o - V_o) - (U_i - V_i)}{(U_0 + V_0) - (U_i + V_i)} \ln\left(\frac{U_0 + V_o}{U_i + V_i}\right)$$

(constant concentration equation)   (131)

where

$$U_o = \sum_j U_j^+ C_{jo}^+ \qquad \text{(for positive ions only),}$$

$$V_o = \sum_j U_j^- C_{jo}^- \qquad \text{(for negative ions only),}$$

$$U_i = \sum_j U_j^+ C_{ji}^+ \qquad \text{(for positive ions only),} \qquad (132)$$

$$V_i = \sum_j U_j^- C_{ji}^- \qquad \text{(for negative ions only).}$$

This equation is often used to describe the liquid junction potential for multi-univalent ions. However, this equation is not often used to describe the membrane potential of biological cells because of difficulties in obtaining the mobilities of the ionic species within the membrane.

The biological membrane contains complex molecular components such as proteins and lipids. These have intrinsic electrolyte properties which include the presence of fixed ions. Consequently, the assumption of an electrically neutral membrane, which was made in the Planck model, requires modification. In the following section, the properties of membranes containing a uniform distribution of fixed charges will be discussed.

3.1.1.3. Mixtures of Equilibrium and nonequilibrium Potentials

3.1.1.3.1. Fixed Charge Theory. As a result of combining the above two theories, a fixed charge membrane theory was developed by Teorell[120,131] and Meyer and Sievers.[132] The theory includes the equilibrium of the electrochemical potentials of ions at the membrane boundaries and the diffusion of ions in the membrane. Therefore, the total membrane potential E is composed of three separate potential differences: E_o corresponds to the first transition region between the solution (o) and membrane phases, E_{diff} corresponds to the ion diffusion potential in the membrane, and E_i corresponds to the other transition region between membrane and the solution (i) phases (see Figure 26B). Thus, the transmembrane potential is

$$E = E_o + E_{diff} - E_i \qquad (133)$$

The boundary potentials E_o and E_i are expressed by the Donnan equilibrium potentials which are

$$E_o = \frac{RT}{Z_j F} \ln \frac{[a_j]_o}{(a_j)_o} = \frac{RT}{Z_j F} \ln \frac{2[a_j]_o}{-f + [f^2 + 4[a_j]_o^2]^{1/2}}$$

$$E_i = \frac{RT}{Z_j F} \ln \frac{[a_j]_i}{(a_j)_i} = \frac{RT}{Z_j F} \ln \frac{2[a_j]_i}{-f + [f^2 + 4[a_j]_i^2]^{1/2}} \qquad (134)$$

respectively, where $E_0 = \bar{\psi}_o - \psi_o$, $E_i = \bar{\psi}_i - \psi_i$, and ψ_o and ψ_i are the electrical potentials at the outside and inside bulk phases away from the membrane, respectively, $\bar{\psi}_o$ and $\bar{\psi}_i$ are the electrical potentials in the membrane adjacent to the outer and inner membrane boundaries, respectively, $[a]_o$ and $(a)_o$ are the activity of ions of the bulk solution and the solution in the membrane near the membrane surface, and f is the positive fixed charge density which is homogeneous in the membrane.

Equilibrium is assumed to prevail here at the boundaries of the membrane (e.g., the membrane is assumed to be thick or else it is a very thin film having a low ionic conductance so that the main resistance to the ionic fluxes is localized in the membrane proper and the equilibrium on the surfaces has enough time to be practically established) so that the boundary potential may be expressed by the Donnan ratios for the ions, and these ratios are determined from the fixed charge density in the membrane and the surrounding electrolyte concentrations. The diffusion potential, E_{diff}, has a somewhat complicated expression. With the conditions of electroneutrality, zero electrical current, and a quasi-stationary state for each ion species, the expression for E_{diff} is

$$E_{\text{diff}} = \frac{RT}{F} \ln \xi \tag{135}$$

where ξ is a solution of

$$\frac{\xi U_i - U_o}{V_i - \xi V_o} = \frac{\ln \bar{K} - \ln \xi}{\ln \bar{K} + \ln \xi} \frac{\xi C_i - C_o}{C_i - \xi C_o} \tag{136}$$

and

$$U_o = \sum_j u_j (C_j^+)_{x=0} \qquad U_i = \sum u_j^+ (C_j^+)_{x=d} \qquad V_o = \sum_j u_j (C_j^-)_{x=0}$$

$$V_i = \sum u_j^- (C_j^-)_{x=d} \tag{137}$$

where d is the thickness of the membrane. The parentheses denote ion concentrations within the membrane; in addition,

$$(C^+)_i = \sum (C_j^+)(x = d) \qquad (C^+)_o = \sum (C_j^+)(x = 0)$$

$$(C^-)_i = \sum (C_j^-)(x = d) \qquad (C^-)_o = \sum (C_j^-)(x = 0) \tag{138}$$

The parameter $\bar{K}$ is defined implicitly by the equation

$$\bar{K} = \frac{(\bar{C})_i^- - 0.5[\ln(\bar{K}/\xi)/\ln\bar{K}]f}{(\bar{C})_o^- - 0.5[\ln(\bar{K}/\xi)/\ln\bar{K}]f} = \frac{(\bar{C})_i^+ + 0.5[\ln(\bar{K}\xi)/\ln\bar{K}]f}{(\bar{C})_o^+ + 0.5[\ln(\bar{K}\xi)/\ln\bar{K}]f} \tag{139}$$

where $(\bar{C})$ is the average concentration of ion in the membrane, f is the fixed ion concentration in equivalents per unit volume. For $f = 0$, it can be verified that Eqs. (139) and (136) lead to Eq. (123), as expected.

In solving for the membrane potential it is necessary to determine ξ from the transcendental equation [Eq. 136]. This can be done by trial and error.

In order to relate the ionic concentrations at the inner and outer edges of the membrane to that in the bulk solutions, the Donnan equation is utilized [Eq. (134)].

A theory similar to the fixed charge membrane theory, presented by Polissar,[133] rests on the surface potentials at the membrane interfaces instead of the membrane boundary potentials. Ohki[134] and MacDonald and Bangham[135] were the first to show the large contribution of surface potential to the observed transmembrane potential of phospholipid membranes. Colacicco,[136] Kamo and Kobatake,[137] and McLaughlin and Harary[138] also suggested the contribution of surface potentials to the membrane potential for model membrane systems.

Gilbert[139] has discussed the transmembrane potential of squid axon membranes in terms of inner and outer surface potentials estimated from various experimental data. Aono and Ohki[140] analyzed the membrane potential of squid axons in terms of the surface potential, considering ion adsorption processes.

None of the above work, however, has satisfactorily and explicitly explained each contribution of surface potential and ion diffusion potential to the transmembrane potential.

Recently, Ohki[141] has showed theoretical as well as experimental analyses for such a transmembrane potential profile for lipid membrane systems.

3.1.1.3.2. Surface/Diffusion Potential Theory.[141] The transmembrane potential, E, is expressed as a difference between the electrical potential, E_i, and E_o of the two bulk phases in the two aqueous compartments separated by a membrane or as a sum of phase boundary potentials produced at the membrane–electrolyte interfaces and the diffusion potential within the membrane arising from the movement of ionic species through the membrane (Figure 28).

$$E = E_i - E_o = V_o - V_i + E_D \qquad (140)$$

where V_o and V_i are the surface potentials at the two membrane interfaces and E_D is the diffusion potential in the membrane. The surface potential is expressed (ie, Eq. (87)) by

$$V = \psi + 4\pi n \mu_D \qquad (141)$$

where the first term is the Gouy–Chapman double layer potential at the membrane surface, and the second term is the polarization (permanent dipole) potential ($4\pi n \mu_D = \psi_D$) of the molecules of the surface membrane, in which n is the number of dipoles per unit area and μ_D is the dipole moment of a membrane molecule.

In the case of a membrane symmetrical with respect to its molecular constituents, the terms due to the polarization potential of the surface membrane in Eq. (140) may cancel each other and the equation becomes:

$$E = \psi_o - \psi_i + E_D \qquad (142)$$

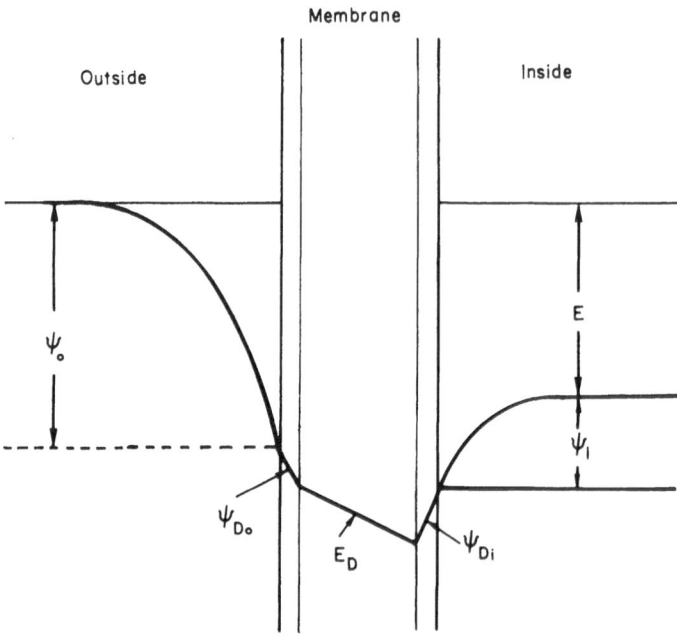

Figure 28. A schematic transmembrane potential profile E = transmembrane potential; ψ_0 = the outer diffuse double layer potential; ψ_i = the inner diffuse double layer potential; ψ_D = polarization potential due to membrane molecular dipoles; $\psi_{Do} \neq \psi_{Di}$ = asymmetrical polarization potentials; E_D = diffusion potential.

Let us suppose that the two electrolyte solutions (the same uni–univalent electrolyte), with concentrations denoted by $[C]_o$ and $[C]_i$, are separated by a membrane† having the same net charge density, σ, on its outside and inside surfaces, and the surrounding ions do not bind to the charged sites of the membrane. Then ψ_o and ψ_i are expressed as

$$\psi_o = 2kT/e \cdot \sinh^{-1}(2\pi\sigma e/\varepsilon\kappa_o kT), \tag{143a}$$

$$\psi_i = 2kT/e \cdot \sinh^{-1}(2\pi\sigma e/\varepsilon\kappa_i kT) \tag{143b}$$

where $\kappa = (8\pi n^0 e^2/\varepsilon kT)^{1/2}$ is the Debye constant of the solution; $\varepsilon = 80$, the dielectric constant of water; n^0 equals the number density of the univalent positive or negative ions in the bulk solution. Then the concentrations of univalent cation $[C^+]_o^s$ and univalent anion $[C^-]_o^s$ at the outer surface of the membrane are given by

$$[C^+]_o^s = [C^+]_o \exp(-e\psi_o/kT) \quad \text{and} \quad [C^-]_o^s = [C^-]_o \exp(+e\psi_o/kT) \tag{144}$$

† The membrane potential arising under these conditions is called the "concentration potential."[269]

and, similarly, the concentrations of univalent cation and anion at the inner surface of the membrane are

$$[C^+]_i^s = [C^+]_i \exp(-e\psi_i/kT) \quad \text{and} \quad [C^-]_i^s = [C^-]_i \exp(+e\psi_i/kT)$$
$$(145)$$

The diffusion potential within the membrane may be expressed by the Goldman and Hodgkin and Katz equation [Eq. (126)] as follows:

$$E_D = RT/F \cdot \ln\{(P^+[C^+]_o^s + P^-[C^-]_i^s)/(P^+[C^+]_i^s + P^-[C^-]_o^s)\} \quad (146)$$

For precision, the concentration $[C]$ should be replaced by the activity of each ionic species, where P is the permeability coefficient of ionic species through the membrane. The permeability coefficient is defined as $P^+ = RTbU^+/h$, where b is the partition coefficient of the ion at the membrane and aqueous phases, U^+ is the mobility of a cation, and h is the thickness of the membrane.

Figure 29 shows the theoretical values of membrane potential, surface potential, and diffusion potential with respect to various surface charge densities (surface charges are negative in sign) for the following three cases of relative ionic permeabilities ($f = P^-/P^+$): (1) $f = 0$, (2) $f = 2/3$, (3) $f = 3.0$. Only the cases where f is not a large number (say $f < 10$) will be dealt with in the following discussion.

As seen from Figure 29, when the surface charge density is small, the diffusion potential contributes greatly to the membrane potential, the magnitude and sign of which strongly depend upon the relative permeabilities of positive and negative ions. This is also seen from Eq. (142) together with Eq. (143).

However, when the surface charge density is large enough (as when σ is greater than one electronic charge per 100 Å^2), the main factor contributing to the membrane potential is the difference in surface potentials at the two phase boundaries, not the diffusion potential. In this situation the membrane potentials hardly depend on the relative ionic permeabilities of the membrane and are the same for all cases ($f = 0, 2/3,$ or 3.0).

This situation can be described as follows. When the argument of the inverse hyperbolic sine in Eq. (143) is much greater than 1 [e.g.,

$$|2\pi\sigma e/\varepsilon\kappa kT| = |-2\pi e l^{1/2}/(8\pi\varepsilon kTNC)^{1/2}A| \gg 1, \quad (147)$$

where C is the concentration expressed as moles per liter (l), A is the area (in Å^2) per electronic charge ($\sigma = -e/A$)], which means that the surface charge density is high enough for a certain bulk electrolyte concentration, then the surface potential is expressed by

$$\psi = 2kT/e \cdot \sinh^{-1}(2\pi\sigma e/\varepsilon\kappa kT)$$
$$\cong -2kT/e \cdot \ln\lambda + kT/e \cdot \ln C \quad (147)$$

where $\lambda = 4\pi e l^{1/2}/(8\pi\varepsilon kTN)^{1/2}A$ is a constant when the surface charge

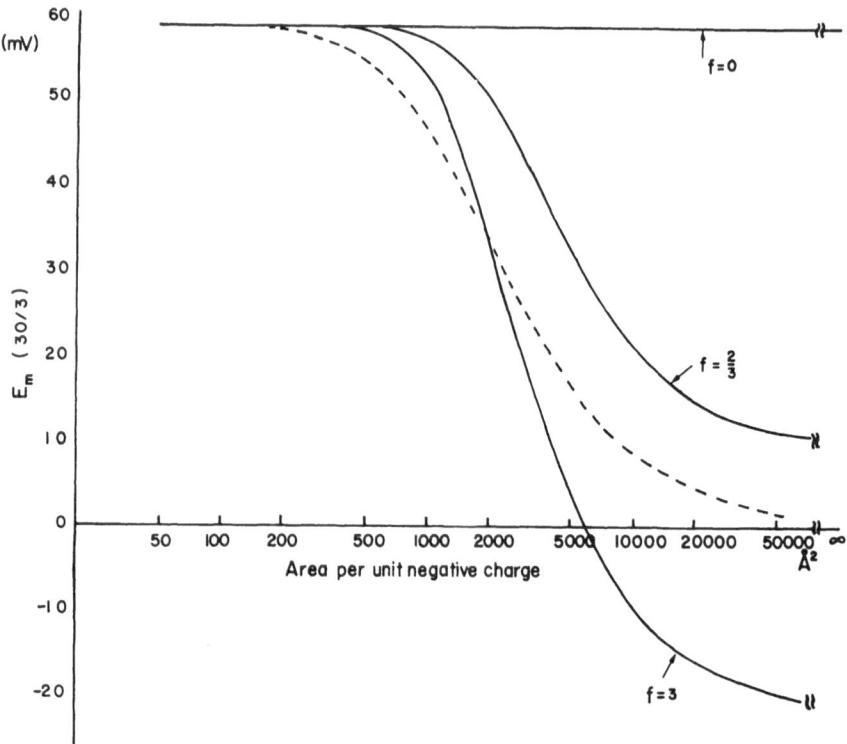

Figure 29. Calculated and observed values of concentration potential E_m (30/3), where E_m(30/3) is the transmembrane potential for 30 mM salt in the outside compartment and 3 mM salt in the inside compartment of the membranes with various surface charge densities. Those densities are indicated as the area (Å^2) per negative electronic charge. Solid lines are the theoretical values for the cases of the following three relative permeabilities: (1) $f = p^-/p^+ = 0$; (2) $f = 2/3$; and $f = 3.0$. Dotted line is the calculated value of the difference in surface potentials (diffuse double layer potentials) in the bulk solutions of 3 mM and 30 mM with respect to variation of surface charge densities. All calculations were done by assuming no ion binding with the membrane at a temperature of 24°C (Reference 141).

density is a constant. Then Eqs. (143a) and (143b) become

$$\psi_o \cong -2kT/e \cdot \ln \lambda + kT/e \cdot \ln C_o \qquad (149a)$$

$$\psi_i \cong -2kT/e \cdot \ln \lambda + kT/e \cdot \ln C_i \qquad (149b)$$

Therefore the difference in the two surface potentials is

$$\psi_o - \psi_i \cong kT/e \cdot \ln (C_o/C_i). \qquad (150)$$

As for the diffusion potential, Eq. (146) can be rewritten with the use of Eqs. (144) and (145):

$$E_D = RT/F \cdot \ln \{([C^+]_o Y_o P^+ + [C^-]_i Y_i^{-1} P^-)/([C^+]_i Y_i P^+ + [C^-]_o Y_o^{-1} P^-)\} \qquad (151)$$

where $Y_o = \exp(-e\psi_o/kT)$ and $Y_i = \exp(-e\psi_i/kT)$.

When ψ_o and ψ_i are both large and negative values ($|\psi_o|, |\psi_i| \gg kT/e$), the diffusion potential, E_D, is approximated as

$$E_D \cong (RT/F) \ln \frac{[C]_o}{[C]_i} e^{-e(\psi_o - \psi_i)/kT} = 0 \qquad (152)$$

Therefore when the surface charge density is large enough, the transmembrane potential, Eq. (142), becomes:

$$E_m = \psi_o - \psi_i + E_D \cong \psi_o - \psi_i$$

$$= RT/F \cdot \ln (C_o/C_i). \qquad (153)$$

The expression of the transmembrane potential is in the same form as the Nernst equation. However, in this case the transmembrane potential is due mainly to the difference in the two surface potentials, and the diffusion potential is approximately zero.

The above theory was tested and confirmed as a valid theory by performing concentration potential measurements on lipid bilayers and surface potential measurements on lipid monolayers.[141]

Although the Hodgkin–Katz[128] and Goldman[127] equation [Eq. (126)] has been widely used to explain the membrane potential of many biological cells, the equation has been found not to be adequate to account for changes in resting membrane potential of squid axons, when the internal salt solution is altered in a certain manner,[142–144] although attempts have been made[145] to explain the observed membrane potential by retaining the Goldman–Hodgkin–Katz equation and introducing modified ionic permeability factors. Recently, Ohki[308] has measured the membrane potential of squid axons as a function of the nature of the ionic species (ion substitution) and the ionic strength (dilution) of the extracellular medium, and has analyzed the observed membrane potentials using both the Goldman–Hodgkin–Katz equation and the surface/diffusion potential equation, described above. He concluded that the membrane potential theory including the surface potential of the membrane (the surface/diffusion potential theory) is better able to analyze the observed membrane potential of squid axon (Figure 30A, B). It is also found that the external surface of squid axon membranes possesses a relatively high surface charge density ($-e/200 \text{ Å}^2$). In the calculation, the relative ionic permeability $P_K : P_{Na} : P_{Cl} = 1 : 0.04 : 0.45$ for the squid axon was used and zero surface charge density for the inner membrane surface of the squid axon was assumed.

Whenever the membrane surfaces possess fixed charges, contributions from both the surface potential and the diffusion potential should be involved in the observed membrane potential. The extent of their contributions depends upon the surface charge densities, the magnitude of the surface potential, and also the relative permeabilities of cations and anions involved in the transport

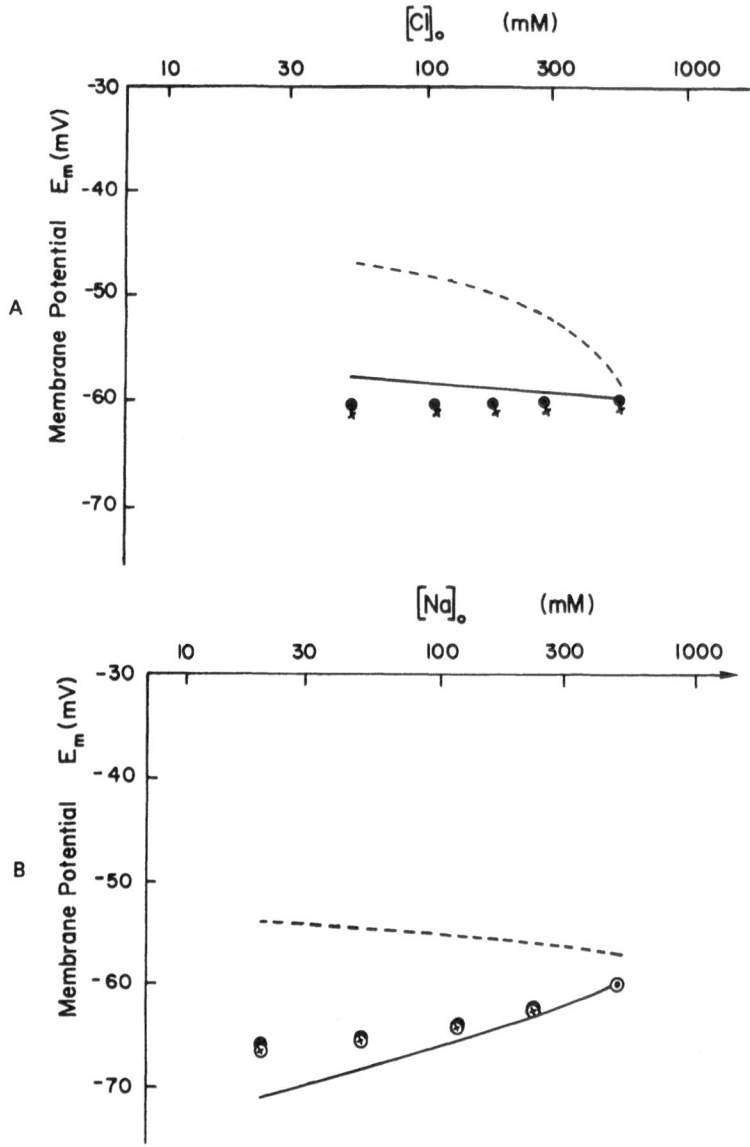

Figure 30. (A) Membrane potential of squid axon versus extracellular Cl⁻ concentration. Various amounts of extracellular Cl⁻ were replaced with either glutamate O or isethionate X, while the other ionic concentrations were kept constant ($[Na^+]_o = 488$ mM, $[K^+]_o = 10$ mM, and $[Ca^{2+}]_o = 20$ mM, 1 mM Tris; pH = 8.0).- - -: Calculated membrane potential by use of the G–H–K equation with the relative ionic permeabilities ($P_K : P_{Na} : P_{Cl} : P_{glutamate(isethionate)} = 1 : 0.04 : 0.45 : 0$) (Reference 128); ——: Calculated membrane potential by use of the S–D (surface/diffusion potential) equation with the same relative ionic permeabilities as the above. (B) Membrane potentials of squid axon as a function of various dilutions of extracellular NaCl replaced by isotonic (to sea water) nonelectrolytes while $[K^+]_o = 10$ mM and $[Ca^{2+}]_o = 20$ mM were kept constant. ●: Sucrose; O: Glucose; ×: Manitol; ···: Calculated membrane potential by use of the G–H–K equation with the relative ionic permeabilities ($P_K : P_{Na} : P_{Cl} = 1 : 0.04 : 0.45$); ——: Calculated membrane potential by use of the S–D potential equation with the same ionic permeabilities as the above. (Ohki, ref. 308).

process. Since membrane surfaces of most biological cells (such as axons, muscles, etc.) are highly charged,[146,147] the experimentally observed membrane potential should therefore involve both a surface potential and an ion diffusion potential. However, many investigators have so far used solely the diffusion potential equation proposed by Hodgkin and Katz to explain observed membrane potentials for such highly charged membranes.

It should be noted, however, that membrane potentials observed in biological membrane systems involve various complex factors of polyvalent ions and their binding with the charge sites of membrane surfaces. As we have discussed earlier in this chapter, individual ions, especially divalent or polyvalent ions, have their own characteristic constants for binding with membrane molecules at their surfaces, and the degree of binding can vary from membrane to membrane. It has been shown that the surface potential is intimately related to the degree of ion binding with the charged groups of membrane surfaces (in Section 2). Furthermore, when an ion binds strongly with the polar groups of the membrane, the orientation of polar groups at the membrane surface and also the dielectric constant of the polar group region of the membrane surface can be altered. Consequently, the net dipole moment of the membrane molecule can be altered, which would change the surface potential V [Eq. (141)]. If the membrane molecular dipole is asymmetric with respect to the membrane, another factor of polarization potential will contribute to the transmembrane potential through the difference in surface potentials. It is known that in many biological membranes, the molecular constituents of the outer surface membrane are different from those of inner membrane surfaces.[148,149] Therefore, the difference in the molecular dipoles between the two sides (outer and inner surfaces) of the membrane would contribute to the measured value of the membrane potential (as discussed in Section 2).

In addition, biological membranes are believed to be composed of an inhomogeneous distribution of molecular assemblies, such as specific ionic channels, and molecular segregation with respect to the two-dimensional surfaces of the membranes occurs, which is quite a different situation from that in simple phospholipid membranes.

In the lipid bilayer systems, since the membrane molecules are arranged in such a way that the charged groups face a water phase and the interior of the membrane is a hydrocarbon phase, the contribution of surface potential to the membrane potential is important. It should be mentioned that the contribution of surface potential to the membrane potential, as discussed above, is generally a transient one in these systems. However, since the electrical conductance due to ion permeation across the lipid bilayer membrane is very low, we can observe the transient potential difference as a quasi-steady state phenomenon. However, if a constant ion distribution is restored by a transport process with a nonelectrical current (active transport) and maintained continuously, the above membrane potential process could become a steady state process.

For biological membranes and polymer-resin membranes, which may contain charged groups within the membrane matrix, the above argument with regard to the contribution of the surface potential to the transmembrane potential may not be valid. In particular, when there are highly conductive ionic channels composed of lipoprotein molecular assemblies within the membrane (where channels may be mostly hydrophilic and probably main ionic transport pathways across the membranes), the observed transmembrane potential is a very complicated mixture of ion diffusion and Donnan equilibrium potentials. The idealized surface potential concept may not be a good approach to analyze such a transmembrane potential.

In these situations (biological membranes and polymer-resin membranes), the fixed charge membrane theory with various aqueous pore sizes may be a better theoretical model to explain the mechanism of the observed transmembrane potential. Ohki[150] attempted to analyze the transmembrane potential for a membrane having an aqueous pore the surface of which possesses fixed charges. He used a number of assumptions to simplify the mathematical treatment. Further development of such a membrane potential theory therefore seems to be necessary in order to describe the observed membrane potential for biological cell systems.

The other extreme case of membrane potential theory based on the diffusion potential approach is the thermodynamic adsorption potential theory of Ling,[151] based on the idea that the entire cell is a fixed charge system and the membrane potential is expressed by the Donnan potential between the plasma and extracellular phases. The intracellular fixed charge system has various affinities for adsorbing ions. In this theory, the plasma membrane plays a small role in the origin of membrane potential. Although this situation may occur under certain idealized conditions for biological cells, at normal physiological environmental conditions the membrane and membrane boundaries interacting with ions seem to have a major responsibility for observed membrane potentials of cells.

However, macromolecular moieties[152] (cytoskeletal protein-like) located at the inner surface of nerve membranes seem to have an important role in stabilizing the membrane as well as maintaining membrane potential, especially for maintaining excitability of nerve fibers.[152-154] According to Tasaki *et al.*,[153,154] the maintaining capability (survival time) of excitability of squid axons depends greatly upon the species of intracellular anions present. The sequence of anions arranged in accordance with their favorability for maintaining excitability under intracellular perfusion is:

F > HPO_4 > glutamate, aspartate > citrate > tartrate > propionate > SO_4 > acetate > Cl > NO_3 > Br > I > SCN.

The above sequence corresponds to that of the lyotropic series of anions (Table 10). This series is known as the sequence of anions for salting out of various proteins in water.

Therefore, this implies that the proteins in the intracellular phase may have an important role in maintaining the production of action potential. Also, Tasaki has stressed the importance of the outer surface of the axon membrane in producing the membrane potential of an ion exchange membrane (Ca^{2+}, K^+, or other monovalent cations.[153,154]). However, the explicit formulation for the membrane potential in terms of these views has not been put forward yet.

3.1.2. Potential Due to Electron Transfer across Membranes

Until recently the possibility of electron conduction in living systems was not seriously considered, with perhaps a few notable exceptions to be mentioned below. Traditionally, the origin of electrical potentials observed in living systems has been attributed almost exclusively to ionic permeability. The measured electrical events were considered to be the result of the movement of ions in an aqueous medium. This traditional view was apparently influenced by the fact that electron conduction could only take place in metallic conductors. However, the development of inorganic and organic semiconductors has conclusively shown that electron conduction is not restricted to metals. Thus, in recent years the possibility that a mechanism analogous to electron conduction in metals might play a critical role in living systems has been considered by many investogators. The two major areas involving such electronic processes are respiration (mitochondria) and photosynthesis (chloroplasts). Several reviews dealing with various aspects of electronic processes in biology are available.[155–159]

Photosynthesis is the most likely process to involve electronic conduction. It is now generally believed that two types of mechanisms are possible in the primary process of light conversion. First, in the energy migration mechanism light absorbed by the pigments in the thylakoid membrane is rapidly transferred without significant loss to a so-called "reaction center." By a process still unknown, the excitation energy is separated upon arrival at the reaction center into two parts; one part is of a reducing nature (as electrons) and the other part is of oxidizing character (as positive holes). The electrons and holes thus separated are prevented from recombining immediately, owing possibly to the hydrophobic lipid phase in the interior of the membrane. The second mechanism is called the charge separation model which postulates that electrons and holes may be transferred between different reaction centers. These ideas were first conceived by Van Niel[160] who formulated an overall view of photosynthesis in terms of the oxidant (OH) and reductant (H). Later, Katz[160a] translated these chemical terms into "electrons" and "holes." According to Katz, it is assumed that chlorophyll molecules are arranged in two-dimensional "crystals" which are semi- and photoconductive.

Experimental evidence in support of Katz's hypothesis was provided by the work of Arnold and Sherwood[161] and others,[162] who studied the optical and electrical properties of dried chloroplasts and found that the electrical conductivity of these preparations measured as a function of temperature followed the familiar Arrhenius equation for semiconductors.

From a historical prospective, Lund[163] had suggested as early as 1928 that the electrical potential in cells was the result of redox reactions across the cell membrane. Later, a similar suggestion was made by Stiehler and Flexner.[164] However, the redox concept of membrane potentials in biology was not generally accepted until 1962. In that year, Jahn revived and extended his earlier ideas.[165] In particular, Jahn proposed that on either side of biological membranes are located two redox enzyme systems connected electronically by means of conjugated lipoid materials. The possibility of electron conduction by molecules such as carotenoids had been mentioned earlier by Dartnall.[166] More recently, in another biological system, evidence for the occurrence of redox reaction (or electronic processes) at the surface of crustacea has been obtained.[167]

The electron transport system in mitochondria is another example.[29,168] Here, the redox reaction systems are so arranged (as shown in Figure 31) that electron transfers occur across the membrane: electrons flow from a low potential state to a higher potential state.

However, as has been mentioned above, the redox potential E' also depends on the concentration ratio of oxidants and reductants [Eq. (68)]. Therefore, the order of flow of electrons is not necessarily that of E_0' but that of the total free energy.

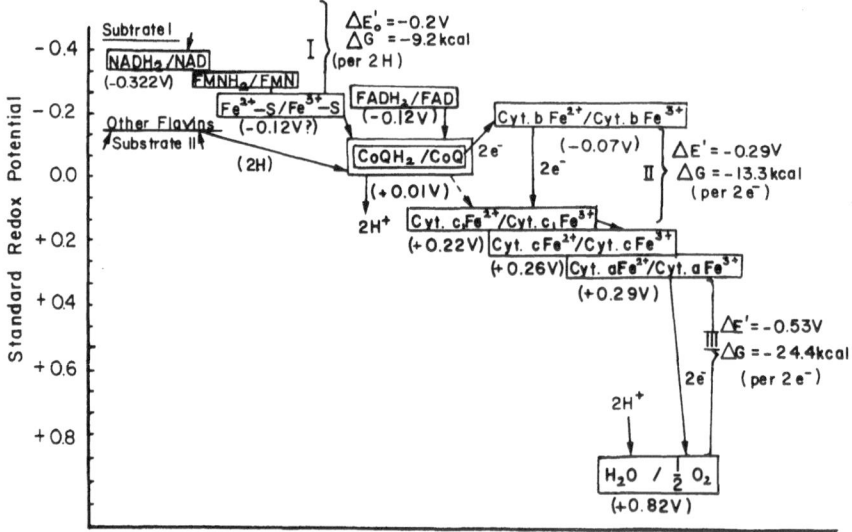

Figure 31. An electron transport system and redox potentials in mitochondria. FMN refers to Flavin mononucleotide in $NADH_2$ dehydrogenase, FAD refers to Flavin adenine dinucleotide in succinate dehydrogenase, I, II, and III correspond to the reaction processes which may be involved in phosphorylation, Fe—S: non-heme iron, Cyt: Heme in cytochromes (after ref. 171).

It is well known that the energy released during the flow of electrons is often utilized to synthesize ATP (electrical energy is stored as chemical energy). The electronic conduction in photosynthetic systems is induced by light energy while the latter electronic conduction is purely due to chemical reaction. Both systems contain redox reaction systems that act when the electrical potential across the membrane is observed.

It is not certain that an electronic flow exists across membrane systems involved in redox systems, since there is no direct measurement of electronic current; nevertheless conduction is very likely, on *a priori* grounds, during such processes in the membrane.[169]

3.1.2.1. Photoelectrical Potential

The molecular mechanism by which a photoelectrical potential across the membrane arises has been investigated for model membrane systems with incorporated photosensitive pigments (such as chlorophyll) where the membrane is immersed in solution. This solution contains a reducing agent (D) on one side of the membrane and an oxidizing agent (A) on the other (Table 21). According to Tien[115,170] photochemical redox reactions in such systems can give rise to an electrochemical gradient for electrons. Namely, pigment molecules excited by photons generate unidirectional electron (or hole) currents in the lipid membrane by pumping electrons from donor molecules to the side of the membrane where acceptors are located [e.g., if $FeCl_3$ is present on one side of the solution, the excited pigment molecule may donate an electron to the ferric (Fe^{3+}) ion since the chlorophyll molecule excited by light has a reduction potential more negative than that of the Fe^{3+}/Fe^{2+} reduction–oxidation couple, and the oxidized chlorophyll is subsequently reduced by substrate, most likely H_2O (Figure 32). These coupled oxidation–reduction reactions allow a net flow of electrons to the $FeCl_3$–Chl-incorporated membrane phase]. Although the explicit nature of electron flow across the membrane cannot be measured by instruments, one can, from various time constant analyses of the induced transmembrane potential, deduce that electron flow most likely occurs.

The observed photoelectric effect must therefore result from an imbalance of ionic charge carriers across the membrane. On one side of the bathing solution, the formation of an electrical double layer (after reduction of Fe^{3+}) at the solution–membrane interface occurs, comprising a layer of anions in the solution and a layer of Chl^+ (ion-radicals) in the membrane. The most likely positive ion carriers on the other side of the bathing solution are assumed to be H^+. The observed photoelectric voltage E_{op} would be given by

$$E_{op} = E^0 + \left(\frac{U_h - U_e}{U_h + U_e}\right)\frac{RT}{F}\ln\frac{(Chl^+)(A^-)(D)}{(Chl^-)(A)(D^+)} \tag{154}$$

where U_h and U_e are the mobilities of holes and electrons, respectively, and E^0 is a constant which is characteristic of the system.

Table 21
Effect of Organic Compounds on the Ch1–BLM Photo-emf^a

The Cell Arrangement:
Salt bridge | Outer solution^b | Ch1–BLM^c | Inner solution^d | Salt bridge

Inner solution (M)	Photo-emf (mV)	Remarks
Ascorbic acid (4×10^{-3})	125–188	240 mV with an applied field
Tannic acid (4.5×10^{-7})	113–133	Also observed with Oolong tea
Hydroquinone (7×10^{-4})	79–96	
Gallic acid (10^{-2})	72–84	
Quinhydrone (7×10^{-4})	45–64	
Benzoquinone (2.8×10^{-4})	−4–+16	10–15 mV dark potential
Flavin mononucleotide (FMN) (7×10^{-3})	138–167	White ppt. disappearing with time
Oolong tea (4×10^{-3} g/mL)	132	
Catechin (3.5×10^{-3})	92–121	
Riboflavin (8×10^{-5})	79–99	
Ferrozine	85	A complexing agent for Fe^{2+}
β-NAD (6×10^{-4} g/mL)	67	Gave a negative dark membrane potential
Cytochrome c (2×10^{-5} g/mL)	48	Generated a dark membrane potential
Vitamin K (2.4×10^{-2} g/mL)	37	
Methyl viologen (4×10^{-2})	73	$K_4Fe(CN)_6$ used in place of $FeCl_3$
Cystine/cysteine	108	
2-Hydroxyl-1,4-naphthoquinone	140	(see footnote e)
1,4-Naphthoquinone-2-sulfonic acid (K salt)	80	(see footnote e)

^{a} Reference 115.
^{b} Outer solution = 0.1 M acetate at pH 5 + lm M $FeCl_3$.
^{c} Ch1–BLM was formed from spinach chloroplast extracts.
^{d} Inner solution = 0.1 M acetate at pH 5 + the compound indicated.
^{e} The cell arrangement used is as follows: Fe^{3+}, Fe^{2+}/Ch1–BLM/Redox couple. The other conditions were the same.

A number of photopigments incorporated into lipid membranes, similar to the above mentioned systems, have been studied. Table 21 gives the photoelectric voltage for various photopigments and for the solutions on both sides of the membranes.

Similar mechanisms with more complicated arrangements of pigments involved in conversion of light energy to electrical energy may be relevant to photosynthetic cells or organelles (such as photosynthetic enzymes in chloroplasts, etc.).

3.1.2.2. Electron Transfer Processes in Mitochondria[29]

When a substance A and a substance B react by an oxidation–reduction reaction, the reaction will proceed according to the difference in oxidation-reduction reaction potential between the two substances ($\Delta E' = E'_A - E'_B$).

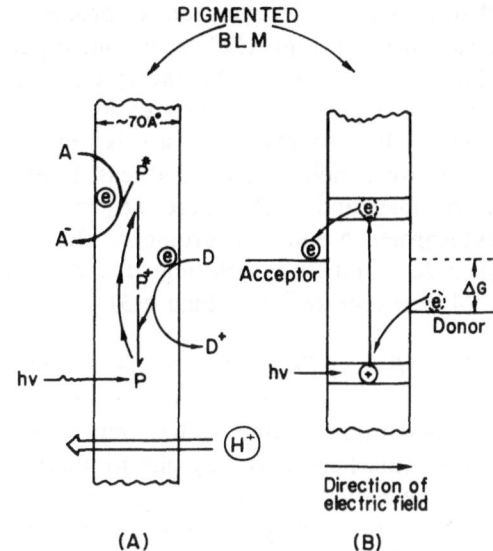

Figure 32. (A) A pigmented BLM (black lipid membrane) separating two aqueous solutions with an electron acceptor on one side and a donor on the other side of the membrane. P = pigment molecule; P* = excited pigment molecule after adsorption of a photon (hν); P$^+$ = oxidized pigment molecule; A = electron acceptor; A$^-$ = reduced electron acceptor; D = electron donor; ⓔ = electron; ⊕ = positive hole; H$^+$ = proton. (B) Corresponding energy diagram illustrating ground and excited states of a pigment molecule in the membrane. ΔG = nFE, theoretically available electrochemical free energy.

The free energy change per mole of reactant or product is proportional to this electrical potential difference in the standard state (both substances are present at unit activity):

$$\Delta G_0' = -nF\Delta E_0' \tag{155}$$

or a general relation for the free energy difference is:

$$\Delta G' = -nF(E_A' - E_B') = -nF(E_{0A}' - E_{0B}') - RT \ln\left[\frac{(A_{Ox})(R_{Red})}{(A_{Red})(B_{Ox})}\right] \tag{156}$$

If several redox reaction systems are arranged across the membrane, electrons will flow from the low potential state of the oxidation–reduction (redox) potential to a high potential state.

In mitochondria, the first redox couple in the electron transfer chain is NADH$_2$/NAD for which the standard redox potential, $E_0' = -0.322$ V, is lowest (see Figure 31). The last step of the electron transfer system is the H$_2$O/$\frac{1}{2}$O$_2$ couple for which the standard redox potential, $E_0' = +0.816$ V, is highest. It is generally considered that electrons flow from the low potential for a standard redox E_0' to a higher potential. However, since the oxidation–reduction potential is also dependent on the concentration ratio of the two substances involved in the electron transfer reaction [A$_{Ox}$ and A$_{Red}$, see Eq. (156)], the direction and order of the electron transfer in various redox systems are not necessarily those given by the standard redox potential (E_0'), when the standard redox potentials (E_0') are close to each other. It is understood that, during the electron transfer process, ATP (adenosine triphosphate) will be synthesized from ADP (adenosine diphosphate) utilizing the energy of the electron transfer when the energy difference between the two redox systems involved is greater than that required to synthesize ATP, i.e., at the stage where the energy release is more than 7–10 Kcal/mol ($\Delta G_0' = -7$–10 Kcal/mol). There are at least three main such electron transfer stages: 1) $NADH_2$ and FMN where $\Delta E_0' = -0.2$ V, $\Delta G = -9.2$ Kcal (per 2H), 2) cytochrome b and cytochrome c, where $\Delta E_0' = -0.29$ V, $\Delta G = -13.3$ Kcal (per $2e^-$) and 3) cytochrome a and O_2 where $\Delta E_0' = -0.53$ V and $\Delta G = -24$ Kcal (per $2e^-$) (see Figure 31).

3.1.2.3. Energy Production Systems and H^+ Transport

Wherever the activity of life exists, energy is necessary to sustain it, whether it may be due to the influx or efflux or dissipation of energy in the system (mechanical energy due to muscle contraction, electrical energy due to nerve activity, thermal energy for body temperature, and chemical energy for the process of synthesis and secretion of substances). The energy for all these activities is derived from the chemical energy liberated by the hydrolysis of energy-rich phosphate compounds such as ATP. Other substances like ATP are GTP, CTP, UTP (nucleotide polyphosphates), acetylphosphate compounds, and creatine phosphate, etc.

When the synthesis of ATP from ADP and inorganic phosphorus is done by an oxidation reaction, it is called oxidative phosphorlylation. This rection occurs in many respiratory chain systems of animals. In this reaction, the electron from a hydrogen ($H^+ + e^-$) in a substrate having high energy plays a role in the electron transport system. Another type of ATP synthesis is due to photophosphorylation, as occurs in green plants (chloroplasts) and bacteria (chlomatophores), where most organic substances and O_2 necessary for all livings are produced by this reaction. In this reaction, the electrons liberated from chlorophyll molecules, having high energy, participate in the electron transport systems.

In each cell system, the proteins (cytochrome, flavin proteins, etc.) containing redox reaction systems responsible for electron transport processes are located on their membranes, and, often in such systems, H^+ is simultaneously transported across the membrane utilizing the energy of electron flux which is produced by the redox reactions [also see Eq. (156)]. As a result, an electrochemical potential gradient of H^+ will be created across the membrane. Mitchell[168] proposed a unified theory for these electron transfer systems relating H^+ transport and ATP synthesis in the energy production systems. For a long time it was thought that the ATP synthesis is brought about by intermediate high-energy compounds which are produced during electron transport due to redox reactions. Mitchell's proposal was that the energy liberated during the electron transport reaction is also used to transport H^+ across the membrane (electron and proton transports are coupled) and is stored in the form of an electrochemical potential difference with respect to H^+ across the membrane ($\Delta\mu_{H^+}$). The H^+ flux due to this proton chemical potential gradient operates H^+-transporting ATPase ($F_i + F_o$) in the reverse direction and its energy will be utilized to synthesize ATP.

This hypothesis was supported by several experiments. One of them was to demonstrate ATP synthesis in a vesicle, which contains the proton-transporting ATPase and has a pH gradient across the vesicle cell membrane, where the pH difference corresponds to an energy (electrochemical potential difference $\simeq \Delta pH = 3.5$) comparable to that required to produce ATP.[172] Another experimental support was demonstration of ATP synthesis by applying an external electric potential difference ($\sim 210\,mV$) across the membrane of a cell containing ATPase ($F_o + F_i$) under the condition of no pH gradient.[173]

3.1.3. Active Transport and Electrogenic Potential

A perfect semipermeable membrane (at least inpermeable to one ionic species) will maintain a constant membrane potential indefinitely without expenditure of energy, because it is in a state of thermodynamic equilibrium. Most biological membranes, however, are permeable, to various degrees, to all ions. Thus, if the ionic concentrations across the cell membranes are not restored, in a certain time, the ionic concentration differences will be diminished at equilibrium and consequently there will be no membrane potential difference across the membrane. However, for living cells a definite ionic concentration difference for each ionic species is always maintained (Table 19). The high internal $[K^+]$ and low internal $[Na^+]$ seen in most biological cells result from the continual active transport of Na^+ out of the cell by way of an ion transfer enzyme present in the membrane, ATPase, which utilizes the energy of the hydrolysis of a high-energy phosphate bond in the ATP molecule.[174-177]

In some Na pumps, the exchange of K^+ for Na^+ is coupled, and inward movement of K^+ in exchange for Na^+ is obligatory.[178-181] Even without coupled K^+ uptake, potassium ions, to which the membrane is relatively permeable, move into the cell passively to replace the Na^+ pumped out so as to maintain electroneutrality of the cytoplasm.

When metabolically energized transport is eliminated by general inhibitors of oxidative metabolism, such as cyanide or azide, or by specific inhibitors of Na transport, such as ouabain, a net influx of Na^+ results, the internal K^+ is gradually displaced, and the resting potential shows a corresponding slow decay as the ratio $[K^+]_i/[K^+]_o$ gradually decreases. Over the long term, it is the metabolically energized extrusion of Na^+ that keeps the Na^+ and K^+ concentration gradients from running downhill to equilibrium. By continual maintenance of the potassium concentration gradient, the sodium pump plays an important indirect role in maintaining the resting potential, although it may not directly affect the main transmembrane potential by generating electric current.

Active transport has been shown to contribute directly as well as indirectly to the resting potential of some cells.[182,183] This occurs when Na^+ is transported from a cell interior to the exterior without one-to-one coupled exchange with K^+ or other monovalent cations (Figure 33). The pump is then said to be electrogenic rather than neutral because it is directly responsible for a steady transfer of positive charge (i.e., Na^+) out of the cell. The pump is also electrogenic if the ratio of coupled Na^+ to exchanged K^+ is greater than one, as in the active pump in the squid axon membrane which is considered to exchange three sodiums for two potassiums.[177] The potential contributed by an electrogenic pump depends upon the rate at which sodium can leak back into the cell. The tendency to leak back increases as the resting potential is further hyperpolarized by the pump.

The electrogenic potential may be determined by measuring the change in potential due to temperature change of the system or introduction of specific inhibitors (such as ouabain) for Na transport. Since that part of the membrane potential due to passive ionic processes across the membrane (as in the squid axon membrane) is less sensitive to temperature than that due to an active transport process, one can deduce from the temperature coefficient of the membrane potential the contribution due to an active transport process; or since ouabain is supposed to inhibit the active transport process, one can measure the active transport contribution to the membrane potential as the difference in membrane potential in two environments (one, a normal physiological solution; the other, a normal physiological solution containing ouabain).

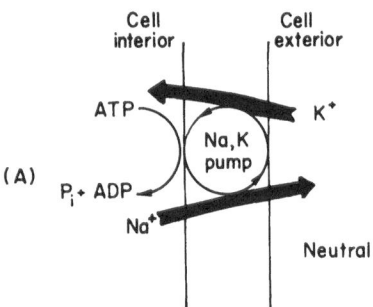

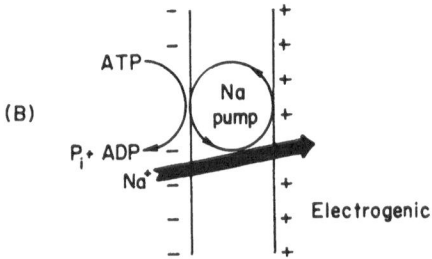

Figure 33. (A) Idealized neutral pump in which 1 K^+ is exchanged for 1 Na^+. In this case, the Na pump plays an indirect role in the resting potential by maintaining low internal Na^+ and high internal K^+ concentrations. (B) Electrogenic pump. The pump is electrogenic if it produces a net movement of charge across the membrane. The most common Na pump exchanges 3 Na^+ for 2 K^+, and is therefore electrogenic.

Mullin and Noda have derived an equation for the membrane potential in the presence of electrogenic pumping under a steady state condition.[185] If sodium and potassium are the only ions undergoing active transport, then it can be shown that the membrane potential is given by the equation:

$$E = \frac{RT}{F} \ln \frac{rP_K[K]_o + P_{Na}[Na]_o}{rP_K[K]_i + P_{Na}[Na]_i} \tag{157}$$

where r is the Na:K transport coupling ratio. The contribution of the electrogenic sodium pump to the resting potential is given approximately by the difference between Eqs. (126) and (157) (since the terms $P_{Na}[Na]_i$ in the denominators are negligible):

$$E_{pump} = \frac{RT}{F} \ln \left[\frac{1}{r} \frac{rP_K[K]_o + P_{Na}[Na]_o}{P_K[K]_i + P_{Na}[Na]_o + P_a[Cl]_i} \cdot \left(1 + \frac{P_a[Cl]_o}{P_K[K]_i} \right) \right] \tag{158}$$

The maximum value which Eq. (158) can acquire is equal to $(RT/F) \ln (1/r)$, which is 10 mV for a coupling ratio of 3:2. The electrogenic potential depends upon the Na:K coupling ratio. There have been several arguments as to the numerical value of the ratio. It seems that the Na:K coupling ratio for the squid axon membrane is reasonably well accepted to be 3:2. Some of the value of the so-called electrogenic potentials measured for various cells are given in Table 22. As seen in the table, the magnitudes of most electrogenic potentials are small (<10–15 mV). It may be said, therefore, that for cells having a high resting potential (such as nerve cells, muscle cells, and some nonexcitable cells) this electrogenic potential is not important as the source of the electrical potential, but for those cells having a low resting potential,[147] the electrogenic potential itself may become an important component of the electrical potential. Besides the Na active pump,[186,271] there are various other

Table 22
Electrogenic Potential of Various Cells at a Steady State

Cells	Electrogenic potential (mV)	Reference
Helix neurons	2–3	(279)
Crayfish stretch receptor neurons	10	(280)
Leech sensory neurons	5	(281)
Barnacle muscle	6	(282)
Limulus photoreceptor	5	(283)
Squid axon	1.4	(284)
Aplysia neurons	22	(285)
Anisodoris neurons	15	(286)
Squid stellon ganglion	15	(287)

ion pump systems: Ca^{2+},[187–189,271] Cl^-,[189,271] and H^+ pumps.[189–191,271] It has been reported recently that the rates of the Na pump in squid axon depend on the membrane voltage. Further studies seem to be needed on the molecular mechanism of the electrogenic potential.

3.2. Excitation Potential

Living systems respond to various external stimuli. The response to an external stimulus is called "excitation." A local response caused by a certain stimulus sometimes travels in the body as a signal and causes a response at a distant part of the body. There are, in principle, two mechanisms by which such signals may be transmitted within a living body. One method depends on the relase of a specific chemical agent as a result of a local stimulus at point A. After circulating in the body, the substance reacts at a remote site B, where some cell responds to the substance. Certain kinds of such messenger substances, such as hormones, play an important role in regulating metabolic effects in the organism. Another way to send a stimulus signal is done by way of an electrical signal. A long extension of a nerve cell, called an axon, is able to send such signals, by producing an "action potential," along the axon to another cell at a distance. In most cases, however, an electrical signal is not transmitted directly from one cell to another, but a special chemical transmitter released by the electrical stimulus mediates the transmission of the electrical signal between the two cells. The chemical transmitter substances interact with the receptors located at the second cell surface and produce receptor potentials (synaptic potentials), which often induce an action potential in the second cell. Receptor potentials are produced not only at the junction between the two nerve cells, but also at sensory cell surfaces by external stimuli, such as light, mechanical displacement, or chemical substances. These receptor potentials will be passed on by action potentials or chemical substances released by such stimuli to various cells and organs in the body (Figure 34).

3.2.1. Action Potentials

Action potentials are produced at membranes of neurons and muscles, as well as by some receptor cells, secretory cells, and protozoa.[192,193]

A certain stimulus, either the application of a brief current from the nerve cell interior to the exterior (depolarizing the resting membrane potential) or the application of high K^+ concentration in the exterior phase or removal of Ca^{2+} from the external solution will initiate an action potential. There are two features associated with the action potential: (1) all-or-none character, with a threshold potential and (2) propagative character (regenerative nature of the action potential).

In the resting state, the membrane is predominantly permeable to K^+, e.g., for squid axon $(P_K : P_{Cl} : P_{Na} = 1 : 0.45 : 0.04)$[194]; hence the resting mem-

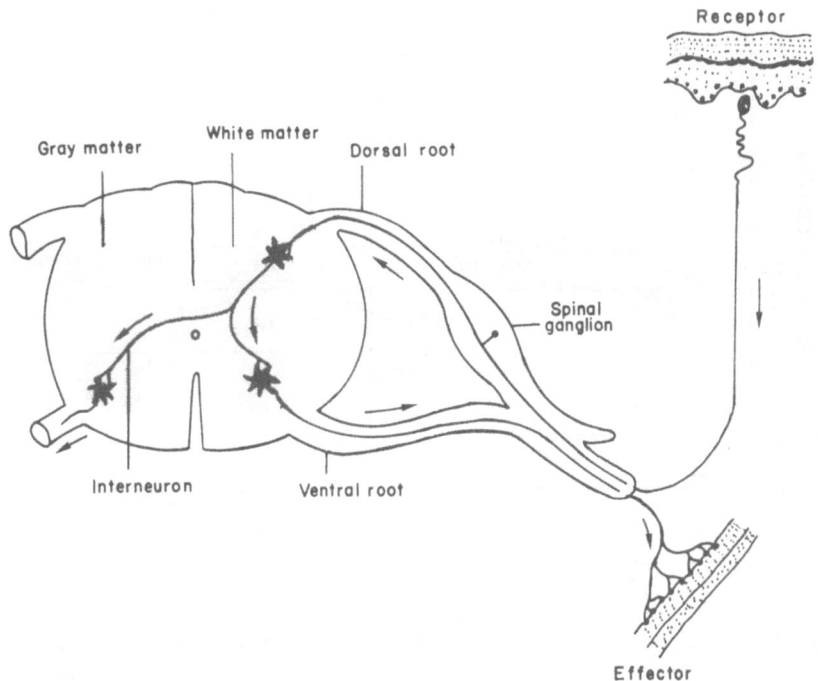

Figure 34. Organization of the vertebrate spinal cord and its segmental roots (after Montagna[274]).

brane potential is approximately equal to the potassium equilibrium potential, $E_K = (RT/F) \ln ([K^+]_o/[K^+]_i)$, i.e., potassium ions are nearly in electrochemical equilibrium across the resting membrane. Sodium, more concentrated outside the cell than inside because of active transport, has an electrochemical gradient opposite to that of K^+, and an equilibrium potential of opposite polarity to that of potassium. The Na potential, $E_{Na} = (RT/F) \ln ([Na^+]_o/[Na^+]_i)$, is positive, whereas E_K is negative (see Table 20).

The sodium equilibrium potential is far from the resting potential since the membrane is much less permeable to Na^+ than to K^+. Therefore, there is a large electrochemical potential gradient ($E_m - E_{Na}$; E_m is the membrane potential) acting on Na^+ in the outside solution. This emf is a substantial source of potential energy for the production of an action potential, which is released in some cells by stimuli that increase the permeability of the membrane ($P_K : P_{Cl} : P_{Na} = 1 : 0.45 : 20$)[194] so as to permit a transient inward sodium current (the net influx of Na^+ is the main component of the action current) through the membrane (Figure 35).

A temporary large increase in membrane permeability to Na^+ will permit the extracellular Na^+ to flow into the cell, and the membrane potential will be depolarized and be close to the Na equilibrium potential (see Figure 35). When the Na conductance is restored to its previous low level (at the resting

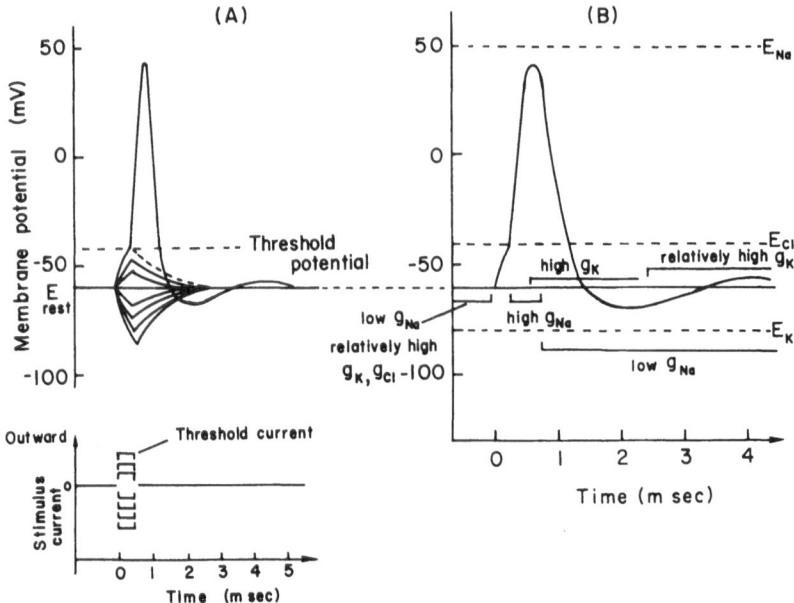

Figure 35. (A) An action potential produced by a nerve cell membrane in response to a depolarization (above the threshold) stimulus. (B) A rough sketch of time-dependent conductance change of a nerve axon membrane (e.g., squid axon).

state) after a brief time (a few milliseconds for most nerve and muscle membranes) the membrane potential returns to its original value near E_K. More precisely, the efflux of K^+ (flow from the cell interior to the exterior) will start in a delayed fashion during the action potential change which mainly contributes to the falling phase of action potential. The quantitative analysis of each ionic current during action potential of squid axon membranes has been done by Hodgkin and Huxley[195] by using the action currents obtained under a voltage-clamped axon. The net number of ion exchanges (Na^+, K^+, etc.) during an impulse will be the same and only the concentration gradient of each ionic species will drain down. When a number of action potentials are produced continuously, therefore, the restoration of ionic environments by an active transport process cannot be kept up because of the large amount of Na/K exchange across the membrane.[196]

It seems to be evident that the ionic pathway for active transport is different from those for action currents which are predominantly an exchange of Na^+ and K^+ due to their electrochemical potential gradients (Figure 36). It is very likely that there is more than one ionic path responsible for the action currents, and that these paths are to some extent specific for each ionic species (such as Na^+ channels and K^+ channels and Ca^{2+} channels, etc.), since the ionic current through each channel can be almost completely suppressed

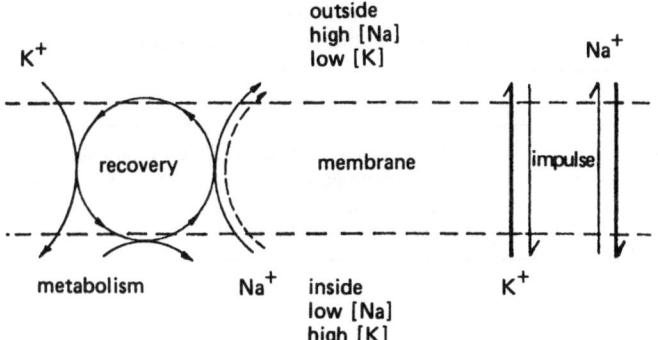

Figure 36. Diagram illustrating movements of ions through a nerve cell membrane. The downhill movements which occur during the impulse are shown on the right; uphill movements during recovery are shown on the left. The broken line represents the component of the sodium efflux which is not abolished by removing external potassium ions. (From Reference 196).

by an application of a specific ion channel inhibitor.[197-199] It should be noted that each channel is not necessarily exclusively specific for one particular ion[197,198] (Table 23).

For example, tetrodotoxin (TXX) and saxitoxin (STX) are inhibitors of the Na channel of squid axons when applied from the extracellular side.[199] On the other hand, the specific inhibitors for K channels are tetraethylammonium (TEA), Cs^+, etc. when applied from the intracellular phase. Also, Mn^{2+}, Co^{2+}, etc.[199] are known to be Ca^{2+} channel inhibitors (Table 24).

Table 23
Relative Selectivity of the Sodium Channel: Ratios of Permeability Coefficient with Respect to Na^+ of Ions Passing through the Early Na Channel of Axons

Squid axons[a] Ions	Li^+	Na^+	K^+	Rb^+	Cs^+		choline$^+$	
Pc/P_{Na}	1.1	1.0	0.083	0.025	0.016		<0.014	
Nodes of Ranvier of frogs[b] Ions	Li^+	Na^+	K^+	Rb^+	Cs^+	Ca^{2+}	Mg^{2+}	
Pc/P_{Na}	0.93	1.0	0.086	<0.012	<0.013	0.043	0.1	
Ions		hydroxylamine		hydrazine		ammonium		formamidine
Pc/P_{Na}		0.94		0.59		0.16		0.14
		guanidine		hydroxyguanidine				aminoguanidine
		0.13		0.12				0.006

[a] Reference 198, 315.
[b] Reference 197, 316, 317.

Table 24
Ion Channel Inhibitors in Squid Axons[a]

Ion channels	Inhibitors
Na channels	Tetrodotoxin (TTX), saxitoxin (STX), local anesthetics
K channels	Tetramethylammonium (TEA), Cs^+, Ba^{2+}, 9-aminoacridine, DDT
Ca channels	Various divalent and polyvalent cations (Co^{2+}, Mn^{2+}, Ni^{2+}, Mg^{2+}, La^{3+}, etc.), verapamil, (D600)

[a] Reference 199.

The molecular mechanism operative in the change in ion channel conductances is not yet clearly understood. However, it is likely that these ion channels are composed of lipid–protein complexes. There are at least two theories for the opening of the sodium channel: (1) Voltage-dependent ion channel conductance: When the depolarization potential is greater than the threshold value, the sodium channel opens.[200] (2) Ca^{2+} removal from the membrane outer surface (probably, from the outer part of the sodium channel): Such Ca^{2+} removal acts as a trigger for opening the sodium channel.[201,202] It seems to this author that the latter mechanism is the primary cause for the opening of the sodium channel and the former (1) is a property associated with the sodium channel molecule and the membrane.

The observed action currents, which are measured via action potentials, are due to the ion flux exchanges but not due to the electronic phenomena.

The regenerative nature of the action potential can be explained in the following manner. Once the Na channel opens up, under normal physiological conditions, the sodium ions rush into the axon and in return the potassium ions will flow from the internal phase of the axon to the external phase. According to the traditional theory of the excitation, these ionic flows create a situation where the local surface charges are replaced by the charges of opposite signs at the membrane regions (A) of excitation. Therefore, the local circuit currents flow through between the neighbouring membrane region (B) and the membrane region being excited (A). This local circuit current is normally in a direction so as to depolarize the neighbouring membrane and, especially at the peak of the action potential, the depolarization current at the neighbouring membrane region is strong enough (more than the threshold current) to produce an action potential at the neighbouring membrane region (Figure 37). Thus, the action potential is propagated along the axon in a regenerative manner. The question which then arises is by what mechanism does the neighbouring membrane open up the sodium channel, i.e., is it depolarized membrane potential or depolarizing membrane current that is effective? Here, again either of the two previously mentioned theories can be invoked. This author is in favor of the depolarizing membrane current. By the externally directed current, due mainly to an outward potassium current,

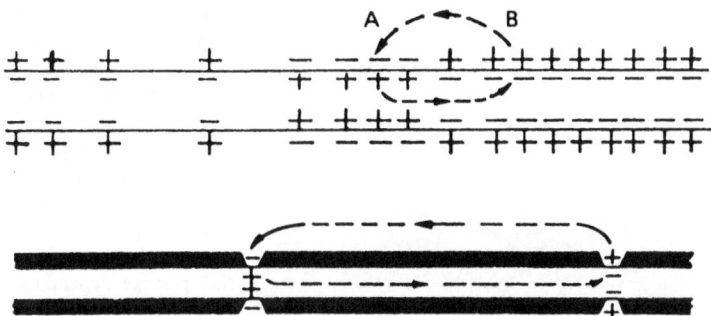

Figure 37. Diagrams illustrating the local circuit theory: The upper sketch represents an unmyelinated nerve fiber, the lower a myelinated nerve fiber. (From Reference 196.)

the calcium ions bound to the external surface at the Na channel will be removed and thus the Na channel lipoprotein molecule will change conformation (Figure 38B). The point which this author would like to stress here is that the opening of the Na channel may not be due to the membrane potential itself but may be due to the specific type of ions driven by the membrane potential or its associated membrane current from the axon interior to the exterior. This is indicated by the fact that the action potential can be produced by reducing the external Ca^{2+} concentration only while the membrane potential and other ionic conditions are not altered appreciably. The opening of Na channel is thus postulated to be due to the removal of Ca^{2+} from the channel molecular matrix introducing a conformational change in the Na channel molecule.[202] The other view of how the Na channel is opened involves a membrane depolarization which gives rise to a voltage-dependent conductance change.[195] It has also been postulated[200] that there is a mechanical barrier to sodium ions in the resting state and that this barrier transiently opens and then closes when the membrane is depolarized, i.e., that the membrane potential controls a voltage-dependent gate (Figure 38A). The major evidence favoring a physical reordering or conformational change is the recent finding of minute gating currents associated with the opening and closing of the Na channels.[203,204] These are interpreted as arising from movements of charged groups associated with the gating mechanism of the Na channel due to the applied voltage. These charge movements could be attributed due to a local polarization of electrons in moieties contained in the Na channel molecule.

Except in certain minor details, the action potentials of vertebrate skeletal muscle resemble those of nerve cells. Action potentials of cardiac muscle or barnacle muscle, however, differ in certain respects from those of nerve and skeletal muscle. The most striking feature of the action potential in the ventricle of the heart is its long duration (see Figure 39), lasting up to 1 sec in amphibians. The upstroke of these action potentials is due primarily to inward current

(A)

Extracellular Phase

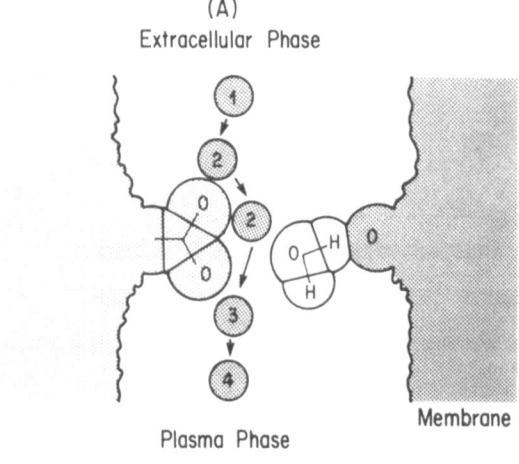

Plasma Phase

Membrane

(B)

Extracellular Phase   Ion Exchange
                           Polymer Phase

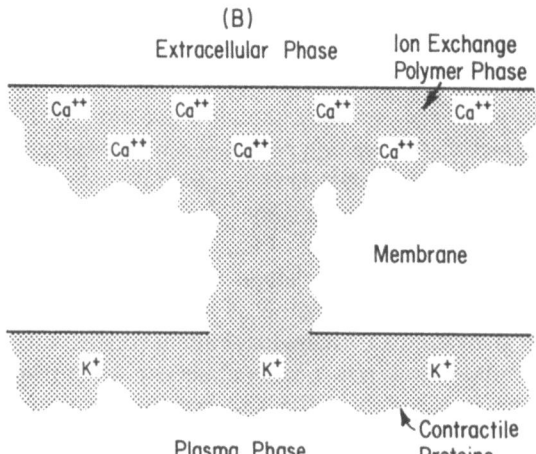

Membrane

Plasma Phase   Contractile
                    Proteins

Figure 38. (A) A model of the Na channel, showing four stages (1, 2, 3, 4) in the passage of a Na^+ (shaded circle) through the channel. A water molecule is hydrogen bonded to a strongly electrogenic oxygen at the right. The two oxygens at the left are part of a carboxyl group. These sites are voltage sensitive and act as a gate for the Na^+ passage depending upon the direction and the magnitude of membrane voltage.[200] (B) A schematic diagram of a squid axon membrane composed of ion-exchange phases. External Surface: when the extracellular Ca^{2+} concentration is relatively high, the external ion-exchange phase is less hydrophilic (low conductance); when $[Ca^{2+}]_o$ is low, the phase becomes hydrophilic and highly conductive. The internal surface is also covered with contractile proteins. These protein phases contract as K^+ is replaced by either Na^+ or Ca^{2+} entering from the extracellular space upon the change in the extracellular ion-exchange phase to the high conductance state. The excitation is initiated by the removal of the bound Ca^{2+} by either the depolarization current (or the outward flowing K^+) or the application of Ca^{2+} exchange substances (e.g., K^+) in the extracellular phase (after ref. 304).

carried by Na^+, as in nerves. However, the prolonged plateau of the action potential results largely from a high Ca^{2+} conductance. The rapid repolarization following the plateau appears to be due to a decrease in Ca^{2+} conductance together with an increase in K^+ conductance.

3.2.1.1. Pacemaker Potentials

Rhythmic firing of action potentials occurs in a number of excitable tissues. In some receptor and nerve cells a steady stimulus produces a steady train of impulses. Certain pacemaker cells are spontaneously active, firing at a steady rate without input from any external source. There are several possible mechanisms based on the ionic hypothesis, for such autorhythmicity. A familiar example of autorhythmicity is known from the pacemaker tissues of vertebrate

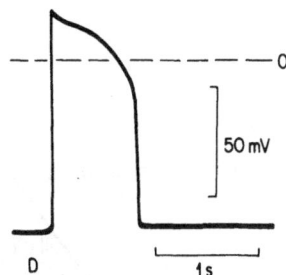

Figure 39. Action potential of a cell in the ventricle of the frog. (After Reference 205).

hearts. The membranes of the cells in the pacemaker region (sinoauricular node) have no fixed resting membrane potential, but undergo a steady depolarization termed a pacemaker potential. The interval between cardiac action potentials depends in part on the rate of depolarization during the pacemaker activity. A slower depolarization brings the membrane to the firing level later, and thus decreases the frequency of spontaneous discharge (action potential). The action potential initiated in the pacemaker cells is conducted and transmitted electrically through the myocardium to excite the remaining cells of the heart.

In the atrial cells of the frog heart, the pacemaker depolarization begins immediately after the previous action potential, when the potassium conductance of the membrane is very high. Then the potassium conductance gradually drops, and the membrane shows a corresponding depolarization due to a moderately high steady conductance for sodium. The pacemaker depolarization continues until the excited sodium conductance is activated, and rapid regenerative upstroke of the cardiac action potential takes over.

Acetylcholine, which slows the heart when released by activity of the vagus nerve, does so by increasing the potassium conductance of the pacemaker cells. This keeps the membrane potential near E_K for a longer time, thereby slowing the pacemaker depolarization and thus delaying the onset of the next action potential. A recent proposal for the rhythmic generation of pacemaker waves in spontaneously "bursting" neurons is as follows.[206] During the slow depolarization, the Ca conductance g_{Ca} is turned on, allowing an influx of Ca^{2+}. As the $[Ca^{2+}]_i$ rises, it gradually activates the potassium conductance, g_K. Thereupon the membrane repolarizes (i.e., is hyperpolarized) toward E_K, causing the calcium conductance g_{Ca} to turn off. As a result, $[Ca^{2+}]_i$ drops, g_K drops, and the membrane potential slowly shift away from E_K, turning on g_{Ca} and initiating a new cycle of pacemaker potential. The slow depolarizing pacemaker wave gives rise to trains of action potentials as the wave reaches and exceeds the firing (action potential) level (Figure 40).

3.2.2. Receptor Potentials

Sense organs provide the only channels of communication into the nervous system from the external world. The processes of sensory reception begin in

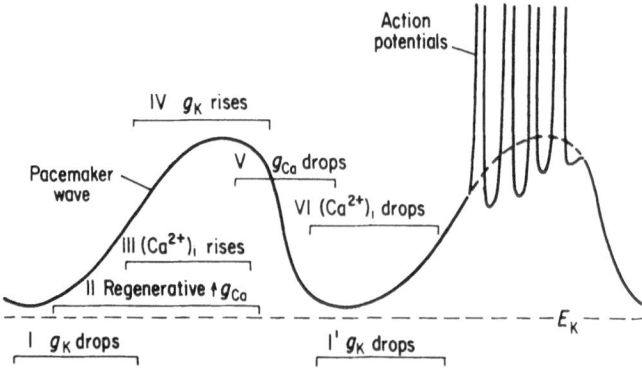

Figure 40. Proposed events (I to VI) responsible for pacemaker waves in spontaneously "bursting" neurons. The wave at the left is shown without action potentials for the sake of simplicity. See text for description of processes involved. (Reference 193, p. 143.)

sensory organs.[207] The receptor cells in the sensory organs have the capability to respond to specific external stimuli by producing electrical potentials termed receptor potentials. Receptor potentials can evoke action potentials in nerve cells by which stimuli can be transmitted to a distant part of the body and induce responses. The receptors can be classified into at least five types: chemoreceptors, mechanoreceptors, photoreceptors, thermoreceptors, and electroreceptors.

3.2.2.1. Chemoreceptors

The sensitivity of cells to specific molecules is manifold. Such sensitivity includes the metabolic response of tissues to chemical messengers as well as the ability of lower organisms such as bacteria or protozoa to detect certain substances in the environment and its motion (chemotaxis). Among chemoreceptors we may mention gustatory (taste) receptors which detect dissolved ions and molecules, and olfactory (smell) receptors, which detect airborne molecules, and as a special case, synaptic receptors for chemical transmitters which propagate an electrical signal carried along an axon from one nerve cell end to another by way of chemical substances.

3.2.2.1.1. Gustatory (Taste) Receptors. One of the examples of taste receptors is the contact chemoreceptors (taste hairs) of insects. These receptor cells send fine dendrites (the extensions of the cell body) to the tips of hollow hairlike projections of the cuticle, called sensilla. Each sensillum has a minute pore that provides access for stimulant molecules to the sensory cells (Figure 41). The sensillum contains several cells, each of which is sensitive to a different chemical stimulus (e.g., water, cations, anions, carbohydrates). Upon intereaction with such chemical substances, the receptor potential is produced at the end of the dendrites that extend to the tip of the sensillum, whereas the action potentials originate near the cell body.

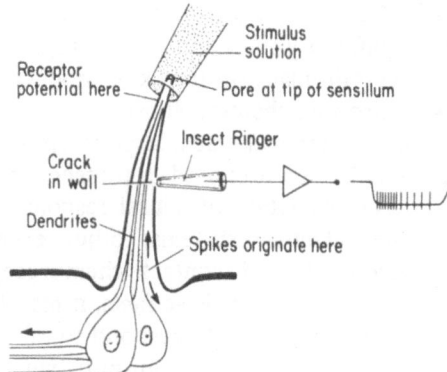

Figure 41. Electrical recording from a contact chemoreceptor sensillum of a fly. The dendrite of each neuron is sensitive to a different class (sugars, cations, anions, water) of substances. Electrical responses (right) are recorded through a crack made in the side of the sensillum. (Reference 193, p. 204).

It is worthwhile to note recent work on the mechanism of generation of taste receptor potential in true slime molds.[208,209] The chemotaxis of the slime mold is closely correlated with a change in the membrane potential, and further the change in membrane potential in response to substances was approximately identical to that in the ζ potential not only for various salts (Figure 42), but also for various sugars. It has been proposed from studies of various receptor cells[208] that the taste receptor potential is related to the change in membrane surface potential due to binding of the substances. Although sugar molecules have no charged groups, it is considered that they are adsorbed or bound to membrane molecules and cause a change in surface charge density. This mechanism does not stress the change in permeability of ions across the membrane but rather the change in surface potential leading to change in the membrane potential.

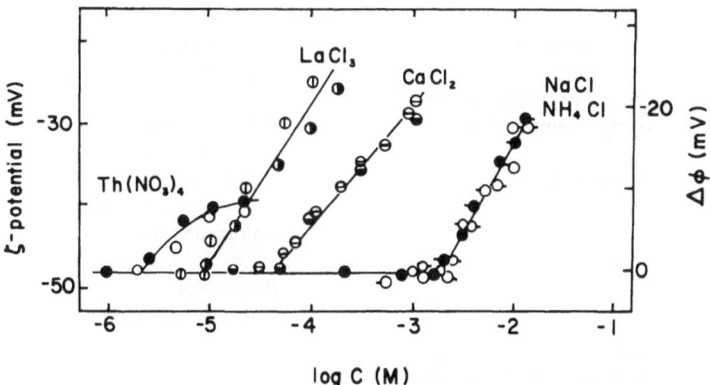

Figure 42. The ζ-potential and the membrane potential, $\Delta\varphi$, of plasmodia of slime mold as a function of salt concentration. ζ: ○ $Th(NO_3)_4$; ⦶ $LaCl_3$: ⊖ $CaCl_2$; —○ NaCl; ○— NH_2Cl. $\Delta\varphi$: ● $Th(NO_3)_4$; ◑ $LaCl_3$; ◐ $CaCl_2$; —● NaCl; ●— NH_4Cl. Temp., 20°C. (Reference 209).

3.2.2.1.2. Olfactory (Smell) Receptors. One example of an olfactory receptor is the olfactory epithelium of the frog.[210] A number of receptor cells, gathered in a certain place on the skin, produce a receptor potential upon interaction with airborne chemical substances (odorants). The receptor potentials can be summed up to make up a large receptor potential which would evoke an action potential when it exceeds a threshold value.

The elicitation of neural responses caused by odorants is not restricted to the olfactory systems; other excitable systems, e.g., the trigeminal, vomeronasal, and contact chemoreceptor, are also able to respond to odorants.[211] From these facts, it may be considered that an excitable membrane in general might respond to odorants.

The molecular mechanism by which such chemoreceptor potentials arise is not yet well understood. However, it is thought that the receptor potentials may arise by processes similar to those (chemosensory membrane) occurring in the postsynaptic nerve or muscle cell.

3.2.2.1.3. Synaptic Potential. A postsynaptic membrane may be considered as a chemosensory membrane. The signal transmissions through most synaptic regions (the region where one nerve ending terminates at a dendrite of another nerve cell) are mediated by chemical transmitters released from the nerve end upon the arrival of an action potential (Table 25). In some synapses, the two membranes are so closely adhered that the electrical impulse (action current) will directly initiate a synaptic potential at the second cell (i.e., an electrical synapse (Fig. 43A)). However, in a chemical synpase, the chemical transmitters released from the presynaptic terminal diffuse and react at a certain part of the dendrite membrane (postsynaptic membrane) where chemical receptors are located. The chemical transmitters react with the

Table 25
Chemical Synapses: Transmitters and Inhibitors

Transmitters	Inhibitors
Excitatory synaptic receptors	
Acetylcholine (ACh)	d-Tubocurarine
	α-Bungarotoxin
Noradrenaline	Yohimbine
	(for α receptors)
	Chlorpromazine
	(for α receptors)
	Propanolol
	(for β receptors)
Glutamic acid etc.	
Inhibitory synaptic receptors	
Glycine	Strychnine
γ-Aminobutyric acid (GABA)	Bicuculline
	Picrotoxin

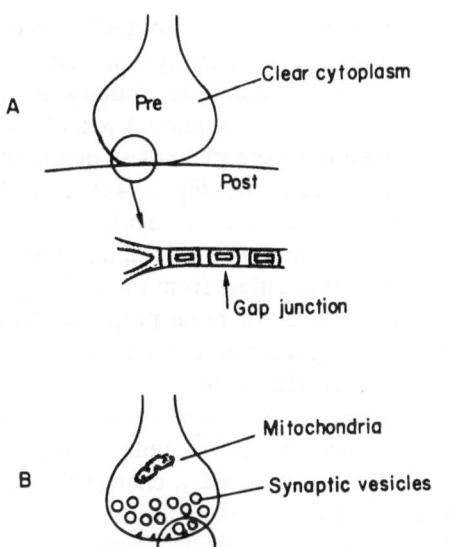

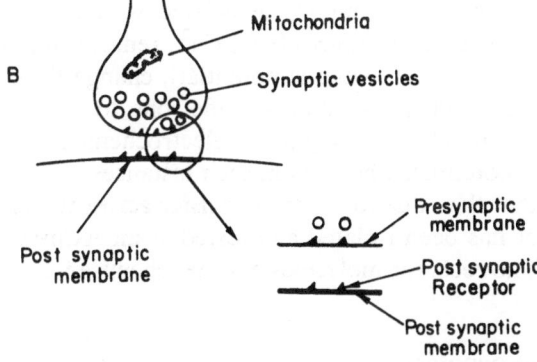

Figure 43. Two kinds of synapses. A. Electrical synapse: Gap junctions between pre- and postsynaptic membranes permit action currents to flow through both membranes, by way of which the excitation signal is transmitted from one nerve cell to other. B. Chemical synapse: No intracellular continuity. Synaptic current flows only across the postsynaptic membrane in response to chemical transmitter-activated opening of membrane channels.

chemical receptors and induce a potential change (the postsynaptic potential) across the postsynaptic membrane.[212-214] The postsynaptic receptor molecules of the vertebrate motor nerve, or muscle, upon binding with ACh (acetylcholine), a chemical transmitter, undergo conformational change that produces an increase in the Na^+ and K^+ conductance of the membrane. As a result, synaptic current (receptor current) flows into the cell, depolarizes the membrane, and generates an action potential (Figure 44).

The excitatory postsynaptic potential is a depolarization potential, because the equilibrium potential of Na^+ and K^+ (reversal potential) is at the level of -20–0 mV whereas the resting potential of the postsynaptic membrane in most nerve and muscle is about -80–-100 mV in a typical physiological environment. On the other hand the inhibitory postsynaptic potential can also be evoked by another type of chemical transmitter which acts by hyperpolarizing the membrane; in this case the receptor responsible for production of the inhibitory postsynaptic current seems to open up anion-selective channels. Since the anion equilibrium potential is at a slightly more hyperpolarized level than the resting potential, the membrane potential produced is the hyperpolarized one. When these two types of receptor (synaptic) potentials are produced simultaneously, the resultant receptor potential is less depolarized than that of the excitatory postsynaptic potential alone (Figure 44). Thus, the excitatory synaptic potential is inhibited. Nervous systems regulate signal transmission by proper arrangements of those excitatory and inhibitoryy synapses at each nerve terminal. The synaptic potential differs from the action potential in the following ways. It does not have an all-or-none property but has a graded nature. Also, it does not travel along the membrane surface in a regenerative fashion as the action potential does. However, it is almost certain that there are specialized lipoprotein complexes associated with each receptor which, when triggered by specific substances (e.g., Ca^{2+} removal for action potential and ACh for neuromuscular synaptic potential), change their conformation. These conformational changes allow specific ions to cross the membrane through the ionic channels according to the electrochemical potential gradient for each ion. The potentials observed in such instances are due to the movements of ions rather than due to electron transfer across the membrane. Only the ACh receptor has been isolated and tested as an ACh-activated channel molecule[215,216]; no receptor molecules of other chemisensory cells have been isolated yet.

3.2.2.2. Mechanoreceptors

The simplest mechanoreceptors are morphologically undifferentiated nerve endings found in the connective tissue of skin. In many mechanoreceptors there have evolved accessory structures whose function is the efficient conversion of mechanical energy to electrical energy in the receptor cell. Examples are Pacinian corpuscles, in which the sensitive ending is covered by a capsule,[217] muscle stretch receptors, in which the mechanically sensitive endings are associated with specialized muscle fibers,[218] and the hair-like sensilla in the exoskeletons of arthropods[219] (Figure 45). Elaborated accessory structures used to detect and analyze sound waves are found in the vertebrate middle and inner ear.[220,221]

2.2.2.2.1. Stretch Receptor Potential. The immediate stimulus thought to act on mechanoreceptors is a stretch or distortion of the surface membrane of the cells. Stretching the membrane of one of the large axons of the crayfish has been shown to increase its permeability to sodium.[218] One hypothesis for mechanoreceptor transduction is that stretching of the receptor membrane slightly enlarges ion-selective channels in the membrane. Such an enlargement of channels may induce a large increase in permeability to a specific ion. Furthermore, it has recently been demonstrated with nerve axons that, upon the spike (action potential) in nerve cells, the axon changes from a swelling phase (the upstroke phase) to a shrinking phase (the refractory phase), produc-

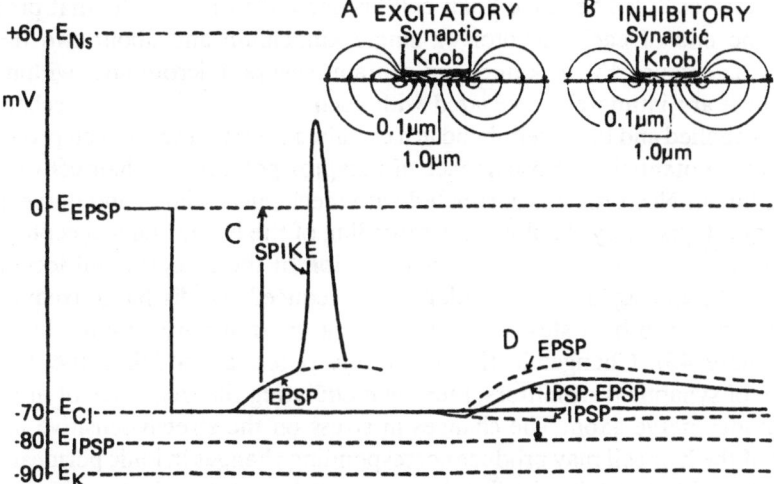

Figure 44. Excitatory and inhibitory synaptic actions in cat brain cells. Synaptic potential and currents are shown in the figures. EPSP—excitatory postsynaptic potential; IPSP—inhibitory postsynaptic potential (after ref. 213).

ing 100 nm changes in the radius of the axon.[222,223,304] This phenomenon may be the result of structural changes in channel proteins, triggered by a certain factor for the impulse production (e.g., Ca^{2+} removal from the outer surface of the membrane and also possibly Na^+ and Ca^{2+} interaction with cytoskeletal proteins at the inner membrane surface of axons).

3.2.2.2.2. Hair Cells. The hair cells of vertebrates are the primary mechanoreceptors in the lateral-line systems of fish and amphibians, in the cochlear nerve of vertebrate hearing organs, and in the organs of equilibrium

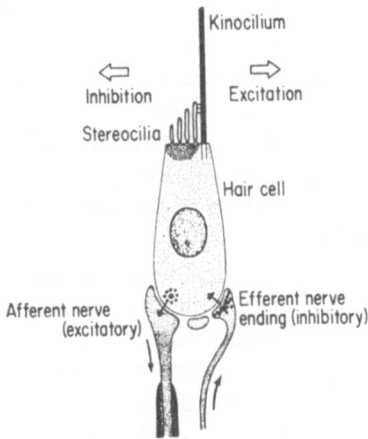

Figure 45. A diagram showing anatomical relation of a hair cell, stimulus direction, and secondary afferent fibers. (Reference 219.)

of vertebrates.[224] The name "hair cell" is derived from the cilia that project from one end of each receptor—a single kinocilium and about two dozen stereocilia (Figure 45). A kinocilium contains several microtubules within the cell which are intimately associated with each other.[225] When a mechanical stress is applied and these cells bend in a certain direction, the hair cell produces a receptor potential. The occurrence of receptor potentials of hair cells in the lateral line of Necturus was correlated with mechanical stimuli and nerve fiber discharges (spikes) by simultaneous recording of the intracellular receptor cell potential.[225] It was found that depolarization of the receptor cell increased discharge frequency and hyperpolarization reduced the discharge frequency. There appears to be a steady release of synaptic transmitter by the hair cell (see Figure 45). Changes in the membrane potential modulate the rate of release of synaptic transmitters, thereby modulating the frequency of firing of the sensory nerve axon. The changes in stress on the stretch-sensitive membrane of the hair cell may produce corresponding changes in ionic permeability which would account for the depolarization and hyperpolarization. However, the detailed molecular mechanism for the production of such receptor (hair cell) potentials is not yet known.

The receptor cells of the cochlea may have a mechanism similar to that of the hair cells, but cochlea receptor cells are more complicated structurally than the single hair cells found in the lateral-line systems of fish and amphibians.

3.2.2.3. Photoreceptors

All photoreceptors have in common photoexcitable pigments associated with the receptor membranes. These photopigment molecules, which are the primary sites of photoreception, are altered by the absorption of light in such a way to change the conductance of photoreceptor cell membranes to a specific ion. Thus, the membrane potential of the receptor cell will be altered, i.e., a "receptor potential" will occur.[226] The vertebrates and the invertebrates have evolved different mechanisms by which the primary transduction process alters the membrane conductance. We will review here primarily the mechanism of photoreceptors in the vertebrates.

In mammals, birds, and many other vertebrates, the receptor cells (rods or cones) are most closely packed in the fovea, which is the small (1 mm^2) central part of the retina. In humans and other mammals with color vision, the fovea contains only cones, whereas the remainder of the retina contains rods as well as cones (Figure 46). In mammals the cones are responsible for color vision and the more sensitive rods are restricted to achromatic vision.

Each receptor cell consists of the outer segment, which contains the photoreceptor membranes and the inner segment, which contains the nucleus, mitochondria, and synaptic contacts. The receptor membranes consist of lamellar flattened membrane sacks (see Figure 46). There are slight differences in their morphologies in cones and rods. The photopigments (e.g., rhodopsin) are embedded in these sack membranes.

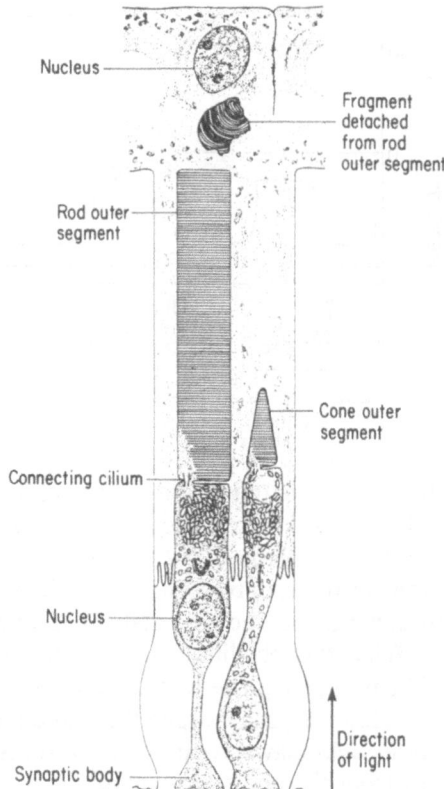

Figure 46. Rod and cone of the frog. Note that the outer segments are pointed away from the source of light, toward the pigment epithelium at the back of the eye. As the visual cells grow, fragments detach from the apical ends and a new membrane reseals the outer segment. (After Reference 226.)

In vertebrate rods and cones, light produces a hyperpolarized receptor potential instead of a depolarization as seen in the photoreceptors of the invertebrates (such as Limulus)[227] (see Figure 47).

It is known that, in the dark, the membrane of the photoreceptor is nearly equally permeable to both Na^+ and K^+. Consequently, the resting potential is about halfway between the sodium and potassium equilibrium potentials, E_{Na} and E_K. Sodium ions leak into the cell at the outer segment and are continually pumped out at the same rate across the membrane of the inner segment. When light is absorbed by the photopigment in the sack membrane, the sodium conductance of the outer segment decreases; thus the membrane is hyperpolarized, i.e., goes toward the E_K value given in Eq. (117). When the light ceases, the sodium conductance of the membrane increases to its high value (in the dark) and the membrane potential goes back to the resting level between E_{Na} and E_K.

The molecular mechanism by which illumination produces a decrease in Na conductance is not well understood yet. One proposal for this mechanism was given by Yoshikami and Hagins[228]: the photochemical alteration of a photopigment molecule in a disk membrane opens up a calcium channel in

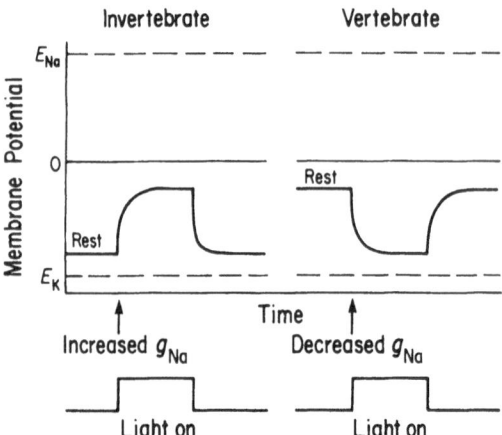

Figure 47. Electrical responses to light by invertebrate and vertebrate visual cells. Most invertebrate photoreceptors undergo an increase in Na (and perhaps Ca) conductance in response to light. The membrane potential shifts toward E_{Na}, producing a depolarization. Vertebrate photoreceptors respond with a decrease in Na conductance. This causes a shift away from E_{Na} toward E_K, producing a hyperpolarization. (After Reference 227).

the membrane. It is known that calcium ions are stored at a high concentration inside the disk. Upon the opening of a calcium channel, calcium ions leak out into the cytoplasm of the rod outer segment and diffuse to the surface membrane of the rod, blocking the sodium channels. This hypothesis is supported by experiments showing that the local application of Ca^{2+} or of certain drugs (caffeine, theophylline) which release internally stored Ca^{2+} mimics the effect of light in giving rise to a hyperpolarized photoreceptor cell.

However, some argue that although the diffusion of Ca^{2+} from the membrane sacks may account for the fast response time from stimulus to response, i.e., from the onset of illumination to the occurrence of a hyperpolarization potential, a great amplification of Ca^{2+} flux by a mechanism other than Ca^{2+} diffusion must be involved in order to account for the electrical activity produced by the absorption of a single quantum of light. Such theories of the mechanism are based on amplification by some chemical changes in enzymatic systems occurring upon illumination. One of these schemes is that a photolyzed rhodopsin molecule activates an enzyme, which in turn produces or degrades many transmitters and other enzymes. Sodium channels are considered to be kept in their open conformation in the dark by protein phosphorylation mediated by cyclic GMP. Illumination activates a phosphodiesterase enzyme, which lowers cyclic GMP levels, protein is dephosphorylated, and the channels close.[229,230]

Another speculation is that the initiation of such potential change is due to the transfer of molecular excitation (exciton theory) whereby a two-dimensional array of photopigments will transmit a signal produced by a single proton by electronic excitation.

It is evident, however, that the photopigment in the mammalian retina is rhodopsin composed of opsin (a protein molecule), which is a single polypeptide and spans several times across one sack membrane and forms a type of molecular oligomer, and 11-*cis*-retinal which is a chromophore surrounded by the polypeptide chain assembly of the opsin (Figure 48). When this chromophore absorbs a certain wavelength of light, its conformation will change and it will become all-*trans*-retinal.[231,232] During this conformational change an ionic channel may be opened, through which Ca^{2+} may flow from the sack membrane to the cytoplasm, or the binding affinity to a certain enzyme (e.g., GTP) may change and subsequently, an enzymic reaction cycle would be accelerated.

The details of the molecular mechanism by which the hyperpolarization potential (receptor potential) is produced, are not well established yet.

Other receptor potentials (e.g., thermoreceptors and electroreceptors) will not be described here.

4. Molecular Interaction in Biological Systems

When two molecules interact with each other, several types of electrostatic interactions or forces may be involved, some of which have been described in the preceding sections (e.g., charge–charge, charge–dipole, dipole–dipole interactions). Here, we would like to mention two other kinds of electrical interactions which were not described above: namely, short-range repulsive interactions and the London dispersion interaction. The latter interaction plays an especially important role in biological systems.

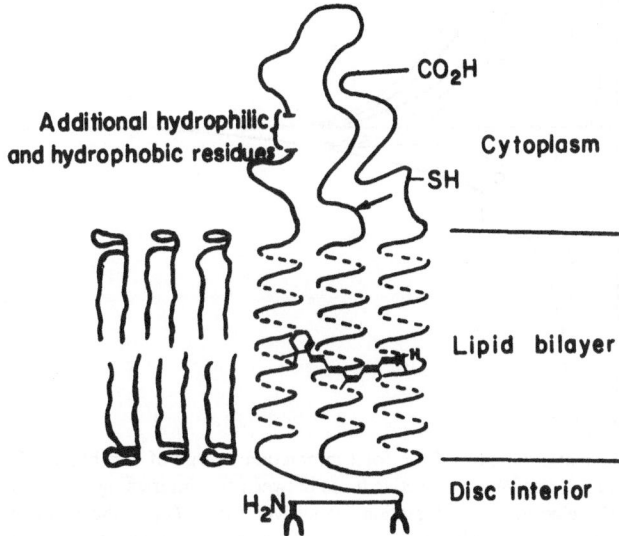

Figure 48. Schematic representation of rhodopsin and phospholipids in the disc membrane bilayer. Three of the $7 \pm 2\alpha$ helices of rhodopsin are shown. The 11-cis retinylidene chromophore is parallel to the membrane surface and is presumed to lie between some of the α helices. (After Reference 275).

4.1. Short-range Repulsive Interactions

When two molecules approach each other, the electron clouds surrounding each molecule would be expected to interact and create a repulsive force tending to push them apart. This repulsive force would arise from simple coulombic repulsion of charges. However, because of the differences in shape and size of molecules, no definite theoretical treatment of this short-range repulsion force has been established. This repulsive interaction has been described by the following general formula:

$$U_R = +B/r^b \tag{159}$$

where B and b are constants and the sign is positive. The exponent b takes a value somewhere between 9 and 15, indicating that the repulsive force is extremely strong if r (distance between the two molecules) is small (Figure 49). Usually, a value of 12 is used for b (such as in the Lennard-Jones potential

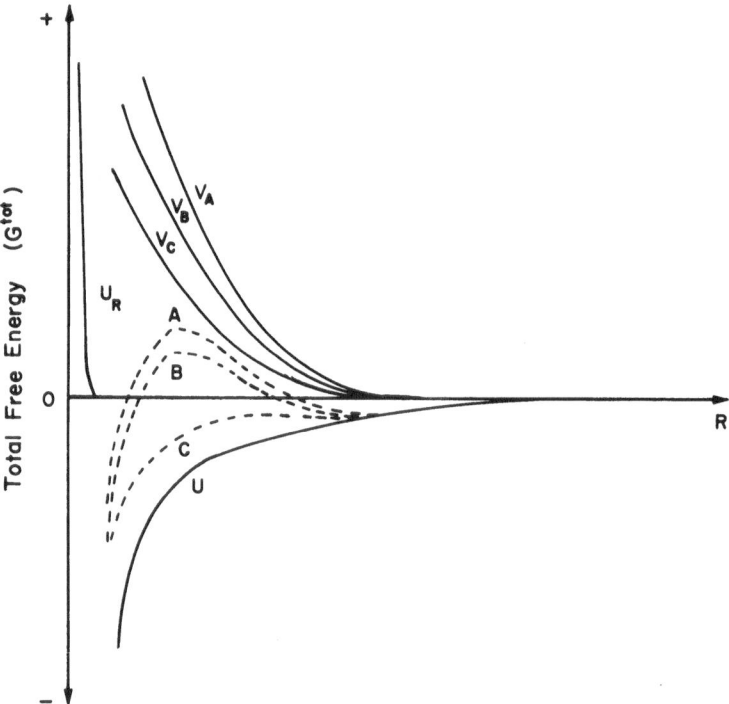

Figure 49. A schematic diagram of the total interaction energy of two bodies. U refers to the energy due to the van der Waals attractive force between two interacting bodies, V refers to the energy due to the electrostatic interaction (repulsive) force, U_R is the short-range repulsive interaction energy. R *is* the distance between the two bodies. The total interaction energy $G^{tot} = U_R + U + V$ (dotted curves). Curves A and B are the cases where there is a maximum (primary maximum) between two areas of minimum (primary minimum and secondary minimum). Curve C is the case where there is no maximum (no barrier) so that the two bodies could come in close contact with each other at the primary minimum region.

[233]). Because of this repulsive interaction factor, other interaction energy expressions have to be corrected for this short-range repulsive contribution. If we use $b = 12$, the correction factors for the various types of interactions considered in the preceding sections are given in Table 26.

4.2. London Dispersion (van der Waals) Interactions

The so-called London dispersion or van der Waals interactions are those between molecules that have neither a net charge nor a permanent dipole moment. This interaction is essentially due to the interactions between a transient electrical dipole in one molecule and its induced electrical dipole in the other molecule. This type of electrical interaction plays an important role in biological systems (e.g., in surface tension, stability of biological membranes, condensation properties, adhesion and fusion of biological cell membranes, enzyme–substrate recognition, etc.).

According to the second-order perturbation theory,[234,235] the dispersion energy between two nonpolar molecules is given by:

$$W_{\alpha\alpha} = -\sum_{\rho \neq 0} \sum_{\sigma \neq 0} \frac{|\langle O(1)O(2)|J(1,2)|\rho(1)\sigma(2)\rangle|^2}{\delta_\rho(1) + \delta_\sigma(2)} \tag{160}$$

$$J(1,2) = -\mathbf{p}(1)\frac{3\mathbf{e}_{12}\mathbf{e}_{12} - \tilde{1}}{(r_{12})^3}\mathbf{p}(2) = -\mathbf{p}(1)\frac{\tilde{T}}{(r_{12})^3}\mathbf{p}(2) \tag{161}$$

where $\delta_\rho(1) = E_\rho(1) - E_0(1)$ and $\delta_\sigma(2) = E_\sigma(2) - E_0(2)$; $\mathbf{p}(1)$ is the electric dipole moment of molecule 1; $O(1)$ is the ground state wave function of molecule 1; $J(1,2)$ is the dipole–dipole interaction energy between dipole 1 and dipole 2; $\rho(1)$ is the ρth excited state wave function of molecule 1; E_0 is the energy of the ground state; E_ρ is the energy of the ρth excited state;

Table 26
Correction of Interaction Energy Expressions for
Short-range Repulsive Contribution

Interaction	Noncorrected form[a]	Correction factor
Charge–charge	$-\dfrac{331.9 n_1 n_2}{\varepsilon r}$ (kcal/mole)	0.92
Charge–fixed dipole	$-\dfrac{69.1 n\mu \cos\theta}{\varepsilon r^2}$	0.83
Rotating dipole–rotating dipole	$-\dfrac{7.0 \times 10^4 \mu_1^2 \mu_2^2}{\varepsilon^2 r^6 T}$	0.5

[a] n: the number of electronic charge on q.

$\mathbf{r}_{12}$ is the distance vector between molecules 1 and 2:

$$\mathbf{r}_{12} = \mathbf{r}_2 - \mathbf{r}_1 = r_{12}\mathbf{e}_{12} \qquad (162)$$

where r_{12} is the distance between molecules 1 and 2, and $\mathbf{e}_{12}$ is the unit vector of $\mathbf{r}_{12}$ and $\tilde{T} = 3\mathbf{e}_{12}\mathbf{e}_{12} - \tilde{1}$ is a tensor.

Equation (160) can be written in an approximate form as (see Ohki and Fukuda,[236]):

$$W_{\alpha\alpha} = -\frac{1}{4(r_{12})^6} \frac{\bar{\delta}_1\bar{\delta}_2}{\bar{\delta}_1 + \bar{\delta}_2} \mathrm{Tr}\,[\tilde{T}\alpha_1\tilde{T}\alpha_2] \qquad (163)$$

where $\bar{\delta}_1$ and $\bar{\delta}_2$ are the ionization potentials of the molecules 1 and 2, respectively, α_1 and α_2 are the polarization tensors of molecules 1 and 2, respectively, and Tr refers to the trace of a tensor. When a molecule has a symmetrical polarizability, the expression for the dispersion interaction [Eq. (163)] becomes:

$$W_{\alpha\alpha} = -\frac{3}{2(r_{12})^6} \frac{\bar{\delta}_1\bar{\delta}_2\alpha_1\alpha_2}{\bar{\delta}_1 + \bar{\delta}_2} \qquad (164)$$

It is seen from Eq. (164) that the dispersion interaction force ($F_\alpha = -\partial W_{\alpha\alpha}/\partial r$) is a short-range force and weak. However, when a molecular assembly consists of a large number of nonpolar molecules, then the dispersion interaction force (or energy) between two bodies of such molecular assemblies becomes a long-range force (or interaction energy) (Verywey and Overbeek).[237]

The dispersion attractive potential varies as r^{-6} [as seen in Eq. (164)]. Accordingly:

$$W_{\alpha\alpha} = -\frac{\lambda_{12}}{r^6} \quad \text{and} \quad \lambda_{12} = \frac{3\alpha_1\alpha_2}{2} \frac{\bar{\delta}_1\bar{\delta}_2}{\bar{\delta}_1 + \bar{\delta}_2} \qquad (165)$$

in which λ_{12} is a constant depending on the properties of the atoms (or molecules) under consideration.† For two identical molecules having a symmetric polarizability, λ_{12} is expressed as:

$$\lambda_{12} = \tfrac{3}{4}\alpha^2\bar{\delta} = \lambda \qquad (166)$$

Hamaker[238] introduced the following quantity A:†

$$A_{12} = \pi^2\rho_1\rho_2\lambda_{12} \qquad (167)$$

where ρ is the number of atoms contained in a unit volume of the substance considered. Then, the dispersion interaction energy between two parallel

---

† Strictly speaking, λ_{12} or A (Hamaker constant) is not constant but depends on the distance between the two bodies and on temperature,[270] which will be mentioned in the section on Retardation Effect (Section 4.2.1). Depending on the nature of interacting bodies, this factor (λ or A) could be negative in sign, although in most cases it has a positive sign.

plates, separated by distance R, is:

$$\bar{W}_{\alpha\alpha} = -\int\int \frac{\lambda_{12}}{r_{12}^6} \, d\tau_1 \, d\tau_2 = -\frac{\lambda_{12}\pi^2\rho_1\rho_2}{12\pi}\left[\frac{1}{R^2} - \frac{2}{(R+h)^2} + \frac{1}{(R+2h)^2}\right]$$

$$= \frac{-A_{12}}{12\pi}\left[\frac{1}{R^2} - \frac{2}{(R+h)^2} + \frac{1}{(R+2h)^2}\right] \tag{168}$$

where h is the thickness of the plates.

(Example 1)

When the thickness of the two plates made of the same molecular assembly is infinite, then Eq. (168) becomes

$$\bar{W}_1 = -\frac{A}{12\pi R^2} \qquad \text{where } A = (3\pi/4)\alpha^2\bar{\delta}\rho^2. \tag{169}$$

(Example 2)

When two identical spherical bodies of radius a interact with each other, the dispersion energy W_2 becomes:

$$\bar{W}_2 = -\frac{Aa}{12R} \tag{170}$$

When two interacting bodies are in a solution of a certain dielectric constant $\varepsilon > 1$, the magnitude of the dispersion interaction is smaller than the case in vacuum. In such a case, the dispersion energy W is expressed by:

$$W = W_{12} - W_{00} - [(W_{10} - W_{00}) + (W_{20} - W_{00})]$$

$$= W_{12} + W_{00} - W_{10} - W_{20} \tag{171}$$

where W_{12} refers to the interaction between the two bodies, 1 and 2, W_{10} is the interaction between body 1 and the medium 0, which is replaced by an equal volume of the other body 2. W_{00} is the interaction between the two media replaced by the two bodies.

Then, the Hamaker constant corresponding to the expression of Eq. (171) is:

$$A = A_{12} + A_{00} - A_{10} - A_{20} \tag{172}$$

For example, the Hamaker constant for the Decane|Vacuum|Decane system is $A = 50 \times 10^{-14}$ ergs, and for the Decane|H_2O|Decane system $A = 2.8 \times 10^{-14}$ ergs.[239]

4.2.1. Retardation Effect

The electric field which is responsible for the dispersion interaction between nonpolar molecules propagates itself between the particles with the speed of light. Thus, if a pair of molecules are widely separated, a time lag or a phase difference develops between vibration at the two locations. Generally, this effect decreases the London dispersion energy from that expected when this retardation is not considered. The effect of retardation is expected to be significant if the distance between the particles is comparable to the electromagnetic wavelengths (λ) corresponding to main absorption bands in the ultraviolet region (of the order of a few hundred angstroms). According to Casimir and Polder,[240] the additional retardation interaction contribution to the dipole-dipole term of Eq. (160), is given by:

$$J_{ret,1} = \sum_j \left[-\frac{e}{mc}(\mathbf{P}_{j1} \cdot \mathbf{A}_1) + \frac{e^2}{2mc^2}A_1^2 \right] \tag{173}$$

where a refers to a particle 1, $\mathbf{P}_j$ is the operator for the momentum of an electron and $\mathbf{A}$ is the vector potential of the electromagnetic field.

For the case of two interacting bodies with a large separation ($r \gg \lambda$), the energy of the retardation correction is

$$W_{12(dispersion)} = -\frac{23\hbar c \alpha_1 \alpha_2}{4\pi r^7} \tag{174}$$

Overbeek[241] gave a simple expression for the retardation correction for the case $\alpha_1 = \alpha_2$:

$$W_{\alpha\alpha(dispersion)} = -\frac{3\hbar\omega_L \alpha^2}{4r^6}f(P) \tag{175}$$

where f is the retardation correction factor and

$$P = \frac{2\pi r}{\lambda_L},$$

$$f = \begin{cases} 1.01 - 0.14P & \text{for } 0 < P < 3 \\ \dfrac{2.45}{P} - \dfrac{2.04}{P^2} & \text{for } 3 < P < \infty \end{cases}$$

For most substances, the retardation effect in the dispersion interaction energy becomes significant at distances greater than 100 Å. The important point is that the Hamaker constant A defined in Eq. (167) is no longer constant but depends on the separation of interacting bodies.

Further studies of the van der Waals interaction among molecules having permanent dipole moments have been carried out using quantum mechanical methods by Dzyaloshinskii et al.[242] Their approach includes the retardation effect as well as a polarization effect due to molecular dipoles. Then, the total dispersion interaction force (F) acting on a unit area of each of the slabs is given by

$$F = \frac{A}{6\pi R^3} = \frac{\hbar\bar{\omega}}{8\pi^2 R^3} \tag{176}$$

where $\bar{\omega}$ is:

$$\bar{\omega} = \frac{2kT}{\hbar} \sum_{n=0}^{\infty} \frac{[\varepsilon_1(i\xi_n) - \varepsilon_3(i\xi_n)][\varepsilon_2(i\xi_n) - \varepsilon_3(i\xi_n)]}{[\varepsilon_1(i\xi_n) + \varepsilon_3(i\xi_n)][\varepsilon_2(i\xi_n) + \varepsilon_3(i\xi_n)]}$$

$$\times \exp\left[\frac{-2R\xi_n\varepsilon_3(i\xi_n)}{c}\right] \times \{[1 + 2R\xi_n\varepsilon_3(i\xi_n)^{1/2}/c + 2[R\xi/c]^2\varepsilon_3(i\xi_n)\}$$

(177)

where ε_1, ε_2, ε_3 are dielectric constants for substance 1 and 2 and medium 3, $\xi = 2\pi nkT/\hbar$, and c is the velocity of light.

4.2.2. Orientation Effect

In the framework of the Lifshitz Theory, the additional term in the dispersion force, $A/6\pi d^3$, per unit area of the two slabs is

$$A_{n=0} = A - A \text{ (dispersion)}$$

$$= \tfrac{3}{4}kT\frac{[\varepsilon_1(0) - \varepsilon_3(0)][\varepsilon_2(0) - \varepsilon_3(0)]}{[\varepsilon_1(0) + \varepsilon_3(0)][\varepsilon_2(0) + \varepsilon_3(0)]}$$

(178)

which arises from the term $n = 0$ in the Lifshitz equation [Eq. (177)].

The correction term refers to the permanent dipole–induced dipole interaction. In the limit of low density in media 1 and 2 and when medium 3 is vacuum, this expression corresponds to the permanent dipole interaction of Keesom.[243]

4.3. Electrical Double Layer Interaction and DLVO Theory

It is well known that Verwey and Overbeek[237] and Derjaguin and Landau[244] (DLVO) have given a quantitative treatment of the interaction of electrical double layers. According to them, the free energy of a double layer may be expressed as a difference between the surface energy G_s of the system in its equilibrium state and the surface free energy G_s^0 of a standard state in which no double layer is present:

$$G_{\text{double layer}} = G_s(\sigma) - G_s^0(\sigma = 0)$$

(179)

Although this expression presents the basis of the free energy of the double layer, it is not truly expressive of the parameters that must be considered. According to Verwey and Overbeek[237] and Ikeda,[245] the free energy of a system of reversible double layers is determined by the following;

$$G = -\sum_{k} dS_k \int_0^{\psi_K} \sigma \, d\psi$$

(180)

where ψ is the potential difference between the two phases separated by the

kth interface of which dS_k is a surface element and ψ_k is the equilibrium surface potential. It should be noticed that the free energy of the double layer system is a negative quantity.

Verwey and Overbeek formulated a second expression for the free energy of the double layer based on the Gouy–Chapman model.

$$G = G(0) + \int_0^1 \frac{d\lambda}{\lambda} v(\rho'\psi')\psi_K \, dv \tag{181}$$

where λ is a parameter describing the stage of the discharging process, which varies from 1 to 0 as the charge of the ith ion varies from $z_i e$ to 0, the second term on the right represents the reversible work done in discharging all ions at constant surface potential ψ'_K, the quantities ρ' and ψ' represent the space charge density and the potential in the Gouy diffusion layer at any stage λ, and dv is a volume element in the solution. By use of Ikeda's basic derivation, $G(0)$ may be substituted into the above equation to give the free energy per unit area of a system composed of two large parallel plates immersed in solution:

$$G = -\frac{\varepsilon}{8\pi d}(\psi_0 - \psi_R)^2 + \int_0^1 \int_0^R \frac{\lambda'\psi'}{\lambda} \, dx \, d\lambda \tag{182}$$

This expression may be reduced using the Poisson equation; a detailed analysis may be found in Devereux and de Bruyn.[246]

The equations used are for the free energy per square centimeter of two flat double layers at a fixed distance of separation R. Therefore, the reversible work required to bring the two parallel surfaces together from infinity to a distance R apart is determined by the expression:

$$V = G - G_\infty \tag{183}$$

where G_∞ is the free energy per square centimeter of surface when the distance of separation is infinite. V determines the free energy of interaction of the double layer system. This interaction may be of a repulsive or attractive nature depending on whether the double layers are dissimilar.

The free energy of interaction of two interacting bodies may be expressed as a sum of electrical double layer interaction energy (in general, repulsive) and the van der Waals attraction energy (in general, attractive):

$$G^{\text{tot}} = V + U \tag{184}$$

where V and U refer to the electrical double layer interaction and the van der Waals interaction energies, respectively.

By using the above equations, Derjaguin and Landau, and Verwey and Overbeek (DLVO) proposed a system for the process of adhesion. According to the work of Derjaguin and Landau, the interaction of two diffused electrical double layers would cause the formation of an energy barrier between the two interacting particles. This energy barrier would lie between two minima.

For the two particles to come into close contact they would have to cross this energy barrier. Figure 49 shows a schematic drawing of the system. Curves A and B represent cases where the minima are separated by a maximum (energy barrier). Curve C represents a special case where the disappearance of the force barrier has taken place, and no energy barrier exists between the two minima. Derjaguin and Landau[244] derived the following criterion for the adhesion reaction to occur

$$\frac{1}{\gamma n_1} = cf(\beta)\frac{A^2 e^6 Z_1^6}{\varepsilon^3 (kT)^5} \tag{185}$$

where A is the van der Waals constant, e the charge of an electron, γ the concentration of the electrolyte in moles per cm^3, n_1 the number of dominating ions in one molecule, ε the dielectric constant, β the ratio of the valency of the auxiliary ion Z_2 to that of the dominating ion Z_1, C a constant and $f(\beta)$ a function of the "asymmetry of the electrolyte", $f(1) = 1$, retaining the same order of magnitude for the other value of β. While the theoretical analysis in the field of colloid science had not by any means been completed, these early papers that presented the basic ideas were necessary for the theoretical biologists and the biophysicists to proceed further.

The electrostatic interaction energy between diffused double layers based on the linear Debye–Hückel approximation [Eq. (18)] for two parallel plates having constant surface charge are given by Usui[247] in the form:

$$V_R^\sigma = \frac{\varepsilon \kappa}{8\pi}[(\psi_{01}^2 + \psi_{02}^2)(\coth \kappa R - 1) + 2\psi_{01}\psi_{02} \operatorname{cosech} \kappa R] \tag{186}$$

where ψ_{01} and ψ_{02} are the surface potentials of plane 1 and plane 2 at the infinite separation distance, respectively.

The linear approximation and an employment of the boundary condition of constant surface potentials yields[314]:

$$V_R^\psi = \frac{\varepsilon \kappa}{8\pi}[(\psi_{01}^2 + \psi_{02}^2)(1 - \coth \kappa R) + 2\psi_{01}\psi_{02} \operatorname{cosech} \kappa R] \tag{187}$$

Under the most physiological conditions, the Debye–Hückel length $(1/\kappa)$ is about 10 Å. At distances greater than 2 to $3\kappa R$, the difference between V_R^σ and V_R^ψ becomes negligibly small.

The expression for the electrostatic interaction energy between the double layers of two spherical bodies[248] is given by:

$$V_R^\sigma = \frac{4\pi^2}{\varepsilon \kappa^2}\left(\frac{a_1 a_2}{a_1 + a_2}\right)\left\{(\sigma_1 + \sigma_2)^2 \ln\left[\frac{1}{1 - \exp(-\kappa R)}\right]\right.$$
$$\left. + (\sigma_1 - \sigma_2)^2 \ln\left[\frac{1}{1 + \exp(-\kappa R)}\right]\right\} \tag{188}$$

where a_1 and a_2 are the respective radii of the two bodies 1 and 2.

The van der Waals energy for two flat plates separated by distance R is given by Eqs. (168) and (169), and that for two identical spherical bodies of radius a is given by Eq. (170). For the case of two spherical shells, the expression for the dispersion interaction energy is given elsewhere.[265]

4.4. DLVO Theory Applied to Biological Systems

Although the DLVO theory has many shortcomings, many attempts have been made to apply it to interaction phenomena in biological systems.[250] Since all mammalian cells examined thus far carry a net negative charge at their surfaces, the contact between these cells can be considered as the interaction of two electrical double layers. The DLVO theory tells us that electrostatic forces of double layers tend to keep cells apart, while attractive forces, principally van der Waals forces, tend to favor close approach of two cells would be hindered by a considerable energy barrier between the cells due to electrostatic repulsive force. For close contact to take place, this energy barrier of repulsion must be overcome. In a state where the membranes are separated by a few angstroms, they are held together by attractive forces with an energy equal to that of any chemical bond which could be formed between them. In a primary energy minimum, close contact of the cells can occur. It should be noted, however, that if the interacting bodies are large enough, it is also possible that cell adhesion occurs at the secondary minimum because the depth of this minimum could be large compared to the thermal energy kT.

While it has been stated that there are inherent limitations in the DLVO theory, it does appear to be useful in explaining adhesion and contact in some biological cell systems. An early paper by Van den Tempel in 1958[251] showed that the theory could be used to analyze systems that were similar to colloidal ones. His work on emulsified oil globules, in relation to contact phenomena, enabled him to set up equations of repulsion and attraction resulting from the double layer. These equations, which are a direct result of the DLVO theory, have been applied with great success to biological systems. Van den Tempel was able to measure the thickness of the double layer and he confirmed that the secondary minimum predicted in the DLVO theory does exist.

Puck and others[252,253] used the theories of colloid stability to explain the adsorption of bacteriophage by bacteria. Their findings appeared to be compatible with those of the basic colloid theory. This work and that of Valentine and Allison[254] led to the novel result that initial contact between virus particles and bacterial surfaces was regulated by electrostatic repulsion and the forces that govern the double layer. While these early studies were not as successful as one might have wished, they did show a correlation between the DLVO theory and the experimental data. Though the evidence is not conclusive, the work by Curtis[255] and Pethica[256] also tends to show that the DLVO theory is relevant to some biological systems. Other studies done at the time tended to show little or no correlation between the DLVO theory and the experimental data. Using other types of cellular systems Wilkins, Ottewill, and Bangham[307] were able to attain partial agreement between electrophoretic studies and flocculation data on sheep leucocytes.

Recent work has suggested that probes with low radius of curvature are important in crossing the primary maximum. [249,257-259] It is also believed that approach and contact can be made by microvilli with a low radius of curvature (<0.1 μm). Microvilli of these dimensions encounter significantly less electrostatic repulsion [Eq. (289)] than larger pseudopodial type projections, and are therefore better able to overcome the potential energy barrier that opposes close cell contact. The formation of microvilli appears to be related to the thickness of the cell coat, and work with virus-induced fusion has played a large role in this work. [258,260]

However, most biological cell membranes are not perfect two-dimensional molecular arrays. Most surfaces are composed of heterogeneous molecular assemblies and have stereolitical molecular roughness. In some cases, some molecular components of membranes extend their molecular residues into the extracellular phase at a considerable distance from the membrane surface (e.g., glycoprotein residues) as shown in Figure 50. [261-263] Such extensions of polymer molecules may exert a repulsive force between two closely opposed membranes because of steric effects resulting from the exclusion volume of such molecular extensions (based on the principle of steric exclusion in colloid chemistry). [264] In some cells, the plasma membranes are covered by a so-called non-structured cell-coating material (mucopolysaccharides). In such cases, the straight application of the DLVO theory obviously does not hold. The adhesion

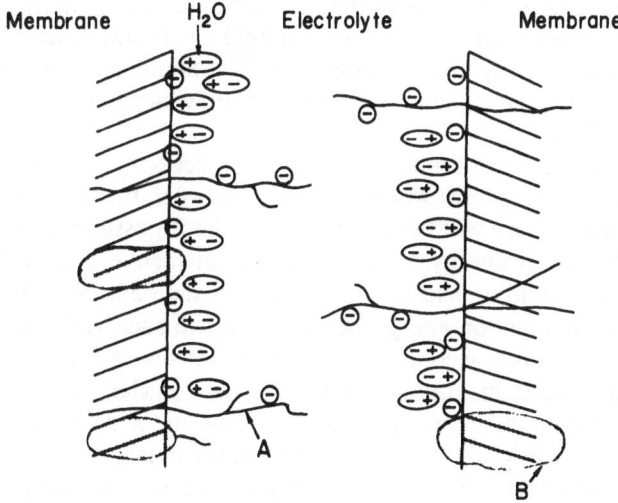

Figure 50. A diagram of two opposed cell membrane surfaces, showing heterogeneous surfaces (oriented adsorbed water H_2O, molecular extensions of membrane protein moieties (A), globular protein molecules (B), etc.).

between cell surfaces may be mediated by a quite different process. There may be specific interactions between specific molecular components. Moreover, polar surfaces of membranes are believed to be surrounded by oriented water molecules whose physical state is quite different from that of free water molecules in the bulk aqueous phase. The degree of hydration and the degree of orientation (water structure) may vary from one membrane surface to another. At the present time the analysis of these problems is quite difficult. The presence of such structured water would greatly modify the interaction energy (dispersion energy) so that strong repulsive interaction would be present when cells approach each other closely. These situations are, in practice, difficult to handle with the simple DLVO theory, although all effects, including polarization of molecules, are included in the formal expression for the dispersion interaction [Eq. (177)].

Red blood cells and many other biological cells do not adhere in their normal physiological environments.[265] On the other hand, lipid vesicles (e.g., phosphatidylserine vesicles) having the same surface charge density as that of many biological cell surfaces do adhere easily.[249] However, phosphatidylcholine vesicles do not adhere to each other in spite of having zero net surface charge density in a similar physiological environment. It should be noted that the surfaces of red cell membranes and other biological cell membranes indeed have many hydrophilic molecular extensions which originate from the membrane molecules (e.g., shown in Fig. 50) and also, that a phosphatidylcholine membrane possesses a greater surface hydration shell than that of a phosphatidylserine membrane surface.

In order to elucidate adhesion phenomena occurring among biological cells, a number of studies on membrane adhesion using model membrane systems (such as lipid membranes as well as lipid membranes with incorporated proteins) have recently been carried out.[296] Some of these aggregation phenomena have been analyzed from the viewpoint of the DLVO theory[249,288–290].

In order for two membranes to adhere, it is necessary to reduce the energy barrier (primary maximum) which may exist between the two membranes as they come close. There will be two ways to make it possible: 1) reducing the repulsive interaction energy, or 2) increasing attractive interaction energy. When the size of lipid vesicles is small (e.g., less than 500 Å in diameter) the van der Waals interaction energy between the two such vesicles is relatively small, there is no stable secondary minimum point where the two vesicles can stay apart in a stable manner. In such case, two vesicles would aggregate at the primary minimum depending on the situation of repulsive interaction energy ("true adhesion"). On the other hand, for larger size lipid vesicles (>1000 Å in diameter), it is possible that two such vesicles can stay at the secondary minimum point because the van der Waals attractive interaction energy becomes relatively large to make a deep secondary minimum in the interaction energy profile. These have been recently studied using various phospholipid vesicles.[290] The effect of hydration on membrane surface on membrane aggregation has also been considered where the repulsive interaction due to the hydration shells on a certain lipid membrane is so large that even two interacting neutral charged membranes have a great repulsive maximum (primary maximum) between them[99,266,277,291] and consequently they do not easily adhere to each other.

The condition for the "true adhesion" is not sufficient for the two interacting bodies to fuse and become one. As long as the force of adhesion of aqueous molecules onto the surfaces of the bodies is stronger than the attractive forces between the interior phases of the two hydrocarbon bodies at a certain distance or the nature of the surface hydrophilic layers, which have a certain thickness, is similar to aqueous phase, the two bodies would be separated apart like two different phases and stay side by side.

Recently, Ohki proposed a physical principle underlying membrane fusion processes[292] in terms of molecular interaction. For simplicity, two interacting membranes are considered as two flat hydrocarbon bodies having hydrophilic layers on their surfaces, separated by an aqueous solution at a certain distance R. The thickness of the hydrophilic layer is h which may be different from those of the flat body as well as the aqueous phases in their molecular nature and molecular density. Since the electrostatic interaction gives a repulsive force and is not a main factor for membrane fusion, we assume the electrostatic energy term to be a constant contribution to the total interaction energy at the "true adhesion," and the van der Waals interaction energy would contribute mainly to membrane fusion. Then, the van der Waals interaction between the two bodies will be expressed as a function of the separation distance R, the thickness h, and the Hamaker constant A in each phase:[293]

$$U = -\sum_i \sum_j A(i, j) h(x_i x_j) \tag{189}$$

where A_{ij} is the Hamaker constant of interaction between the ith and jth phases and $H(x_i x_j) = 1/r_{ij}^2$ where r_{ij} is the distance between the two interfaces of either the surface hydrophilic layer or the bulk hydrocarbon phase. The Hamaker constant is approximately given in terms of the molecular polarization, α, ionization potential, I, and the molecular density, ρ, of each phase as follows:

$$A_{ij} \cong \frac{3\pi^2}{2} \frac{I_i I_j}{I_i + I_j} \alpha_i \rho_i \alpha_j \rho_j \tag{190}$$

Since the ionization potential of the hydrocarbon unit C_2H_4 and H_2O are similar quantities,[294,295] Eq. (190) can be expressed approximately by

$$U \equiv -\frac{3\pi^2 I}{4} \left\{ (\lambda_p - \lambda_w)^2 \frac{1}{R^2} + (\lambda_p - \lambda_w)(\lambda_H - \lambda_p) \right.$$

$$\left. \times \frac{2}{(R + h)^2} + (\lambda_H - \lambda_p)^2 \frac{1}{(R + 2h)^2} \right\}, \tag{191}$$

where $\lambda_i = \alpha_i \rho_i$ and suffixes W, P and H refer to the aqueous, the polar hydrophilic layer, and the hydrocarbon phases, respectively. In Eq. (191), the first term is mainly due to the interaction of the two hydrophilic layers separated by an aqueous phase. The third term is mainly due to the interaction between the two hydrocarbon bodies through the hydrophilic interface layer as a medium. The second term is the cross interaction term between the aqueous/hydrophilic interface layer interaction and the hydrocarbon body/hydrophilic interface layer interaction. The equation [Eq. (191)] shows that if the interaction coefficient λ_p (molecular polarizability × its density) of the surface layer is nearly the same as that of the aqueous phase, the contribution from the first and second terms to the total interaction energy is negligibly small and the third term contributes mainly to the interaction energy. In this case if the thickness of the surface layer is relatively large, the interaction energy of the two bodies will be small. On the other hand, if the interaction coefficient of the surface layer is nearly equal to that of the interior hydrocarbon bodies, the first term will contribute predominantly to the interaction energy, since $\lambda_w = \lambda_H$, regardless of the thickness of the surface layer. The interaction energy will approach to that of the hydrocarbon phase at the molecular distance R. Since the separation distance R is small, the attractive interaction force of the two bodies would become large enough to overcome the aqueous adhesion interaction force and thus cohesion of the two bodies will take place which may mean fusion of the two bodies. Therefore, depending on the nature of the surface layer and its thickness, the total attractive energy varies. If the total attractive energy becomes larger than a certain value, the two bodies which are in consideration may be able to fuse and become one body.

The above argument was based on an average quantity of the membrane. However, some of the fusion processes could be a dynamical or a local process. In such a situation, the above-mentioned theory should be modified as a local event, but the principal base for the theory is unchanged.

The above interpretation appears to explain many membrane fusion events observed for model membrane systems;[296] the divalent or polyvalent cation-induced membrane fusion is directly related to the surface energy increase (to become more hydrophobic) of the membrane caused by binding of these ions to the membranes.[297,298] The increased temperature or osmotic pressure-induced membrane fusion[298–300] is related to the membrane expansion whereby the surface energy of the surface hydrophilic layers increases.[298] The boundary region between two different phases in the membrane caused by different membrane components[301] or metal ion binding to the membrane[302] or phase defect points[303] in the membrane could correspond to a higher energy (more hydrophobic) state than the other parts of the homogeneous membrane phases.

These fusion events all have a common feature of increase in surface energy at the site of membrane fusion. In this sense, these membrane fusions can be explained by the theoretical interpretation presented here and it will also accommodate most of the existing membrane fusion theories. So far, the results are encouraging, namely, the DLVO theory has some relevance to such membrane systems. However, when one faces the adhesion problem for even model membranes which are heterogeneous and have rough membrane surfaces like those of most biological cells, one might expect to encounter difficulty in theoretical analyses. Nevertheless, constant attempts[99,266,267,293,298] and great success to elucidate the biophysical mechanism of cell–cell interaction are hoped and awaited for by cell biologists.

Acknowledgements

The author would like to express his sincere thanks to Drs R. J. Kurland and H. Oshima for their reading and valuable commenting on the manuscript. He also wishes to express many thanks to Mrs. C. B. Ohki for her art work, to Ms. M. Lawrence for her typing of the manuscript and to Miss J. Duax for her help in preparing the manuscript. This work was supported partially by a grant from the U.S. National Institutes of Health (GM24840).

References

1. J. Edsall and J. Wyman, in *Biophysical Chemistry*, Vol. 1, Academic Press, New York (1958).
2. N. Davison, in *Statistical Mechanics*, McGraw-Hill, New York (1962), p. 485.
3. R. G. Bates, *Electrometric pH Determinations*, John Wiley & Sons, New York (1954).
4. H. C. Brown, D. H. McDaniel, and O. Häfliger, in *Determination of Organic Structure by Physical Methods*, E. A. Brande and F. C. Nachod, eds., Academic Press, New York (1955).
5. N. Bjerrum, *Z. Plupik. Chem.* **106**, 219 (1923).
6. J. G. Kirkwood and F. H. Westheimer, *J. Chem. Phys.* **6**, 506 (1938).
7. F. H. Westheimer and M. W. S. Shookhoff, *J. Am. Chem. Soc.* **61**, 555 (1939).
8. C. Tanford and J. Kirkwood, *J. Am. Chem. Soc.* **79**, 5333 (1958).
9. W. H. Orttung, *J. Phys. Chem.* **72**, 4066 (1968).
10. W. H. Orttung, *Biochemistry* **9**, 2394 (1970).
11. S. J. Shire, G. I. H. Hanania, and F. R. N. Gurd, *Biochemistry* **13**, 2974 (1974).
12. L. H. Botelho, S. H. Friend, J. B. Matthew, L. D. Lehman, G. I. H. Hanania, and F. R. N. Gurd, *Biochemistry* **17**, 5197 (1978).
13. J. B. Matthew, G. I. H. Hanania, and F. R. N. Gurd, *Biochemistry* **18**, 1919 (1979).
14. M. S. Fernandez and P. Fromherz, *J. Phys. Chem.* **81**, 1755a (1977).
15. S. Ohki and R. Kurland, *Biochim. Biophys. Acta* **645**, 170 (1981).
16. J. A. Marinsky, *Coord. Chem. Rev.* **19**, 125 (1976).
17. J. Kirkwood, *J. Chem. Phys.* **2**, 351 (1934).
18. P. Debye, *Z. Physik. Chem. Cohen Festband*, 56 (1927).
19. A. Vöet, *Chem. Rev.* **20**, 169 (1939).
20. J. T. G. Overbeek and H. G. Bungenberg De Jong, in *Colloid Science*, Vol. II, H. R. Kruyt, ed., Elsevier, Amsterdam (1949), pp. 184.
21. S. N. Vinogradov and R. H. Linnell, *Hydrogen Bonding*, Van Nostrand Reinhold Co, New York (1971).
22. J. Bernal and R. Fowler, *J. Chem. Phys.* **1**, 515 (1933).

23. G. Bradley and M. Kasha, *J. Am. Chem. Soc.* **77**, 4462 (1955).
24. C. Coulson, *Research* **10**, 149 (1957).
25. H. Tsubomura, *Bull. Chem. Soc. Japan* **27**, 445 (1954).
26. N. Sokolov, *Usp. Phys. Nauk* **57**, 205 (1955).
27. M. V. Volkenstein, in *Molecular Biophysics*, Academic Press, New York (1977), Chapter 4.
28. R. Gabler, in *Molecular Biophysics*, Academic Press, New York (1978), p. 162.
29. A. L. Lehninger, *Biochemistry*, Worth Publishers Inc. (1970).
30. M. N. Jones, in *Biological Interfaces*, Elsevier, Scientific Publish. Co., Amsterdam (1975).
31. J. T. Davies and E. K. Rideal, *Interfacial Phenomena*, Academic Press, New York (1963).
32. D. A. Haydon, *Recent Progr. Surface Sci.*, Vol. 1, J. F. Danielli, K. G. A. Pankhurst, and A. C. Riddiford, eds., Academic Press, New York (1964), Chapter 3, p. 94.
33. M. C. Phillips, *Progress in Surface and Membrane Sciences*, **5**, 139 (1972).
34. A. Goldup, S. Ohki, and J. F. Danielli, *Recent Progr. Surface Sci.*, Vol. 3, pp. 193–260, J. F. Danielli, K. G. A. Pankhurst, and A. C. Riddiford, eds., Academic Press, New York (1970).
35. H. T. Tien, *Bilayer Lipid Membrane; Theory and Practice*, Marcel Dekker, New York (1974).
36. A. D. Bangham, *Progr. Biophys. Mol. Biol.* **18**, 29 (1968).
37. B. Abraham-Shrauner, *J. Math. Biol.* **2**, 333 (1975).
38. A. L. Loeb, J. T. G. Overbeek, and P. H. Wiersema, *The Electrical Double Layer Around a Spherical Particle*, MIT Press, Cambridge, Mass. (1961).
39. B. Abraham-Shrauner, *J. Colloid Interface Sci.* **44**, 79 (1973).
40. S. L. Brenner and R. E. Roberts, *J. Phys. Chem.* **77**, 2367 (1973).
41. L. R. White, *J. Chem. Soc., Faraday Trans. 2*, **73**, 577 (1977).
42. J. Bentz, *J. Colloid Interface Sci.* **80**, 179–192 (1981).
43. H. Müller, *Kolloidchem. Beihefte* **26**, 257 (1928).
44. S. G. A. McLaughlin, G. Szabo, and G. Eisenman, *J. Gen. Physiol.* **58**, 667 (1971).
45. S. Nir, C. Newton, and D. Papahadjopoulos, *Bioelectrochem. Bioenerg.* **5**, 116 (1978).
46. S. Ohki and R. Sauve, *Biochim. Biophys. Acta* **511**, 377 (1978).
47. M. Eisenberg, T. Gresalfi, T. Riccio, and S. McLaughlin, *Biochemistry* **18**, 5213 (1979).
48. A. Lau. A. McLaughlin, and S. McLaughlin, *Biochim. Biophys. Acta* **645**, 279 (1981).
49. M. M. Hammoudah, S. Nir, J. Bentz, E. Mayhew, T. P. Stewart, S. W. Hui, and J. Kurland, *Biochim. Biophys. Acta* **645**, 102 (1981).
50. S. McLaughlin, N. Mulrine, T. Gresalfi, G. Vaio, and A. McLaughlin, *J. Gen. Physiol.* **77**, 445 (1981).
51. J. A. Cohen and M. Cohen, *Biophys. J.* **36**, 623 (1981).
52. F. Oosawa, *Polyelectrolytes*, Marcel Dekker, New York (1971), p. 160.
53. G. S. Manning, *Ann. Rev. Phys. Chem.* **23**, 117 (1972).
54. H. Wennerström, B. Lindman, G. Lindblom, and G. J. Tiddy, *J. Chem. Soc., Faraday Trans. 1*, **75**, 663 (1979).
55. S. Engström and H. Wennerström, *J. Phys. Chem.* **82**, 2711 (1978).
56. R. Kurland, S. Ohki, and S. Nir, in *Solution Behavior of Surfactants*, Vol. 2, K. L. Mittal and E. J. Fendler, eds., Plenum Press, New York (1982).
57. A. N. Frumkin, *A. Phys. Chem.* **109**, 34 (1924).
58. I. Prigogine, P. Mazur, and R. Defay, *J. Chim. Phys.* **50**, 146 (1953).
59. D. A. Haydon and F. H. Taylor, *Phil. Trans.* **A253**, 255 (1960).
60. D. C. Grahame, *J. Chem. Phys.* **18**, 903 (1950).
61. F. Booth, *J. Chem. Phys.* **19**, 391, 1327 (1951).
62. J. J. Bikerman, *Phil. Mag.* **33**, 384 (1942).
63. A. L. Loeb, *J. Colloid. Sci.* **6**, 75 (1951).
64. W. E. Williams, *Proc. Phys. Soc.* **66**, 372 (1953).
65. S. Levine, S. Mingins, and G. M. Bell, *J. Phys. Chem.* **67**, 2095 (1963).
66. F. P. Buff and F. H. Stillinger Jr., *J. Chem. Phys.* **39**, 1911 (1963).
67. C. A. Barlow Jr. and J. R. McDonald, *J. Chem. Phys.* **40**, 1535 (1964).
68. S. Levine, K. Robinson, G. M. Bell, and J. Mingins, *J. Electroanal. Chem.* **38**, 253 (1972).

69. D. Stigter, *J. Phys. Chem.* **68**, 3603 (1964).
70. R. H. Brown Jr., *Progr. Biophys. Mol. Biol.* **28**, 341 (1974).
71. A. P. Nelson and D. A. McQuarrie, *J. Theor. Biol.* **55**, 13 (1975).
72. P. A. Forsyth Jr., S. Marcelja, D. J. Mitchell, and B. W. Ninham, *Biochim. Biophys. Acta* **469**, 335 (1977).
73. R. Sauve and S. Ohki, *J. Theor. Biol.* **81**, 157 (1979).
74. S. Levine, *Proc. Phys. Soc.* **39**, 1897 (1953).
75. J. O'M. Bockris, B. E. Conway, and E. Yeager, eds., *Comprehensive Treatise of Electrochemistry*, Vol. 1, Plenum Publishing Co., New York (1980).
76. D. O. Shah and J. H. Schulman, *J. Lipid Research* **8**, 227 (1967).
77. A. D. Bangham and D. Papahadjopoulos, *Biochim. Biophys. Acta* **126**, 181 (1966).
78. D. Papahadjopoulos, *Biochim. Biophys. Acta* **163**, 240 (1968).
79. G. Colacicco, *Chem. Phys. Lipid* **10**, 66 (1973).
80. S. A. Rice and M. Nagasawa, *Polyelectrolyte Solution*, Academic Press, N.Y. (1961), p. 118.
81. V. S. Vaidhyanathan, *Colloid Surf.* **6**, 291 (1983).
82. H. Oshima, T. W. Healy, and L. R. White, *J. Colloid Interface Sci.* **90**, 17 (1982).
83. B. Hille, *J. Gen. Physiol.* **51**, 221 (1968).
84. D. L. Gilbert and G. Ehrenstein, *Biophys. J.* **9**, 447 (1969).
85. C. L. Schauf, *J. Physiol. (London)* **248**, 613 (1975).
86. B. Hille, A. Wood, and B. I. Shapiro, *Phil. Trans. Royal Soc. London*, **B270**, 301 (1975).
87. S. Ohki, *Bioelectrochem. Bioenerg.* **5**, 204 (1978).
88. D. A. Haydon and V. B. Myers, *Biochim. Biophys. Acta* **307**, 429.
89. J. D. Castle and W. L. Hubbell, *Biochemistry* **15**, 4818.
90. A. D. Bangham, M. W. Hill, and N. G. A. Miller, in *Methods in Membrane Biology*, Vol. 1, E. D. Korn, ed., Plenum Press, New York (1974), p. 1.
91. G. V. F. Seaman and V. F. Cook, in *Cell Electrophoresis*, E. J. Ambrose, ed., Little, Brown, Boston (1965), p. 78.
92. H. Alpes and W. G. Pohl, *Naturwissenschaften* **65**, 652 (1978).
93. M. S. Fernandez, *Biochim. Biophys. Acta* **646**, 23 (1981).
94. O. Stern, *Z. Elektrochem.* **30**, 508 (1924).
95. D. C. Grahame, *Z. Elektrochem.* **62**, 264 (1958).
96. F. Franks, ed., *Water, A Comprehensive Treatise*, Vol. 4, Plenum Press, New York (1975).
97. S. Marcelja, S. and N. Radic, *Chem. Phys. Lett.* **42**, 129 (1976).
97a. D. M. LeNeveu, R. P. Rand, and V. A. Parsegian, *Nature* **259**, 601 (1976).
98. R. P. Rand, *Ann. Rev. Biophys. Bioeng.* **10**, 277 (1981).
99. G. Cevc and D. Marsh, *J. Phys. Chem.*, **87**, 376 (1983).
100. G. Cevc, A. Watts, and D. Marsh, *Biochemistry* **20**, 4955 (1981).
101. M. V. Smoluchowski, *Bull. Intern. Acad. Sci.* (Cracovie), 184 (1903).
102. E. Hückel, *Physik. Z.* **25**, 204 (1924).
103. D. J. Shaw, *Introduction to Colloid and Surface Chemistry*, Butterworths, London (1979).
104. D. C. Henry, *Trans. Faraday Soc.* **44**, 1021 (1948).
105. F. Booth, *Proc. Roy. Soc. (London)* **A203**, 514 (1950).
106. E. J. Ambrose, ed., *Cell Electrophoresis*, J. and A. Churchill, London (1965).
107. G. V. F. Seaman, in *The Red Blood Cell*, D. M. Surgenor, ed., Academic Press, New York (1974), p. 1136.
108. G. V. Sherbet, *The Biophysical Characterisation of the Cell Surface*, Academic Press, New York (1978).
109. Z. Samec, V. Mareck, and J. Weber, *J. Electroanal. Chem.* **96**, 245 (1977).
110. Z. Samec, V. Mareck, and J. Weber, *J. Electroanal. Chem.* **103**, 1 (1979).
111. J. Koryta, P. Vanýsek, and M. Březina, *J. Electroanal. Chem.* **75**, 211 (1977).
112. L. L. Boguslavsky, A. G. Volkov, and Kandelaki, *FEBS Letters* **65**, 155 (1976).
113. L. I. Boguslavsky, A. G. Volkov, I. A. Kozlov, and A. M. Malyan, *Biofizika* **21**, 286 (1976).
114. S. S. Anderson, I. G. Lyle, and E. Paterson, *Nature* **259**, 147 (1976).

115. H. Ti Tien, *Photochemistry and Photobiology* **24**, 95 (1976).

116. F. G. Donnan, *Chem. Rev.* **1**, 73 (1924).

117. J. Bernstein, *Elektrobiologie, Braunschweig, Fr. Vieweg* (1912).

118. P. J. Boyle and E. J. Conway, *J. Physiol. (London)* **100**, 1 (1941).

119. R. H. Adrian, *J. Physiol. (London)* **133**, 631 (1956).

120. A. L. Hodgkin and P. Horowicz, *J. Physiol (London)* **145**, 405 (1959).

121. G. N. Ling and R. W. Gerard, *Nature (London)* **165**, 113 (1950).

122. R. D. Keynes and P. R. Lewis, *J. Physiol. (London)* **113**, 73 (1951).

123. A. L. Hodgkin and R. D. Keynes, *J. Physiol. (London)* **119**, 513 (1953).

124. W. Nernst, *Z. Phys. Chem. Stoechiom. Verwandschaftslehre*, **2**, 613 (1888).

125. M. Planck, *Am. Phys. Chem.* **39**, 161 (1890).

126. D. A. MacInnes, *The Principle of Electrochemistry*, Reinhold, New York (1939).

127. D. E. Goldman, *J. Gen. Physiol.* **27**, 37 (1943).

128. A. L. Hodgkin and B. Katz, *J. Physiol. (London)* **108**, 37 (1949).

129. P. Henderson, *Z. Phys. Chem. Stoechiom. Verwandschaftslehre* **59**, 118 (1907).

130. T. Teorell, *Proc. Soc. Exp. Biol. Med.* **33**, 282 (1935).

131. T. Teorell, *Progr. Biophys. Chem.* **3**, 305 (1953).

132. K. H. Meyer and J. F. Sievero, *Helv. Chim. Acta* **19**, 649, 665, 987 (1936).

133. M. J. Polissar, in *Kinetic Basis of Molecular Biology*, F. H. Johnson, H. Eyring, and M. J. Polissar, eds., Wiley, New York (1954), p. 515.

134. S. Ohki, *Biochim. Biophys. Acta* **282**, 55 (1972).

135. R. C. MacDonald and A. D. Bangham, *J. Membrane Biol.* **7**, 29 (1972).

136. G. Colacicco, *Nature (London)* **207**, 936 (1965).

137. N. Kamo and Y. Kobatake, *J. Colloid Interface Sci.* **46**, 83 (1974).

138. S. McLaughlin and H. Harary, *Biophys. J.* **14**, 200 (1974).

139. D. L. Gilbert, in *Biophysics and Physiology of Excitable Membranes*, W. J. Adelman Jr., ed., Van Nostrand Reinhold, New York (1971), p. 264.

140. O. Aono and S. Ohki, *J. Theor. Biol.* **37**, 273 (1972).

141. S. Ohki, *Physiol. Chem. and Physics* **13**, 195 (1981).

142. I. Tasaki, A. Watanabe, and T. Takenaka, *Proc. Natl. Acad. Sci. U.S.A.* **48**, 1177 (1962).

143. I. Tasaki and T. Takenaka, *Proc. Natl. Acad. Sci. U.S.A.* **50**, 619 (1963).

144. S. Ohki and O. Aono, *Jap. J. Physiol.* **29**, 373 (1979).

145. P. F. Baker, A. L. Hodgkin, and T. I. Shaw, *J. Physiol. (London)* **164**, 355 (1962).

146. N. Lakshminarayanaiah, *Bull. Math. Biol.* **39**, 643 (1977).

147. S. Ohki, in *Progress in Surface and Membrane Sciences*, Vol. 10, D. A. Cadenhead and J. F. Danielli, eds., Academic Press, New York (1976), p. 117.

148. R. F. Zwaal, B. Roelofsen, and C. M. Colley, *Biochim. Biophys. Acta* **300**, 159 (1973).

149. L. D. Bergelson and L. I. Barsukov, *Science* **197**, 224 (1970).

150. S. Ohki, *J. Phys. Soc. Japan* **20**, 1674 (1965).

151. G. N. Ling, in *A Physical Theory of the Living State; The Association–Induction Hypothesis*, Blaisdell, Waltham, Massachusetts (1962).

152. J. Metuzals and I. Tasaki, *J. Cell Biol.* **78**, 597 (1978).

153. I. Tasaki, I. Singer, and A. Watanabe, *Proc. Natl. Acad. Sci. U.S.A.* **54**, 763 (1965).

154. I. Tasaki, *Physiology and Electrochemistry in Nerve Fibers*, Academic Press, New York (1982).

155. J. M. Olson, G. Hind, H. Lyman, and H. W. Siegelman, eds., *Energy Conversion by the Photosynthetic Apparatus*, Brookhaven National Lab., New York (1967).

156. D. D. Eley, in *Horizons in Biochemistry*, M. Kasha and B. Pullman, eds., Academic Press, New York (1962), p. 341.

157. H. Kallmann and M. Silver, eds., *Electrical Conductivity in Organic Solids*, Wiley, New York (1961).

158. P. D. Boyer, *Annual Rev. Biochem.* **46**, 955 (1977).

159. F. M. Harold, *Current Topics in Bioenergetics* **6**, 84 (1977).

160. C. B. Van Niel, *Adv. Enzymol.* **1**, 263 (1941).

160*a*. E. Katz, in *Photosynthesis in Plants*, Iowa State College Press, Ames, 1949, p. 287.

161. W. Arnold and H. K. Sherwood, *Proc. Natl. Acad. Sci. U.S.A.* **43**, 105 (1957).

162. R. C. Nelson, *J. Chem. Phys.* **27**, 864 (1957).

163. E. J. Lund, *J. Exp. Zool.* **51**, 265 (1928).

164. R. D. Stiehler and L. B. Flexner, *J. Biol. Chem.* **126**, 603 (1939).

165. T. L. Jahn, *J. Theor. Biol.* **2**, 129 (1962).

166. H. J. A. Dartnall, *Nature* **162**, 122 (1948).

167. P. S. B. Digby, *Proc. Roy. Soc.* **161B**, 504 (1965).

168. P. Mitchell, *Biol. Rev.* **41**, 445 (1966).

169. H. Metzner, ed., *Progress in Photosynthesis Research*, Vol. 3, Tubingen, Germany (1969).

170. H. Ti Tien, *Photochem. Photobiol.* **16**, 271 (1972).

171. Y. Kagawa, *Biological Membrane*, Iwanami Book Store, Tokyo (1978), p. 228.

172. A. J. Jagendorf and E. Uribe, *Proc. Natl. Acad. Sci. U.S.A.* **55**, 170 (1966).

173. H. T. Witt, E. Schlodder, and P. Gräber, *FEBS Letters* **69**, 272 (1976).

174. J. C. Skou, *Biochim. Biophys. Acta* **23**, 394 (1957).

175. P. C. Caldwell, A. L. Hodgkin, R. D. Keynes, and T. I. Shaw, *J. Physiol. (London)* **152**, 561 (1960).

176. E. J. Harris, in *Active Transport and Secretion*, R. Brown and J. F. Danielli, eds., Cambridge Univ. Press, Cambridge (1954), p. 228.

177. P. DeWeer, *MTP International Review of Science*, Physiology Series One, Vol. 3, C. C. Hunt, ed., Butterworths, London (1975).

178. R. D. Keynes and R. Rybova, *J. Physiol. (London)* **168**, 58P (1963).

179. R. L. Post, C. D. Albright, and K. Dayani, *J. Gen. Physiol.* **50**, 1201 (1967).

180. P. J. Garrahan and I. M. Glynn, *J. Physiol. (London)* **192**, 217 (1967).

181. R. A. Sjodin and L. A. Beaugé, *J. Gen. Physiol.* **51**, 152 (1968).

182. P. G. Kostyuk, O. A. Krishtal, and V. I. Pidoplichiko, *J. Physiol. (London)* **226**, 373 (1972).

183. R. C. Thomas, *J. Physiol. (London)* **220**, 55 (1972).

184. R. C. Thomas, *Physiol. Rev.* **52**, 563 (1972).

185. I. J. Mullins and K. Noda, *J. Gen. Physiol.* **47**, 117 (1963).

186. S. L. Bonting and L. L. Caravaggio, *Arch. Biochem.* **101**, 37 (1960).

187. M. W. Weiner and R. H. Maffly, in *Physiology of Membrane Disorders*, T. E. Andreoli, J. F. Hoffman, and D. D. Fanestil, eds., Plenum Press, New York (1978), p. 287.

188. M. Chiesi and A. Martonosi, in *Membrane Biochemistry*, E. Carafoli and G. Semenza, eds., Springer-Verlag, New York–Berlin (1979), p. 120.

189. D. C. Tosteson, ed., in *Membrane Transport in Biology*, Vol. 2, Springer-Verlag, New York–Berlin (1979).

190. M. Wikström and E. Sigel, in *Membrane Biochemistry*, E. Carafoli and G. Semenza, eds., Springer-Verlag, New York–Berlin (1979), p. 82.

191. M. Yoshida, H. Okamoto, N. Sone, H. Hirata, and Y. Kagawa, *Proc. Natl. Acad. Sci. U.S.A.* **74**, 936 (1977).

192. B. Katz, *Nerve, Muscle and Synapse*, McGraw-Hill, New York (1966).

193. R. Eckert and D. Randall, in *Animal Physiology*, W. H. Freeman and Co., San Francisco (1978).

194. A. L. Hodgkin, *Biological Rev. Cambridge Phil. Soc.* **26**, 339 (1951).

195. A. L. Hodgkin and A. F. Huxley, *J. Physiol. (London)* **117**, 500 (1952).

196. A. L. Hodgkin, in *The Conduction of the Nerve Impulse*, Charles C. Thomas Publisher, Springfield, Illinois (1964), p. 72.

197. B. Hille, *Progr. Biophys. Molec. Biol.* **21**, 1 (1970).

198. R. E. Taylor, in *The Neurosciences: A Study Program*, G. C. Quarton, T. Melnechuk, and F. O. Schmitt, ed., Rockefeller University Press, New York (1967).

199. T. Narahashi, *Physiol. Rev.* **54**, 813 (1974).

200. B. Hille, *J. Gen. Physiol.* **66**, 535 (1975).

201. I. Tasaki and I. Singer, *J. Cell Comp. Physiol.* **66**, 137 (1965).

202. I. Tasaki, *Nerve Excitation*, Charles C. Thomas Publ., Springfield, Ill. (1968).

203. C. M. Armstrong and F. Bezamilla, *J. Gen. Physiol.* **62**, 53 (1974).

204. R. Keynes and E. Rojas, *J. Physiol. (London)* **239**, 393 (1974).

205. R. K. Orkand, *J. Physiol. (London)* **196**, 311 (1968).

206. R. Eckert and H. D. Lux, *J. Physiol. (London)* **254**, 129 (1976).

207. W. R. Loewenstein, in *Handbook of Sensory Physiology, Principle of Receptor Physiology*, Springer-Verlag, New York (1971).

208. Y. Kobatake, K. Kurihara, and N. Kamo, in *Surface Electrochemistry*, T. Takamura and A. Kozawa, eds., Japan Scientific Societies Press (1978).

209. M. Hato, T. Ueda, K. Kurihara, and Y. Kobatake, *Biochim. Biophys. Acta* **426**, 73 (1976).

210. R. C. Gesteland, in *Discovery* **27(2)**, Proprietors, Professional and Industrial Pub. Co., London (1966).

211. A. Alvanitaki, H. Takeuchi, and N. Chalazonitio, in *Olfaction and Taste*, Vol. 2, T. Hayashi, ed., Pergamon Press, Oxford (1971).

212. J. G. Eccles, *Scientific American* **212(1)**, 56 (1965).

213. J. G. Eccles, in *The Understanding of the Brain*, McGraw-Hill, New York (1972).

214. S. W. Kuffler and J. Nicholls, in *From Neuron to Brain*, Sinauer, New York (1976).

215. M. Weber, T. David-Pfeuty, and J. P. Changeux, *Proc. Natl. Acad. Sci. U.S.A.* **72**, 3443 (1975).

216. A. Karlin, *Cell Surface Rev.*, **6**, 191 (1980).

217. W. R. Loewenstein, *Scientific American* **203(2)**, 98 (1960).

218. S. Nakajima and K. Onodera, *J. Physiol. (London)* **200**, 161 (1969).

219. G. G. Harris and A. Flock, in *Lateral Line Detectors*, P. H. Cahn, ed., Indiana Univ. Press, Bloomington (1967).

220. W. S. Beck, *Human Design*, Harcourt Brace Jovanovich, New York (1971).

221. H. Davis, *Ann. Otol. Rhinol. Laryngol.* **77**, 644 (1968).

222. K. Iwasa, I. Tasaki, and R. C. Gibbons, *Science* **210**, 338 (1980).

223. I. Tasaki and K. Iwasa, *Biochem. Biophys. Res. Commun.* **101**, 172 (1981).

224. D. E. Hillman, *Brain Res.* **13**, 407 (1969).

225. G. G. Harris, L. S. Fishkopf and A. Flock, *Science* **167**, 76 (1970).

226. R. W. Young, *Scientific American* **223(4)**, 89 (1970).

227. J. Toyoda, H. Nosaki and T. Tomita, *Vision Res.* **9**, 453 (1969).

228. S. Yoshikami and W. A. Hagins, in *Biochemistry and Physiology of Visual Pigments*, H. Langer, ed., Springer-Verlag, New York (1973), p. 245.

229. M. L. Woodruff and M. D. Bownds, *J. Gen. Physiol.* **73**, 629 (1979).

230. W. L. Hubbell and M. D. Bownds, *Ann. Rev. Neurosci.* **2**, 17 (1979).

231. D. F. O'Brian, *Science* **218**, 961 (1982).

232. P. A. Hargrave, *Biochim. Biophys. Acta* **492**, 83 (1977).

233. J. E. Lennard-Jones, *Proc. Phys. Soc. London* **43**, 461 (1931).

234. F. London, *Trans. Faraday Soc.* **33**, 8 (1937).

235. J. O. Hirschfelder, C. F. Curtiss, and R. B. Bird, *Molecular Theory of Gases and Liquids*, John Wiley & Sons, New York (1964).

236. S. Ohki and N. Fukuda, *J. Theoretical Biol.* **15**, 362 (1967).

237. E. J. W. Verwey and J. Th. G. Overbeek, *Theory of the Stability of Lyophobic Colloids*, Elsevier, Amsterdam (1948).

238. H. C. Hamaker, *Physica* **4**, 1058 (1937).

239. S. Nir, *Progr. Surface Sci.* **8**, 1 (1977).

240. H. B. G. Casimir and D. Polder, *Phys. Rev.* **73**, 360 (1948).

241. J. Th. G. Overbeek, in *Colloid Science* Vol. 1, H. R. Kruyt, ed., Elsevier, Amsterdam (1952).

242. I. E. Dzyaloshinskii, E. M. Lifshitz, and L. P. Pitaevskii, *Adv. Phys.* **10**, 165 (1961).

243. W. H. Keesom, *Proc. Amst.* **15**, 250 (1913).

244. B. V. Derjaguin and L. Landau, *Acta Phys. Chim. URSS* **14**, 633 (1941).

245. Y. Ikeda, *J. Phys. Soc. Japan* **8**, 49 (1953).

246. O. F. Devereux and P. L. de Bruyn, *Interaction of Plane Parallel Double Layers*, MIT Press (1963).
247. S. Usui, *J. Colloid Interface Sci.* **44**, 107 (1973).
248. G. R. Wiese and T. W. Healy, *Trans. Faraday Soc.* **66**, 490 (1970).
249. S. Ohki, N. Duzgunes and K. Leonards, *Biochemistry* **21**, 2127 (1982).
250. L. Weiss and J. P. Harlos, *Prog. Surface Sci.* **1**, 335 (1972).
251. M. Van den Tempel, *J. Colloid Sci.* **13**, 125 (1958).
252. T. T. Puck and L. J. Tolmach, *Arch. Biochim. Biophys.* **51**, 229 (1954).
253. J. Dirkx, J. Beumer, and M. P. Beumer-Jochmans, *Ann. Ins. Pasteur* **93**, 340 (1957).
254. R. C. Valentine and A. C. Allison, *Biochim. Biophys. Acta* **34**, 10 (1959).
255. A. S. G. Curtis, *Amer. Nat.* **94**, 37 (1960).
256. B. A. Pethica, *Exp. Cell Res. Suppl.* **8**, 123 (1961).
257. L. Weiss, *The Cell Periphery Metastasis and Other Contact Phenomena*, North-Holland Press, Amsterdam (1967).
258. G. Poste, *Microbios* **2**, 227 (1970).
259. A. D. Bangham and B. A. Pethica, *Proc. Royal Phys. Soc. of Edinburgh* **28**, 43 (1960).
260. G. Poste, *Int. Rev. Cytol.* **33**, 157 (1972).
261. K. Olden, E. L. Hahn, and K. M. Yamada, *J. Cell Biochem.*, **18**, 635 (1982).
262. R. C. Hughes, S. D. Pena, and P. Visher, in *Cell Adhesion and Motility*, A. S. G. Curtis and J. D. Pitts, eds., Cambridge Univ. Press, Cambridge (1980), p. 329.
263. S. Subtelny and N. K. Wessells, *The Cell Surface: Mediator of Developmental Processes*, New York, Academic Press (1980).
264. N. G. Maroudas, *Nature* **254**, 695 (1975).
265. K. Ohsawa, H. Ohshima, and S. Ohki, *Biochim. Biophys. Acta* **648**, 206 (1981).
266. V. A. Parsegian and D. Gingell, in *Recent Advances in Adhesion*, L-H. Lee, ed., Gordon and Breach Sci. Pub., London (1973).
267. D. Gingell and L. Ginsberg, in *Membrane Fusion*, G. Poste and G. L. Nicholson, eds., North-Holland, Amsterdam (1978), p. 791.
268. K. S. Spiegler and M. R. J. Wyllie, in *Physical Techniques in Biological Research*, G. Oster and A. Pollister, eds., Academic Press, New York (1956).
269. F. Helfferich, in *Ion Exchange*, McGraw-Hill, New York (1962), p. 374.
270. E. M. Lifshitz, *J. Exp. Theor. Phys.* **29**, 94 (1955).
271. F. Schuurmans Stekhoven, and S. L. Bonting, *Physiol. Rev.* **61**, 1 (1981).
272. H. J. Curtis, and K. S. Cole, *J. Cell Comp. Physiol.* **19**, 135 (1942).
273. A. F. Huxley and R. Stümpfli, *J. Physiol. (London)* **112**, 496 (1951).
274. W. Montagna, *Comparative Anatomy*, Wiley, New York (1959).
275. P. A. Hargrave, in *Progress in Retinal Research*, N. N. Osborne and G. J. Chader, eds., Pergamon Press, Oxford (1982), p. 1.
276. R. Kurland, C. Newton, S. Nir, and D. Papahadjopoulos, *Biochim. Biophys. Acta* **551**, 137 (1979).
277. A. McLaughlin, C. Grathwohl, and S. McLaughlin, *Biochim. Biophys. Acta* **513**, 338 (1978).
278. N. E. Hill, W. E. Vaughan, A. H. Price, and M. Davies, *Dielectric Properties and Molecular Behaviour*, Van Nostrand Reinhold Co., London (1969).
279. R. C. Thomas, *J. Physiol. (London)* **220**, 55 (1972).
280. P. G. Sokolove and I. M. Cooke, *J. Gen. Physiol.* **57**, 125 (1971).
281. D. A. Baylor and J. G. Nicholls, *J. Physiol. (London)* **203**, 571 (1969).
282. H. Koike, H. M. Brown, and S. Hagiwara, *J. Gen. Physiol.* **57**, 723 (1971).
283. J. E. Brown and J. E. Lisman, *J. Gen. Physiol.* **59**, 720 (1972).
284. P. DeWeer and D. Geduldig, *Science* **179**, 1326 (1973).
285. D. O. Carpenter, *Comp. Biochem. Physiol.* **35**, 371 (1970).
286. M. F. Marmor and A. L. F. Gorman, *Science* **167**, 65 (1970).
287. D. O. Carpenter, *Science* **179**, 1336 (1973).
288. S. Nir and J. Bentz, *Colloid Int. Sci.* **65**, 399 (1978).

289. S. Nir, J. Bentz, and A. R. Portis, Jr., "Bioelectrochemistry Ions, Surfaces, Membranes," in *Advances in Chemistry*, No. 188, M. Blank, ed., Amer. Chem. Soc. (1980), p. 75.
290. S. Ohki, S. Roy, H. Ohshima, and K. Leonards, *Biochem.*, in press (1984).
291. P. Rand, *Ann. Rev. Biophys. Bioeng.* **10**, 277 (1981).
292. S. Ohki, *Phys. Lett.* **103A**, 153 (1984).
293. M. J. J. Vold, *Colloid Sci.* **16**, 1 (1961).
294. *Handbook of Chemistry and Physics*, The Chemical Rubber Co., Cleveland, Ohio, 1965.
295. S. Ohki and N. J. Fukuda, *Colloid Interface Sci.* **27**, 208 (1968).
296. S. Nir, J. Bentz, J. Wilschut, and N. Duzgunes, *Prog. Surface Sci.*, Vol. 13, S. G. Davison, ed., Pergamon Press, New York, 1983, p. 1.
297. S. Ohki, *Biochim. Biophys. Acta* **889**, 1 (1982).
298. S. Ohki, *J. Membrane Biol.* **77**, 265 (1984).
299. W. Breisblatt and S. Ohki, *J. Membrane Biol.* **23**, 385 (1975).
300. C. Miller, P. Avan, J. N. Telford, and E. Racker, *J. Membrane Biol.* **30**, 21 (1976).
301. S. W. Hui, T. P. Stewart, L. T. Boni, and P. L. Yeagle, *Science* **217**, 921 (1981).
302. D. Papahadjopoulos, G. Poste, B. E. Schaffer, and W. J. Vail, *Biochim. Biophys. Acta* **352**, 10 (1974).
303. S. E. Schullery, C. F. Schmidt, P. Felgasr, T. W. Tillack, and T. E. Thompson, *Biochemistry* **19**, 3919 (1980).
304. I. Tasaki and K. Iwasa, in *Structure and Functions in Excitable Cells*, eds. D. Chang, I. Tasaki, W. Adelman, and Leuchtung, Plenum Publishing Corp., New York (1983).
305. B. R. Copeland and H. C. Anderson, *J. Chem. Phys.* **74**, 2549 (1981).
306. H. Oshima and S. Ohki, *J. Colloid Interface Sci.*, in press (1984).
307. D. J. Wilkins, R. H. Ottewill, and A. D. Bangham, *J. Theor. Biol.* **2**, 165 (1962).
308. S. Ohki, in *Charge and Field Effects in Biosystems*, eds. M. J. Allen and P. N. R. Usherwood, Abacus Press, England, 1984, p. 147.
309. D. A. Haydon, *Biochim. Biophys. Acta* **50**, 450 (1961).
310. E. Donath and V. Pastushenko, *Bioelectrochem. Bioenerg.* **6**, 543 (1979).
311. S. Levine, M. Levine, K. A. Sharp, and D. E. Brooks, *Biophys. J.* **42**, 127 (1983).
312. R. W. Wunderlich, *J. Colloid Interface Sci.* **88**, 385 (1982).
313. H. Ohshima, T. W. Healy, and L. R. White, *J. Chem. Soc., Faraday Trans. 2*, **79**, 1613 (1983).
314. R. Hogg, T. W. Healy, and D. Fuerstenau, *Trans. Faraday Soc.* **62**, 1638 (1966).
315. W. K. Chandler and H. Meves, *J. Physiol. London* **180**, 788 (1965).
316. B. Hille, *J. Gen. Physiol.* **58**, 599 (1971).
317. B. Hille, *J. Gen. Physiol.* **59**, 637 (1972).

2

Electrochemistry of Low Molecular Weight Organic Compounds of Biological Interest

GLENN DRYHURST

1. Introduction

The electrochemical behavior of a large number of low molecular weight compounds of biological interest has been studied. Reviews such as those of Brezina and Zuman,[1] Underwood and Burnett,[2] McCreery,[3] and Dryhurst[4] have summarized much of the available information on the electrochemical behavior of biologically significant molecules. However, except in relatively few instances, such reviews have reported the electrochemical results without much attempt to correlate the information with known or suspected biological redox reactions. In this review only a rather small number of low molecular weight biomolecules will be considered. However, they will be discussed in some depth and an attempt will be made to demonstrate that electrochemical studies of biomolecules can provide some unique and valuable insights into their biochemical redox behavior.

There is now a formidable array of sophisticated and powerful electrochemical tools and techniques available to investigate the redox properties of ions and molecules in solution. Probably the most complex and therefore scientifically challenging redox processes occur in living systems. Biochemistry

GLENN DRYHURST • Department of Chemistry, University of Oklahoma, Norman, OK 73019, USA.

textbooks notwithstanding, there are only a very few examples where the mechanisms of such biological redox reactions are understood. Thus, bio-electrochemistry is an area of great promise and provides significant challenges to the electrochemist.

When considering the biological redox reactions of low molecular weight organic compounds there are at least two aspects of the process which must be understood. The first is the reaction pathway and mechanism followed by the species which undergoes the redox reaction. This species might be the substrate in an enzymic reaction or it might be a cofactor. The second aspect which must be understood is the role of the enzyme as it directs or controls the reaction pathway of the substrate or cofactor molecules. Electrochemical investigations can make valuable contributions to an understanding of both of these apsects, but more particularly to unraveling the redox mechanisms of relatively low molecular weight substrate or cofactor species.

Under electrochemical conditions, electrons are added to (reduction) or withdrawn from (oxidation) a molecule or ion of interest at an electrode surface and the consequences of that process are followed by means of many electrochemical, spectral, and other techniques. Naturally, there is no assurance that the redox pathways and mechanisms developed on the basis of electrochemical experiments will bear any relationship to biological redox pathways. Nevertheless, there is now a small but significant body of evidence which supports the view that redox mechanisms deduced for small organic biomolecules using purely electrochemical techniques do give rather unique insights into biological redox processes.

The electrochemistry and biological redox behavior of members of four classes of compounds will be considered. These are catecholamines, phenothiazines, biological quinones, and purines.

2. Catecholamines

Catecholamines are of considerable biological interest because some members of this class of compounds are involved in certain types of chemical neurotransmission processes. Other catecholamines are powerful neurotoxins and are suspected of being involved in mental illnesses such as depression, schizophrenia, and other psychotic disorders. Adams[5] has presented an excellent introduction to the structure of nerve cells or neurons and the funadmentals of chemical neurotransmission processes. Neurons are cells which are designed to communicate with one another. This is accomplished in chemical neurotransmission through regions between them called synaptic gaps. The chemical neurotransmission process is illustrated in Figure 1. The synaptic gap is typically 200–800 Å in width and is bordered on one side by the presynaptic membrane of one neuron and on the other side by the postsynaptic membrane of a second neuron. Inside the presynaptic membrane there are a group of vesicles which

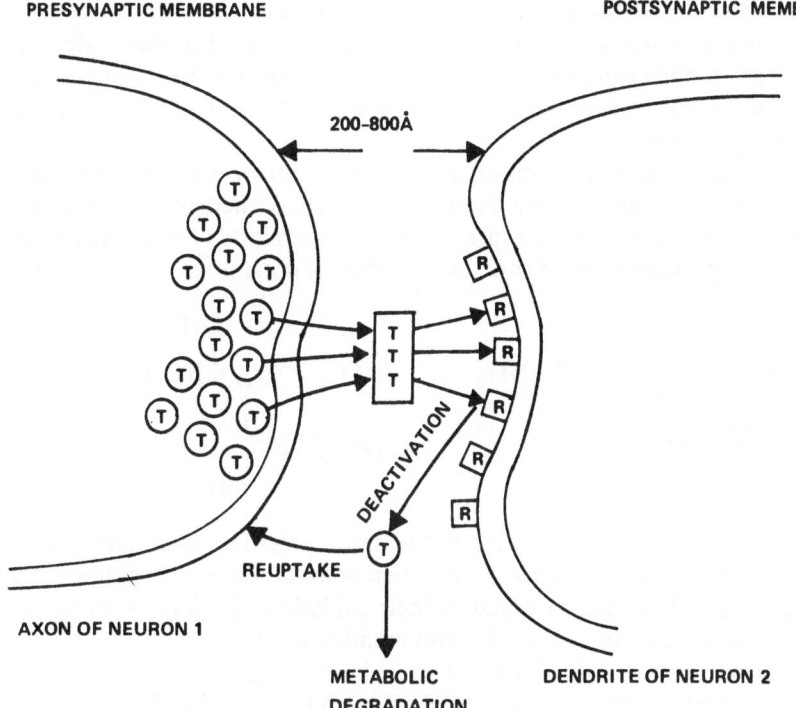

Figure 1. Schematic representation of a chemical synaptic gap: ⓣ represents the chemical neurotransmitter substance, Ⓡ represents a receptor site.

contain the chemical transmitter substance, represented by ⓣ in Figure 1. The chemical neurotransmission process is initiated by an electrical potential pulse or spike, called the action potential (*ca.* 0.1 V for 1–2 ms), in neuron 1 in Figure 1. In some fashion this action potential activates a process which releases some of the vesicle-contained neurotransmitter substance into the synaptic gap. This substance then travels by diffusive and perhaps convective mass transport to receptor sites (denoted Ⓡ in Figure 1) located on the postsynaptic membrane of neuron 2. The interaction between the receptors and the transmitter substance results in sudden permeability changes of the postsynaptic membrane to sodium and other small ions which in turn generates a small postsynaptic charge across the postsynaptic membrane. The postsynaptic potential generally has an amplitude of a few millivolts for milliseconds to seconds duration and is dependent upon the amount of neurotransmitter substance released. The postsynaptic potential may be excitatory or inhibitory. By some unknown process the receiving neuron integrates the receptor postsynaptic potentials and determines whether to initiate a new action potential which then causes the same process to be propagated from neuron 2 to neuron 3 and so on.

Once the neurotransmitter substance has interacted with the receptor sites on the postsynaptic membrane it must be deactivated so that it does not continuously excite or inhibit the membrane. There are at least three deactivation processes—mass transport away from the synaptic gap, reuptake, and enzymic degradation.

There are a number of different types of neutrotransmitter substances associated with various neurons. However, in the so-called adrenergic system the important chemical neurotransmitters are dopamine (**1**) and norepinephrine (**2**). Electrochemical studies of the oxidation of dopamine, norepinephrine,

$$HO\text{—}\bigcirc\text{—}CH_2CH_2NH_2$$
$$HO$$

(**1**)

$$\overset{OH}{\underset{|}{HO\text{—}\bigcirc\text{—}CHCH_2NH_2}}$$
$$HO$$

(**2**)

and similar catecholamines have led to the development of information which could have a bearing on some of the reactions of these compounds, and intermediates and products formed in their oxidation reactions, with various other naturally occurring molecules and membrane sites.

The first in-depth study of the electrooxidation mechanism of a biologically important catecholamine was that of Hawley and co-workers[6] who investigated the bahavior of epinephrine (**3**). This compound, often called adrenaline,

$$\overset{OH}{\underset{|}{HO\text{—}\bigcirc\text{—}CHCH_2NHCH_3}}$$
$$HO$$

(**3**)

is not a neurotransmitter substance but its basic electrooxidation is the same as that for dopamine and norepinephrine. Typical cyclic voltammograms of epinephrine are shown in Figure 2. Below pH 2 a single voltammetric oxidation peak, peak I_a, is observed, followed on the reverse sweep by quasi-reversible reduction peak I_c (Fig. 2A). On subsequent cycles only this single couple could be observed. The peak I_a reaction involves a simple $2e-2H^+$ electrooxidation of the catechol nucleus of epinephrine (I, Figure 3) to an open chain o-quinone (II, Figure 3). At pH 3 and above, however, after scanning through oxidation peak I_a, reduction peak I_c greatly decreases in height and reduction peak II_c appears at more negative potentials (Figure 2B). On the second sweep towards positive potentials two new voltammetric oxidation peaks, peaks II_a and III_a, appear (Figure 2B). This cyclic voltammetric behavior has been rationalized by the mechanism shown in Figure 3. Thus, following the initial $2e-2H^+$

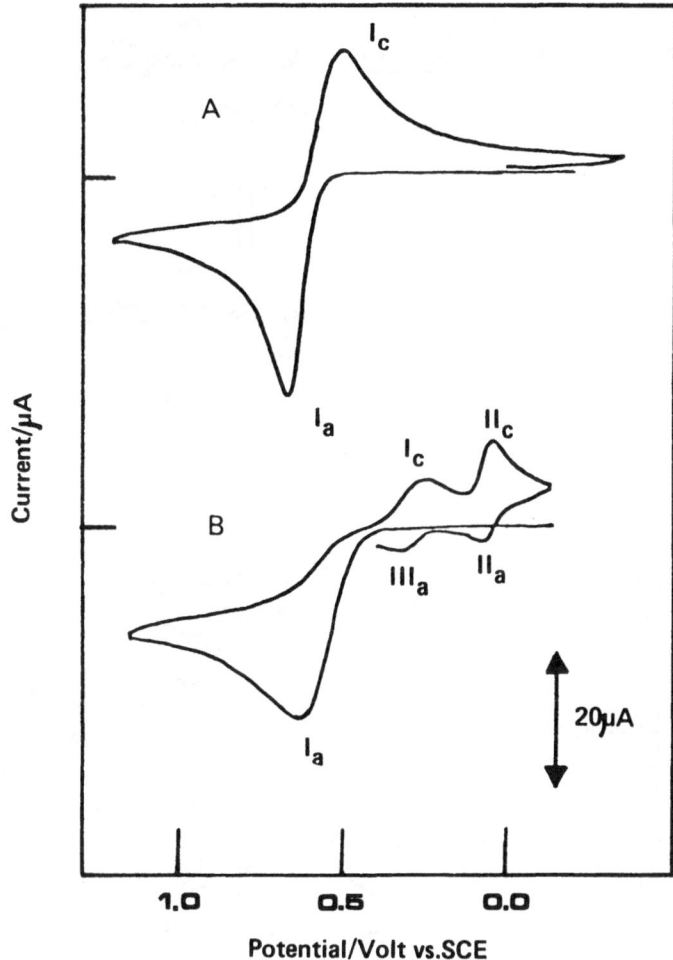

Figure 2. Cyclic voltammograms of (A) 1.3 mM epinephrine in 1 M H$_2$SO$_4$ and (B) 1.7 mM epinephrine at pH 3.0 at a carbon paste electrode. Sweep rate: 40 mV s^{-1}. (Reprinted with permission of the American Chemical Society).[6]

oxidation of epinephrine to the protonated open chain quinone (II, Figure 3), some unprotonated quinone (III, Figure 3) is formed. This unprotonated species undergoes a cyclization reaction to give leucoadrenochrome (IV, Figure 3) characterized by the pseudo first-order rate constant k_2. The value of k_2 is strongly pH-dependent since it depends upon deprotonation of the open chain o-quinone. Leucoadrenochrome (IV, Figure 3) rapidly reacts with the open chain o-quinone (II, Figure 3) giving adrenochrome (V, Figure 3). Reduction peak II$_c$ observed in the cyclic voltammogram shown in Figure 2B is thought to correspond to a 2e–2H$^+$ reduction of adrenochrome to

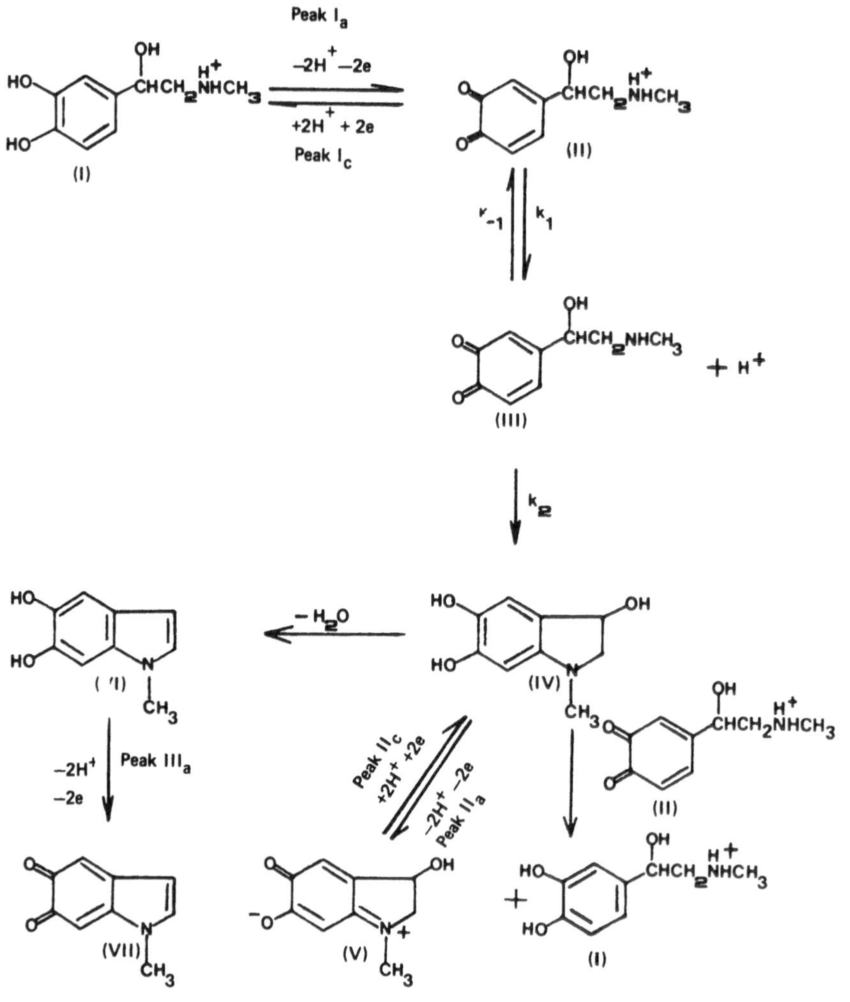

Figure 3. Mechanism proposed by Hawley and co-workers[6] for the electrochemical oxidation of epinephrine.

leucoadrenochrome, while oxidation peak II_a is the reverse reaction. Voltammetric oxidation peak III_a is thought to be caused by oxidation of 5,6-dihydroxyindole (VI, Figure 3), formed by dehydration of leucoadrenochrome (IV, Figure 3), to 5,6-dioxoindole (VII, Figure 3), presumably via a $2e-2H^+$ reaction. The initial cyclized product, leucoadrenochrome, is more easily oxidized than epinephrine and hence is chemically oxidized by the open chain o-quinone. Accordingly, based on the reaction sheme shown in Figure 3, the peak I_a process at intermediate pH values is an *eccc* reaction, i.e., three chemical steps follow the electron transfer process.

The rate of the cyclization reaction (III → IV, Figure 3) has been studied by chronoamperometric methods[6] and using rotating disc electrochemistry.[7] Values of the observed rate constant (k_{obsd}) are shown in Table 1. The relationship between k_{obsd} and the rate constants k_1 and k_2 shown in Figure 3 is given in Eq. (1).

$$k_{obsd} = \frac{k_1 k_2}{k_{-1}[H^+] + k_2} \qquad (1)$$

At low pH this rate law reduces to that shown in Eq. (2)

$$k_{obsd} = \frac{k_2 K_a}{[H^+]} \qquad (2)$$

and k_2 may be estimated from a plot of k_{obsd} vs. $K_a/[H^+]$ assuming that the dissociation constant of the side chain of the open-chain o-quinone (II, Figure 3) is the same as that of the catecholamine.

It is clear from the data presented in Table 1 that the open chain o-quinone is a rather reactive species, particularly around physiological pH, and is susceptible to nucleophilic attack, particularly at the C(6)-position. The major metabolic pathways for removal of the catecholamine neurotransmitters are O-methylation or side chain oxidation,[9,10] i.e., formation of the open chain o-quinone is not a major route for metabolism of these compounds. Nevertheless, electrochemical studies[8,11–13] reveal that the catecholamine neurotransmitters dopamine and norepinephrine are very easily oxidized at physiological pH. It thus seems likely, therefore, that small amounts of the open chain o-quinone oxidation intermediates of these compounds might be formed in central nervous system tissue. In fact, *in vitro* studies have revealed that the enzymic hydroxylation of dopamine to norepineprhine in the presence of dopamine-β-hydroxylase has a minor pathway leading to dopamine o-quinone.[14] In order to gain some insights into the possible fate of the o-quinone intermediates of dopamine or norepinephrine in central nervous system tissue, Tse and co-workers[15] have used electrochemical techniques to examine the fast chemical reactions of these species under conditions similar to those which probably exist in mammalian brain. According to these workers there are three reaction pathways of the o-quinones which might be important: (1) intracyclization yielding aminochromes (see Figure 3), (2) addition of external nucleophiles kown to be present in high concentrations in brain, and (3) reduction to the original catecholamine by endogenous reductants before reactions 1 or 2 have time to occur. The relative rates of these reactions dictate the fate of the oxidized catecholamine neurotransmitter substance. Using chronoamperometric techniques it was found[8,15] that nucleophilic addition of amino acids to the C(6)-position of dopamine o-quinone, or suitable model compounds, proceeded more slowly than the intracyclization reaction (i.e., equilvalent to III → IV, Figure 3). On the other hand, nucleophilic attack of

Table 1

Cyclization Rate Constants for Oxidized Catecholamines[a,b]

pH	Observed rate constant $(k_{obsd})/s^{-1}$ for				
	Epinephrine	Norepinephrine	α-Methyl norepinephrine	Dopamine	Isoproterenol
3.5	0.025				0.020
4.0	0.098				0.055
4.5	0.27				0.15
5.0	0.99		0.019		
5.5			0.043		
6.0		0.66	0.12		
6.3	19*				
6.5		0.15	0.23		
7.0	20	0.36	0.50		
7.15				0.263^d	
7.4					
pK_a	8.88	8.90	8.85	8.92	8.87
k_2^c	7.4×10^3	5×10^1	8.1×10^1		1.2×10^3

[a] Data obtained by chronoamperometry by Hawley et al.[6] except where marked * in which case the data was obtained by rotating disc electrochemistry.[7]
[b] The reaction of interest corresponds to III → IV, Figure 3.
[c] Calculated as discussed in text assuming K_a of the oxidized catecholamine is the same as the reduced form.
[d] From Sternson et al.[8]

aromatic amines such as aniline proceeded faster than the intracyclization, and attack by sulfhydrylamino acids such as glutathione and cysteine proceeded at rates more than three orders of magnitude greater than the intracyclization reaction (Table 2).

Ascorbate, which is present in quite high concentrations in brain, reduces dopamine o-quinone to dopamine at pH 7.4 at a rate faster than the intracyclization reaction such that it could compete seriously with the nucleophilic attack of thiols, although it could not completely eliminate it.

The electrochemical studies of Sternson et al.[8] and Tse et al.[15] allow several conclusions to be drawn regarding the fate of dopamine (or norepinephrine) o-quinone if it is formed in small amounts in central nervous system tissue via some minor or aberrant oxidative pathway. First, the reduction of the o-quinones by endogenous ascorbic acid is sufficiently rapid to prevent formation of aminochromes (see Figure 3), non-sulfhydrylamino acid adducts, or 6-hydroxylated products (*vide infra*). Indeed, all attempts to identify such compounds in brain tissue have been unsuccessful which supports this first conclusion. Second, the fast reaction of sulfhydryl nucleophiles with dopamine o-quinone cannot be totally eliminated by ascorbic acid reduction. Hence, the nucleophilic addition of glutathione, which is present in brain in quite high concentrations, or protein or membrane sulfhydryl groups should be the predominant reaction of the o-quinones in central nervous system tissue. In fact, it has been suggested[16] that if the sulfhydryl function was part of a nerve terminal membrane structure such interactions could result in neuronal damage and hence lead to symptoms of mental disorders.

Sternson et al.[8] have noted that nucleophilic attack of water on dopamine o-quinone leads to 6-hydroxydopamine (**4**). This reaction is very slow (Table

(4)

2) and may only be observed, using electrochemical techniques, in very acid solution. 6-Hydroxydopamine is an interesting compound in neuropharmacology because it has the ability to selectively damage or destroy norepinephrine nerve terminals when injected into small animals.[17] A number of theories have been advanced to explain this property of 6-hydroxydopamine based on the ease of oxidation of this compound and interactions of the oxidation products with adrenergic nerve terminal components.[17-19]

Electrochemical studies[20] have revealed that indeed 6-hydroxydopamine is very easily oxidized. A cyclic voltammogram of 6-hydroxydopamine at a carbon paste electrode at pH 7 is shown in Figure 4. The evidence leads to the conclusion that peak I_a is a $2e-2H^+$ electrooxidation of 6-hydroxydopamine

Table 2
Pseudo First-order Rate Constants for Reaction
of Dopamine o-Quinone with Various
Nucleophiles at pH 7.4[a]

Nucleophile	k_{obsd}/s^{-1}
None (intracyclization)	0.147^b
Aniline	14.5^b
Glutathione	$208^{b,c}$
Cysteine	$315^{b,c}$
H_2O in 2 M $HClO_4$	0.0001^d

[a] Unless otherwise noted.
[b] Data from Tse *et al.*[15]; normally the catecholamine concentration was 0.1 mM and approximately a five-fold excess of nucleophile was used.
[c] Because of the very fast rates of these reactions the values could be low by as much as 50%.
[d] Data from Sternson *et al.*[8]

[I, Eq. (3)] to a *p*-quinone [II, Eq. (3)] which dissociates at around pH 7 to the zwitterion III [Eq. (3)].

At physiological pH and over conventional cyclic voltammetric time periods no evidence for intracyclization of the *p*-quinone was noted. However, after longer time periods a follow-up reaction was noted. This is illustrated in Figure 5. Figure 5A is a cyclic voltammogram of a freshly electrooxidized solution of 6-hydroxydopamine at 25°C. Peak I_c is due to reduction of the simple *p*-quinone and oxidation peak I_a is the reverse reaction. After 40 minutes at 25°C the cyclic voltammogram shown in Figure 5B was obtained where a new small oxidation peak, peak II_a, may be observed. After 40 minutes at 37°C the latter peak is much larger (Figure 5C). The species responsible for oxidation peak II_a is 5,6-dihydroxyindole. It has been proposed[21] that following the initial electrooxidation of 6-hydroxydopamine (I, Figure 6) to the correspond-

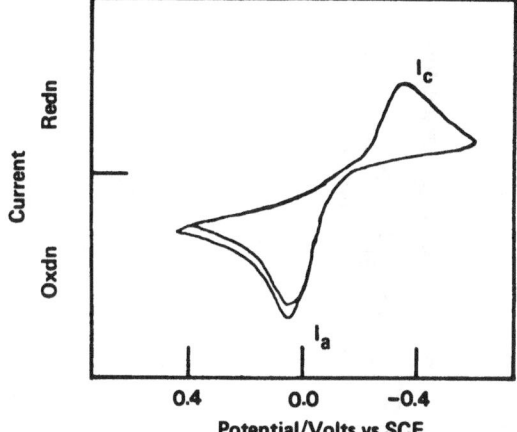

Figure 4. Cyclic voltammogram of 6-hydroxydopamine at pH 7 at a carbon paste electrode at 25°C. (Reprinted with permission of North-Holland Publishing Co., Amsterdam).[20]

ing p-quinone (II, Figure 6), the deprotonated form of the latter species (III, Figure 6) slowly undergoes a 1,2-cyclization reaction to an aminochrome (IV, Figure 6) and hence to 5,6-dihydroxyindole (V, Figure 6). 6-Aminodopamine (VI, Figure 6), which has a similar neurotoxicity to 6-hydroxydopamine,[22]

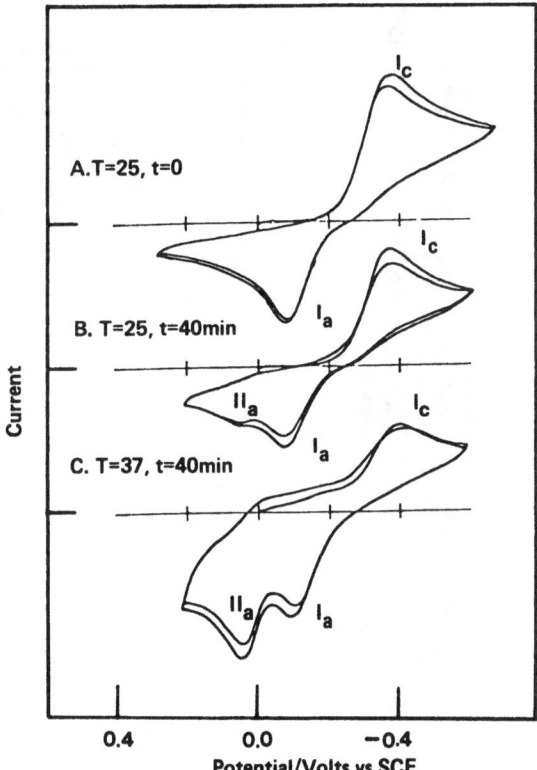

Figure 5. Cyclic voltammograms of electrochemically oxidized *ca.* 1 mM 6-hydroxydopamine at a carbon paste electrode in pH 7.4 McIlvaine buffer: (A) immediately following exhausitive electrooxidation at 25°C; (B) 40 miutes after electrooxidation at 25°C; (C) 40 minutes after electrooxidation at 37°C. (Reprinted with permission of North-Holland Publishing Co., Amsterdam).[20]

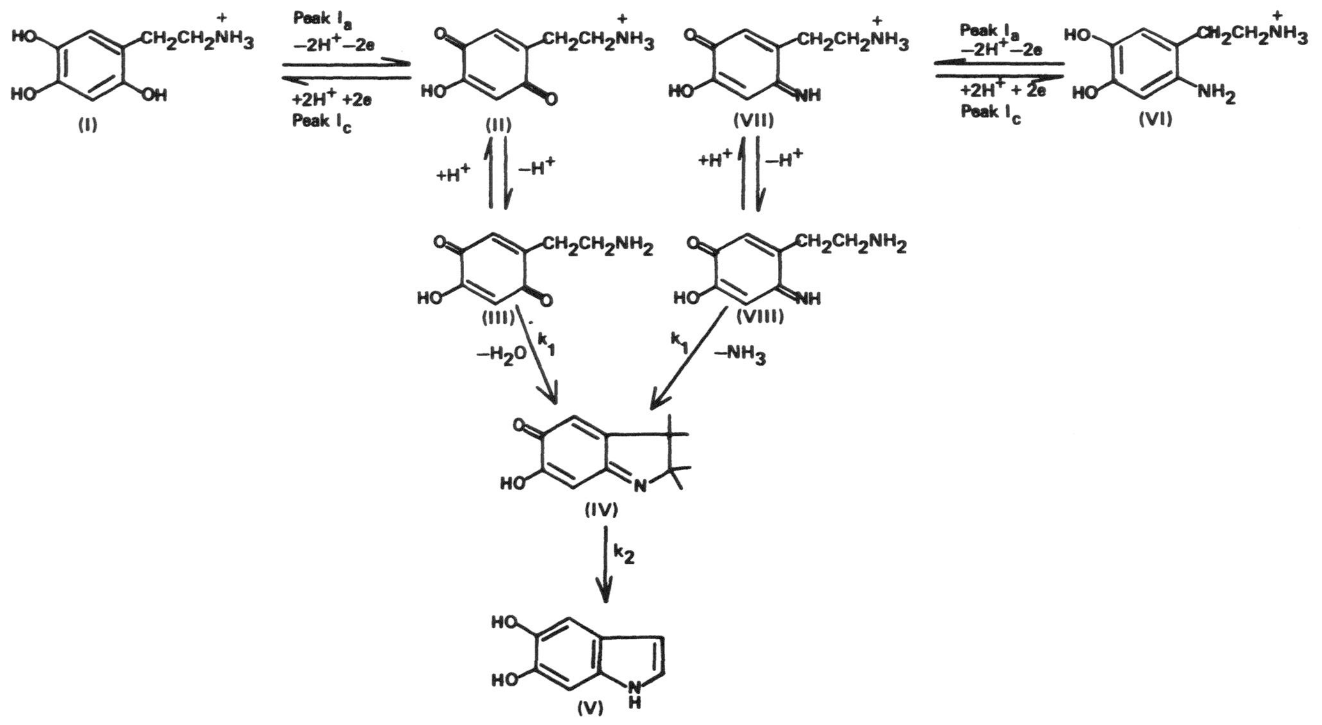

Figure 6. Mechanism proposed by Blank et al.[21] for the electrochemical oxidation of 6-hydroxydopamine and 6-aminodopamine.

undergoes the same basic electrochemical and chemical reactions except that a p-quinoneimine (VII and VIII, Figure 6) is the primary $2e$–$2H^+$ elecrooxidation product.[21]

It has been stated[23] that if 6-hydroxydopamine and 6-aminodopamine are to be selective in their neurotoxic behavior the damaging process must occur intraneuronally following uptake and hence concentration of the neurotoxin within the neuron. *In vivo* electrochemical measurements have shown that about 20% of 6-hydroxydopamine is converted to its p-quinone within a few minutes after injection into rat brain.[24,25] The redox equilibrium between 6-hydroxydopamine and its p-quinone or 6-aminodopamine and its p-quinoneimine is apparently maintained by an active redox buffer, probably ascorbate–dehydroascorbate, which exists in brain tissue.[26] The potential of this redox buffer is apparently close to -0.200 V vs. SCE.

The p-quinone formed on electrooxidation of 6-hydroxydopamine undergoes rapid nucleophilic attack at physiological pH with thiols such as glutathione.[27] This reaction may be readily observed by cyclic voltammetry or by use of chronoamperometric techniques. The product of reaction of 6-hydroxydopamine-p-quinone with glutathione has been isolated and identified as 2,4,5-trihydroxy-6-S-(glutathionyl)phenethylamine (**5**).[28] Liang *et al.*[28] have also found that, following injection of 6-hydroxydopamine into mouse brain, compound **5** may be detected in the brain tissue of the sacrificed animal. Compound **5** is rapidly air and electrochemically oxidized. Using chronoamperometric methods this oxidation has been shown to follow basically the same

$$\text{HOOC}\cdot\text{CH}_2\cdot\text{NH}\cdot\overset{\|}{\underset{\text{O}}{\text{C}}}\cdot\overset{\underset{\text{CH}_2}{|}}{\text{CH}}\cdot\text{NH}\cdot\overset{\|}{\underset{\text{O}}{\text{C}}}\cdot\text{CH}_2\text{CH}_2\overset{\underset{\text{NH}_2}{|}}{\text{CH}}\cdot\text{COOH}$$

(**5**)

reaction pathway as 6-hydroxydopamine, as shown in the simplified scheme in Eq. (4). At pH 7.4 and 37°C, $k_1 = 1.1 \pm 0.04 \times 10^{-3}$ s^{-1} which corresponds to a half life for the p-quinone intermediate [II, Eq. (4)] of about 12 minutes. The intermediate indoline [III, Eq. (4)] is apparently rapidly converted to the final indole [IV, Eq. (4)] and cannot be observed.

Liang and co-workers[27] have proposed that fast chemical interactions of the p-quinone, formed on oxidation of 6-hydroxydopamine, with ubiquitous thiol groups of nerve terminal proteins and membrane constituents could be mainly responsible for the neurotoxicity of 6-hydroxydopamine.

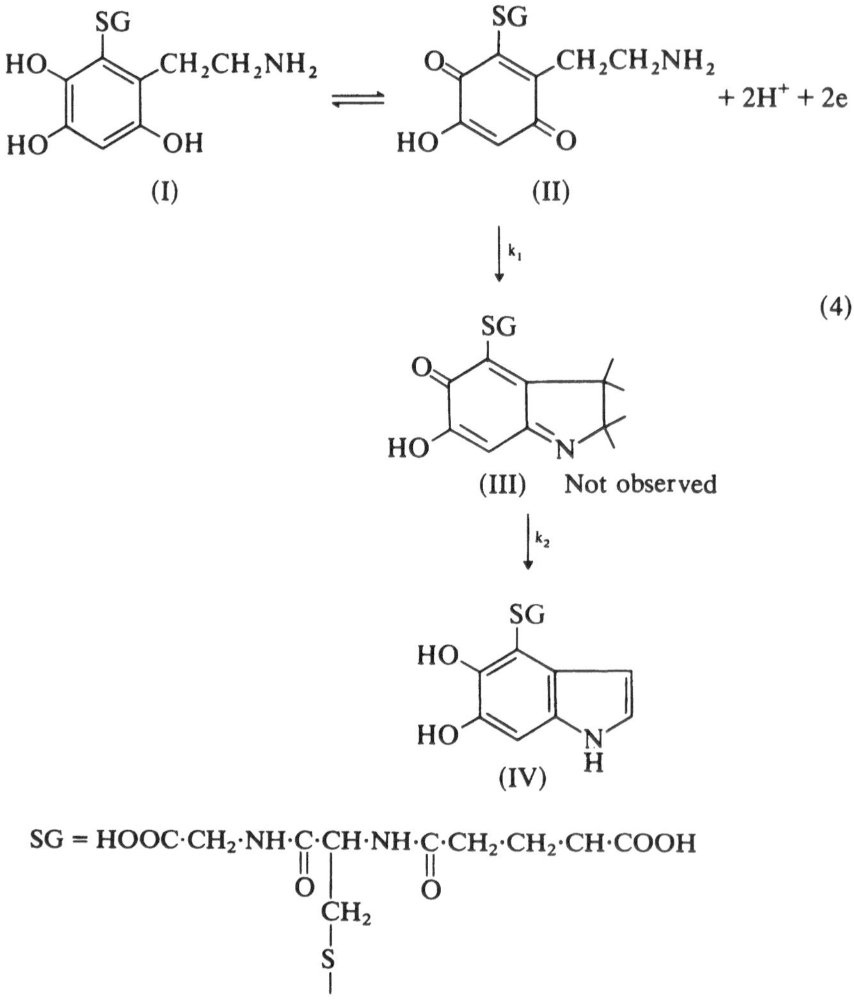

$$+ 2H^+ + 2e \qquad (4)$$

(I)   (II)

(III)   Not observed

(IV)

$$SG = HOOC \cdot CH_2 \cdot NH \cdot \underset{\underset{O}{\|}}{C} \cdot CH \cdot NH \cdot \underset{\underset{O}{\|}}{C} \cdot CH_2 \cdot CH_2 \cdot CH \cdot COOH$$

$$\underset{\underset{\underset{|}{S}}{CH_2}}{}$$

The redox reactions of dopamine, norepinephrine, and 6-hydroxy-dopamine and related compounds discussed above have been elucidated primarily by electrochemical techniques. Such studies clearly provide some real insights into the ways in which these compounds and intermediate species formed upon electrooxidation might be involved in certain types of neurochemical behavior.

Relatively simple electrochemical techniques such as cyclic voltammetry or chronoamperometry provide unequivocal evidence for the formation of *o*-quinone intermediates upon oxidation of the adrenergic neurotransmitters norepinephrine and dopamine. If such intermediates are formed *in vivo*, the electrochemical evidence suggests that any *o*-quinones which escape reduction by ascorbate will react with thiol functions. The consequences of such reactions are highly speculative but if the thiol group is part of a nerve terminal (membrane) structure it would not be surprising if profound neurological effects would result.

The electrochemical studies of 6-hydroxydopamine and 6-aminodopamine reveal that oxidation leads to formation of reactive *p*-quinoid intermediates. The electrochemical studies suggest that reaction of these *p*-quinoid intermediates with nucleophilic groups located on membranes or other protein macromolecules might be related in some way to the degenerative effects of these compounds. It is also worth noting that the redox buffer which exists in central nervous system tissue will maintain both 6-hydroxydopamine and its *p*-quinone in a constant ratio, regardless of their absolute concentrations. Thus, the reactive *p*-quinone species is always available provided that 6-hydroxydopamine is available from some source. 6-Hydroxydopamine is suspected of being the aberrant metabolite of the neurotransmitter catecholamines that causes schizophrenia.[29] According to Senoh and co-workers,[30] 6-hydroxydopamine is an autoxidation product of dopamine and it can form to a significant extent in intact animals.

Using graphite working electrodes, it has been demonstrated that *in vivo* measurements of physiologically important catecholamines and their metabolites, along with 5-hydroxtryptamine (serotonin) and its metabolite 5-hydroxyindole, are possible via their voltammetric oxidation signals. Adams has reviewed these *in vivo* electroanalytical techniques.[5,31–33] An advantage of the *in vivo* electrochemical sensors is that most other neurotransmitter substances are not electrooxidizable. Thus, the implanted sensing microelectrode is blind to acetylcholine, γ-aminobutyric acid, and the other putative amino acid neurotransmitters.

These studies on the electrochemistry of the catecholamine neurotransmitters pioneered by Adams strongly support the contention that not only can biologically relevant mechanistic information be obtained from such studies, but also that uniquely valuable practical analytical techniques can be developed.

3. Phenothiazines

Phenothiazines are not naturally occurring substances but a number of phenothiazine derivatives have useful medicinal applications. For example, phenothiazine and some C-substituted derivatives have anthelmintic properties (i.e., the ability to destroy worms). Various *N*-substituted phenothiazines have important pharmacological properties. For example, promethazine [10-(2-dimethylamino-1-propyl)phenothiazine] possesses antihistaminic activity and chlorpromazine [2-chloro-10-(3-dimethylamino-1-propyl)phenothiazine] has a profound psychotherapeutic activity and is widely used, as are related phenothiazines, for the treatment of various mental illnesses.

No attempt will be made here to present a comprehensive review of the electrochemistry of phenothiazines. Rather the electrochemical behavior of chlorpromazine and some of its metabolites will be discussed, along with the relationship of the observed electrochemistry to the biochemistry of these important drugs.

7,8-Dihydroxychlorpromazine is one of the major metabolites of chlorpromazine in both animals and humans.[34–37] A typical cyclic voltammogram of 7,8-dihydroxychlorpromazine is shown in Figure 7[38] where it is observed that a single voltammetric oxidation peak is formed and, on the reverse sweep, an almost reversible reduction peak. The behavior shown in Figure 7 is typical of that observed between pH 1–8.[38] Additional electrochemical studies[38] reveal that 7,8-dihydroxychlorpromazine [I, Eq. (5)] is electrochemically oxidized in an uncomplicated $2e$–$2H^+$ process giving 7,8-dioxochlorpromazine [II_a, Eq. (5)] which can also exist as a tautomeric quinoneimine spedies [II_b, Eq. (5)].

A precusor of 7,8-dihydroxychlorpromazine in the metabolism of chlorpromazine is 7-hydroxychlorpromazine.[34–37] The latter compound gives two pH-dependent voltammetric oxidation peaks at a carbon paste electrode below ph 2,[39] while at higher pH values a single peak is observed. A cyclic voltammogram of 7-hydroxychlorpromazine in 1 M HCl is shown in Figure 8. Controlled potential electrolysis and coulometry at peak I_a potentials at very low pH gives a purple solution in a $1e$ reaction. Below pH 2 the peak I_a processes are due to the $1e$ electrooxidation of the protonated form of 7-hydroxychlorpromazine (I, Figure 9) to a protonated cation radical (II, Figure 9). The latter species is stable for many hours in strongly acid solution (e.g., 5 M HCl), but in less acidic conditions (i.e., $H_0 = -0.2$ to <2) it disproportionates to the

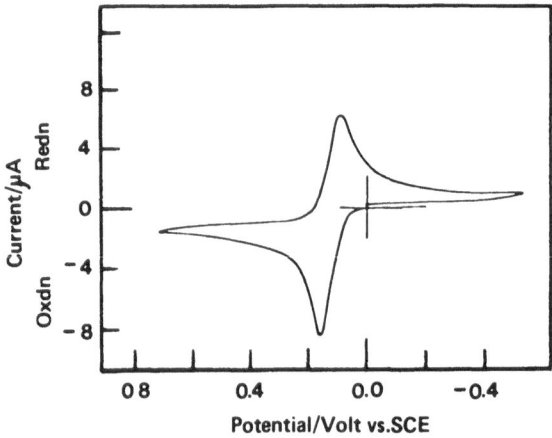

Figure 7. Cyclic voltammogram of 0.5 mM 7,8-dihydroxychlorpromazine in pH 4 McIlvanine buffer at a carbon paste electrode. Sweep rate: 100 mV s^{-1}. (Reprinted with permission of the American Pharmaceutical Association).[38]

The chemical structures with reaction equation (5):

Structure (I): 7-hydroxychlorpromazine with substituent $CH_2CH_2CH_2N(CH_3)_2$

$$\xrightleftharpoons[+2H^+ + 2e]{-2H^+ - 2e}$$

Structure (II_a): quinone form with $CH_2CH_2CH_2N(CH_3)_2$

$$\updownarrow \qquad (5)$$

Structure (II_b): with $CH_2CH_2CH_2N(CH_3)_2$

quinoneimine cation III (Figure 9) and 7-hydroxychlorpromazine (I, Figure 9). The quinoneimine then undergoes a series of reactions generating 7,8-dioxochlorpromazine (IV, Figure 9). These reactions will be discussed later. Peak II_a is thought to correspond to further $1e-1H^+$ electrooxidation of the protonated cation radical (II, Figure 9) to the quinoneimine III (Figure 9) which again generates 7,8-dioxochlorpromazine (IV, Figure 9).

At pH 2, 7-hydroxychlorpromazine exhibits only a single voltammetric oxidation peak.[39] Based on a number of remarkably incisive experiments, Neptune and McCreery[40] deduced that at pH 2 the oxidation of 7-hydroxychlorpromazine (I, Figure 10A) is a $2e$ reaction on a short time scale generating a reactive (half-life $\approx$ 100 ms) quinoneimine cation (II, Figure 10A) which reacts to form a yellow intermediate (λ_{max} = 424 nm) stable for a few tens of seconds. An internal Michael addition then takes place giving 6,9-dihydroxychlorpromazine (IV, Figure 10A). Based on the suspected potential required to oxidize the latter compound, it has been proposed[40] that 6,9-dihydroxychlorpromazine (IV, Figure 10B) is then chemically oxidized by the

Figure 8. Cyclic voltammogram of 0.75 mM 7-hydroxychlorpromazine in 1 M HCl at a carbon paste electrode. Sweep rate: 200 mV s^{-1}. (Reprinted with permission of the American Chemical Society).[39]

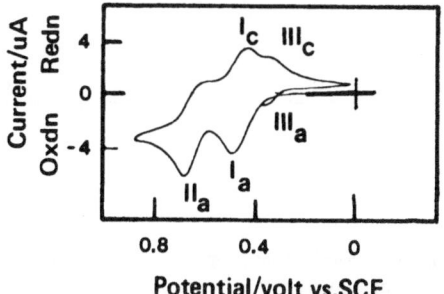

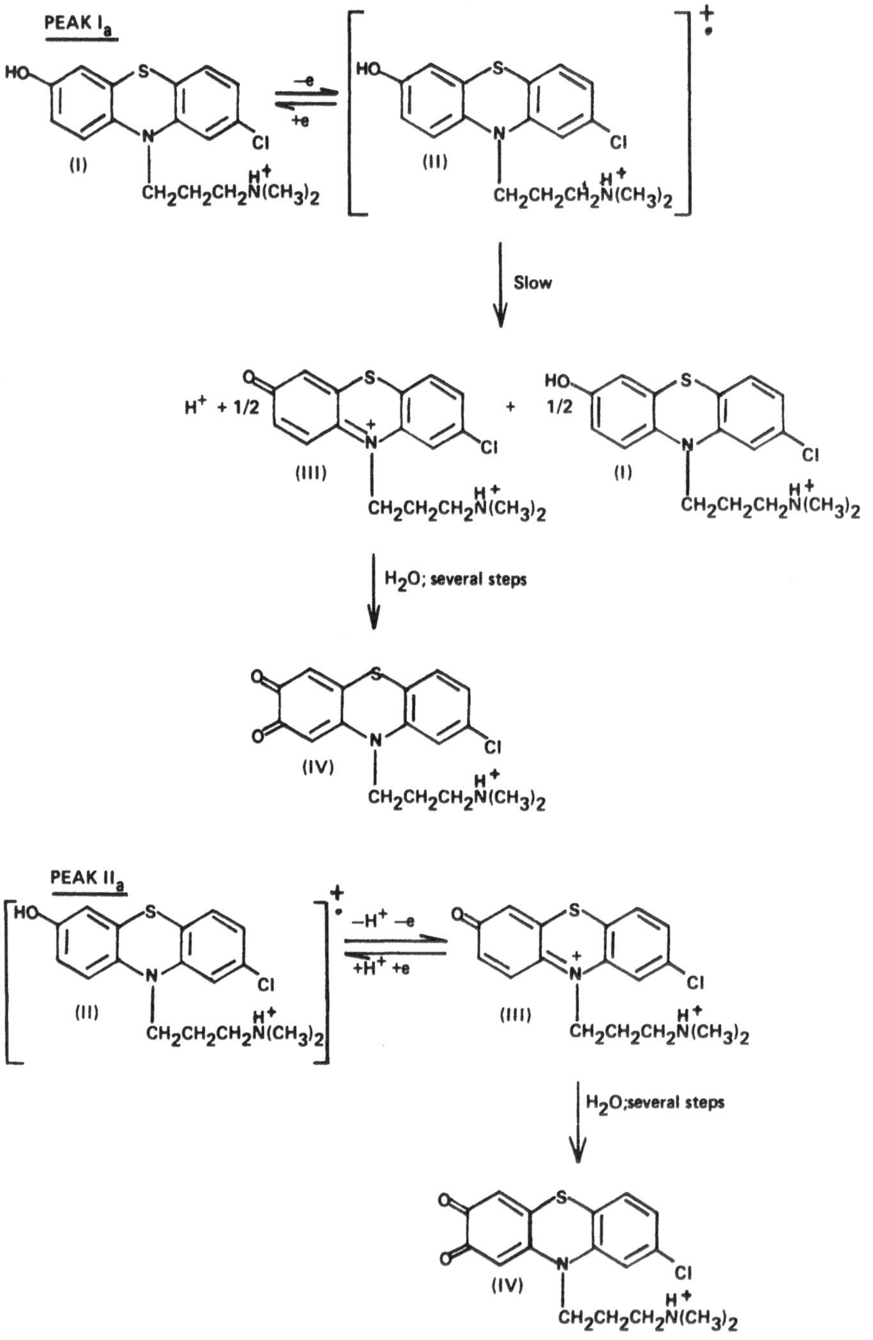

Figure 9. Mechanism proposed[39] for voltammetric oxidation peaks I_a and II_a observed below pH 2 for 7-hydroxychlorpromazine at a carbon paste electrode.

quinoneimine intermediate (II, Figure 10B) giving 7-hydroxychlorpromazine (I, Figure 10B) and 6,9-dioxochlorpromazine (V, Figure 10B). The sequence of reactions shown in Figure 10A and B results in 1 mol of the 1,4-benzoquinone derivative (III, Figure 10A) being converted into 0.5 mol each of 7-hydroxychlorpromazine (I, Figure 10B) and 6,9-dioxochlorpromazine (V, Figure 10B), the former being a stable product. 6,9-Dioxochlorpromazine (V, Fig. 10C), however, is not stable and reacts with water to give 6,8,9-trihydroxychlorpromazine (VI, Figure 10C), followed by a reverse Michael

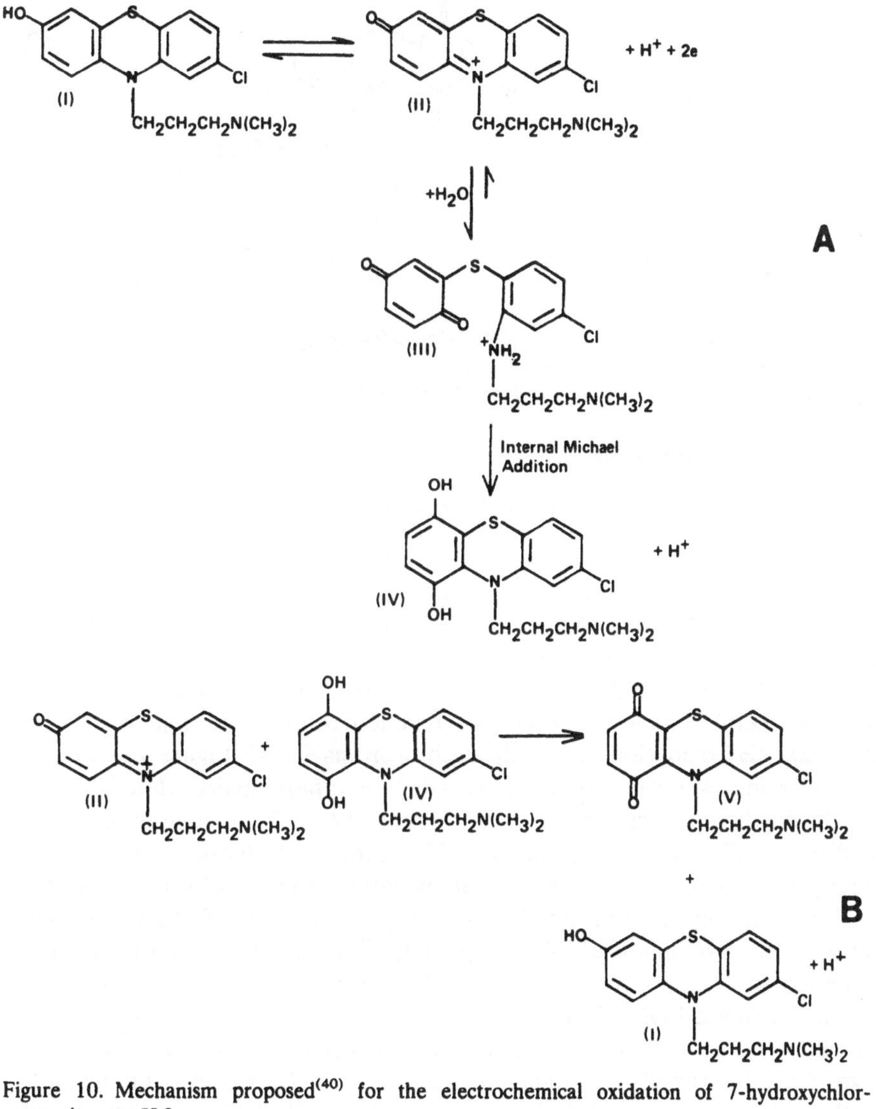

Figure 10. Mechanism proposed[40] for the electrochemical oxidation of 7-hydroxychlorpromazine at pH 2.

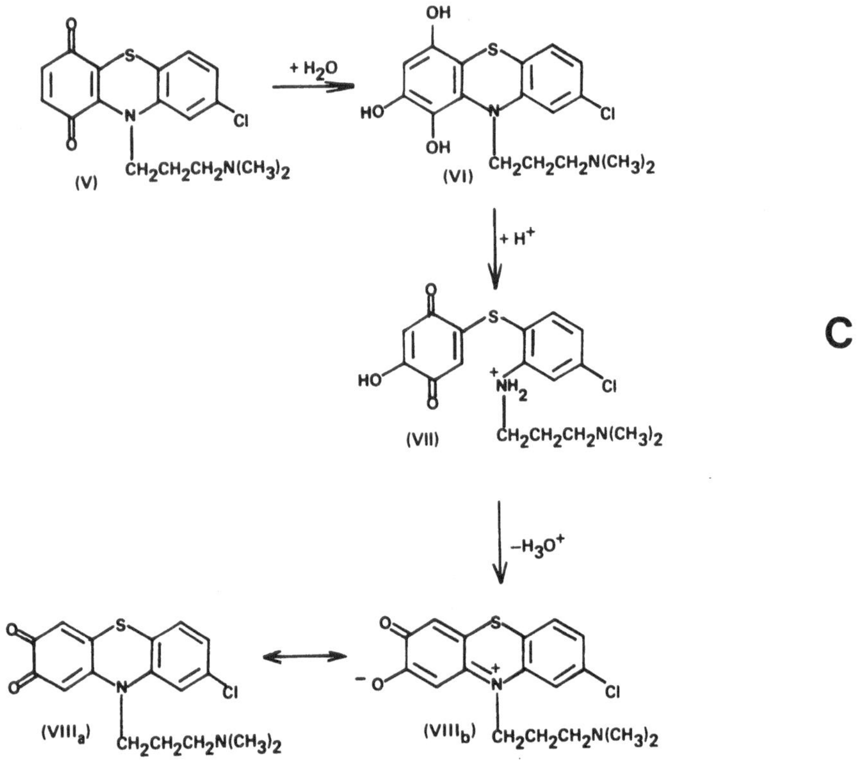

Figure 10 (*continued*)

reaction to form a substituted hydroquinone (VII, Figure 10C) which rearranges to give 7,8-dioxochlorpromazine (VIII_a–VIII_b, Figure 10C).

   The electrochemical oxidation of 7-hydroxychlorpromazine at pH 7 results in the very rapid formation of 7,8-dioxochlorpromazine.[40] Neptune and McCreery[40] have proposed two reaction routes. The first involves an initial $2e-2H^+$ oxidation of 7-hydroxychlorpromazine (I, Figure 11A) to the quinoneimine cation (II, Figure 11A) which is rapidly hydroxylated at pH 7 to 7,8-dihydroxychlorpromazine (III, Figure 11A). This is then immediately oxidized to 7,8-dioxochlorpromazine (IV, Figure 11A). Because colored intermediates could be observed during the oxidation reaction which could not be accounted for by the scheme shown in Figure 11A, an additional reaction pathway must be involved. This is thought[40] to involve an initial $2e-1H^+$ oxidation of 7-hydroxychlorpromazine (I, Figure 11B) to the quinoneimine cation (II, Figure 11B) which rapidly hydrolyzes to a 1,4-benzoquinone intermediate (V, Figure 11B). An internal Michael reaction of this intermediate gives 6,9-dihydroxychlorpromazine (VI, Figure 11B) which is immediately further oxidized to purple 6,9-dioxochlorpromazine (VII, Figure 11B). Several

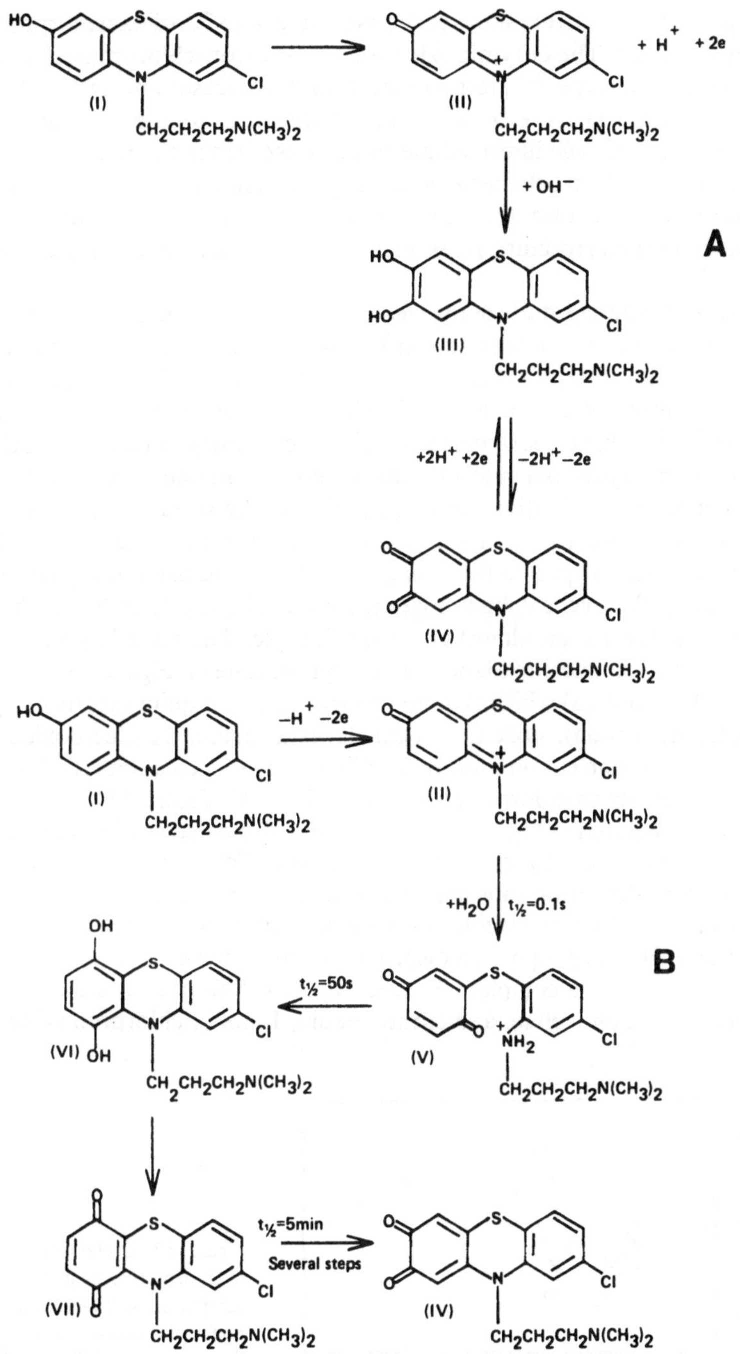

Figure 11. Mechanism proposed[40] for the electrochemical oxidation of 7-hydroxychlor-promazine at pH 7.

subsequent and relatively slow reactions then lead to 7,8-dioxochlorpromazine (IV, Figure 11B). The enzymic oxidation of 7-hydroxychlorpromazine at pH 7 with peroxidase (type VI from horseradish peroxidase) and H_2O_2 also gives 7,8-dioxochlorpromazine as a major product. In addition, under suitable circumstances a purple intermediate can be observed in the reaction.[40] These studies of the electrochemical and enzymic oxidation of 7-hydroxychlorpromazine demonstrate the fundamental importance of the quinoneimine intermediate (i.e., structure III in Figure 9 and structure II in Figures 10 and 11).

The first detailed study of the electrochemical oxidation of chlorpromazine itself was carried out by Merkle and Discher.[41] However, these studies were carried out in $1\,N\,H_2SO_4$ and give relatively little insight into the biological redox chemistry of this compound. More recently, McCreery and his co-workers[38,42–44] have reexamined the electrochemistry of chlorpromazine. At around physiological pH chlorpromazine, under suitable circumstances, gives two voltammetric oxidation peaks. A typical cyclic voltammogram at pH 6.5 in a 2-(N-morpholino)ethanesulfonic acid buffer is shown in Figure 12. In a McIlvaine buffer at pH 6 only a single voltammetric oxidation peak may be observed.[38] The cyclic voltammogram shown in Figure 12 indicates that peak I_a and peak I_c form an almost reversible couple. The peak I_a process is due to a $1e$ electrochemical oxidation of chlorpromazine (I, Figure 13) to a cation radical (II, Figure 13) which is quite reactive so that it quite rapidly disappears (see later discussion). Oxidation peak II_a is a further $1e$ elecrooxidation of the chlorpromazine cation radical (II, Figure 13) to a dication (III, Figure 13) which hydrolyzes to chlorpromazine sulfoxide (IV, Figure 13).

It may be noted in Figure 13 that the chlorpromazine cation radical reacts at physiological pH to give other products. Tozer and co-workers[45,46] observed that the cation radical hydrolyzes in aqueous solution to yield chlor-promazine and chlorpromazine sulfoxide in a roughly second-order reaction. Thus, these authors proposed a disproportionation mechanism although it was based on a rather incomplete kinetic analysis. Subsequent studies[42] have revealed that in phosphate and citrate media, 1 mol of chlorpromazine cation

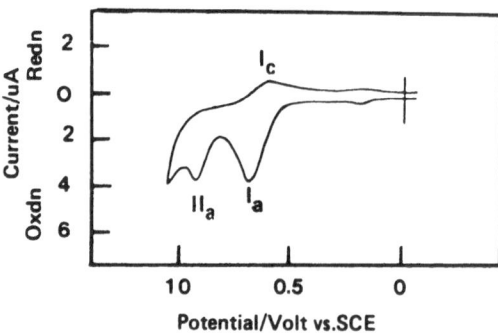

Figure 12. Cyclic voltammogram of chlorpromazine in pH 6.5 2-(N-morpholino)ethanesulfonic acid buffer at a carbon paste electrode. Sweep rate: $100\,mV\,s^{-1}$. (Reprinted with permission of the American Chemical Society).[43]

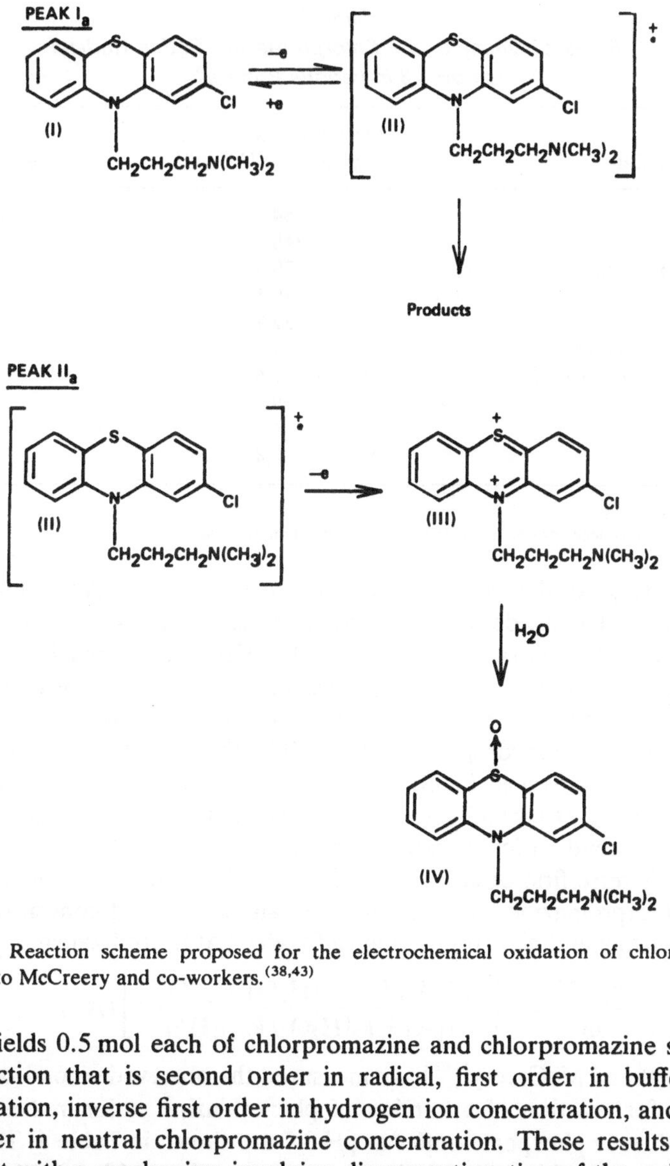

Figure 13. Reaction scheme proposed for the electrochemical oxidation of chlorpromazine according to McCreery and co-workers.[38,43]

radical yields 0.5 mol each of chlorpromazine and chlorpromazine sulfoxide in a reaction that is second order in radical, first order in buffer anion concentration, inverse first order in hydrogen ion concentration, and inverse first order in neutral chlorpromazine concentration. These results are not consistent with a mechanism involving disproportionation of the radical, but indicate a direct reaction of the cation radical with buffer constituents, i.e., nucleophilic attack by a buffer anion. The resulting adduct then reacts to form chlorpromazine sulfoxide or hydroxylated derivatives and the original nucleophile is regenerated. The products and kinetics of the reaction appear to strongly depend on the nucleophile species. The relative rates of reaction of the chlorpromazine cation radical with different buffers are shown in Table 3. These results are shown for pH 3.8 because comparisons at pH 7 were not

Table 3
Rates of Decay of the Chlorpromazine Cation Radical
at pH 3.8 in various Buffers[a,b]

Buffer	Observed second-order rate constant/$M^{-1} s^{-1}$	Relative rate
Citrate	484	1052
Succinate	340	738
Oxidized glutathione	70.4	153
Maleate	40.1	87
Phosphate	32.2	70
Adenosine-5'-triphosphate	27.6	60
Adenosine-5'-monophosphate	21.6	47
Acetate	9.2	20
Glycine	1.4	3
Monochloracetate	0.46	1

[a] From Reference 43.
[b] Chlorpromazine concentration, 1.9 mM; buffer concentration, 0.02 M.

possible owing to the very fast reaction at the latter pH. In sulfhydryl-containing media the decay of the chlorpromazine cation radical could be measured at pH 1 where it was second order. However, at pH 7 the decay of the radical in glutathione-containing solutions was essentially instantaneous.

The observed kinetics for reaction of the chlorpromazine cation radical in the presence of dihydrogen phosphate ($H_2PO_4^-$) may be accounted for by the mechanism shown in Figure 14.[44] The rate law for this mechanism is presented in Eq. (6) which clearly accounts for the observed facts that the reaction is second order in cation radical (A^+), first order in buffer anion ($H_2PO_4^-$), inverse first order in hydrogen ion (H^+), and inverse first order in neutral chlorpromazine concentration (A) and that equal concentrations of chlorpromazine and chlorpromazine sulfoxides (AO) are formed.

$$\frac{d[A^+]}{dt} = -\left[\frac{2K_1(k_2/k_{-2})k_3[H_2O][H_2PO_4^-]}{([A] + k_3[H_2O]/k_{-2})[H^+]}\right](A^+)^2 \qquad (6)$$

Sackett and McCreery[44] have measured the observed second-order rate constants for reaction of a number of phenothiazine cation radicals. Some typical values in phosphate buffer at pH 3.7 are shown in Table 4. The cation radicals shown in this table all give rise to equimolar yields of sulfoxide and parent phenothiazine. However, the kinetics of decay of the cation radicals of chlorpromazine, promazine, triflupromazine, nor$_1$ chlorpromazine, and nor$_2$ chlorpromazine are qualitatively the same whereas in the case of the cation radicals of methoxypromazine and acepromazine the decays are qualitively different. The cation radicals generated from promethazine (**6**), fluphenazine (**7**), perphenazine (**8**), and chlorphenethazine (**9**) did not give rise to equal yields of the sulfoxide and parent phenothiazine, but give lower yields of the

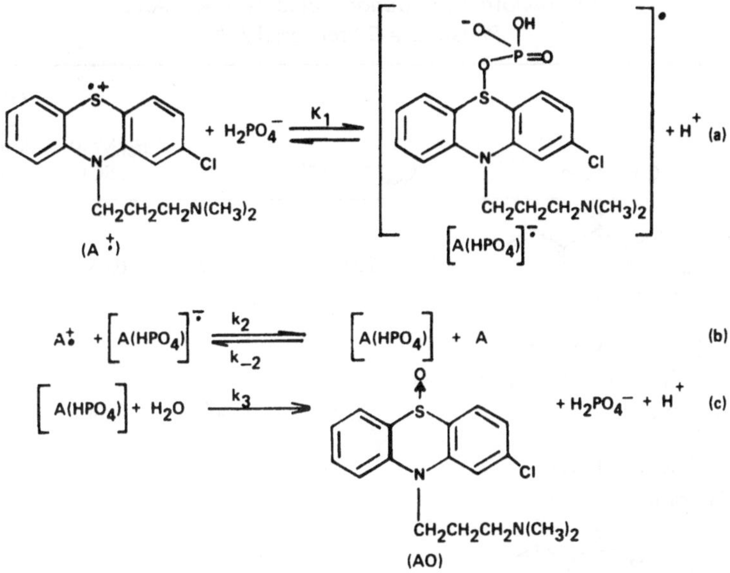

Figure 14. Reaction mechanism proposed to occur between dihydrogen phosphate and chlor-promazine cation radical.[44]

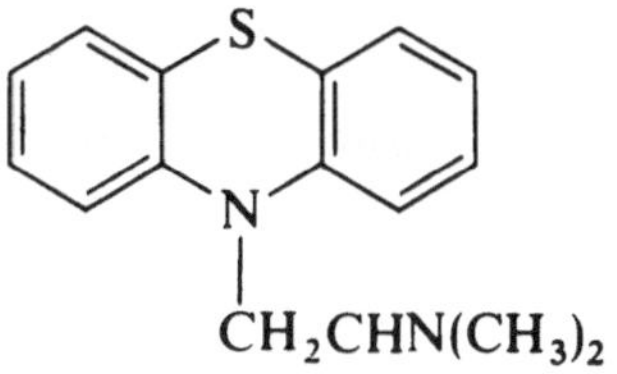

sulfoxide and a variety of unidentified products, possibly ring hydroxylated compounds.[44] It may be concluded, therefore, that an unbranched aminopropyl chain is necessary to produce a 50:50 mixture of neutral phenothiazine derivative and sulfoxide. A shorter carbon chain or piperazine ring contained in the side chains leads to the formation of different products.

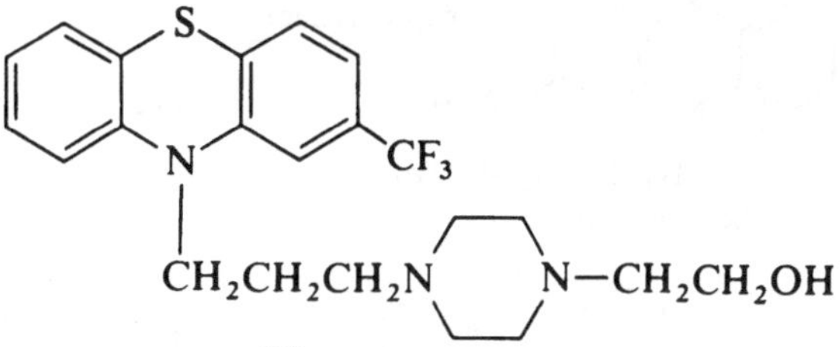

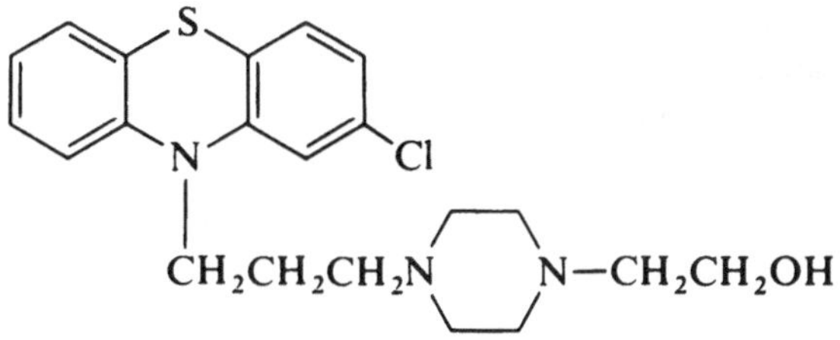

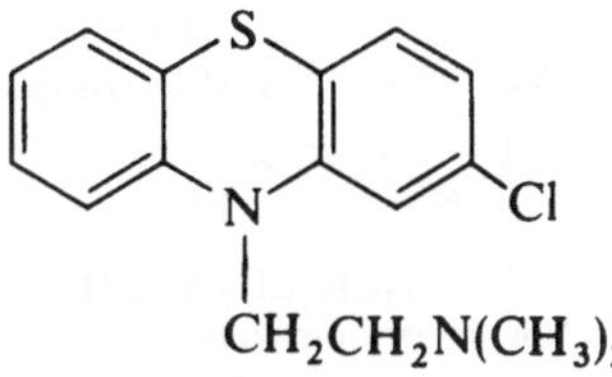

Table 4
Observed Second-order Rate Constants for Decomposition
of Phenothiazine Cation Radicals in Aqueous
Phosphate Buffer, pH 3.7[a,b]

Compound	$k_{obs}/M^{-1} s^{-1}$	σ_p	$E_p/$Volt[c] vs. SCE	Average[d] clinical dose/μmol kg^{-1}
(phenothiazine cation radical, 2-Cl, N–CH$_2$CH$_2$CH$_2$N(CH$_3$)$_2$) (Chlorpromazine)	22.9	+0.227	0.68	12.0
(phenothiazine cation radical, N–CH$_2$CH$_2$CH$_2$NH$_2$) (Promazine)	0.89	0.00	0.58	33.0
(phenothiazine cation radical, 2-CF$_3$, N–CH$_2$CH$_2$CH$_2$N(CH$_3$)$_2$) (Triflupromazine)	70.5	+0.540	0.75	4.6
(phenothiazine cation radical, 2-COCH$_3$, N–CH$_2$CH$_2$CH$_2$N(CH$_3$)$_2$) (Acepromazine)	57.4	+0.502	0.67	5.0
(phenothiazine cation radical, 2-OCH$_3$, N–CH$_2$CH$_2$CH$_2$N(CH$_3$)$_2$) (Methoxypromazine)	14.7	−0.268	0.55	30.1
(phenothiazine cation radical, 2-Cl, N–CH$_2$CH$_2$CH$_2$NHCH$_3$) (Nor$_1$ chlorpromazine)	6.7			
(phenothiazine cation radical, 2-Cl, N–CH$_2$CH$_2$CH$_2$NH$_2$) (Nor$_2$ chlorpromazine)	9.0			

[a] Data from Reference 44.
[b] 0.08 M H$_2$PO$_4^-$.
[c] First voltammetric oxidation peak potential observed at a carbon paste electrode.
[d] From References 47 and 48.

The data shown in Table 4 summarize not only kinetic results but also voltammetric oxidation peak potentials, Hammett sigma values, and clinical doses. As might be expected, the rate of the cation radical reaction for the first three compounds in Table 4 increases with increase in the electron-withdrawing properties of the C(2)-position substituent as indicated by the σ_p values. The voltammetric peak potentials corresponding to oxidation of the parent compound to the cation radical follow the trend of the Hammett σ_p values.

The electrochemical and kinetic work of Sackett and McCreery[44] clearly indicates that the products of phenothiazine cation radical decay are dependent on the structure of the radical, particulrly the N(10)-side chain substituents. There have been several reports that the cation radical is an intermediate in phenothiazine metabolism[49,50-52] but the results of Sackett and McCreery[44] give the first definitive insights into the mechanism and structural factors which might influence the formation of metabolites from the radical. It has also been established that the mechanism and rates of reaction of phenothiazine drug cation radicals depend strongly on the identity of the nucleophile which attacks the radical. In addition, it is noteworthy that, for those phenothiazine drugs which have a common mechanism (chlorpromazine, promazine, and tri-flupromazine), the compounds giving the more reactive radical have lower clinical doses (Table 4). Indeed, it has been proposed[44] that a cation radical–nucleophile interaction could be the molecular form of receptor binding for these drugs. Only chlorpromazine, promazine, and triflupromazine cation radicals can be compared directly because they react by the same mechanism to give the same type of products (see Figure 14). The other compounds listed in Table 4 react by somewhat different mechanisms. These results, however, do implicate the rate of radical reactions with nucleophiles as being important to clinical activity, rather than overall radical stability.

As was mentioned earlier, major metabolites of chlorpromazine are 7-hydroxychlorpromazine and 7,8-dihydroxychlorpromazine.[34-37] It is also known that 7-hydroxychlorpromazine is comparable to chlorpromazine as a psychoactive drug.[53,54] Indeed, the blood levels of 7-hydroxychlorpromazine, but not chlorpromazine, have been correlated with the improvement of psychotic symptoms of mentally ill patients.[55,56] In addition, both 7-hydroxychlorpromazine and 7,8-dihydroxychlorpromazine have been linked with some of the observed side effects of long term chlorpromazine treatment such as corneal opacity and skin pigmentation.[57-59] The electrochemical studies of Neptune and McCreery[40] on 7-hydroxychlorpromazine (see Figures 10 and 11), however, make it possible to make reasonable suggestions regarding the pharmacology of this compound. Thus, the electrochemical oxidation of 7-hydroxychlorpromazine leads, in part, to cleavage of the phenothiazine ring system giving a 1,4-benzoquinone intermediate (V, Figure 11). Although there are at least 77 metabolites of chlorpromazine in humans, the majority of these remain unidentified.[37] However, it is known that many of these metabolites are not phenothiazine derivatives and could, therefore, readily arise from the ring-opened 1,4-benzoquinone intermediate. It is also observed that at least two intermediates in the oxidation of 7-hydroxychlorpromazine are quinones and it has therefore been suggested[40] that such intermediates could be involved in the formation of "pseudo melanin" and hence the skin pigmentation effect noted with chronic chlorpromazine patients.

The electrochemical (and chemical and enzymic) oxidation of 7-hydroxychlorpromazine leads to the eventual formation of 7,8-dioxochlorpromazine which, according to the mechanism proposed by Neptune and McCreery,[40] proceeds through a trihydroxylated intermediate (VI, Figure 10). It has been suggested[40] that the oxidized form of the latter compound (i.e., VII, Figure 10) could react with neuromembrane proteins, particularly at thiol sites, in the same way described for the p-quinone of 6-hydroxydopamine (see p. 143) and hence cause some neurophysiological effect.

It has been noted in earlier discussion (Figures 10 and 11) that quinone-imine intermediates are formed on electrochemial oxidation of hydroxylated chlorpromazine derivatives. Neptune and McCreery[39] have concluded that since hydroxylated chlorpromazine metabolites are important in the beneficial and side effects of chlorpromazine then intermediates such as quinoneimines may be important *in vivo*.

It is also of interest to note that tyrosine hydroxylase catalyzes the hydroxylation of 7-hydroxychlorpromazine to 7,8-dihydroxychlorpromazine and then to 7,8-dioxochlorpromazine.[60] Under similar conditions the electrochemical oxidation of 7-hydroxychlorpromazine progresses through a similar reaction seequence (see Figure 11A) to the same product observed enzymically. A comparison of the information obtained from the enzymic and electrochemical oxidations strongly supports the conclusion that electrochemical techniques yield considerably more mechanistic information and allow the ready detection of transient intermediate species. The fact that the same products are formed electrochemically and enzymatically lends support to the view that the electrochemically generated intermediates may be real possibilities for involvement in the *in vivo* metabolic processes.

Investigations into the electrochemistry of chlorpromazine and the behavior of the cation radical of this drug[43,44] also make it possible to draw some conclusions regarding the involvement of this radical in the pharmacology of chlorpromazine. Thus, depending on the type of nucleophiles present, the chlorpromazine cation radical in neutral aqueous solution may have a lifetime ranging from a few milliseconds to several minutes. However, this range of lifetimes is such that it can survive long enough to interact with the physiological environment. Kinetic and mechanistic studies indicate that the cation radical forms a covalent bond with a variety of physiologically occurring nucleophiles (e.g., phosphate, sulfhydryl, and carboxyl groups, see Table 3). The cation radical nucleophile adduct (e.g., see structure $[A(HPO_4)]^{\tau}$ in Figure 14) eventually breaks down to the nucleophile in its original form. Thus, if the reaction observed *in vitro* resembles a drug/receptor site interaction, the interaction would be strong but not irreversible. The mechanisms proposed by McCreery *et al.*[43,44] suggest that the covalent bond formed between the chlorpromazine cation radical and a nucleophile is stable until attacked by another radical or another solution component. In addition, if the chlorpromazine cation radical is formed *in vivo*, interaction with surrounding nucleophiles is much more likely than a reaction with water or another molecule of chlorpromazine. The work of McCreery *et al.*[43,44] does not establish that the cation radical of chlorpromazine is an active pharmacological entity. However, this work clearly indicates that the redox properties of chlorpromazine are such that the cation radical is likely to be generated *in vivo*, and gives valuable insights regarding the probable routes of its degradation. The finding[44] that the more clinically active antipsychotic phenothiazine drugs give rise to the most reactive cation radicals also supports the hypothesis that the interaction of the phenothiazine cation radical with nucleophiles associated with nerve terminal receptor sites may be related to receptor binding.

It is, however, quite clear from studies of the electrochemical and other oxidations of chlorpromazine and some of its hydroxylated metabolites that electrochemical techniques provide powerful tools to generate and detect transient intermediates and products and provide valuable insights into reaction mechanisms. It has yet to be demonstrated whether the electrode and related chemical mechanisms will prove to be the same as or similar to *in vivo* processes. Nevertheless, the present evidence supports the conclusion that there is a very likely correlation between the the biological and electrochemical processes.

4. α-Tocopherol and α-Tocopherylquinone

α-Tocopherylquinone (**10**) and α-tocopherol (**11**) are the oxidized and reduced form, respectively, of vitamin E although the vitamin is now generally considered to be the tocopherol (**11**) form. Replacement of ring-substituted methyl groups by hydrogens leads to other tocopherols or tocopherylquinones (i.e., β, γ, δ). The tocopherol structure **11** is often referred to as a chromanol.

The chemistry and biochemistry of the tocopherylquinones and tocopherols has been recently reviewed in some detail.[61] The first recognized function of vitamin E was as an antisterility factor for the laboratory rat.[62] However, it is now known to have multiple functions *in vivo*. For example, it is required for the maintenance of structural and functional integrity of skeletal, cardiac, and smooth muscle and, in some animals, the peripheral vascular system. Tocopherols also function as intracellular antioxidants, particularly with respect to the stabilization of ingested fats and perhaps products arising in the metabolic synthesis and degradation of lipids.[61] It has also been proposed[63] that the tocopherylquinones may have some functional role in electron transport systems.

$$\text{(Structure 10: quinone ring with } R_5, R_6, R_3 \text{ substituents and side chain)}$$

CH$_2$·CH$_2$·C·CH$_2$ [CH$_2$·CH$_2$·CH·CH$_2$]$_3$ H

(with OH and CH$_3$ on central carbon; CH$_3$ on repeating unit)

(10)   where $R_3 = R_5 = R_6 = CH_3$

$$\text{(Structure 11: chroman ring with } R_8, R_7, R_5 \text{ substituents)}$$

CH$_3$ CH$_3$ CH$_3$ CH$_3$

(CH$_2$)$_3$·CH(CH$_2$)$_3$· CH(CH)$_2$)$_3$·CH·CH$_3$

(11)   where $R_5 = R_7 = R_8 = CH_3$

The relationship between the electrochemistry of tocopherols and toco-pherylquinones and their biological redox reactions is not as obvious as with the other systems discussed in this chapter. However, the electrochemistry serves to show the complexity of their redox behavior and to point out some important points in the redox mechanisms of quinones.

The first reports dealing with the electrochemical behavior of vitamin E-type compounds are those of Smith and co-workers[64,65] who studied the polarographic oxidation waves of α-tocopherol and a number of tocopherol model compounds in methanol–water (1:1, v/v). These workers concluded that the electrooxidation of tocopherols consisted of an initial, reversible $2e$–$2H^+$ reaction of the tocopherol [(I, Eq. (7)] to give an unstable quinonoid intermediate [(II, Eq. (7)] which was rapidly transformed into the corresponding tocopherylquinone [(III, Eq. (7)]. These first reports pointed out the possibility of an unstable species intermediate between tocopherol and toco-pherylquinone in the oxidation reaction. Controversy still remains regarding the nature of this putative intermediate species. Among the proposed alternative structures for this intermediate are α-tocopheroxide (**12$_a$** or **12$_b$**)[66] and 9-hydroxy-α-tocopherone (**13**).[67,68] The latter compound found rather general acceptance as the partner of α-tocopherol in the primary reversible oxidation–reduction reaction.[69] A more recent electrochemical investigation by Marcus and Hawley[70] has allowed a more detailed understanding of the oxidation of tocopherol. Using acetonitrile as solvent, a typical cyclic voltammogram of α-tocopherol is shown in Figure 15. Electrochemical studies reveal that the peak I$_a$ process is a $2e$–$1H^+$ electrooxidation of α-tocopherol (I, Figure 16). The product of this reaction has been proposed[70] to be a carbonium ion (II, Figure 16). Cyclic voltammetric peak I$_c$ (Figure 15) is the reverse reaction. The putative carbonium ion was clearly short-lived because controlled potential electrooxidation of α-tocopherol at peak I_a potentials gives α-tocopherylquinone (IV, Figure 16) in quantitative yield. As might be expected, the carbonium ion is very unstable in the presence of nucleophiles such as water and cyclic voltammetric reduction peak I_c decreases in height. Marcus and Hawley[70] have demonstrated that 9-hydroxy-α-tocopherone (III, Figure 16) gives rise to no voltammetric oxidation or reduction peaks

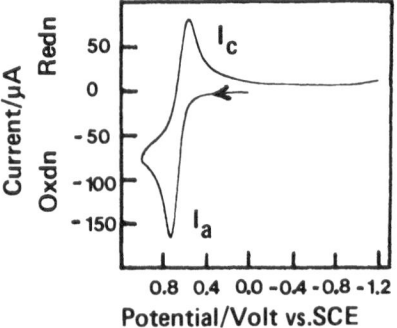

Figure 15. Cyclic voltammogram at a platinum electrode of 1.1 mM α-tocopherol in acetonitrile containing 0.1 M tetraethylammonium perchlorate. Sweep rate: 83.5 mV s^{-1}. (Reprinted with permission of Elsevier Publishers, Amsterdam).[70]

between +1.0 and −2.0 V vs. SCE. Thus, if reduction of 9-hydroxy-α-tocopherone is to occur, and it can be reduced to α-tocopherol with chemical reductants,[66–68,71] then formation of the electroactive carbonium ion must occur. The failure to observe a reduction peak for the carbonium ion (II, Figure 16) with solutions of 9-hydroxy-α-tocopherone (III, Figure 16) in acetonitrile is thought[70] to indicate that not only is the equilibrium constant

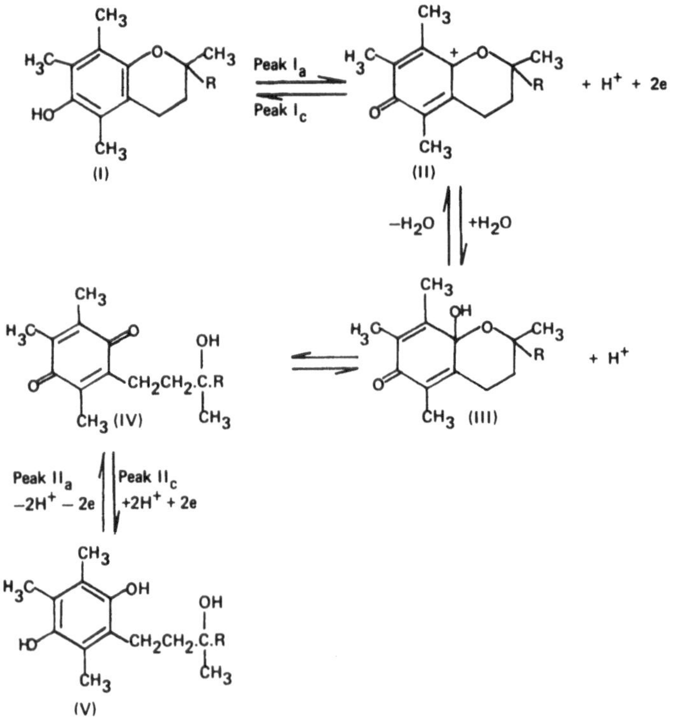

Figure 16. Mechanism proposed for the electrochemical oxidation of α-tocopherol in acetonitrile.[70] When R=C$_{16}$H$_{33}$, I is α-tocopherol; when R=CH$_3$, I is 2,2,5,7,8-pentamethyl-6-hydroxychroman.

for the hydration reaction (II → III, Figure 16) large, but also that the rate of the dehydration reaction (III → II, Figure 16) is slow. The rate of hydration of the putative carbonium ion (II → III, Figure 16) in aqueous acetonitrile has been measured by a double potential step chronoamperometry method using the model compound 2,2,5,7,8-pentamethyl-6-hydroxychroman (**14**) because of the very low solubility of α-tocopherol. The second-order rate constant was found to be 2.5 $M^{-1}\,s^{-1}$.

$$
\begin{array}{c}
\text{CH}_3 \\
\text{H}_3\text{C} \underset{\text{HO}}{\overset{\text{O}}{\bigcirc}} \begin{array}{c}\text{CH}_3\\\text{CH}_3\end{array} \\
\text{CH}_3
\end{array}
$$

(14)

The cyclic voltammetric behavior of the α-tocopherol model compound **14** in aqueous acetonitrile containing perchloric acid is shown in Figure 17. Again oxidation peak I_a is the $2e$–$1H^+$ oxidation of the chromanol to the carbonium ion (I → II, Figure 16). No peak I_c corresponding to electrochemical reduction of the carbonium ion may be observed on the reverse sweep. However, a peak at more negative potentials may be observed (peak II_c, Figure 17) and on the next positive-going sweep a new oxidation peak (peak II_a, Figure 17) is formed. Under such acid conditions apparently what happens is that a rapid ring-opening of 9-hydroxy-2,2,5,7,8-pentamethylchromanone (III, Figure 16) occurs giving the tocopherylquinone species (IV, Figure 16). Peak II_c is the $2e$–$2H^+$ reduction of the latter quinone to the corresponding hydroquinone (V, Figure 16) while oxidation peak II_a is the reverse reaction.

Marcus and Hawley[72] used a chronoamperometric method to study the kinetics and mechanism of the ring opening of 7-hydroxy-2,2,5,7,8-pentamethyl-6-chromanone (III → IV, Figure 16), a model of the hemiketal intermediate in the oxidation of α-tocopherol to α-tocopherylquinone. In

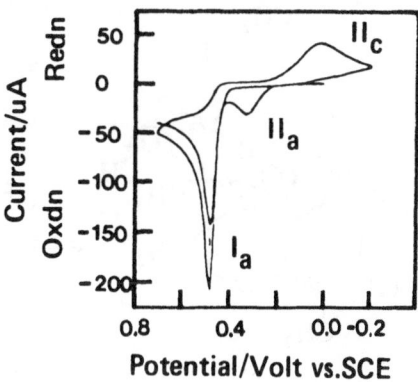

Figure 17. Cyclic voltammogram at a platinum electrode of 0.95 mM 2,2,5,7,8-pentamethyl-6-hydroxychroman in water–acetonitrile (75:25, v/v) containing 0.87 M perchloric acid. Sweep rate: 167 mV s^{-1}. (Reprinted with permission of Elsevier Publishers, Amsterdam).[70]

water–acetonitrile solution in the presence of strong acids the O(1)-oxygen of 9-hydroxy-2,2,5,7,8-pentamethyl-6-chromanone is protonated to form the corresponding oxonium salt [I, Eq. (8)]. Removal of the proton from the hydroxy group in the C(9)-position by the solvent then results in formation of the corresponding quinone [II, Eq. (8)]. The system also exhibits both

general acid and general base catalysis. The mechanism proposed for general acid catalysis involves proton transfer from the acid [HA in Eq. (9)] to the O(1)-position followed by removal of a proton from the hydroxyl group at the C(9)-position by the solvent, water [Eq. (9)]. In the case of general base catalysis, the reaction proceeds by removal of the proton from the hydroxy group at C(9) by the base [A$^-$ in Eq. (10)] and transfer of a proton from the solvent to the O(1) position [Eq. (10)].

Marcus and Hawley[73] have also studied the electrochemical reduction of α-tocopherylquinone and 2,3,5-trimethyl-6-(3'-methyl-3'-hydroxybutyl)quinone (2,3,5-TMHQ, **15**) in nonaqueous solvents. The electrochemical

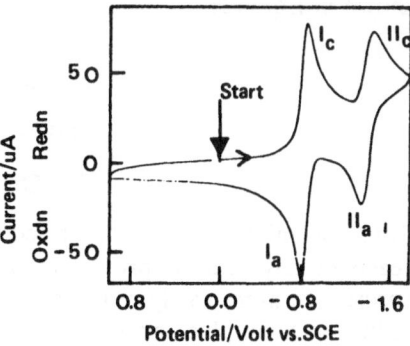

(**15**)

behavior of 2,3,5-THMQ is essentially identical to that of α-tocopherylquinone but is not complicated by adsorption effects associated with the large alkyl side chain of the latter species. Thus, mechanistic studies were carried out on 2,3,5-THMQ.

A cyclic voltammogram of 2,3,5-TMHQ in acetonitrile is shown in Figure 18. Peak I_c is a reversible $1e$ reduction of α-tocopherylquinone or its model quinone (Q) to an anion radical (Q$^{\overline{\cdot}}$) [Eq. 11a)] and peak II_c the further reversible $1e$ reduction of the anion radical to a dianion [Q^{-2}, Eq. (11b)].

$$Q + e \underset{\text{Peak } I_a}{\overset{\text{Peak } I_c}{\rightleftharpoons}} Q^{\overline{\cdot}} \qquad (11a)$$

$$Q^{\overline{\cdot}} + e \underset{\text{Peak } II_a}{\overset{\text{Peak } II_c}{\rightleftharpoons}} Q^{-2} \qquad (11b)$$

This behavior is typical of simple quinones in aprotic solution.[74] However, the electrochemical behavior of 2,3,5-TMHQ and α-tocopherylquinone in acetonitrile is altered considerably by the addition of a weak acid such as ethyl malonate (Figure 19). Thus, the peak I_c process remains unchanged but reduction peak II_c broadens and shifts towards more positive potentials. In addition peak II_a, corresponding to electrooxidation of the quinone dianion

Figure 18. Cyclic voltammorgram at a platinum electrode of 1.45 mM 2,3,5-trimethyl-6-(3'-methyl-3'-hydroxybutyl)-quinone in acetonitrile containing 0.1 M tetraethylammonium perchlorate. Sweep rate: 83.5 mV s^{-1}. (Reprinted with permission of Elsevier Publishers, Amsterdam).[73]

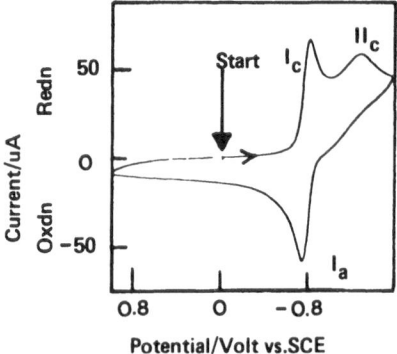

Figure 19. Cyclic voltammogram at a platinum electrode of 1.4 mM 2,3,5-trimethyl-6-(3'-methyl-3'-hydroxybutyl)quinone and 2.8 mM ethyl malonate in acetonitrile containing 0.1 M tetraethyl-ammonium perchlorate as supporting electrolyte. Sweep rate: 83.5 mV s⁻¹. (Reprinted with permission of Elsevier Publishers, Amsterdam).[73]

(Q^{-2}) to the anion radical (Q^{-}), is absent. The probable reduction mechanism involved is shown in Eq. (12), where HA denotes the proton source.

$$Q + e \overset{\text{Peak I}_c}{\underset{\text{Peak I}_a}{\rightleftharpoons}} Q^{-} \tag{12a}$$

$$Q^{-} + e \rightleftharpoons Q^{-2} \tag{12b}$$

$$Q^{-2} + HA \rightleftharpoons QH^{-} + A^{-} \tag{12c}$$

$$QH^{-} + Q \rightleftharpoons 2Q^{-} + H^{+} \tag{12d}$$

The protonation of the dianion, which is fast compared to the rate of potential sweep, causes the reduction peak for the anion radical (peak II$_c$, Figure 19) to shift to more positive potentials and the electrooxidation peak for the dianion (peak II$_a$) to disappear. Since no peak is observed for electrooxidation of the monoprotonated dianion (QH^{-}), the latter species must react and disappear, probably by reducing quinone species diffusing towards the electrode surface to give the quinone anion radical [Eq. (12d)].

In the presence of a stronger proton donor such as benzenethiol a cyclic voltammogram of 2,3,5-TMHQ or α-tocopherylquinone shows some profound changes (Figure 20). Thus, peak I$_c$ is moved to more positive potentials and peak II$_c$ and I$_a$ disappear. A scheme consistent with these observations is shown in Eq. (13).

$$Q + e \rightleftharpoons Q^{-} \tag{13a}$$

$$Q^{-} + HA \rightleftharpoons QH\cdot + A^{-} \tag{13b}$$

$$QH\cdot + e \rightleftharpoons QH^{-} \tag{13c}$$

$$Q^{-} + QH\cdot \rightleftharpoons Q + QH^{-} \tag{13d}$$

Thus, the anion radical, Q^{-}, formed by the initial $1e$ reduction of the quinone reacts rapidly with the proton donor (HA) to give neutral radical QH·. Such species are more easily reduced than the parent quinone[75] so a further $1e$ reduction takes place giving QH⁻. Thus, the peak I$_c$ process in Figure 20 is a

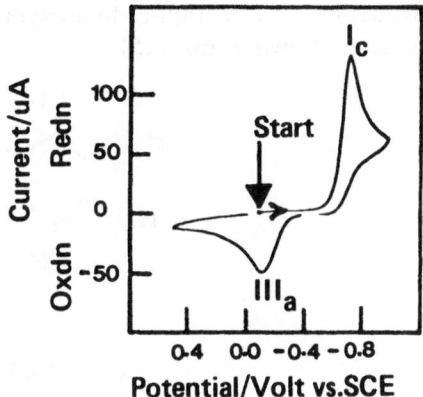

Figure 20. Cyclic voltammogram at a platinum electrode of 1.47 mM 2,3,5-trimethyl-6-(3'-methyl-3'-hydroxybutyl)-quinone and 29 mM benzenethiol in acetonitrile containing 0.1 M tetraethylammonium perchlorate as supporting electrolyte. Sweep rate: 83.5 mV s^{-1}. (Reprinted with permission of Elsevier Publishers, Amsterdam).[73]

$2e$ reaction. Oxidation peak III$_a$ in Figure 20 is a $2e$ process corresponding to electrooxidation of the hydroquinone monoanion (QH$^-$) formed in the peak I$_c$ reaction.

In the presence of a very strong proton donor such as H$_2$SO$_4$, 2,3,5-TMHQ gives a new reduction peak III$_c$ (Figure 21). The electrode reaction responsible for this peak is shown in Eq. (14).

$$Q \underset{-H^+}{\overset{+H^+}{\rightleftharpoons}} QH^+ \underset{-e}{\overset{+e}{\rightleftharpoons}} QH\cdot \underset{-e}{\overset{+e}{\rightleftharpoons}} QH^-$$

$$\scriptstyle -H^+ \uparrow\downarrow +H^+ \qquad -H^+ \uparrow\downarrow +H^+$$

$$QH_2^+\cdot \underset{-e}{\overset{+e}{\rightleftharpoons}} QH_2 \tag{14}$$

The protonated quinone, QH$^+$, is the species undergoing the initial reaction to give the neutral radical QH· which is then reduced to the hydroquinone species by one of the pathways shown in Eq. (14). The overall peak III$_c$ process clearly involves $2e$ and 2H$^+$. The electrode reaction responsible for peak IV$_a$ (Figure 21) is a $2e$ electrooxidation of the hydroquinone species, e.g., 2,3,5-

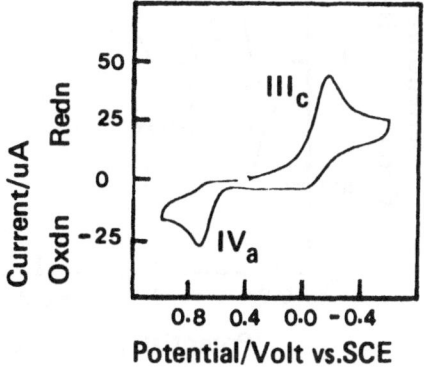

Figure 21. Cyclic voltammogram at a platinum electrode of 1.4 mM 2,3,5-trimethyl-6-(3'-methyl-3'-hydroxybutyl)-quinone and 1 mM H$_2$SO$_4$ in acetonitrile containing 0.1 M tetraethylammonium perchlorate. Sweep rate: 83.5 mV s^{-1}. (Reprinted with permisssion of Elsevier Publishers, Amsterdam).[73]

trimethyl-6-(3'-methyl-3'-hydroxybutyl)hydroquinone (**16**) according to the scheme shown in Eq. (15).

$$QH_2 \rightleftharpoons QH_2^{\ddagger} + e$$

$$QH_2^{\ddagger} \rightleftharpoons QH\cdot + H^+$$

$$QH\cdot \rightleftharpoons QH^+ + e \qquad\qquad (15)$$

$$QH\cdot + QH_2^{\ddagger} \rightleftharpoons QH^+ + QH^2$$

$$QH^+ \underset{fast}{\overset{fast}{\rightleftharpoons}} Q + H^+$$

In the presence of $HClO_4$ the cyclic voltammetric behavior of 2,3,5-TMHQ or α-tocopherylquinone in acetonitrile becomes even more complex (Figure 22). Peak III_c is due to the electrochemical reduction of QH^+ [Eq. (14)] and peak P_c to reduction of protons. The electrooxidation peak at *ca.* 0.7 V is due to cyclization of 2,3,5-TMHQ [I, Eq. (16)] to give 9-hydroxy-2,2,5,7,8-pentamethyl-6-chromanone [II, Eq. (16)] and subsequent dehydration to give the electroactive carbonium ion [III, Eq. (16)]. Application of the initial potential of 0.3 V (Figure 22) reduces the carbonium ion [III, Eq. (16)]

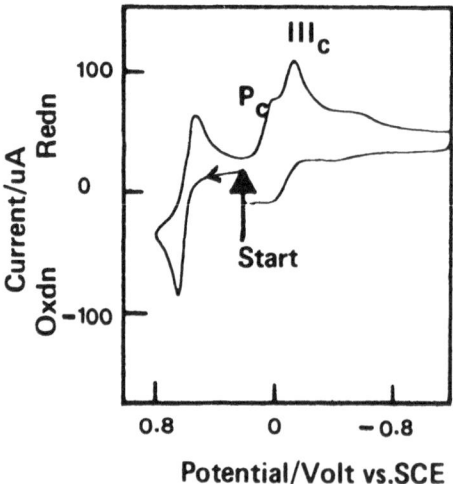

Figure 22. Cyclic voltammogram at a platinum electrode of 2.0 mM 2,3,5-tri-methyl-6-(3'-methyl-3'-hydroxybutyl)-quinone and 4.0 mM $HClO_4$ in acetonitrile containing 0.1 M tetraethyl-ammonium perchlorate. Sweep rate: 83.5 mV s^{-1}. (Reprinted with permission of Elsevier Publishers, Amsterdam).[73]

to the chromanol [IV, Eq. (16)]. Then, on the first sweep towards positive potentials the latter species is oxidized giving the characteristic oxidation peak at 0.7 V.

The actual biological significance of the electrode reaction scheme for oxidation of α-tocopherol and related model compounds shown in Figure 16 are not clear at this time. A perusal of the relevant literature,[76] however, reveals that electrochemical techniques such as cyclic voltammetry and double potential step chronoamperometry are ideally suited to unraveling the complex redox chemistry of these molecules.

The electrochemical behavior of α-tocopherylquinone and related model compounds points out a number of interesting facts. For example, in aprotic media in the absence of protons or proton donors essentially all biological quinones, including α-tocopherylquinone, are electrochemically reduced in stepwise $1e$ processes giving first an anion radical and then a dianion. Both these processes are electrochemically reversible. However, addition of protons or a proton donor to the α-tocopherylquinone or other bioquinones in an aprotic solvent has a profound effect on the second $1e$ reduction process. With increasing concentration of the proton donor the second $1e$ process is shifted towards more positive potentials and ultimately merges with the first $1e$ process so that a single $2e$ reaction occurs. The resulting $2e$ process is no longer electrochemically reversible, indeed it is totally irreversible.

There is a significant amount of biochemical evidence that bioquinones, such as ubiquinone, are located within a lipidic membrane and that the quinone function is not exposed on the membrane surface.[77] Because of the very low solubility of α-tocopherylquinone in water it is almost certainly also membrane bound. Nevertheless, studies strongly support the view that membrane-bound biological quinones are held in a very hydrophobic environment. This environment must have a profound effect on their redox reactions, and it might be expected that such reactions might be similar to those observed electrochemically in aprotic solvents and illustrated in this chapter with α-tocopherylquinone and its model compounds. This would suggest that radicals would participate as intermediates in the redox reactions of these molecules. This in turn suggests that radicals shoud be formed in the electron transport chain where ubiquinone (coenzyme Q) and possibly other quinones including α-tocopherylquinone are involved. There is some evidence that radicals are participants in the electron transport chain. For example, EPR spectroscopy of mitochondria during electron flux gives evidence for radical formation.[78–82] However, the identity of the detected radicals appears to be uncertain and have been variously attributed to the radical forms of flavins, NADH, and ubiquinone.[81]

The existence of acidic compounds, such as phosphatides and cardiolipids,[83] within the lipidic inner membrane of the mitochondria and the proposal that these compounds could function as a proton pool[84] might support the view that $2e$–$2H^+$ reductions of biological quinones might be likely. However, even in nonaqueous medium with a relatively plentiful proton supply where, under electrochemical conditions, an overall $2e$–$2H^+$ reduction of quinones may occur, the resulting hydroquinone is not reversibly (and often not simply) oxidized back to the quinone. These sorts of electrochemical findings seem to bring into question the generally held biochemical view, often widely disseminated in textbooks,[85] that membrane-bound biological quinones such as ubiquinone and perhaps α-tocopherylquinone are simply and reversibly reduced.

5. Purines

Purines occur extensively in all living organisms. Adenine and guanine are the usual purine bases found in nucleic acids. Adenine also occurs as a component of a number of coenzymes such as nicotinamide adenine dinucleotide (NAD$^+$), nicotinamide adenine dinucleotide phosphate (NADP$^+$), and flavin adenine dinucleotide (FAD). N-methylated xanthines such as caffeine, theophylline, and theobromine are central nervous system stimulants.[86]

The electrochemical oxidation and reduction of purine bases has been studied quite extensively and has recently been reviewed.[87] In order to point out the biochemical significance of recent electrochemical studies, the electrochemical and biological oxidation of uric acid (**17**) and some of its derivatives will be considered.

(17)

A rather widely occurring enzyme which catalyzes the oxidation of uric acid is uricase (urate:oxygen oxidoreductase, EC 1.7.3.3). This enzyme is highly selective for uric acid and, except under ususual circumstances, it catalyzes the conversion of uric acid to allantoin (**18**). The absence of uricase

(18)

from the liver of man has lead to the generally accepted view that uric acid is the final product of purine metabolism. Experimental data, however, do not support this generalization. For example, it has been reported that about 20% of labeled uric acid injected into human subjects is broken down.[88–90] Degradation of uric acid by intestinal flora[91] might account for part of this degradation but a number of peroxidatic enzymes have been suggested as being involved in uric acid catabolism. For example, catalase,[92] peroxidase,[92–94] and methemoglobin[95] can catalyze the oxidation of uric acid in the presence of H_2O_2. Peroxidases are widely distributed in mammalian tissues and include lactoperoxidase,[96] liver peroxidase,[97] erythrocyte peroxidase,[98] peroxidase of the adrenal medulla,[99] and myeloperoxidase (verdoperoxidase).[93] Soberon and Cohen[100,101] have demonstrated the peroxidatic oxidation of uric acid by leucocytes _in vitro_. Among the products formed from uric acid upon enzymic oxidation with peroxidase are allantoin, alloxan,[94,102] carbamyldiurea, cyanuric acid, parabanic, oxaluric acid, oxonic acid, and alloxanic acid.[92] Alloxan produces necrosis of the β-cells of the islands of Langerhans[103] and hence causes experimental diabetes. It has also been found that intraperitoneal injection of uric acids in rabbits causes hyperglycemia.[104]

Many of the other products of uric acid noted above are toxic and are not considered to be normal metabolites in the human. However, the possibility certainly exists that conditions associated with peroxidatic oxidation of uric acid such as large accumulations of leucocytes (localized abscesses, leucocytic exudates, leukemia, etc.) could give rise to toxic substances. Thus, there is a real need to understand the basic mechanism of enzymic oxidations of uric acid. Our understanding of this mechanism has been drawn from chemical, enzymic, and electrochemical studies. Because this review is primarily concerned with the electrochemistry of biomolecules, the electrochemical behavior of uric acid will be discussed first.

Uric acid is electrochemically oxidized at a graphite electrode by way of a single, pH-dependent oxidation peak.[105] A cyclic voltammogram of uric acid at pH 7 is shown in Figure 23. The peak I_a process is a $2e-2H^+$ reaction between pH 1–12. Using a pyrolytic graphite electrode having a relatively rough surface (RPGE), after scanning oxidation peak I_a two voltammetric reduction peaks, peaks I_c and II_c (Figure 23), are observed. Reduction peak I_c, which forms a quasi-reversible couple with peak I_a, may only be observed at relatively fast sweep rates. At slow sweep rates peak I_c is absent although peak II_c may still be observed.[106] Double potential step chronoamperometry experiments reveal that the $2e-2H^+$ electrooxidation product of uric acid peak I_a disappears in a first- (or pseudo first-) order reaction characterized by a rate constant at pH 8.0 of 32.5 s^{-1}.[106] This rate constant corresponds to a half-life of 21 ms. At lower and higher pH the rate constant seems to become considerably larger.[106] Thin-layer spectroelectrochemical experiments reveal that a uv-absorbing intermediate is formed on electrochemical oxidation of uric acid at peak I_a potentials.[106] Some typical thin-layer spectroelectrochemical data are presented in Figure 24. Curve 1 in Figure 24 is the spectrum of uric acid at pH 7.5 before electrolysis and shows the characteristic bands at $\lambda_{max} = 292$ nm and 235 nm. Upon application of a potential corresponding to oxidation peak I_a, both uric acid absorption bands decrease and, correspond-

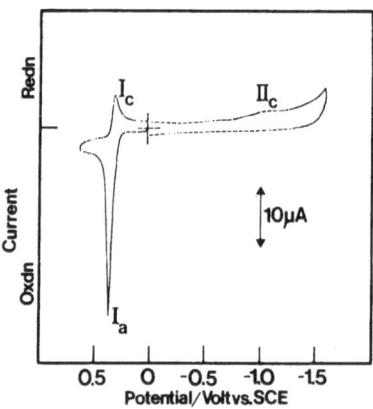

Figure 23. Cyclic voltammogram of 1 mM uric acid in phosphate buffer, pH 7, at a pyrolytic graphite electrode. Sweep rate: 200 mV s⁻¹.

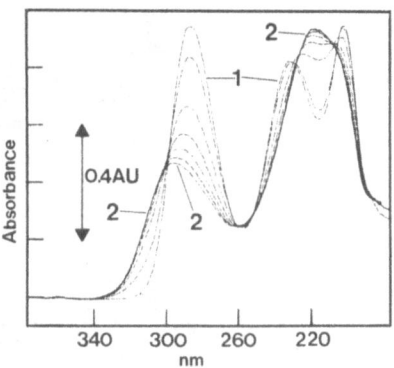

Figure 24. Spectrum of 1.0 mM uric acid electrolyzing at 0.65 V vs. SCE in 0.5 M NaCl + 0.005 M Na$_2$HPO$_4$, pH 7.5, at a reticulated vitreous carbon electrode in a thin-layer cell. Curve (1) is the spectrum of uric acid before electrolysis. Curve (2) is the spectrum of the uv-absorbing intermediate species. Repetitive sweeps of 19 s are shown.

ingly, new absorption bands grow in at longer wavelength (λ_{max} = 302 nm) and at shorter wavelength (λ_{max} = 223 nm). These two bands grow to a maximal value, represented by curve 2 in Figure 24, and then decrease and disappear. Thus, a uv-absorbing intermediate is generated electrochemically. The decay of the uv-absorbing intermediate follows first-order kinetics and in phosphate (Na$_2$HPO$_4$ + NaH$_2$PO$_4$) buffers between pH 7–9 having an ionic strength of 0.5 M, for example, the observed rate constant is 0.0035 s^{-1} which corresponds to a half-life of about 3 minutes.[107] This intermediate is thus much longer-lived than that responsible for reduction peak I$_c$ (see Figure 23). If, after electrochemically generating the uv-absorbing intermediate (i.e., curve 2 in Figure 24), the potential applied to the optically transparent reticulated vitreous carbon electrode[108] is shifted to a value at or more negative than reduction peak II$_c$ observed on cyclic voltammetry of uric acid (Figure 23), then the uv-absorbing intermediate disappears much more rapidly than it does in the absence of an applied potential. This behavior indicates that the uv-absorbing intermediate is reducible at peak II$_c$ potentials.[107] The major organic product formed upon electrochemical oxidation of uric acid is allantoin (**18**).[105,107] On the basis of such studies, the electrode reaction proposed for electrochemical oxidation of uric acid (I, Figure 25) at around physiological pH is an initial 2e–2H$^+$ reaction of its anion (pK_a = 5.75) to a very unstable anionic quinonoid compound (II, Figure 25) which will be referred to as a diimine. This is thought to be the unstable species responsible for cyclic voltammetric reduction peak I$_c$ (Figure 23). The first-order disappearance of this species is thought to be due to attack of water giving the second anionic intermediate imine–alcohol species (III, Figure 25). There are a number of resonance structures possible for the latter species which effectively give it a very conjugated chromophore extending along O$\cdots$C(2)$\cdots$N(3)$\cdots$C(4)$\cdots$N(9)$\cdots$C(8)$\cdots$O. This chromophore is more delocalized than that of the anion of uric acid (O=C(6)—C(5)=C(4)—$\bar{\text{N}}$—) and hence accounts for the uv absorption band of the imine–alcohol occurring at a longer wavelength than that of uric acid (see Fig. 24). Reduction peak II$_c$

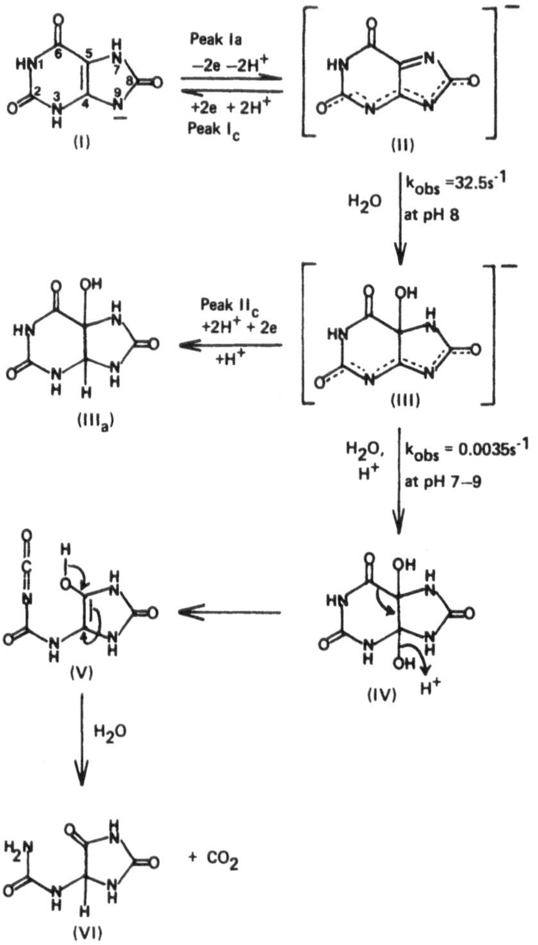

Figure 25. Proposed mechanism for the electrochemical oxidation of uric acid at around physiological pH according to Wrona et al.[107]

observed on cyclic voltammetry of uric acid has been proposed[105–107] to be caused by reduction of the putative imine–alcohol in a $2e$–$2H^+$ reaction across its $C(4)=N(9)$-double bond to give a dihydro species (III_a, Figure 25). That the uv-absorbing intermediate is electrochemically reducible is quite certain, but whether it gives rise to III_a (Figure 25) is certainly open to question. The imine–alcohol (III, Figure 25) has been proposed to decay in a second pseudo first-order hydration reaction giving the 4,5-diol species (IV, Figure 25) which breaks down to give allantoin and CO_2.

The diimine and imine–alcohol intermediates (II and III, Figure 25) are of great interest in the mechanism outlined in Figure 25 and there is some substantial electrochemical, spectral, and analytical data to support their structures. Consider first the putative diimine intermediate which is too short-lived to be detected by spectroelectrochemical techniques but can be detected via its reduction peak I_c in cyclic voltammetry (Figure 23). A number of *N*-methylated uric acid derivatives also give reduction peak I_c after scanning peak I_a. These derivatives are shown as the group I compounds in Figure 26. Thus, provided that the *N*-methyl substituents allow formation of an *ortho* and/or *para* quinonoid diimine in the $2e$ peak I_a process, peak I_c can be observed on the reverse sweep.

Mono-, di- or trisubstitution of uric acid with methyl groups where at least one of the methyl groups is substituted at the N(7)–position always eliminates cyclic voltammetric peak I_c (Figure 26), regardless of the sweep

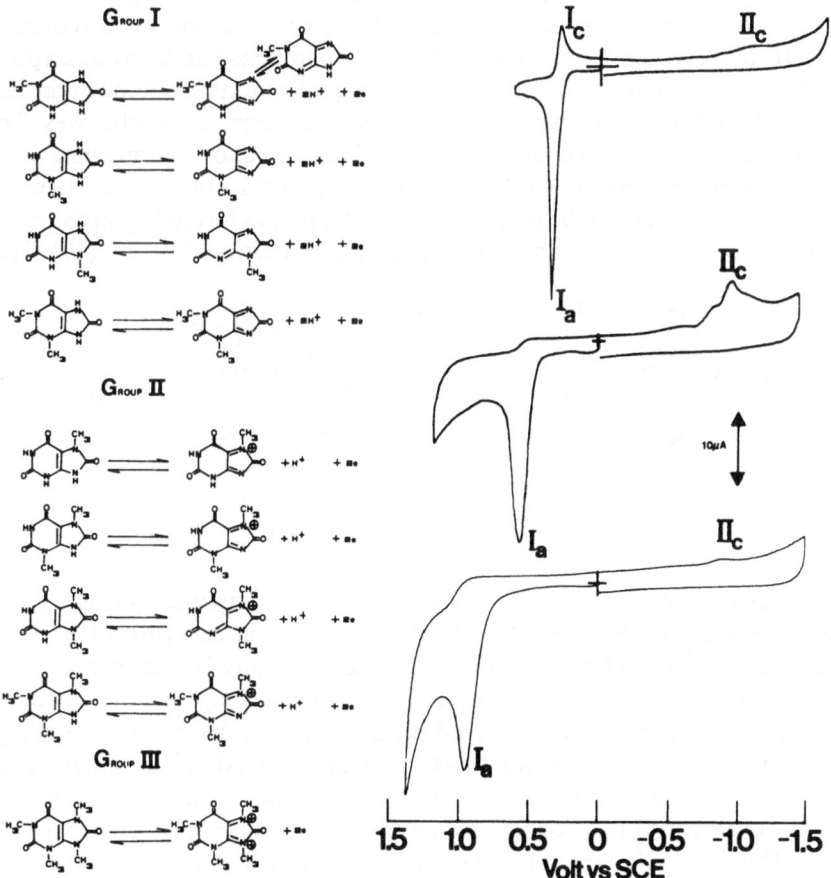

Figure 26. Proposed primary electrode reactions for the peak I_a electrochemical oxidation of *N*-methylated uric acids. The cyclic voltammograms are, from top to bottom, uric acid, 1,3,7-trimethyl uric acid and, 1,3,7,9-tetramethyl uric acid in phosphate buffer, pH 7, at a RPGE at $200\,mV\,s^{-1}$.

rate employed.[107] These compounds constitute the group II uric acids shown in Figure 26. The absence of peak I_c with this group of uric acids is easily rationalized based on the putative diimine intermediate. Thus, $2e$ oxidation of the group II uric acids must form a diimine carrying a positive charge at the N(7)-position regardless of whether the original uric acid derivative is a neutral species or a monoanion. The expected facile and rapid nucleophilic attack of water, or perhaps other solution nucleophiles, on the positively charged $-C(5)=\overset{\oplus}{N}(CH_3)(7)-$ double bond explains the failure to observe a diimine reduction peak. The single group III compound, 1, 3, 7, 9-tetramethyl ruric acid, undergoes a $2e$ reaction to give a dication and, not surprisingly, peak I_c cannot be observed on cyclic voltammetry of this compound (Figure 26).

The product of nucleophilic attack of water on the diimine primary product of electrochemical oxidation of uric acids is an imine–alcohol (see Figure 25). This species is characterized in the case of uric acid in terms of its reduction peak II_c observed under cyclic voltammetric conditions and its uv absorption spectrum under thin-layer spectroelectrochemical conditions. Reduction peak II_c may be observed on cyclic voltammetry of all uric acid derivatives. The general reaction involved in forming the imine–alcohol intermediate from neutral or anionic diimines in the case of the group I uric acids is shown in Figure 25. In the case of the positively charged diimines formed upon oxidation of the group II uric acids, the reaction scheme is illustrated in Eq. (17). That

$$\text{R}_1\text{-N} \cdots =O \xrightarrow[\text{Fast}]{\text{H}_2\text{O}} \text{R}_1\text{-N} \cdots =O + \text{H}^+ \tag{17}$$

where R_1 and R_3 may be CH_3 and/or H

an imine–alcohol structure is reasonable for the compound giving rise to reduction peak II_c is attested to by the fact that in the case of 1, 3, 7, 9-tetramethyl uric acid, peak II_c is extremely small (Figure 26) under conditions where all other compounds give a relatively large peak II_c. This is so because the diimine primary product formed on $2e$ electrooxidation of tetramethyl uric acid carried a double positive charge (Figure 26). Attack by water would give a positively charged imine–alcohol which in turn would be significantly more susceptible to nucleophilic attack by water to form a diol species [Eq. (18)]. Thus, peak II_c should be much smaller than in the case of all other uric acid derivatives.

That the imine-alcohol species is the uv-absorbing intermediate is also supported by thin-layer spectroelectrochemical studies of *N*-methylated uric acids.[107] For example, uric acid and its 1- and 7-methyl derivatives upon

$$(18)$$

electrooxidation give rise to a uv-absorbing intermediate with one band having λ_{max} (*ca.* 302–310 nm, pH 6–9) at longer wavelengths that the parent uric acid (λ_{max} *ca.* 292 nm at pH 6–9). This behavior suggests that these intermediates have a more extensively conjugated structure than the parent uric acids, resulting in the observed bathochromic·shift. Clearly, the uv-absorbing imine–alcohol anion (III, Figure 25) can exist in several resonance forms even if the N(1) and/or N(7) positions are methylated. Such a structure, as noted earlier, would be expected to exhibit an absorption band at longer wavelengths than the parent uric acids. All other N-methylated uric acids studied (Figure 26) which give a uv-absorbing intermediate were substituted at N(3) or N(9) and hence cannot form an imine–alcohol having the many resonance forms shown in structure III (Figure 25). Accordingly, such imine–alcohol intermediates should and do absorb at shorter wavelengths.

In the case of 1, 3, 7, 9-tetramethyl uric acid, it was not possible to detect a uv-absorbing imine–alcohol intermediate.[107] This is to be expected because the first hydration of the diimine dication [Eq. (18)] formed on 2e oxidation of this compound gives a positively charged imine–alcohol [Eqn. (18)] which would be rapidly attacked by water to give an unstable uv-inactive 4,5-diol [Eq. (18)].

In some very recent[109] work the uv-absorbing intermediate generated upon electrochemical oxidation of uric acid was trapped and identified by gas chromatography–mass spectrometry. This was accomplished by carrying out the experiment shown in Figure 24. At the point where the uv-absorbing intermediate reached its maximum concentration (i.e., curve 2 in Figure 24), the solution in the thin-layer cell was rapidly ejected into a vial maintained at −78°C. The frozen mixture was then lyophilized and the resulting dry product containing NaCl, Na_2HPO_4, and the trapped uv-absorbing inter-mediate was treated with N,O-bis-trimethylsilylacetamide or N,N-tri-methylsilyl-trifluoroacetamide which converts N–H or O–H groups to the corresponding trimethylsilyl $[-Si(CH_3)_3]$ derivative. GC/MS studies on the derivatized product were in accord with the imine–alcohol structure (III, Figure 25). Thus, electrochemical, spectral, kinetic, and mass spectroscopic evidence favor the imine–alcohol structure as the uv-absorbing intermediate generated on electrochemical oxidation of uric acid. A natural, first-order decay reaction of this intermediate would result in formation of the 4,5-diol species (IV, Figure 25). This has never been trapped or detected; hence, if it is formed, it must undergo a rapid decomposition to the products allantoin and CO_2 at around pH 7.

The enzymic oxidation of uric acids with peroxidase (type VIII from horseradish peroxidase) has recently been studied and critically compared to the electrochemical oxidation.[110] Peroxidase was studied because this enzyme will oxidize not only uric acid but also most of its N-methyl derivatives.[94,111] Typical spectra of uric acid obtained during the peroxidase-catalyzed oxidation of uric acid are shown in Figure 27. Curve 1 is the spectrum of uric acid. Addition of hydrogen peroxide and peroxidase causes the initial rapid decrease of the spectrum because of dilution. However, it is quite clear from Figure 27

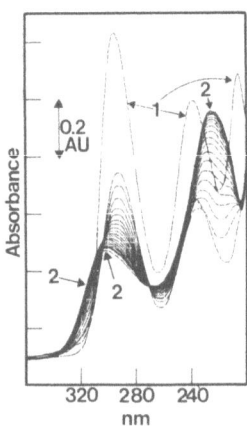

Figure 27. Spectrum of 200 μM uric acid undergoing oxi-dation in the presence of 200 μM H_2O_2 and 0.7 μM peroxi-dase in 0.5 M NaCl puls 5 mM Na_2HPO_4, pH 7.5. Curve (1) is the spectrum of uric acid before electrolysis. Curve (2) is the spectrum of the uv-absorbing intermediate species. Repetetive scans of 10 s.

Table 5

Observed First-order Rate Constants for Reaction of
the UV-Absorbing Intermediate Formed on Enzymic
(Peroxidase) and Electrochemical Oxidation of Uric Acid
Derivatives at pH 7[a,b]

Compound	λ_{max} for uv-absorbing intermediate/ nm	k_{obs}/s^{-1}	
		Electro-chemical	Enzymic
Uric acid	302	0.0035	0.0037
1-Methyl uric acid	308	0.0018	0.0017
3-Methyl uric acid	<270	0.0041	0.0043
7-Methyl uric acid	310	0.0024	0.0024
9-Methyl uric acid	<280	0.015	0.017
1,3-Dimethyl uric acid	<300	0.0019	0.0019
3,7-Dimethyl uric acid	<280	0.0050	0.0050
3,9-Dimethyl uric acid[c]	290	0.18[c]	0.11[c]
7,9-Dimethyl uric acid	<300	<0.001	0.0008
1,3,7-Treimethyl uric acid	<280	0.0010	0.0010
1,3,7,9-Tetramethyl uric acid	N.D.[d]		

[a] Data obtained in phosphate buffer, pH 7, having an ionic strength of 0.5 M.
[b] Results from Reference 110 unless otherwise noted.
[c] Result from Reference 112.
[d] This compound is not oxidized in the presence of peroxidase and gives no detectable uv-absorbing intermediate when electrochemically oxidized.

that with time the uric acid uv bands decrease and new bands grow in at longer and shorter wavelengths. Curve 2 in Figure 27 is the spectrum observed when all uric acid has been oxidized. The two bands at λ_{max} = 302 nm and 223 nm then decrease and disappear. Thus, a uv-absorbing intermediate is formed upon enzymic oxidation of uric acid. Comparison of curves 2 of Figures 24 and 27 reveals that the electrochemically and enzymically generated intermediates are spectrally identical. Further studies[110] revealed that at around physiological pH the spectra of the uv-absorbing intermediates formed on enzymic and electrochemical oxidation of numerous N-methylated uric acid derivatives were identical. In addition, the decay of the uv-absorbing intermediate generated enzymically followed first-order kinetics and the observed rate constant was indistinguishable from that measured for the electrochemically generated intermediate in all cases. Spectral and kinetic results are presented in Table 5.

Further evidence for formation of the same intermediate both upon enzymic and electrochemical oxidation of purines was obtained by cyclic voltammetry. Voltammogram A in Figure 28 was obtained at a pyrolytic graphite electrode immersed in a solution containing 200 μM uric acid, 200 μM H_2O_2 in 0.5 M NaCl plus 5 mM Na_2HPO_4, pH 7.5. The initial sweep in this voltammogram (labeled 1 in Figure 28) is towards negative potentials and no

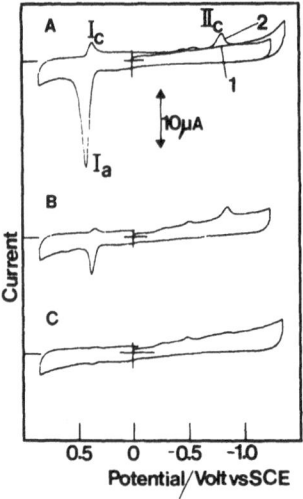

Figure 28. Cyclic voltammograms at a rough pyrolytic graphite electrode of (A) 200 μM uric acid in a solution contining 200 μM H_2O_2, 0.5 M NaCl puls 5 mM Na_2HPO_4, pH 7.5. Curve 1 shows initial sweep towards negative potentials. (B) Cyclic voltammogram of solution in (A) except 0.7 μM type VIII peroxidase enzyme added. (C) Same as (B) except several minutes later. Sweep rate: 200 mV s^{-1}.

reduction peaks are observed. After scanning oxidation peak I_a, however, both the diimine (peak I_c) and imine–alcohol (peak II_c) peaks may be observed on the second negative-going sweep. However, with addition of peroxidase to the solution, the imine–alcohol peak II_c may be observed on the first sweep to negative potentials without initially sweeping oxidation peak I_a (Figure 28B). This implies, therefore, that the reducible intermediate responsible for cyclic voltammetric peak II_c is generated by both enzymic and electrochemical oxidation of uric acid. After almost complete enzymic oxidation of uric acid the voltammogram shown in Figure 28C is observed where all oxidation and reduction peaks have disappeared. Similar cyclic voltammetric behavior has also been noted with various other N-methylated uric acid derivatives.

Such results reveal that upon electrochemical and enzymic (peroxidase) oxidation of uric acid and uric acid derivatives a uv-absorbing intermediate is formed. The spectral, kinetic, and electrochemical behavior of this intermediate whether generated enzymically or electrochemically are identical. Analytical data on the trapped and derivatized electrochemical intermediate are in accord with the uv-absorbing intermediate having an imine–alochol structure (i.e., III, Figure 25).

It has been concluded[107,110], therefore, that since the imine–alcohol intermediate and the same products are formed in both the electrochemical and the enzymic reaction then the enzyme-catalyzed reaction must, in a chemical sense, follow the same reaction pathway as the electrochemical process. This strongly implies, therefore, that the primary product of the enzyme reaction is the putative diimine (i.e., structure II, Figure 25). This species is too short-lived to be detected in the enzymic process and it can only be detected in the electrochemical oxidation reaction. Nevertheless, the over-whelming evidence supports the view that the diimine species precedes formation of the imine–alcohol intermediate in both the electrochemical and enzymic oxidation of uric acids.

For comparitive purposes, it is worthwhile reviewing the information available from purely biochemical studies of the peroxidase-catalyzed oxidation of uric acids. In fact, more detailed studies have been carried out on the uricase-catalyzed oxidation of uric acid.[113-117] However, this enzyme is virtually specific for uric acid and its mechanism has not been studied using the electrochemical approach outlined earlier for the case of the peroxidase enzymes.

Paul and Avi-Dor[94] found that 1-methyl uric acid is oxidized in the presence of horseradish peroxidase and H_2O_2. This reaction is most rapid when the uric acid molecule is uncharged. According to these workers the primary oxidation product obtained at pH 3–5 could be reversibly transformed into two other spectrophotometrically active forms by changing the pH to >6 or >1. Depending on the pH, the primary oxidation product was proposed to undergo a nonenzymic decomposition to allantoin (pH 3–6) or alloxan (pH ≤1). The reaction was represented as shown in Eq. (19). Paul and

$$1\text{-Methyl uric acid} \xrightarrow[\text{H}_2\text{O}_2]{\substack{\text{Horseradish} \\ \text{peroxidase}}} B_1 \rightarrow C_1 (\rightarrow ? \rightarrow \text{allantoin}) \qquad (19)$$

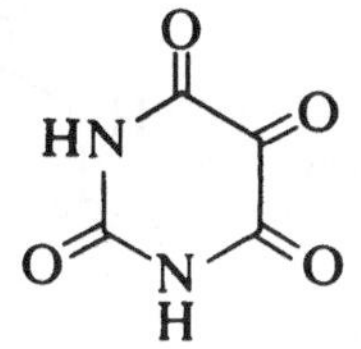

$$B_3 \rightarrow \text{alloxan}$$
$$\updownarrow$$
$$B_1 \rightarrow C_1$$
$$\updownarrow$$
$$B_2 \rightarrow C_2 \text{ (not examined)}$$

Avi-Dor[94] speculated that the various intermediate species represened as B in Eq. (19) might be various ionic forms of the diimine species (**19**). Uptake of water would then give the diol **20**.

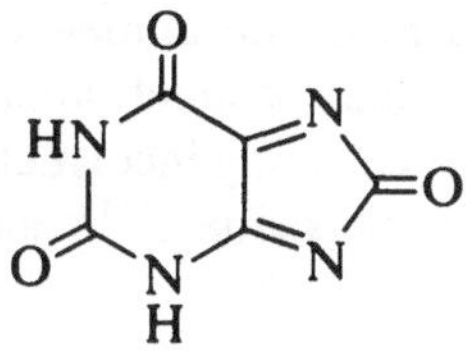

(19)

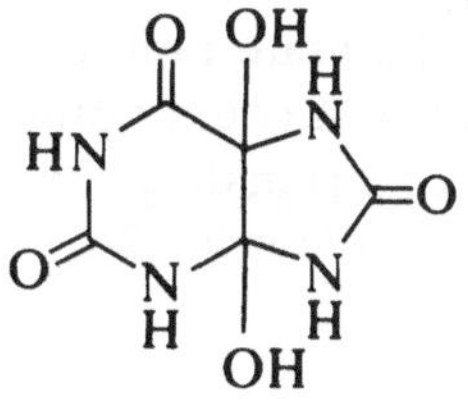

(20)

Canellakis and co-workers[92] examined the oxidation of uric acid with the catalase–H_2O_2, lactoperoxidase–H_2O_2, verdoperoxidase–H_2O_2, and horseradish peroxidase–H_2O_2 systems. On the basis of this investigation it was proposed that at around pH 7 uric acid (I, Figure 29) is oxidized, via one or

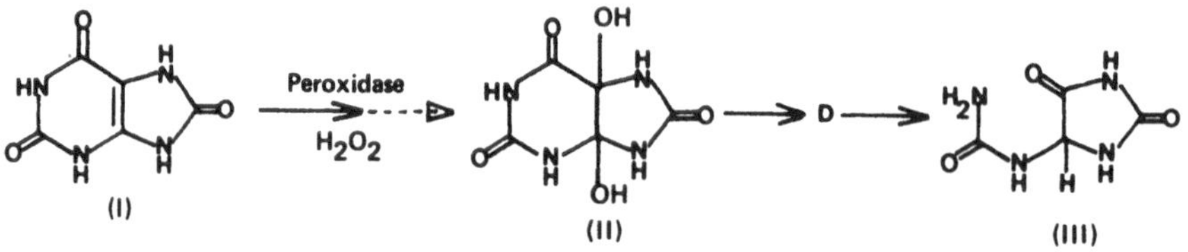

Figure 29. Proposed generalized mechanism at around pH 7 of oxidation of uric acid by peroxidase enzymes according to Canellakis and co-workers.[92]

more unknown intermediate species, to uric acid-4, 5-diol (II, Figure 29). Then, possibly via another intermediate D, the diol decomposes to give allantoin (III, Figure 29).

The reaction schemes shown in Eq. (19) and Figure 29 were thrown into some doubt when Howell and Wyngaarden[111] found that 7-methyl (**21**) and 9-methyl uric acid (**22**) were oxidized in the presence of methemoglobin (a hemeprotein or peroxidase enzyme)–H_2O_2 or horseradish peroxidase–H_2O_2 but that 1,3,7,9-tetramethyl uric acid (**23**) was not. These workers thus

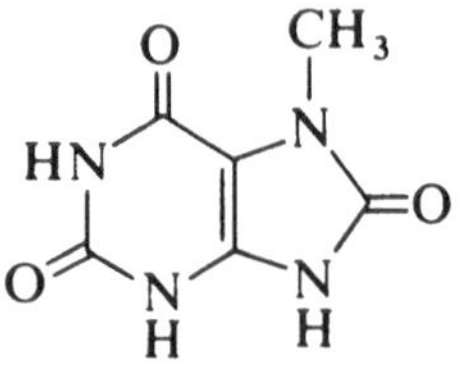

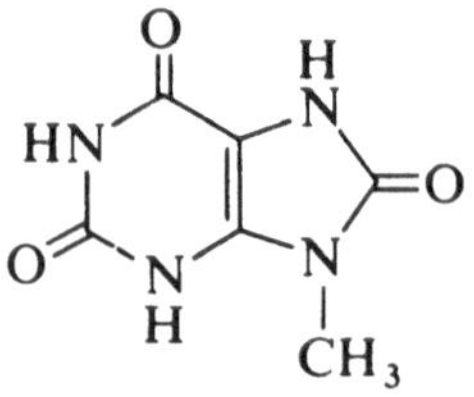

concluded that an unsubstituted N(7) or N(9) position is necessary for the enzyme-catalyzed reaction to occur. Thus, an alternative mechanism was proposed in which the initial oxidation step involved a dehydrogenation of the uric acid [I, Eq. (20)] to give a radical intermediate [II, Eq. (20)]. Loss of an electron to a hydroxyl radical, formed from peroxide, could then give a carbonium ion [III, Eq. (20)] which could then break down to give allantoin at around pH 7. Alternatively, it was suggested[111] that attack of a hydroxyl radical on the urate radical [II, Eq. (20)] would form the imine–alcohol IV [Eq. (20)] which could again break down to the final product, allantoin [V, Eq. (20)]. Unfortunately, this mechanism is almost certainly incorrect because 7,9-dimethyl uric acid (**24**) is readily oxidized by peroxidase[110] and hence

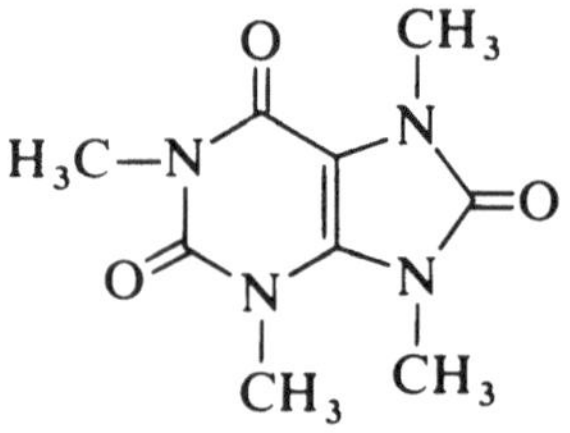

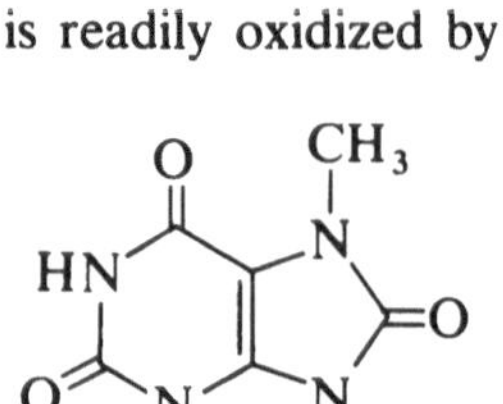

$$(20)$$

an unsubstituted N(7) or N(9) position is not a prerequisite for susceptibility of uric acid to oxidation by peroxidase enzymes.

The mechanisms obtained in these enzyme studies are, in fact, based on rather insubstantial chemical evidence. However, by parallel investigation of the enzymic and electrochemical oxidations, significant mechanistic insights are obtained. Spectral, kinetic, and electrochemical information confirm that under both enzymic and electrochemical oxidation conditions the same uv-absorbing intermediate is formed upon oxidation of uric acid or its N-methylated derivatives. The GC/MS data on the trapped and derivatized intermediate generated electrochemically is that expected for the imine–alcohol structure. This intermediate is almost certainly derived by partial hydration of the initial 2e electrooxidation product of the uric acid, i.e., a diimine species. This species is too short-lived to be detected in homogeneous solution if generated in the enzymic oxidation of uric acids. However, because of its peculiar adsorption at a rough pyrolytic graphite electrode[106] it is easily detected as an electrooxidation product. There is no reason why the imine–alcohol species detected upon enzymic oxidation of uric acids should not be derived from an enzymically generated diimine species. Quite conventional reaction pathways leading from the imine–alcohol intermediate to the ultimate products can then be described (Figure 25).

The electrochemical and enzymic (peroxidase) oxidations outlined in this section clearly support the view that electrochemical techniques can provide useful information regarding the pathways and mechanisms of enzymic redox reactions. Work currently underway in the author's laboratory on the electrochemical and enzymic oxidation of guanine, adenine, hypoxanthine, xanthine, and various nucleoside and nucleotide species indicates that there is a considerable parallelism between the electrochemical and enzyme-catalyzed processes.

6. Conclusions

The reader should not be left with the impression that electrochemical methods and approaches can be used in all instances to unravel the biological redox chemistry of biologically interesting organic molecules. Indeed, not only are the examples shown in this section the most illustrative in the literature, they are almost the only examples. This does not imply that electrochemical techniques have only a very limited usefulness as tools to help unravel the complexities of biological redox processes. It simply reflects the rather small amount of work that has been carried out by electrochemists in this area. It may be noted from the examples quoted in this chapter that the complex reaction processes discussed are largely oxidation reactions. It is certainly true in electrochemistry that oxidation reactions are generally far more complex and difficult to unravel than are reduction reactions. Unfortunately most, although certainly not all, biological redox reactions are oxidation processes. Thus, it is necessary for electrochemists not only to study primarily oxidation reactions but also to study the mechanisms of such reactions for rather complex biomolecules which present formidable problems not only in electrochemistry but also in kinetics, analytical chemistry, and structure elucidation. Thus, investigations of the redox chemistry of biological molecules by electrochemical techniques provide some major challenges to electrochemists. The results of such studies might have profound and far-reaching effects in our understanding of the biological or *in vivo* redox processes associated with such biomolecules. This is a challenging area and it is to be hoped that many electrochemical scientists can be enticed to take up the challenge.

Acknowledgements

The author would like to acknowledge the generous financial support of his research on biological electrochemistry by the National Institutes of Health through Grant Nos. GM-21034 and GM-25842 and the Research Council of the University of Oklahoma. The excellent work of my coworkers, Barbara H. Hansen, James L. Owens, Michael T. Cleary, Henry A. Marsh, Monika Z. Wrona, Anna Brajter-Toth, R. N. Goyal, and Wedad R. Hussein on the electrochemistry and enzyme chemistry of purines is gratefully acknowledged.

References

1. M. Brezina and P. Zuman, *Polarography in Medicine, Biochemistry and Pharmacy*, Wiley (Interscience), New York (1958).
2. A. L. Underwood and R. W. Burnett, in *Electroanalytical Chemistry*, Vol. 6, A. J. Bard, ed., Marcel Dekker, New York (1973), pp. 1–85.
3. R. L. McCreery, *CRC Critical Review in Analytical Chemistry*, Vol. 7 (1978), pp. 88–119.
4. G. Dryhurst, *Electrochemistry of Biological Molecules*, Academic Press, New York (1977).
5. R. N. Adams, *Anal. Chem.* **48**, 1126A (1976).
6. M. D. Hawley, S. V. Tatawawadi, S. Piekarski, and R. N. Adams, *J. Am. Chem. Soc.* **89**, 447 (1967).
7. P. Malachesky, L. Marcoux, and R. N. Adams, *J. Phys. Chem.* **70**, 4068 (1966).
8. A. W. Sternson, R. L. McCreery, B. Feinberg, and R. N. Adams, *J. Electroanal. Chem.* **46**, 313 (1973).
9. J. Axelrod, in *Adrenergic Mechanisms*, J. R. Vane, G. E. W. Wolstenholme, and M. O'Conner, eds., Little, Brown and Co., Boston, Mass. (1960).
10. J. J. Kapin, in *Actions of Hormones on Molecular Processes*, G. Litwack and D. Kritchevsky, eds., Wiley, New York (1964).
11. E. G. Ball and T. T. Chen, *J. Biol. Chem.* **102**, 691 (1933).
12. G. Satori and C. Cattaneo, *Gazz. Chim. Ital.* **72**, 525 (1942).
13. S. Senoh and B. Witkop, *J. Am. Chem. Soc.* **81**, 6231 (1959).
14. S. Kaufman and S. Friedman, *Pharmacol. Rev.* **17**, 71 (1965).
15. D. C. S. Tse, R. L. McCreery, and R. N. Adams, *J. Med. Chem.* **19**, 37 (1976).
16. R. N. Adams, *Bull. Menninger Clin.* **38**, 57 (1974).
17. H. Thoenen, J. P. Tranzer, and G. Hausler, in *New Aspects of Storage and Release of Catecholamines*, H. J. Shuman and G. Kronberg, eds., Springer-Verlag, Berlin (1970), p. 130.
18. A. Saner and H. Thoenen, *Mol. Pharmacol.* **7**, 147 (1970).
19. J. De Champlain and R. Nadeau, *Fed. Proc.* **30**, 877 (1971).
20. R. N. Adams, E. Murrill, R. L. McCreery, C. L. Blank, and M. Karolczak, *Eur. J. Pharmacol.* **17**, 287 (1972).
21. C. L. Blank, P. T. Kissinger, and R. N. Adams, *Eur. J. Pharmacol.* **19**, 391 (1972).
22. A. Rotman, J. Lundstrom, E. McNeal, J. Daly, and C. Creveling, *J. Med. Chem.* **18**, 138 (1975).
23. C. L. Blank, R. L. McCreery, R. M. Wightman, W. Chey, R. N. Adams, J. R. Reid, and E. E. Smissman, *J. Med. Chem.* **19** 178 (1976).
24. R. McCreery, R. Dreiling, and R. N. Adams, *Brain Res.* **73**, 15 (1974).
25. R. McCreery, R. Dreiling, and R. N. Adams, *Brain Res.* **73**, 22 (1974).
26. R. L. McCreery, Ph.D. Dissertation, University of Kansas (1974).
27. Y. O. Liang, R. M. Wightman, P. Plotsky, and R. N. Adams, in *Chemical Tools in Catecholamine Research*, Vol. I, G. Jonsson, T. Malmfors, and C. Sachs, eds., North-Holland Publishing Co., Amsterdam (1975), p. 15.

28. Y. O. Liang, P. M. Plotsky, and R. N. Adams, *J. Med. Chem.* **20**, 581 (1977).

29. L. Stein and C. D. Wise, *Science* **171**, 1032 (1971).

30. S. Senoh, C. R. Creveling, S. Udenfriend, and B. Witkop, *J. Am. Chem. Soc.* **81**, 6236 (1959).

31. R. N. Adams, *Trends in Neurosciences* (Pers. Ed.) **1**, 160 (1978).

32. R. Keller, I. Mefford, A. Oke, E. Strope, J. Conti, R. M. Wightman, P. Plotsky, and R. N. Adams, *Modern Pharmacology–Toxicology* **10**, 761 (1977).

33. R. N. Adams, *J. Pharm. Sci.* **58**, 1171 (1969).

34. V. Fishman and H. Goldenberg, *Proc. Soc. Exp. Biol. Med.* **112**, 501 (1963).

35. J. W Daly and A. A. Manian, *Biochem. Pharmacol.* **16** 2131 (1967).

36. A. A. Manian, D. H. Efron, and S. R. Harris, *Life Sci.* **10**, (part I), 679 (1971).

37. P. Turano, W. J. Turner, and A. A. Manian, *J. Chromatogr.* **75**, 277 (1973).

38. R. L. McCreery, *J. Pharm. Sci.* **66**, 357 (1977).

39. M. Neptune and R. L. McCreery, *J. Org. Chem.* **43**, 5006 (1978).

40. M. Neptune and R. L. McCreery, *J. Med. Chem.* **21**, 362 (1978).

41. F. H. Merkle and C. A. Discher, *J. Pharm. Sci.* **53**, 620 (1964).

42. H. Y. Cheng, P. H. Sackett, and R. L. McCreery, *J. Am. Chem. Soc.* **100**, 962 (1978).

43. H. Y. Cheng, P. H. Sackett, and R L. McCreery, *J. Med. Chem.* **21**, 948 (1978).

44. P. H. Sackett and R. L. McCreery, *J. Med. Chem.* **22**, 1447 (1979).

45. T. N. Tozer and D. Tuck, *J. Pharm. Sci.* **54**, 1169 (1965).

46. L. Levy, T. N. Tozer, L. D. Tuck, and D. B. Loveland, *J. Med. Chem.* **15**, 898 (1972).

47. I. Creese, D. R. Burt, and S. H. Snyder, *Science* **192**, 481 (1976).

48. R. Byck, in *The Pharmacological Basis of Therapeutics*, 5th Ed., L. S. Goodman and A. Gilman, eds., Macmillan, New York (1975).

49. I. S. Forrest and D. E. Green, *J. Forensic Sci.* **17**, 592 (1972).

50. L. H. Piette, G. Bulow, and I. Yamazaki, *Biochim. Biophys. Acta* **88**, 120 (1964).

51. G. M. Gooley, H. Keyser, and F. Setchell, *Nature* **223**, 80 (1969).

52. I. S. Forrest, F. M. Forrest, and M. Berger, *Biochim. Biophys. Acta* **29**, 441 (1958).

53. J. Buckley, M. Steenberg, H. Barry, and A. A. Manian, *J. Pharm. Sci.* **62**, 715 (1973).

54. A. A. Manian, D. H. Efron, and M. Goldberg, *Life Sci.* **4**, 2425 (1965).

55. A. V. P. Mackay, A. F. Healey, and J. Baker, *Brit. J. Clin. Pharmacol.* **1**, 425 (1974).

56. G. Sakalis, T. L. Chan, S. Gershon, and S. Park, *Psychopharmacologia* **32**, 279 (1973).

57. T. L. Perry, C. A. Culling, K. Berry, and S. Hanson, *Science* **146**, 82 (1964).

58. M. H. Van Woert, *Nature* **219**, 1054 (1968).

59. H. R. Adams, A. A. Manian, M. L. Steenberg, and J. P. Buckley, in *Phenothiazines and Structurally Related Drugs*, I. S. Forrest, C. J. Carr, and E. Usdin, eds., Raven Press, New York (1974), p. 281.

60. T. A. Grover, L. H. Piette and A. A. Manian, in *The Phenothiazines and Structurally Related Drugs*, I. S. Forrest, C. J. Carr, and E. Usdin, eds., Raven Press, New York (1974), p. 561.

61. W. H. Sebrell and R. S. Harris, eds., *The Vitamins*, Vol. V, Academic Press, New York (1972), Chap. 16.

62. H. M. Evans, *Vitamins and Hormones* **20**, 379 (1962).

63. F. L. Crane, in *Biological Oxidations*, T. P. Singer, ed., Interscience, New York (1968), p. 533.

64. L. I. Smith, I. M. Kolthoff, S. Wawzonek, and P. M. Ruoff, *J. Am. Chem. Soc.* **63**, 1018 (1941).

65. L. I. Smith, L. J. Spillane, and I. M. Kolthoff, *J. Am. Chem. Soc.* **64**, 447 (1942).

66. P. D. Boyer, *J. Am. Chem. Soc.* **73**, 733 (1951).

67. W. Dürckheimer and L. A. Cohen, *Biochem. Biophys. Res. Commun.* **9**, 262 (1962).

68. W. Dürckheimer and L. A. Cohen, *J. Am. Chem. Soc.* **86**, 4388 (1964).

69. J. Green and D. McHale, in *Biochemistry of Quin*

70. M. F. Marcus and M. D. Hawley, *Biochim. Biophys. Acta* **201**, 1 (1970).

71. W. H. Harrison, J. E. Gander, E. R. Blakley, and P. D. Boyer, *Biochim. Biophys. Acta* **21**, 150 (1956).

72. M. F. Marcus and M. D. Hawley, *J. Org. Chem.* **35**, 2185 (1970).

73. M. F. Marcus and M. D. Hawley, *Biochim. Biophys. Acta* **222**, 163 (1970).
74. S. Wawzonek, R. Berkey, E. W. Blaha, and M. E. Runner, *J. Electrochem. Soc.* **103**, 456 (1956).
75. G. J. Hoijtink, J. van Schooten, E. de Boer, and W. Y. Aalbersberg, *Recl. Trav. Chim. Pays-Bas* **73**, 355 (1954).
76. G. Dryhurst, K. Kadish, and F. Scheller, *Biological Electrochemistry*, Vol. I, Academic Press, New York (1982), pp. 67–85.
77. F. L. Crane, *Ann. Rev. Biochem.* **46**, 439 (1977).
78. D. D. Tyler, J. Gonze, R. W. Estabrook, and R. A. Butow, in *Non-Heme Iron Proteins*, A. San Pietro, ed., Antioch Press, Yellow Springs, Ohio (1965), pp. 447–60.
79. D. Bäckström, B. Norling, A. Ehrenberg, and L. Ernster, *Biochim. Biophys. Acta* **197**, 108 (1970).
80. S. Odashima and H. Abe, *Saibo Siebutsugaku Shimpojiumu* **22**, 217 (1972).
81. L. P. Kayushin and O. N. Brzhevskaya, *Biofizika* **23**, 1024 (1978); *Chem. Abstr.* **90**, 67919h.
82. O. N. Brzhevskaya, O. S. Nedelina, E. M. Sheksheev, and L. P. Kayushin, *Dokl. Akad. Nauk. SSSR* **232**, 221 (1977): *Chem. Abstr.* **86**, 84865x.
83. A. L. Lehninger, *The Mitochondrion*, Benjamin, New York (1964).
84. M. Klingenberg and A. Kroger, in *Current Topics in Bioenergetics*, Vol. 2, D. R. Sanadi, ed., Academic Press, New York (1967), p. 186.
85. R. C. Bohinski, *Modern Concepts in Biochemistry*, Third Edition, Allyn and Bacon, Boston, Mass. (1979), p. 446.
86. L. S. Goodman and A. Gilman, eds., *The Pharmacological Basis of Therapeutics*, 3rd Ed., Macmillan, New York (1967), p. 355.
87. G. Dryhurst, *Electrochemistry of Biological Molecules*, Academic Press, New York (1977), Chap. 3.
88. J. D. Benedict, P. H. Forsham, and D. Stetten, *J. Biol. Chem.* **181**, 183 (1949).
89. J. Buzard, C. Bishop, and J. H. Talbott, *J. Biol. Chem.* **196**, 179 (1952).
90. J. B. Wyngaarden and D. Stetten, *J. Biol. Chem.* **203**, 9 (1953).
91. L. B. Sorensen, *Metabolism* **8**, 687 (1959).
92. E. S. Canellakis, A. L. Tuttle, and P. P. Cohen, *J. Biol. Chem.* **213**, 397 (1955).
93. K. Agner, *J. Exp. Med.* **92**, 337 (1950).
94. K. G. Paul and Y. Avi-Dor, *Acta Chem. Scand.* **8**, 637 (1954).
95. J. B. Wyngaarden, *J. Biol. Chem.* **235**, 3544 (1960).
96. B. D. Polis and H. W. Shmukler, in *Methods of Enzymology*, Vol. II, S. P. Colowick and N. O. Kaplan, eds., Academic Press, New York (1955), p. 813.
97. M. J. Hunter, in *Methods in Enzymology*, Vol. II, S. P. Colowick and N. O. Kaplan, eds., Academic Press, New York (1955), p. 791.
98. G. C. Mills, *J. Biol. Chem.* **234**, 502 (1959).
99. S. Huszak, cited by R. Lemberg and J. W. Legge, in *Hematin Compounds and Bile Pigments*, Wiley (Interscience), New York (1947), p. 430.
100. G. Soberon and P. P. Cohen, *Rev. Soc. Chim. Mexico* **6**, 45 (1962).
101. G. Soberon and P. P. Cohen, *Arch. Biochem. Biophys.* **103**, 331 (1963).
102. A. L. Tuttle, Ph.D. Thesis, University of Wisconsin (1953), cited in reference (101).
103. J. S. Dunn, H. K. Shelhan, and M. C. B. MacLetchie, *Lancet* **1**, 484 (1943).
104. M. Griffiths, *J. Biol. Chem.* **172**, 853 (1948).
105. G. Dryhurst, *J. Electrochem. Soc.* **119**, 1559 (1972).
106. J. L. Owens, H. A. Marsh, and G. Dryshurst, *J. Electroanal. Chem.* **91**, 231 (1978).
107. M. Z. Wrona, J. L. Owens, and G. Dryhurst, *J. Electroanal. Chem.* **105**, 295 (1979).
108. V. E. Norvell and G. Mamantov, *Anal. Chem.* **49**, 1470 (1977).
109. A. Brajter-Toth and G. Dryhurst, *J. Electroanal. Chem.* **122**, 205 (1981).
110. M. Z. Wrona and G. Dryhurst, *Biochim. Biophys. Acta* **570**, 371 (1979).
111. H. R. Howell and J. B. Wyngaarden, *J. Biol. Chem.* **235**, 3544 (1960).

112. M. Z. Wrona, R. N. Goyal, and G. Dryhurst, *Bioelectrochem. Bioenerg.* **7**, 433 (1980).

113. R. Bentley and A. Neuberger, *Biochem. J.* **52**, 694 (1952).

114. E. S. Cancellakis and P. P. Cohen, *J. Biol. Chem.* **213**, 385 (1955).

115. H. R. Mahler, H. M. Baum, and G. Hübšcher, *Science* **124**, 705 (1956).

116. O. M. Pitts and D. G. Priest, *Biochemistry* **12** 1358 (1973).

117. G. P. A. Bongaerts and G. D. Vogels, *Biochim. Biophys. Acta* **567**, 295 (1979).

3

Electrochemistry of Biopolymers

HERMANN BERG

1. Introduction

The development of the electrochemistry of biopolymers has paralleled
research on their structures. In the case of proteins the basic research by the
polarography school of Heyrovsky[1] and the electrophoresis school of Tiselius
and Theorell[2] started in the thirties, whereas research on nucleic acids and
polysaccharides was the focus of interest about 20 years later at Jena[3,4] and
Brno.[5,6] The study of the behavior of proteins and nucleic acids as polyelectrolytes in solution is now a very broad field and the main topics of research
are

- separations by sophisticated variations of electrolysis,[2,7]
- purification by salt precipitation, ion exchange, and other methods,
- ionic equilibria (isoelectric points) and conformational transitions,
  investigated by potentiometric acid–base titrations (conductometry)
  and spectroelectrochemistry,[7,8]
- complex formation, measured by electrochemical and spectrometric
  methods,[4,7]
- electron transfer reactions of proteins,[9] studied by kinetic techniques.

All these topics are discussed very briefly here. The emphasis of this chapter
is on the voltammetric and polarographic behavior of proteins, polysaccharides,
and nucleic acids. In contrast to most of the other physicochemical methods,

HERMANN BERG • Academy of Sciences of the GDR, Central Institute of Microbiology and
Experimental Therapy, Department of Biophysical Chemistry, DDR-69 Jena, GDR.

the electrochemical method permits the adsorbed state of biopolymers to be established, influenced, and measured simultaneously! The theory, instrumentation, and techniques of polarography and voltammetry are widely developed.[10–12] In addition to classical direct current (d.c.) polarography with the dropping mercury electrode (d.m.e.), the following modern techniques and special electrodes are preferred:

- alternating current (a.c.) polarography[11]
- pulse polarography (normal pulse, n.p., and differential pulse, d.p.).

Stationary hanging mercury drop electrodes (h.m.d.e.) are suitable for evaluating a slow equilibrium of adsorption and subsequent reduction. Solid metal or graphite electrodes are used mainly for oxidation. From the molecular biophysical point of view such measurements are performed for characterization of structural and conformational transitions caused by physical and electrochemical influences, such as heat, light, electrical fields, solvents, ions, and other ligands. In all cases, one can distinguish between reversible (allosteric and conformational modifications) and irreversible (denaturation, strand break, enzyme reactions) processes. Besides these investigations, biochemical analysis, clinical tests, and electrochemical synthesis are fruitful applications.

2. Proteins

2.1. Electrochemical Classification

From an electrochemical point of view, proteins can be classified as follows:

- proteins with disulfide–thiol groups
- proteins with prosthetic groups
- proteins yielding catalytic double waves after complexation.

Prior to electron exchange, strong adsorption[13] of mostly globular molecules takes place which raises the question: what happens to secondary and tertiary structures in the vicinity of the electrode? On the one hand, there is evidence for unfolding, flattening, and even splitting of globuli and, on the other hand, some enzymes are still active in the adsorbed state.

Frequently, proteins are classified according to the results of the Koryta equation:

$$\Gamma = (7.36 \times 10^{-4}) D^{1/2} c_s t^{1/2} \tag{1}$$

(Γ = surface concentration, c_s = concentration in solution, D = diffusion coefficient), or of the capacity measurements by a.c. polarography[14]:

$$\Delta C / C_0 = (5.67 \times 10^{-4}) D^{1/2} c_s \tau^{1/2} N_L A \Delta C_m / C_0 \tag{2}$$

(C_0, C, C_m = capacity of free, partially, and fully covered surface, respectively, τ = drop time, N_L = Loschmidt's number, A = area of 1 adsorbed protein).

From the enhancement of A at the electrode in comparison to its magnitude in crystals it has been concluded that ribonuclease bovine serum albumin (in presence of $CaCl_2$), cytochrome P450, ferredoxin, and others belong to the first group,[15-17] whereas methemoglobin and urease are examples of the stable second group.[15,18,19] There are different opinions[15,19] on the conformational changes of cytochrome c upon adsorption at graphite electrodes. In all cases, slight deformation of protein molecules may occur in the first adsorbed layer.

2.2. Polyelectrolyte Behavior in Solution

In view of the well-known amphoteric character of proteins,[2,7,8] the reactivity for electron transfer is clearly dependent on the environment, particularly on pH and ionic strength. For instance, in the case of horse heart cytochrome c, the transition from the "neutral" to the "alkaline" state involves replacement of the sixth ligand of the heme iron, methionine-80, by lysine-79 and therefore the reduction takes place more reversibly.[21]

The interaction of ions with a protein of N identical binding sites is described by the site-binding model[22,23] for the ligand L. The probability (r) that a given site is occupied by L, which has an activity a_L, is

$$\frac{n_L}{N} = r = \frac{Ka_L}{1 + Ka_L} \tag{3}$$

(n_L = number of occupied binding sites, K = binding constant), which is another form of the well-known Scatchard equation. The dependence of protein conformation on pH, surrounding ions, and other protein molecules has been analyzed in detail by von Hippel and Wong,[24] Feinberg et al.,[25] and Ryan[9] using relations between activity coefficients of polyions, solubility, salting-out effects (lyotropic series), inhibition of enzymes, cooperative association,[9] and tendency for denaturation of the superstructures. Serum albumin, ribonuclease, myosin, α-chymotrypsin, tubulin, hemoglobin, myoglobin, histones, RNA polymerase, and cytochromes have all been extensively examined.[23]

2.3. Polyelectrolyte Behavior at Electrodes

As mentioned at the beginning of this chapter, a prerequisite for all types of electron exchange is at least a partial adsorption of the globuli.[13] From a heuristic point of view one can distinguish four electron exchange groups:

- disulfide or thiol groups,
- prosthetic groups,
- catalytic waves of complexes,
- aromatic residues.

2.3.1. Electron Exchange with Disulfide–Thiol Groups

In principle, a reversible redox reaction occurs

$$R-S-S-R + 2e + 2H^+ \rightleftharpoons 2RSH \tag{4}$$

However, this reaction can be disturbed by steric hindrance or side reactions:

$$RSH + Hg \rightarrow RSHg + e + H^+ \tag{5}$$

Furthermore, polarization time and electrode type are responsible for the number of reducible disulfide groups (Table 1). For instance at the d.m.e. only the two interstrand disulfide bridges A7—B7 and A20—B19 of insulin[4,27] are split at -0.6 V vs. SCE, whereas in the case of albumin[28] only 4 out of the 16 disulfide bridges are broken. At the stationary electrode (h.m.d.e.) about 3 disulfide bridges in insulin and 9 in albumin are broken. Albumin denatured by 8 M urea shows practically the same result, which means that

Table 1
Proteins and Their Groups for Electron Exchange

Protein	MW/10^3	Electron acceptor	Number	n_e	$U_{1/2}$(pH 7) (V vs. SCE)
Insulin (dimer)	6 (12)	S—S	(2×) 3	6 (12)	-0.6
Ribonuclease	13.6	S—S	4	8	-0.8
Lysozyme	14.5	S—S	5	10	-0.8
Trypsin	23.8	S—S	6	12	-0.49
Chymotrypsin	25	S—S	5	10	-0.45
Pepsm	35	S—S	3	6	
β-Lactoglobulin	38	S—S	9	18	
Ovalbumin	40	S—S	6	12	-0.8
Takaamylase[68]	52	S—S/SH	4/1	2	0.47
Human serum albumin	69	S—S	16	32	≈ -0.6
Bovine serum albumin	69	S—S	7	14	-0.63
Cytochrome c	13	Fe^{3+}		1	-0.13
Cytochrome c_3	14	Fe^{3+}	4	4	-0.53
Cytochrome b_5	≈ 14	Fe^{3+}		1	-0.58
Methemoglobin	64	Fe^{3+}	4	4	-0.60
Metmyoglobin	16	Fe^{3+}		1	-1.05
Bacteriorhodopsin		R=CH—CH=R			-0.97
Cytochrome oxidase	200	Fe^{3+}, Cu^{2+}		2	-0.2
Tryptophan oxygenase	67	Fe^{3+}, Cu^{2+}		1	≈ -0.2
Glycogen phosphorylase	200	R—CH=N—R	2	4	-0.82^a
Xanthine oxidase	200	FAD, SH		2	-0.59
Cholesterol oxidase		FAD,		2	-0.33
Glucose oxidase	186	FAD, S—S	2	2	-0.36
Ferredoxin (spinach)	13.5	Fe^{3+}	6	6	-0.6
Ferritin	700	Fe^{3+}	n	nx	-0.38^b

^{a} pH 4.9.
^{b} pH 2.

slow surface denaturation may occur to a similar extent at a h.m.d.e., but this is not the case for the d.m.e. The signals of insulin depend on polarization times and the environment. The positive d.c. double steps merge with increasing temperature, content of organic solvent, and by changing the buffer solution (Figure 1), whereas the cyclic voltammogram (c.v.) shows only one cathodic positive peak and a still unexplained negative one $(-1.2\,\mathrm{V})$ (Figure 2). Fast reoxidation needs $4e$, but after longer times only one disulfide can be restored, whereas the second one exhibits steric hindrance as a result of slow rearrangement. Hence the possibility of reoxidation of SH-groups depends on the spatial position and recognition between the two corresponding S atoms which formed the original S—S bridge. The oxidation could involve Hg, too, as shown in Eq. (5). These findings are also consistent with the studies of the redox reaction of urease[18] adsorbed on a stationary mercury electrode where enzymatic activity could be inactivated and regenerated upon reduction and reoxidation. During such a procedure the volume of the globulus extends and contracts, respectively, as measured by the intrinsic viscosity: for ribonuclease (4 disulfide bridges), between 14.4 and $3.3\,\mathrm{cm^3\,g^{-1}}$.

Sterical differences can be distinguished between immunoglobulins,[29] e.g., between Ig G and Ig A.

In order to reduce more disulfide bridges selectively inside a large globulus (α-chymotrypsin) mediators such as 7-methylpterine[30] or methyl viologen may be used which sometimes render electron transfer more effective. In the case of several SH groups buried in large proteins such as TMV-protein, higher sensitivity is achieved by measuring the catalytic waves of their cobalt complexes (cf. section 2.3.3).

2.3.2. Electron Exchange of Prosthetic Groups

Prosthetic groups tightly linked to proteins have been analyzed polarographically (Table 1), mainly with the acceptors Fe^{3+}, Cu^{2+}, FAD, and others.

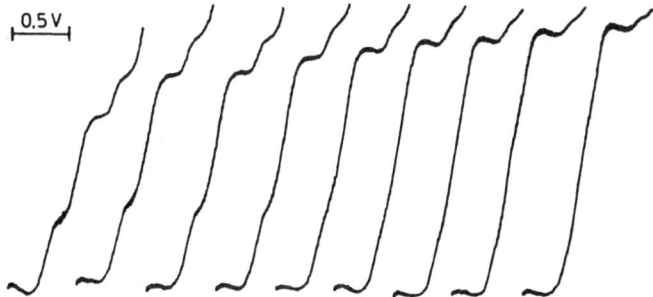

Figure 1. Dependence of the double step in polarography for insulin[4] ($2.5 \times 10^{-5}\,M$) on concentration of ethanol (the third little step is caused by Zn^{++}): 0; 2.5; 4.7; 7; 9; 11; 13; 15; 16.5%, phosphate buffer pH 7, 35°C, starting potential $-0.3\,\mathrm{V}$ (SCE).

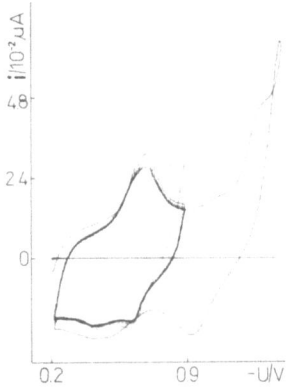

Figure 2. Cyclic d.c. voltammetry of insulin at the h.m.d.e. at pH 7.8, 25°C until equilibrium in the positive region and then to the negative turn. Scan rate: 1 V min^{-1}, starting at −0.2 V (SCE). Cf. Reference 27.

2.3.2.1. The Ferredoxins and Ferritin

Ferredoxin is a small protein,[17] with a molecular weight MW ≈ 5.5×10^3, containing 7 Fe^{3+} linked to an equal number of cysteine residues (in *Clostridia*); however, it lacks histidine, methionine, and tryptophan.

At the d.m.e., with the aid of cyclic a.c. voltammetry, two successive one-electron redox reactions of iron–sulfur clusters were found.[17] Conversely, at the h.m.d.e. a steady state is reached (Figure 3) which is similar to that of apoferredoxin. This observation[17] is explained in terms of a RSH/RSHg redox reaction [cf. Eq. (5)] of two forms of apoferredoxin.

The iron storage protein ferritin[31] has a MW = 700,000 with a Fe^{3+} core buried in a thick protein shell across which electrons cannot tunnel, in contrast to low-molecular weight reducing agents ($FADH_2$). After preparation of the core, one reduction step was measured at −0.38 V (SCE) in 0.01 M HCl solution by normal pulse polarography.

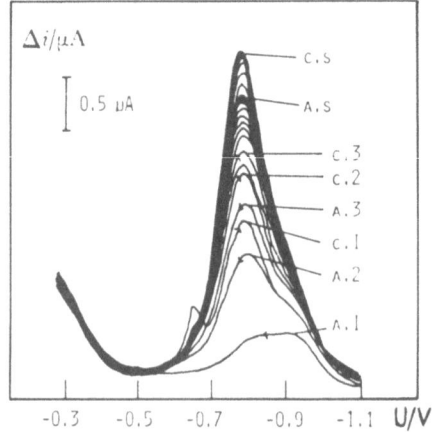

Figure 3. Cyclic a.c. voltammetry[17] (in-phase, $f = 500$ Hz) of ferrodoxin (1.9×10^{-6} M, pH 9.2) after adsorption at h.m.d.e. (−1.1 V). A1, A2, A3: first, second, third anodic scans; C1, C2, C3: first, second, third cathodic scans. The peak at −0.65 V of scan C1 disappears. AS and CS are steady states of the anodic and cathodic scans for apoferredoxin (iron removed).

2.3.2.2. Prosthetic Groups: Flavins, Pyridoxal, and Carotenoids

Oxygenases and dehydrogenases (Table 1) contain flavin residues, whose isoalloxazin ring system is reversibly reduced by $2e$ to 1,5-dihydroflavin. Glucose oxidase[32] (MW = 186,000) contains two isoalloxazin systems and moreover two buried disulfide bridges. The d.p. reduction peak is 0.1 V more positive than the peak of free FAD, due to the protein moiety. Cholesterol oxidase shows similar behavior[33] with nearly the same $U_{1/2}$, −0.32 vs. SCE (pH 7), of a reversible step which was studied extensively at the d.m.e. and the h.m.d.e.

The two pyridoxal cofactors in glycogen phosphorylase b[40] are not reducible in the neutral pH region, because they are hidden in the apolar pocket of the apoenzyme. When the pH is lowered, conformational changes occur and, at pH 5, a reduction step is observed, reaching its maximal height at pH 2. The reason is that the cofactors turn outside the pocket and the bond to the apoenzyme is hydrolyzed below pH 4. As a consequence, the reduction of free pyridoxal is recorded.

There is considerable interest in the reduction of bacteriorhodopsin,[34] the photoreceptor protein of halophilic bacteria, with $U_{1/2} = -1$ V vs. SCE (0.1 M LiClO$_4$), and in its oxidation, with $U_{1/2} = +0.6$ V. This difference of 1.6 V, which is independent of pH, coincides with its fluorescence quantum energy stored. Therefore the carotenoid chromophore may be responsible for both the reduction and oxidation processes. In this respect bacteriorhodopsin is similar to chlorophyll[41] and its excited state permits electron transfer from a weak donor to a weak acceptor.

As a result of basic research on proteins with prosthetic groups, a broad field of applications exists today, especially the use of enzyme electrodes.[54,55]

2.3.2.3. Hemoproteins

The ferric states of hemoproteins are reducible in nearly all cases[4,15,35,36]: ferricytochrome, methemoglobin, deuterohemin, metmyoglobin. However, the electrode processes are more or less irreversible[21] with the exception[37,38] of cytochrome c_3 and c_7 (Table 1). In the case of cytochrome P450 contradictory results have been obtained.[35,39] The pH value and the electrode material[36] may influence the rate of electron exchange and the potential shift.[39] Nevertheless the controlled large-scale electrolysis of methemoglobin and metmyoglobin yields the native hemoglobin and myoglobin with their original biological activity.[35]

The mechanism of electron transfer can be elucidated from the concentration dependence of the diffusion current i_d (d.c. or d.p. measurements) on the one hand and the relative capacity decrease (Figure 4) on the other. The capacity change reaches a limiting value for full coverage of the surface (adsorbed monolayer), whereas the diffusion current further increases as a linear function of concentration. These concentration dependences suggest either a very fast exchange with molecules in the first monolayer or an electron

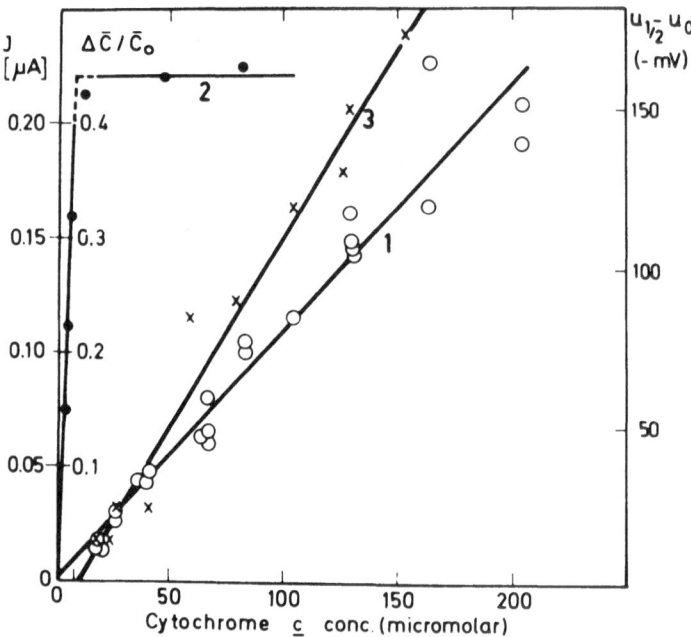

Figure 4. Concentration dependence of limiting current I ($\bigcirc$), capacity decrease $\Delta\bar{C}/\bar{C}_0$ ($\bullet$), and difference between half-wave potential $U_{1/2}$ and normal potential U_0 ($\times$) for cytochrome c. [Courtesy F. Scheller, *Bioelectrochem. Bioenerg.* **4**, 157 (1975).]

transfer through the first layer. Because exchange of adsorbed proteins proceeds slowly, it is more probable that electrons move through already reduced protein molecules of the adsorption layer to reduce proteins approaching from solution by diffusion. The molecules of the first layer may be deformed[19] or unfolded to some extent so that the electron hopping distance is less than 1 nm.

For hemoproteins, the electron exchange rate at the electrode is of the same order as in solution and the porphyrin rings may promote this process.

A second mechanism has been proposed for proteins with disulfide groups (Section 2.3.1) and ferredoxins (Section 2.3.2.1) whose concentration dependence reaches a constant level. Therefore only the adsorption layer can be reduced, whereas at a mercury pool electrode the reduction of all molecules in solution takes place. For this reason, a slow exchange in the adsorption layer is very probable and electron hopping may be inhibited because porphyrin rings are absent.

2.3.3. Catalytic Waves of Complexes

All proteins yield catalytic hydrogen waves ("presodium waves") at the negative end of polarograms by lowering the hydrogen overvoltage as discovered by Heyrovský and Babicka in 1939. Moreover, in 1933 Brdička introduced the catalytic double waves in the presence of cobalt ions which exhibit characteristic shapes, especially for SS/SH-containing proteins (Figure 5). The Brdička currents (i_B) are extremely sensitive for the detection of trace amounts of proteins as low as one ng per ml, and also for the examination of their denaturation and renaturation.[50,51] In spite of numerous studies,[42-50] the mechanism involved is not yet fully understood, but it seems generally accepted that the RSH or RS⁻ groups form complexes with Co^0, which catalyzes hydrogen evolution before the complex is decomposed into Co^0 amalgam and RSH or RS⁻. The wave height is controlled by the number[42] of thiol groups adsorbed and by the recombination rate:

$$RS^- + H^+ \rightleftharpoons RSH \tag{6}$$

For $[Co] \rightarrow 0$ and $[RSH] \rightarrow 0$

$$i_B = Ak_{B'}D_{Co}^{1/2}[Co]D_P^{1/2}[P] \tag{7}$$

(A, electrode surface area; $k_{B'}$, Brdička current activity per mole protein adsorbed at A; D, diffusion coefficients). Ferredoxin,[49] without SS/SH groups, produces a novel Brdička wave originating from Fe—S (S from cysteine).

The proposed reasons for the observation of double waves in the case of TMV-protein with only 1 SH-group per molecule,[44] are presented in Table 2. The explanations of Ivanov and Kuznetzov on the one hand and of Berg and Mairanovsky on the other are rather similar. Remarkable in this respect is the suggestion of Shinagawa[67] who tries to identify the Brdička currents due to an adsorbed n-type semiconductor by their depression in the presence of electron acceptors and their enhancement with electron donors.

Furthermore exceptions have to be taken into account:

- myoglobin contains neither cysteine nor cystine, but gives the Brdička current, which has been attributed to the heme group.[60]
- subtilisin BPN, a proteinase, also without SS/SH, gives no Brdička wave.[60]

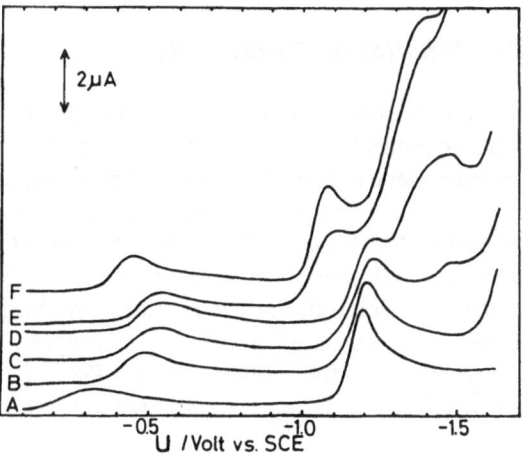

Figure 5. Potential sweep voltammograms[60] of (A) 0.1 M NH₄OH, 0.1 M NH₄Cl and 0.1 M KCl, containing 2×10^{-4} M Co(NH₃)₆Cl with $V = 0.2$ V s⁻¹ at the h.m.d.e., (B) +subtilisin, (C) sperm whale myoglobin, (D) horse heart cytochrome, (E) RNase, (F) bovine serum albumin at 5 μg mL⁻¹.

Table 2
Proposed Explanations for the Double Wave of TMV-Protein

1st wave	2nd wave	Reference
—SH in globulus	—SH in tails	Ivanov[43]
—SH hydrogen bonded	—SH free	Ruttkay-Nedecky[44]
RSH_{ad} chemisorbed	RSH_{ad} desorbing and RSH diffusion	Berg[45]
—SH in hydrophobic parts	—SH in hydrophilic parts	Kuznetzov[46]
adsorption-dependent surface catalysis	diffusion-dependent volume catalysis	Mairanovsky[47]
$\begin{array}{c} S \\ \diagdown \\ \diagup \; Co \\ R_1 \diagup \end{array}$	$\begin{array}{c} S \\ \diagdown \\ \diagup \; Co \\ R_2 \diagup \end{array}$	Kolthoff[48]

2.3.4. Adsorption and Oxidation

Proteins without any prosthetic or SS/SH-groups can be distinguished at the mercury electrode by their different adsorption behavior or at solid electrodes by their oxidation currents. Conformational transitions of poly-L-lysine[40] in solution can be followed in a.c. polarograms: α-helix $\rightarrow$ β-sheet $\rightarrow$ coil. Heating at pH 11.2 to 40°C converts $\alpha \rightarrow \beta$, while changing at 40°C from pH 11.2 to pH 8 causes the transition $\beta \rightarrow$ coil. The oxidation[52,53] of RNase, bovine serum albumin, concanavalin-A, lysozyme, and others has been investigated on gold, platinum, vitreous carbon, carbon paste, paraffin wax-impregnated graphite, and other electrodes using a.c., cyclic, and d.p. voltammetry. Tyrosinyl and trytophanyl residues show well-developed peaks on carbon and graphite surfaces over the range of pH 1–7. When oxidation potentials are more positive, histidine and cystine can also be oxidized in the adsorbed state. During adsorption conformational changes are expected.

2.4. Kinetics of Denaturation

As already pointed out, conformational transitions can be followed by electrochemical techniques in the drop-time range or, using the streaming mercury electrode, to about 0.1 s. In principle, it may be possible, under certain circumstances, to measure kinetic currents caused by fast reactions. In addition to enzymatic cleavage, hydrolytic scissions of peptide bonds under acidic or alkaline conditions have been investigated.[12] Consequently, the adsorption area at the electrode surface increases and buried SS/SH-groups may become more active, but later on are also hydrolyzed. Brdička currents pass through a maximum[12] for proteins with MW > 20,000. A similar pathway was observed during photodegradation of serum albumin by uv light,[56,57] whereas the waves of ribonuclease decrease immediately (Figure 6). Pulse polarography is another sensitive tool for investigating such reactions.

2.5. Complexes with Dyes and Drugs

Besides heavy cations (Zn^{2+}, Cd^{2+}, Cu^{2+}), a number of mainly acidic dyes (methylene blue, methyl orange, eosin) and drugs form strong protein complexes. In order to obtain relations between biological effects and binding data, polarography has been used to record titrations of a ligand with a protein or vice versa.[14,58] This method is based on differences in diffusion coefficients between proteins ($D_P = 0.3-1.2 \times 10^{-6} \, cm^2 \, s^{-1}$) and ligands ($D_L = 3-20 \times 10^{-6} \, cm^2 \, s^{-1}$). The diffusion current during titration is composed of the reduction current of the free ligand (c_f) and the reduction current of the bound ligand (c_b) at independent binding sites[58]:

$$i_d = x(c_f + qc_b) \tag{8}$$

with $c_f + c_b = c_t$, x = constant of Ilkovič equation, $q = i_d/i_{d0} = (D_P/D_L)^{1/2}$ at very high protein concentration, i_{d0} = diffusion current of free ligand, and c_b = total ligand concentration. The value of q may be extrapolated using the equation[56]:

$$\frac{c_a}{1 - (i_d/i_{d0})} = \frac{c_a}{(1 - q)} + \frac{\alpha}{K(1 - q)(1 - \theta)} \tag{9}$$

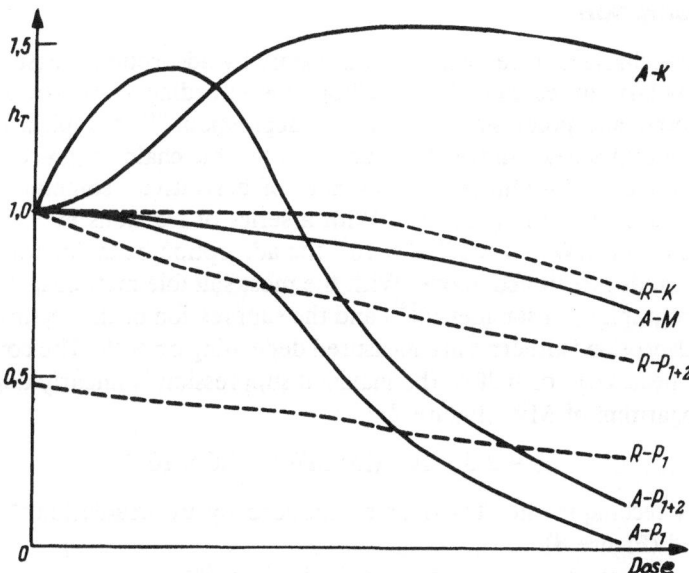

Figure 6. Scheme[57] of relative wave heights (h_r) versus the uv dose for A: serum albumin; R: RNase. K: catalytic presodium wave (peptide wave), M: damping curve for O_2 maximum, P_{1+2}: both Brdička protein waves, P_1: first Brdička protein wave.

where c_a = total concentration of amino acids, α = number of amino acids per binding site, and θ = ratio of covered binding sites n to total number of binding sites N [cf. Eq. (3)].

For $\theta \to 0$ the plot of the term on the left-hand side against c_a yields a straight line with slope $(1 - q)$, the intercept gives K/α. Combining Eqs. (8) and (9),

$$c_f = (i_d/x - qc_t)/(1 - q) \tag{10}$$

and $c_b = c_t - c_f$. From the Scatchard diagrams, $(c_b/c_a)/c_f$ versus (c_b/c_a), α can be obtained as the reciprocal of the intercept, $c_a/c_b = \alpha$.

A competition[59] between electroactive estriol labeled with mercuric acetate, E_sHg, and unlabeled estriol E_s takes place during titration with the corresponding antibody Ab:

$$E_s + Ab \to AbE_s$$

$$E_sHg + Ab \to AbE_sHg$$

The titration curve measured by d.p. polarography has the analogous shape to that obtained by RIA. This electroactive label may be useful for immunoassay.

3. Polysaccharides

3.1. Adsorption

Polysaccharides have been studied mainly by adsorption measurements: electrocapillary curves, dU/dt–U oscillograms according to Heyrovský, cyclic voltammetry, a.c. polarography, maxima depression.[61] A typical example from this group is dextran which forms linear flexible chains of glucopyranose rings in aqueous solution. A large number of derivatives, sometimes linked with dyes, are currently produced. With a series of fractions in a molecular weight range of MW = 4×10^4–2×10^6, the adsorption behavior was studied by all methods mentioned above. With the most suitable methods,[61] such as a.c. polarography (tensammetry[14]) and the suppression of the O_2-maximum, typical adsorption isotherms are measured depending on MW. The concentration (%) necessary for half of the maximal suppression is linearly dependent on the logarithm of MW (Figure 7):

$$c_{1/2} = 2.8 \times 10^{-4}(\log MW) - 7.6 \times 10^{-4} \tag{11}$$

With this technique the strand break induced by uv irradiation[64] can be monitored (Figure 8).

The a.c. data are consistent with the isotherm[65]:

$$\frac{\theta}{1 - \theta} \exp(-2a\theta) = (bc)^{1/\nu} \tag{12}$$

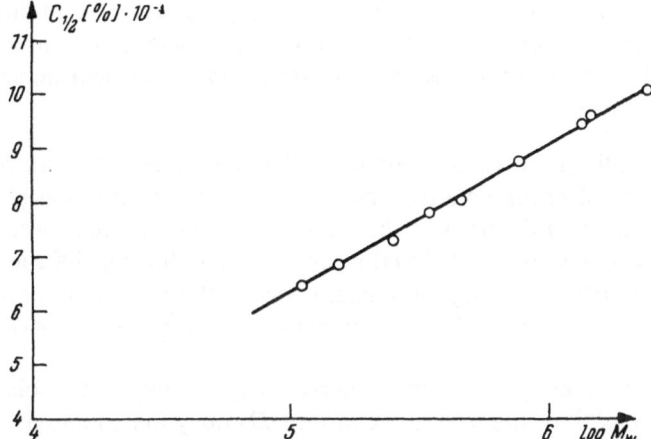

Figure 7. Damping of O_2 maximum by dextran.[61] Dependence of concentration (%) for half-suppression on molecular weight of fractions.

with θ = surface coverage, a = a measure of interaction energy, b = adsorption equilibrium constant, and ν = total number of adsorbed segments. (For $a = 0$, $\nu = 1$, the Langmuir isotherm is obtained.)

3.2. Electron Exchange

Basic research was done with chromophordextran, which exhibits electron exchange with its incorporated dye Cibacron Blue, in order to answer two questions:

1. What is the upper limit of molecular weight and concentration at which the Ilkovič equation can be applied for a direct determination of diffusion constants of linear macromolecules or their electron uptake?

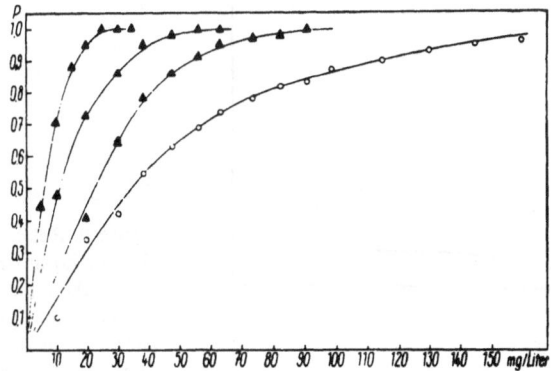

Figure 8. UV splitting of dextran,[64] indicated by damping of the O_2 maximum P (relative): $\bigcirc$ nonirradiated; $\blacktriangle$—$\blacktriangle$ (from right to left) 5, 20, 320 min irradiation at 260 nm.

2. How many reducible groups inside the large globuli or coils of biopolymers are not able to capture electrons because complete access directly to the electrode surface is not possible (electron transfer mechanism and surface denaturation)?

To answer both questions d.c. polarographic measurements were performed with fractions of chromophordextrans, which are condensation products of Cibacron Blue with dextrans in the 2000 to 50,000 molecular weight range. At pH < 7 and low concentration they are reduced, showing diffusion currents i_d. Owing to surface coverage, at higher concentrations the polarographic step height becomes progressively independent of concentration for MW > 10,000 (Figure 9).

Splitting by enzymatic action increases i_d steadily. The polarographic determinations[63] of the diffusion constant D_P obey the equation:

$$D_P = (i_d/\text{const. } n_e c_e)^2 \tag{13}$$

where i_d is the mean diffusion current, n_e, the electron uptake by one depolarizer group, and c_e, the effective concentration of the bound depolarizer.

Comparing the D_P values with D_0 values obtained by optical determinations under the same conditions, these agree fairly well for MW < 500,000.

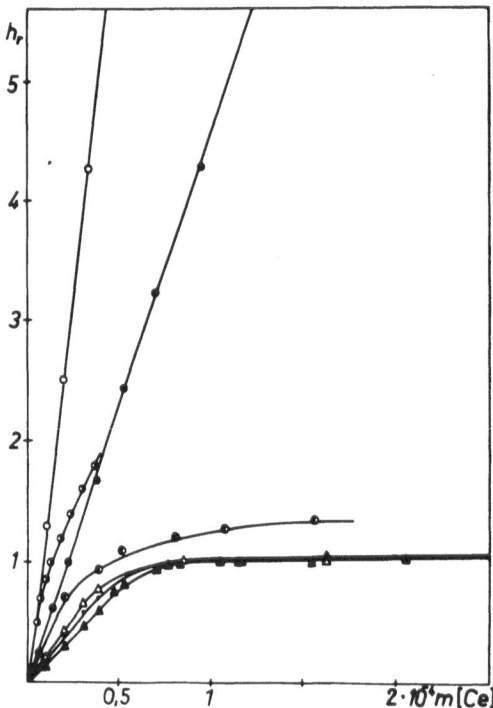

Figure 9. Relative step heights of chromophordextrans[63] depending on the effective chromophore concentration (pH 7) for chromophordextrans of MW: ◑ 4000, ● 10,900 (60°C); ◐ 10,900 (25°C), △ 77,500, ● 110,000, ▲ 500,000; ○ free chromophore.

Deviations become significant for MW $\geq$ 500,000 because the coil dimensions are of the same order as the thickness of the diffusion layer, and depolarizer groups inside are shielded by adsorbed segments and the distances.

These limitations[63] of the classical Ilkovič equation have major implications for analogous measurements, not only with proteins as discussed already (buried depolarizer groups) but also for nucleic acids (extreme coil dimensions).

4. Nucleic Acids

4.1. Electrochemical Classification

In contrast to the variability of proteins, nucleic acids exhibit rather similar primary and secondary helical structures and a limited number of defined conformations, ascertained mainly by x-ray diffraction analysis. The classical Watson–Crick double helical B-form has a variable and flexible chain,[7,8,69] depending on its environment and ligands. Moreover, about 1% of the base pairs (BP) are opened by fluctuation, possibly as soliton excitations,[70] which retain coherence by sharing the energy of a twist deformation among 10 BP, and the hydrogen exchange[71] open state is consistent with this model. Electron exchange[4,72,73] takes place with

- cytosine and adenine: irreversible reduction (mercury electrodes, buffer solution)
- guanine and adenine: irreversible oxidation (graphite electrodes)
- uracil and thymine: irreversible reduction in dimethylsulfoxid[131]

and depends on adsorption and protonation. The electrode mechanisms of monomers and oligonucleotides have been studied carefully under all conditions,[72,73] and they may be similar in polynucleotides and nucleic acids; however, new parameters have to be taken into account. Therefore each polarographic method does not produce signals under all conditions. A simplified scheme, which serves to illustrate the complicated situation, is presented in Table 3. From this distribution of distinct signals, it becomes evident that the timing and mechanism of polarization is very important for comparison of different methods in the literature.

4.2. Polyelectrolyte Behavior in Solution

Nucleic acids are polyanions because each phosphate group has a strong negative charge.[74] On the other hand, the N_1-position in adenosine (pK 3.4) and in guanosine (pK = 9.2), the N_7 in guanosine (pK = 2.2), and the N_3 in cytidine (pK = 4.4), in uridine (pK = 9.2), and in thymidine (pK = 9.8) must be protonated to be able to pick up electrons. Nevertheless, the secondary and tertiary structure in solution is mainly determined by the screening of the

Table 3
Polarographic Responses[a] of DNA obtained with Various Methods

Reduction + adsorption	ds-DNA		ss-DNA		Method
	pH < 8	pH > 8	pH < 8	pH > 8	
d.m.e.	−	−	+	−	d.c., d.p.
	+	−	+	−	n.p.
h.m.d.e.	+	−	+	−	d.c., d.p., n.p.
	(+)	+	+	+	a.c.

[a] +, distinct signal; −, no signal.

negative sugar–phosphate backbone by counterions. For a chain of univalent negative charges with mean distance b, the degree of counterion condensation[75,76] is given by the parameter f, with the reduced charge density, according to

$$f = 1 - (|z|\lambda)^{-1} \tag{14}$$

where

$$\lambda = \frac{e^2}{\varepsilon b k T} \tag{15}$$

and e is the charge of the electron, z is the charge of the counterion and ε is the dielectric constant of the pure solvent.

For the B-form of the double helix, the charge projection to the helix axis for each phosphate group is $b = 0.17$ nm and hence $\lambda = 4.2$ or $f = 0.76$. This means that 76% of one phosphate charge is screened by counterions with a local concentration, $c_L = 1.2$ M according to

$$c_L = 0.0243(\lambda b^3)^{-1} \tag{16}$$

within a cylindrical mantle of 0.7 nm thickness. Because of the greater distance, $b = 0.39$ nm, in single-stranded DNA, the other values decrease to $\lambda = 1.8$, $f = 0.44$, and $c_L = 0.21$ M. At low ionic strength the repulsion between the phosphate charges increases and so does the stiffness of the whole chain. This is also a basic result for the behavior of DNA at surfaces. Further experimental results (melting, precipitation, base pair opening, persistence length) can be explained by this Manning model.[75] The site binding of ions to the polyanion is another theoretical concept, which has also been extended to the binding of larger charged ligands.[77]

4.3. Polyelectrolyte Behavior at Electrodes

Apart from early characterizations of nucleic acids by their oscillopolarographic activity,[3,5,78] or their damping of oxygen maxima,[57] most information has been obtained by a.c. polarography,[79,80] current–time curves,[81,82] (d.m.e.), pulse polarography,[83] linear sweep voltammetry,[84] capacity measurements,[85,86] modulation polarography,[87] and other variations. The vast amount of results which have been published so far, however, are not enough for a quantitative model.

4.3.1. Adsorption

The investigation of electrochemical behavior of nucleic acids involves adsorption and various adsorbed states, since *adsorption determines the electron exchange.* Adsorption depends on *the conformation of nucleic acids in solution,* which is a function of:

- electrode potential
- ionic strength (counterion condensation)
- pH
- temperature
- nature of the solvent
- the ligands (inorganic and organic, outside the double helix or intercalated).

As for other macromolecules,[88] adsorption takes place only with segments of the whole chain molecule, whereas the connecting parts in between form loops into the bulk of the solution. The surface density of segments[88–90] at a given potential depends on the flexibility of the whole chain and on the time necessary to reach saturation, which is a rather irreversible process.[88] Therefore one has to distinguish between the d.m.e. and the h.m.d.e. (cf. Table 3). In this section, only results obtained by a.c. polarography in alkaline medium, where electron uptake is suppressed, will be discussed.

4.3.1.1. Adsorption at the D.M.E.

The decrease in the a.c. polarogram indicates a broad adsorption region of ds-DNA ending with the rearrangement peak 1 (-1.2 V), whereas ss-DNA[91] shows additionally the more negative peak 2 (-1.4 V) (Figure 10). Since its discovery,[79] this sharp second peak has been used as a sensitive tool for detection of single-strand formation during disturbances of the double helix. The position of its potential is evidence for the stronger adsorption of ss-DNA because of its free rotating bases and its lower b-value.[15] At this point, it is important to note that ss-DNA is still adsorbed in the potential region of reduction of its bases C and A, whereas ds-DNA is not! In this respect the use[88] of the h.m.d.e. has important consequences.

4.3.1.2. Adsorption at the H.M.D.E.

Using the h.m.d.e., the slow adsorption kinetics are measurable[88,93] until saturation is reached. The highest surface coverage without denaturation is in

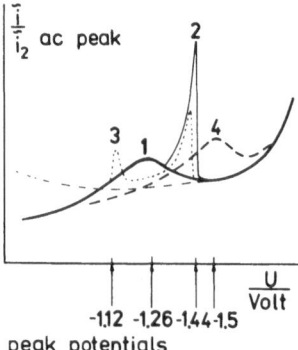

Figure 10. Heights of a.c. peaks[99] relative to peak 2: 1: ds-DNA at d.m.e., 35°C; 2 + 1: ss-DNA at d.m.e., 35°C, or ds-DNA at h.m.d.e., 35°C; 3 + 2: ds-DNA at h.m.d.e., 85°C; 4: ds-DNA at h.m.d.e. potential-induced compaction at 35°C.

the neighborhood of the electrocapillary zero potential. Nevertheless the adsorption energy of a single segment is rather small, but interaction between the segments forms a large network, which can stay adsorbed up to rather negative potentials, at least up to the reduction potential of the bases C and A.[93]

At more negative potentials, only few segments of the same molecule remain adsorbed, which is indicated by the rounded peak 1, and intermolecular interactions of the loops increase, resulting in growing thickness of the adsorption layer. The sharp peak 2 is caused by partial desorption of these adsorbates. For ds-DNA, both peaks are recorded only in the negative direction of polarization (Figure 11), but not in the positive one, in contrast to the higher peak 2 of ss-DNA. The reason is the difference in the potential dependence of adsorption, as mentioned for the d.m.e., and the higher tendency of association for ss-DNA and polynucleotides.[101] A similar hysteresis effect[93] (Figure 11) is seen for tylose and triton X-102, and hence a helix–coil transition

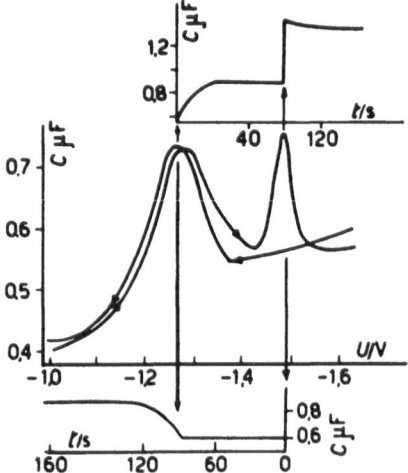

Figure 11. Hysteresis in differential capacity[93] at two h.m.d.e. (a.c. peaks 1 and 2) of ds-DNA (pH 7, 25°C) and time dependence for peak 1 (bottom trace) and peak 2 (top trace). Compare Fig. 13.

region, as sometimes postulated for ds-DNA,[106] is implausible. Both a.c. peaks show a characteristic decrease[93,94] resulting in a final adsorption state. An alternative explanation for the peaks 1, 2, 2a, 3 (peak 3 is the above-mentioned peak 2), where an opening of adsorbed ds-DNA segments may appear), was postulated by Brabec *et al.*[140] Peaks 2 and 2a are recorded only under special conditions.

4.3.1.2.1. The potential-induced state (π-state)[95–100]. The decrease of peak heights can be induced in the potential region of peak 2 either by maintaining a constant potential or by potential sweeps (Figure 12). These fading processes at fixed potentials can be seen schematically in Figure 13, where peak 2 moves to more positive potential values before it disappears. The product must be a more strongly adsorbed network of segments and loops unable to show any response to the alternating field at the electrode. This novel form of adsorption behavior is referred to as the compact π-state of DNA. Such π-states have been discovered also for other biopolymers, e.g., poly A (Figure 14).

The fading process depends on

1. temperature: with increasing temperature, the fading process is accelerated and at least two intermediates can be obtained, because the flexibility increases and the double helix becomes unstable (premelting and melting, cf. Section 4.4),
2. frequency: compaction is favored by lower frequencies,
3. ionic strength: shielding of the phosphate charges enhances the fading process (Figure 15),
4. complex formation: the outside complex in the narrow groove with the peptide netropsin (cf. Section 4.5) causes a greater stiffness of the double helix and consequently a slower fading process.

The adsorption of ds-DNA can be characterized by the scheme

$$\text{bulk} \qquad\qquad \text{peak 1} \underset{\text{electrode}}{} \text{peak 2}$$
$$\text{DNA}_{\text{soln}} \underset{\text{diffusion}}{\rightleftharpoons} \text{DNA}_{\text{ads}} \underset{\text{layer}}{\rightleftharpoons} \text{DNA}_{\text{ass}} \qquad\qquad (17)$$
$$\downarrow$$
$$\pi\text{-DNA}_{\text{ass}}$$
$$(\text{no peak})$$

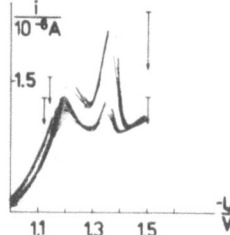

Figure 12. Hysteresis of peak-fading process[96] for ss-DNA by a.c. voltammetry in the potential range of peak 1 and 2 at the same h.m.d.e.: 12 cathodic sweeps (upper trace); 12 anodic sweeps (lower trace).

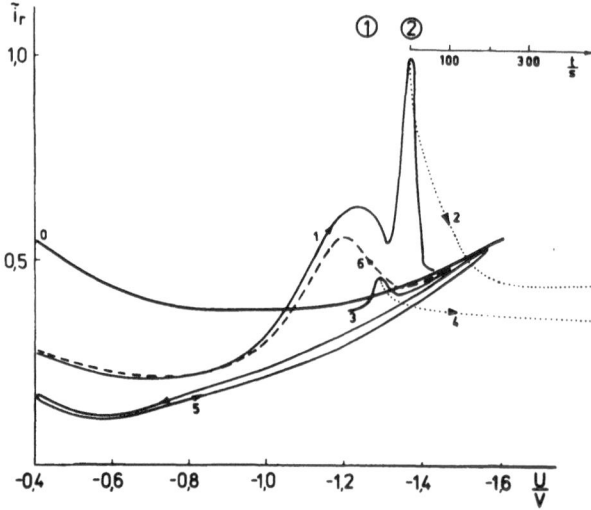

Figure 13. Scheme[96] of peak-fading process and formation of the π-state of DNA, recorded by a.c. polarography. Curve 0: base electrolyte; Curve 1: peak ① and ② of ds-DNA, 65°C; Curve 2: decrease of peak ②, fixed potential; Curve 3: rest peak ② at more positive potential, and Curve 4: its decrease to zero; Curve 5: adsorption region of the π-state of DNA; Curve 6: anodic polarization—only peak ① is recorded.

Once in the π-DNA$_{ass}$ state no disassociation, desorption, and penetration by fresh DNA$_{ad}$ seem to be possible, whereas low-molecular weight electron acceptors can still be reduced. DNA, as well as polynucleotides, tends to aggregate irreversibly in solution under several conditions. The description of a.c. polarograms of ds-DNA must be extended by studies under special conditions.

4.3.1.2.2. Special peaks. Only for ds-DNA at low ionic strength has a tensammetric peak 0 been recorded (d.m.e.) near the electrocapillary zero potential.[103] The desorption of the weakly adsorbed phosphate backbone

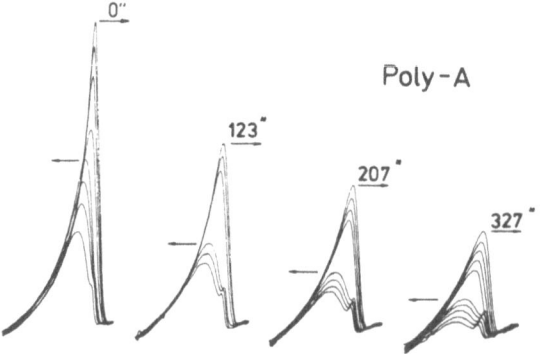

Figure 14. Peak-fading process[96] of poly-A (time in s) also with hysteresis which may be caused by two potential-influenced conformations in the adsorbed state.

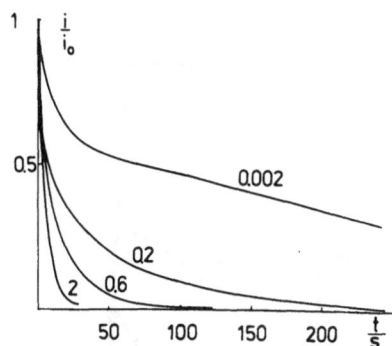

Figure 15. Dependence of peak-fading process (ds-DNA) on the buffer molarity (labeled on the curves).[96]

may be responsible. This peak was not observed for ss-DNA because of the lower f-value and overwhelming base adsorption.

Not so clear is the reason for a further peak (linear ds-DNA) between peak 1 and 2 observed at elevated temperatures[103] or by means of in-phase a.c. polarography (d.m.e.) at room temperature,[102] where it was formerly designated[92] peak 2'. With open-circular DNA, which shows similar adsorption properties to closed duplex DNA and linear ds-DNA, this peak 2' [however, named peak 2 in Reference[104]] should be present at room temperature (pH 8), too.

With the exception of peak 0, it is unfortunate that the interpretations of several authors for the other peaks are different. According to Reynaud,[102] "peak 2 (2') is a capacitive peak in an alkaline medium and a faradaic peak in an acidic medium"! On the other hand, under the same conditions (pH 5.6) his peak 2 for ds-DNA is only 0.04 V more positive than his peak 3 for ss-DNA, but the latter is 7 times higher!

4.3.2. Electron Exchange

As pointed out earlier, reduction of protonated nucleic acids in the adsorbed state takes place at mercury electrodes whereas oxidation is measured mainly at solid electrodes.

4.3.2.1. Reduction

Mostly d.p. and n.p. polarography, exhibiting normally peaks I (-1.2 V) and II (-1.45 V) for ds-DNA or peak III (-1.65 V) for ss-DNA, are used,[106] because of higher sensitivity and broader application[83,107] than d.c. polarography and sweep voltammetry. The peak properties depend on the conditions[106] in acidic as well as in alkaline solutions and on the conformations of DNA. For instance, the height of the asymmetric peak II in d.p. polarography increases by complex formation with spermine indicating that the negative flank of this peak is limited by the adsorption boundary of ds-DNA corresponding to the i_{max}–Upp curve (Figure 16). The height of peak II increases with

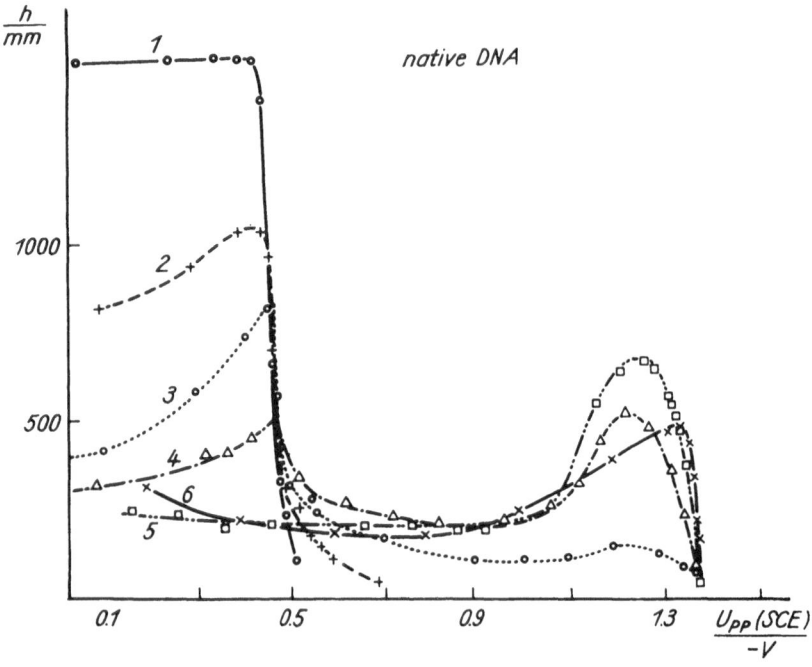

Figure 16. Relative heights of reduction currents[100] (i_{max}, n.p. polarography) of ds-DNA (curve 1) and ss-DNA (curve 5) depending on the difference ΔUpp between the half-wave potential (O) and the prepolarization potential Upp. Curves (2–4) are obtained from complexes of ds-DNA with spermine. Curve (6) n.p. polarogram of DNA reaching the maximal height.

temperature and peak II is converted into peak III by thermal and other denaturations.[107]

The essential influence of adsorption prior to reduction can be shown by n.p. pulse polarography, changing the starting potential ≡ prepolarization potential (Upp).[93,94,107,108] In Figure 16, the relative reduction curve (i vs. ΔU) for C and A is opposite to the maximal currents for a series of Upp of ds-DNA (curve 1), ss-DNA (curve 5), and complexes of ds-DNA with increasing spermine concentrations. It is clearly shown that the adsorption region of ss-DNA overlaps the reduction curve 6, whereas that of ds-DNA does not. This handicap in adsorption is the reason why ds-DNA cannot be reduced by d.c. and d.p. polarography at the d.m.e. In the complex with spermine the phosphate charges are shielded and, in spite of the resulting stabilization of the double helix, the reduction current of the complex increases markedly. An extraordinary effect of ionic strength on the i_{max}–Upp curves has been measured (Figure 17). At low ionic strength (0.01 M Na$^+$) the stiff segments are unshielded and are adsorbed only up to the electrocapillary zero potential.[108] Conversely, at high ionic strength (>0.1 M Na$^+$) the flexible chain

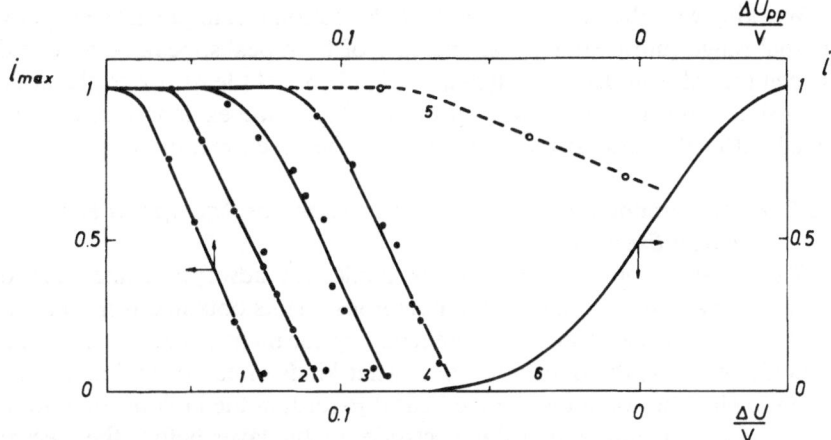

Figure 17. Dependences[108] of the n.p. polarographic step height (h) on the prepolarization potentials Upp for ds-DNA (50 μg mL^{-1}, phosphate buffer, pH 6.2, with NaCl). Ionic strength: 10^{-2} M (curve 1), 2.2×10^{-2} M (curve 2), 4.4×10^{-2} M (curve 3), 8.8×10^{-2} M (curve 4), 0.36 M (curve 5), 0.75 M (curve 6).

enables adsorption until -1.4 V and the surface DNA concentration[93]—not the segment adsorption of each molecule—in the peak region increases.

For intermediate ionic strengths, because of higher repulsion, no current maxima occur, i.e., the current density depends mainly on the amount of DNA molecules screened by counterions and their mobility inside the loose adsorption layer. Therefore surface denaturation,[106,108] favored by the energy of adsorption, seems unlikely, whereas the high field strength ($\approx 10^7$ V cm^{-1}) in 1 M salt solution is effective just in the Helmholtz double layer region (<0.5 nm), i.e., this concerns a small part of the few segments adsorbed and influences merely less than 1% of the whole DNA molecule, which extends over the diffuse double layer into the bulk of solution.[102]

4.3.2.2. Oxidation

Starting with the oxidation mechanisms of G and A bases[72,73,109] and oligonucleotides, the electrochemical activity of nucleic acids is also dependent on the presence of purine residues, as shown by well-separated oxidation peaks on d.p. voltammograms[110] (carbon electrodes) at +0.89 V for G and +1.17 V for A (pH 7.2). Nucleic acids are strongly adsorbed at the rough surface—ss-DNA more than ds-DNA; however no pronounced surface denaturation of ds-DNA takes place at graphite electrodes charged to potentials of 0 to +1.4 V.[110] Distinguishing between G- and A-peaks can enable the determination of the G–C content and, furthermore of conformational changes. The A–T specific netropsin complexes increase the stiffness of ds-DNA, as does spermine, on the one hand, and decrease the negative charge, on the other hand, resulting in a lowering of the A-peak more than the G-peak.

Working with the voltammetric prepolarization technique in the negative potential range (until -0.8 V), an increase of both peaks, peak A $>$ peak G, has been recorded in 0.1 M sodium acetate, pH 6.6. At least two explanations are possible; however, a decision on the validity of the explanations can only be made after the ionic strength dependence has been measured.

4.3.2.3. A Phenomenological Electrostatic Model for Adsorption and Electron Exchange

The construction of a satisfactory model for adsorption and electron exchange which can account for the numerous results obtained is not an easy task.[111,112] The general idea of segment adsorption mentioned before is widely accepted; however, the question arises: what kinds of conformational transitions take place between the double helical B-form in the bulk of the solution and the adsorbed state within the electrode double layer before the electron uptake? Of course, when a DNA chain and its hydrated condensed counterions adsorb, several interferences occur depending on the surface charge. Combining the theory of Manning[75] (Section 4.2), the results of Flemming (Section 4.3), and the adsorption isotherm of Silberberg[113]:

$$\ln \frac{\theta}{P_a P_V} = P\left[\Delta G_s + \Delta G_e + \ln \frac{1-\theta}{1-P_V} + \Delta E_m \theta - (\gamma - f\gamma)^2 \frac{\theta}{P_a c_V}\right] \quad (18)$$

which includes: ΔG_s = the excess free enthalpy per adsorbed segment, and the coulombic term:

$$\Delta G_e = -q^{\pm} z (1-f) \kappa^{-1} \left(\frac{e^2}{\varepsilon kT}\right) \quad (19)$$

where

$$q^{\pm} = q_0^{\pm} + z(1-f)\frac{\theta}{b^2} \quad (20)$$

and the symbols represent the following: θ, fraction of the surface with adsorbed segments; P, total number of segments; P_a, fraction of adsorbed segments; P_V, fraction of polymer in the bulk of solution; ΔE_m, mixing energy; γ, activity coefficient of free counterions; c_V, added salt concentration; q, charge density after segment adsorption; q_0, charge density before adsorption; z, f, and b have the same meaning as in Section 4.2.

Under the premise that f should be the same for P_a and P_V at the positively charged electrode (q^+), ion pairs are formed and ΔG_e of DNA ($z = -1$) enhances ΔG_s and hence θ. Conversely, for q_0^-, repulsion of the polyanion occurs and ΔG_s will be diminished for the phosphate backbone. However, within the distance $\kappa^{-1} = 0.3$–1 nm, the counterion condensation on the attached side of DNA segments is influenced asymmetrically by q^- and by the gradient of the electric field $E \geq 10^6$ V cm^{-1} in the double layer. This field polarizes kinked helical parts perpendicular to the surface, on the one hand, and dissociates counterion pairs attached parallel to the surface to a degree of >0.2 by the dissociation-field effect (M. Wien), on the other hand. This Wien effect is in competition with the enhancement of surface concentration of cations with increasing q_0^-. Here we find an explanation for the influence of spermine and netropsin complexes and also the influence of ionic strength on the width of the adsorption region of nucleic acids. Incidentally, a constant height E in the bulk of the solution can cause a counterion displacement yielding a macrodipole, and hence unwinding from the chain ends may occur by repulsion between both strands.[114]

These concepts are taken into consideration in the following steps leading to an electrostatic model[96,99] according to H. Berg (Figure 18).

Step 1: After diffusion at the surface, partial dehydration[113] of the adsorbed segments within the double layer occurs.

Step 2: As a consequence of step 1, partial transformation of the segments from the B-form into a type of A-form,[8,69] which is of course deformed by the asymmetry of the adsorption forces, occurs. Assuming the geometry of the canonical A-form, the parameters of the B-form change to $b = 0.13$ nm, $\lambda = 5.5$, $f = 0.82$, $c_L = 2.0\ M$. Hence, this transition is accompanied by an additional condensation of nearly $\Delta f = 0.06$ counterions per phosphate.

Step 3: The enhancement of base fluctuation[70] as a consequence of the A-form transition, dipole polarization, and orientation by the field gradient in the double layer further increases ΔG_s. The electrode charge modifies ΔG_e [Eq. (19)] in the following way:

(a) Positively charged electrode q_0^+: ΔG_e increases with P_a and strengthens the total adsorption enthalpy $(\Delta G_s + \Delta G_e)$. Therefore, the condensed counterion concentration c_L is decreased by repulsion of the electrode. Moreover, the enhanced anion concentration leads to release of counterions and the result is a higher P_a and θ at lower and medium c_V. High c_V, i.e. lower $(1 - f)$, decreases κ^{-1} and diminishes q^+ and ΔG_e.

(b) Negatively charged electrode q_0^-: According to Eq. (19), ΔG_e decreases with q^- and so does the total adsorption enthalpy. Hence, P_a decreases and the loops expand steadily as q^- increases. This extension of the adsorption layer from the surface is indicated by the a.c. peaks 1 and 2 as discussed previously. For ds-DNA with $\lambda = 4.2$ the repulsion grows more strongly than for ss-DNA with $\lambda = 1.8$, despite the higher f value.

To explain the potential-induced peak fading and consequently the compaction to the π-state (Section 4.3.1.2.1.) two further processes are invoked:

Step 4: Accumulation[96] of cations in the vicinity of the electrode occurs by shifting of the electrode potential to negative values. In this way the counterion diffusion potential

$$U_D = zRT \ln (c_V/c_L) \tag{21}$$

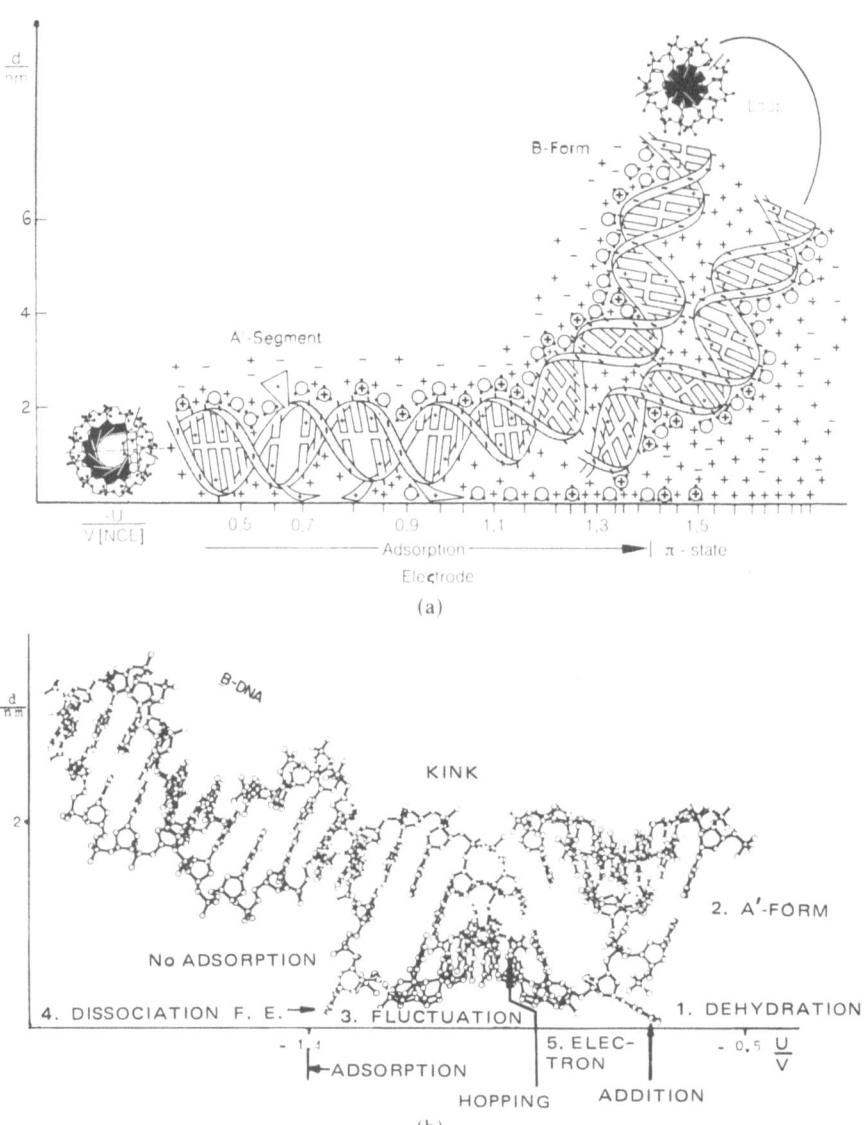

Figure 18a. Electrostatic model[99] according to H. Berg for adsorption and compaction to the π-state. Overall scheme: + Protonated fluctuating bases; Ω water molecules; ⊕ H_3O^+ ions; + counterions. b, Enlarged cut-out of an adsorbed segment (A'-form) and a free part (B-form).

near the electrode is diminished, because c_L increases. If f approaches unity, all phosphate charges will be more and more neutralized and the repulsion is ineffective, i.e., $\Delta G_e \to 0$, whereas ΔG_s brings about strong adsorption. Consequently, phase separation and precipitation spontaneously form a rigid layer.

Such an instability[115] of "neutralized" segments ($f > 0.9$) is supported by the decrease of ε in the double layer, the shortening of b by the $B \rightarrow A$ transition, the lowering of temperature, and by highest values of c_V and z.

As ΔG_e vanishes, the persistence length a:

$$a = a_0 + 339 - 56 \log c_V \tag{22}$$

decreases, too, and the DNA loops are able to contract and water molecules are squeezed out. In this compact mesomorphic phase,[116] the distance between both helix axes may be diminished in the π-state to 2.5–4 nm.

Step 5: The dissociation-field effect can disturb the π-state, too, by disturbing the counterion condensation on the attachment sites of adsorbed segments in such a way that an asymmetrical local charge distribution around the backbone occurs, resulting in more kinks or bending of the double helix. In this way, crystalline precipitates are favored. However, this is not a closed layer, because low-molecular weight electron acceptors are still reduced at such a covered electrode. The π-state of DNA exhibits strong adsorption; neither desorption nor further DNA uptake was observed.

This model takes into account mainly coulombic forces; however, hydrophobic interactions and van der Waals forces (including specific interactions of bases, sugars, or amino acids in nucleoproteins) have a modifying influence.[117] This last general point has a major implication for all conformations of adsorbed biopolymers carrying electron exchange groups.

Step 6: Electrons can be taken up directly by protonated bases (or amino acids) attached immediately to the surface. Additionally a second mechanism is possible, namely, an electron hopping over a distance to bases (or amino acids) inside the dehydrated parts of adsorbed molecules.[4,118] Because water is squeezed out, especially for the semiconductor network of the π-state, the activation energy for electron transfer into the conduction band should be lowered and thus electrons can move inside the double Y helix (or in the globulus[125]).

Under conditions of photoelectrochemistry,[119] DNA adsorbed on or in the vicinity of the electrode acts as an electron scavenger[120,121] over distances between 1–10 nm.

Up to now all experimental facts of reduction and oxidation are consistent with this model and some conclusions can be drawn. However, a quantitative description by an extended polyelectrolyte theory is necessary, in combination with results of surface spectroscopy.[19] From a recent study[123] of surface-enhanced Raman scattering of native CT-DNA (ds) and denatured DNA (ss) at a Ag-electrode the following conclusions may be drawn:

(a) ss-DNA exhibits mainly the same inverse vibrational bands at $734\ \text{cm}^{-1}$ and $1330\ \text{cm}^{-1}$ as poly-A, in contrast to the low response of ds-DNA.

(b) The appearance of the sugar–phosphate symmetric stretching frequency ($818\ cm^{-1}$) of ds-DNA indicates that the helical structure "remains to some extent saved" [123]!

(c) The apparently weak scattering of the other bases may be a confirmation of the favored fluctuation of A–T clusters.[70]

Further spectroscopic research is essential to clarify the conformation of double helical DNA adsorbed at the d.m.e. in buffer solution of pH~7.

4.4. Kinetics of Denaturation

Generally the helix–coil transition[4,8] or denaturation can be caused by low ionic strength:

$$DNA_h \cdot r_h M^+ \rightleftharpoons DNA_c \cdot r_c M^+ + (r_h - r_c)M^+ \tag{23}$$

i.e., $f_h > f_c$, (r = number of cations), or by low pH values:

$$(r_c - r_h)H^+ + DNA_h \cdot r_h H^+ \rightleftharpoons DNA_c \cdot r_c H^+ \tag{24}$$

i.e., the coil (c) binds more protons than does the helix (h) (pH 3.5). On the other hand, the destabilization occurring at alkaline pH is due to the fact that guanine and thymine are readily deprotonated in the coil form. The alkaline cooperative denaturation of DNA, recorded by a.c. peak 2, is in accordance with spectroscopic measurements.[80] The same agreement of the T_m values has been found for the melting curve recorded[79] by a.c. polarography (peak 2) (Figure 19) after the premelting region.[78,83] Melting starts with nucleation of loops and their growth in the A–T clusters. Then these loops merge, forming

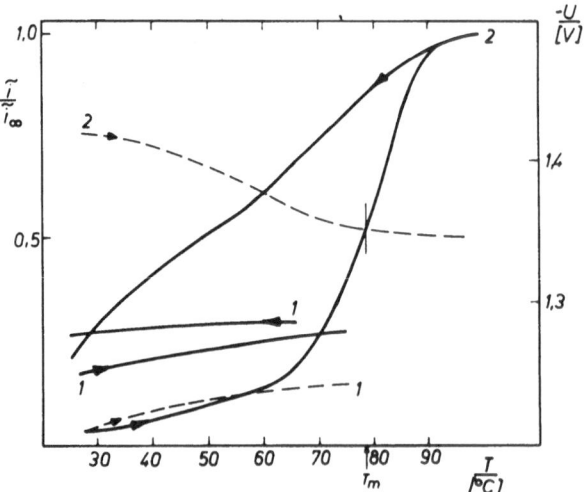

Figure 19. Hysteresis[100] of ds-DNA melting measured by a.c. polarography: —— relative peak height ① and ②, ----- peak potentials of ① and ②.

a large coil region at the expense of the intervening G–C clusters, leading to strand separation. By fixing the electrode potential on the a.c. peak 2, we were successful in following the overall melting and unwinding process after a temperature jump by rapid streaming of the DNA solution (60°C) into the hot buffer (90°C) in a special mixing cell. The relaxation response (Figure 20) at the d.m.e. shows an instantaneous rise of the peak height, followed by relatively slow processes in the range of minutes. At the beginning, hydrogen bridges become weak, the stacking interaction is disturbed, and the nucleation of loops starts. The slow relaxation process is limited by the unwinding and strand separation, depending on molecular weight. The unwinding relaxation time τ_u can be calculated from

$$\frac{i(t)}{i(t \to \infty)} = 1 - e^{-t/\tau_u} \tag{25}$$

For complexes, τ_u increases markedly,[4] depending on the dissociation kinetics of the ligand. The retardation of unwinding and the slow rate of dissociation is essential for the biological activity of inhibitors of polymerases. Local denaturations are caused by ionizing or uv irradiation[124] and reductive denaturation[132] at −1.8 V, which can be recorded with high sensitivity.

4.5. Complexes with Dyes and Drugs

As in the case of proteins[125] (Section 2.5), the complexation with dyes and drugs[56,126,127] is also an important field in nucleic acid research. The

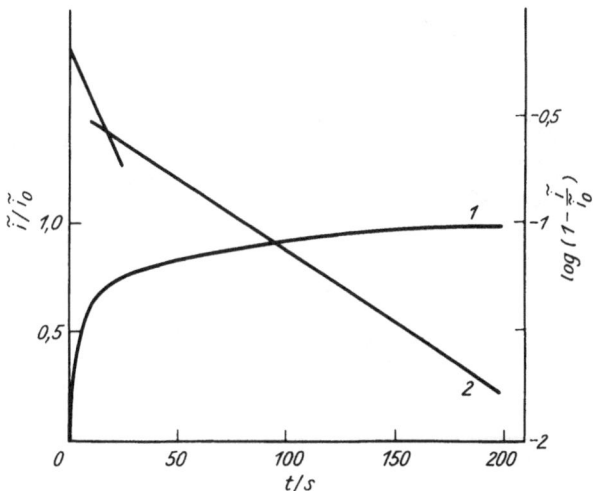

Figure 20. Relative relaxation curve (curve 1) of unwinding[100] of ds-DNA after the temperature jump 60° → 90°C (pH 7), measured by a.c. polarography at peak potential ②. Logarithmic linearization (curve 2).

equations given for proteins [Eqs. (8)–(10)] are valid for nucleic acids, but with a lower q-value, because the diffusion constant of DNA is only 10^{-8} cm^2 s^{-1}. The equilibrium constants of complexes with biologically active ligands, e.g., antibiotics,[122] are of the order of 10^6 M^{-1} and, therefore, their determination requires high sensitivity, e.g., d.p. polarography. Titration technique 1, starting with the polarogram of the ligand, was performed for a series of anthracyclines[128] (Figure 21). From the binding isotherms, r vs. c_f, the affinities for DNA can be estimated: violamycin BI > aclacinomycin > adiramycin > daunomycin > iremycin > steffimycin. Because $q \to 0$ for the quinone wave, whereas the second wave, caused by ring system D, is still measurable, an intercalation model—with the shielded quinone part inside and ring D plus sugar residue outside—is favored.

Another example of intercalation is thiopyronine,[129] which photooxidizes guanine by a photodynamic radical mechanism.[130] A relation between oxidation potentials of bases and their photooxidizability has been discovered,[129,130] not only for numerous complexes with organic ligands but also for complexes with metal ions, e.g., *cis*-Pt,[131] O$_s$O$_4$,[132] Cu^{2+}, Hg^{2+}. Osmium tetroxide should be able to form complexes only with single-stranded nucleic acids. This complex thus turns out to be a very sensitive tool[132] for measurement of radiation damage of ds-DNA.

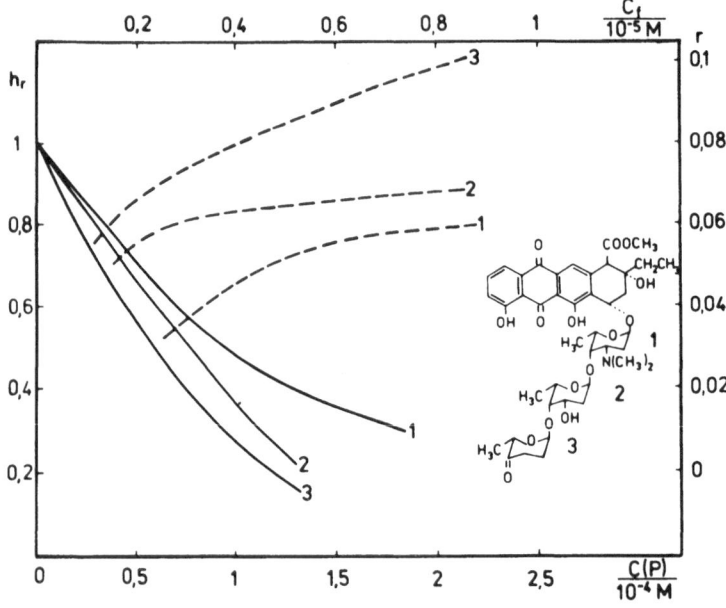

Figure 21. Titration curves[128] (———): relative step height vs. total phosphate concentration c(P) for aclacinomycin A (3) and derivatives (2) and (1) with two and one sugar groups, respectively. Parts of binding isotherms r vs. c_f: – – – –.

The other titration method seems to be more suitable for high-molecular weight DNA: Identical small volumes of a stock solution of the ligand are gradually added to the buffered solution of DNA. Polarograms are recorded after 25 min of deaeration after the first addition and 5 min after each successive addition. In this case the step height of the ligand increases from zero and—after saturation of DNA—becomes parallel to the straight line obtained in calibration without DNA (Figure 22). However, in general it is worthwhile to point out that, in comparison to spectrophotometric titrations, polarography gives approximate results and, moreover, the calculation of binding data needs another theory,[4,126] which includes cooperative interaction.[133]

5. Nucleoproteins

In general all animal and viral DNA and RNA (with the possible exception of tRNA) occur *in vivo* in more or less specific association with structural or functional proteins. For instance, the low-molecular weight basic histones (and sometimes the protamines) appear to be strongly and stoichiometrically complexed to the DNA in cells. Other superstructures with nucleic acids are built up by complexes with enzymes, repressors, ribosomal proteins, capsid proteins (capsomeres), etc.

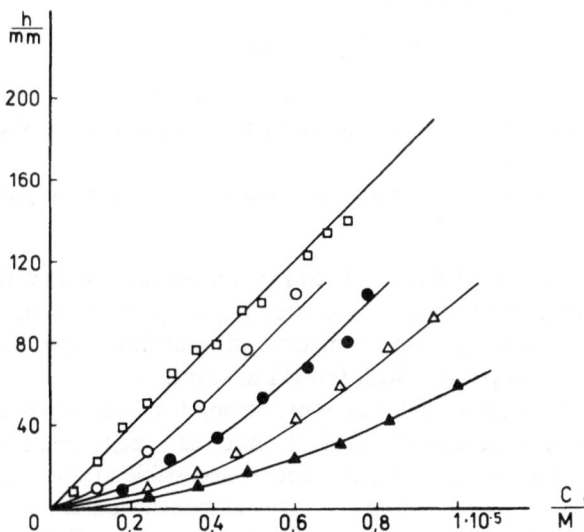

Figure 22. Alternative titration method[128]: for 0 □, 0.15 ○, 0.3 ●, 0.6 △, 1.12 ▲ (×10⁻⁴) M DNA concentrations the step heights with increasing total ligand concentration c_t are shown.

5.1. Principles of Interactions

The electrostatic interaction of anionic and cationic polyelectrolytes are accompanied by changes in hydration, dissociation, and counterion condensation, for example[23] [cf. Eqs. (3) and (23)] in:

$$DNA + P \overset{K_L}{\rightleftharpoons} DNA \cdot P + \Delta r[Na^+]$$

$$\Delta r = r_{DNA} - r_{DNA \cdot P} \tag{26}$$

counterions are pushed away and n_L phosphate groups are bound by the protein, hence:

$$\Delta r = n_L(1 - (|z|\lambda)^{-1}) \tag{27}$$

and the following relation has been evaluated[23]:

$$\ln K_L = \ln K_L^0 + n_L\lambda^{-1} \ln \bar{\gamma}\frac{1}{3}b - n_L(1 - 2\lambda^{-1}) \ln [Na^+] \tag{28}$$

For $[Na^+] = 1\ M$ (mean activity coefficient $\bar{\gamma}$) characteristic data are: DNA–ribonuclease, $n_L = 4, \ln K_L = -3$ (pH 7.7); DNA–RNA polymerase, $n_L = 12, \ln K_L = -4.9$ (pH 7.8); DNA–*lac* repressor (unspecific), $n_L = 12, \ln K_L = -8.5$ (pH 8); DNA–*lac* repressor (operator), $n_L = 8, \ln K_L = 15.9$ (pH 7.4).

5.2. DNA–Histone Complexes

One has to distinguish between the behavior of

- complexes of DNA with several single histones,
- mononucleosomes as organized particles of a short DNA superhelix and 8 histones
- chromatin as mononucleosomes joined by linker DNA region associated with the H1 histone.

Depending on the ratio of histone H1 to DNA these complexes with H1 mostly diminish the peaks in d.p. and a.c. polarography (Figure 23). It may be a result of compaction including also formation of micelles depending on ionic strength[134] and competition with the H1 adsorption.

In a.c. polarography of lower ionic strength nucleosomes are strongly adsorbed without exhibiting any peak[135] like that due to π-state DNA. It is assumed that the undisturbed nucleosome is accessible at the surface only by the flat part consisting of histones.[134] If this conclusion is true, there must be a very large difference in adsorption enthalpy between DNA and histones, which should be investigated in more detail. Chromatin itself shows a marked concentration dependence (Figure 24) and at low concentrations the d.p. peaks become more and more similar to that of DNA.[135] The same was found for

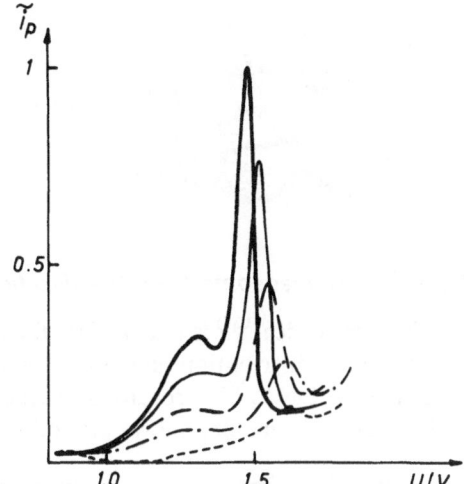

Figure 23. A.c. polarograms of peaks ① and ② of ss-DNA $(0.2 \text{ mg mL}^{-1}; \text{pH } 7,$ 50°C) titrated by histone H1: 0, 0.017, 0.050, 0.084, 0.1 mg mL^{-1}.

polysomes with H1 by dp polarography[134] that means, that the linker DNA between the nucleosomes is responsible for the reduction peak. Further investigations are necessary to clarify the electrochemical behavior of chromatin and its subunits.

5.3. An Electrostatic Model

The nucleosome core particle consists of two histone dimers $(H2a-H2b)_2$, one tetramer $(2H3-2H4)$ and a short DNA (145 BP) wrapped around as -1.5 superhelical turns leading to an asymmetry of the negative charge on the peripheral surface (Figure 25),[137] 12 nm in diameter. The stabilization of this particle is warranted by electrostatic forces.

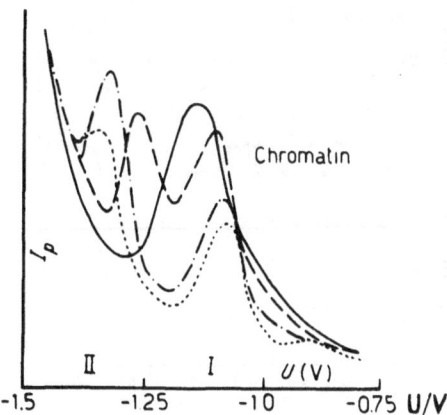

Figure 24. Influence of concentration[136] on the d.p. polarogram of calf thymus chromatin $(0.1 M$ Tris-buffer, $0.1 M \text{ NH}_4\text{Cl}, \text{pH } 8)$. Relative concentrations: —— 1 (corresponds to $E_{260 \text{ nm}} = 3.75$); —— 0.5; –·–·– 0.25; ---- 0.125.

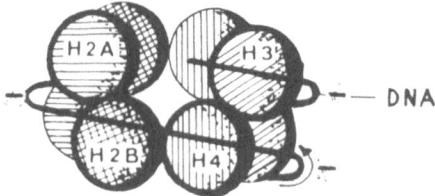

Figure 25. Model of a core particle according to Zinke[137]: DNA helix around the histone heads; asymmetric charge distribution outside.

5.3.1. The Nucleosome Core Stabilization

The basic idea[137] is to correlate the electrostatic and hydrophobic amino acid sequence distribution function of each histone with the adjoining negative backbone of DNA. The resulting charge density for the region of every 10 amino acids (Figure 26) shows the following characteristic properties:

- the N-terminal regions are predominantly basic
- the histone heads (SIS) are mostly hydrophobic and of low effective polarity
- the C-terminal regions are of different nature, whereas H1 differs essentially from the other histones.

A likely position of these parts of histones with respect to DNA is shown in Figure 25 and Figure 27. According to this arrangement of contacts to DNA the positive charge density distribution of the histone core has a maximum in the rear region (DNA bases number 60–80) and a minimum in the front region where both DNA strands have a distance of about 0.7 nm.

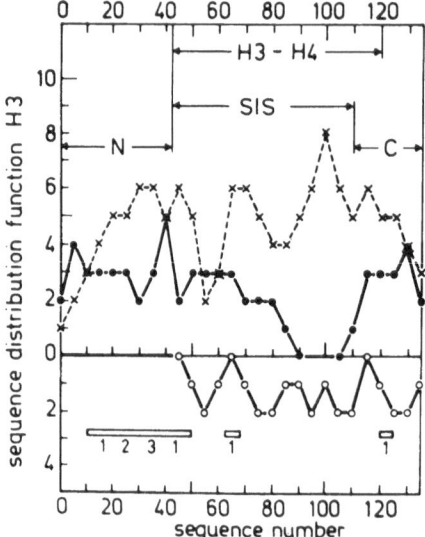

Figure 26. Sequence distribution function[137] of histone H3: × hydrophobic groups (Val, Met, Leu, Ile, Tyr, Phe, Pro, Ala, Gly, Cys); ● basic groups (Arg, Lys, His); ○ acidic groups (Glu, Asp); ▭ prolines.

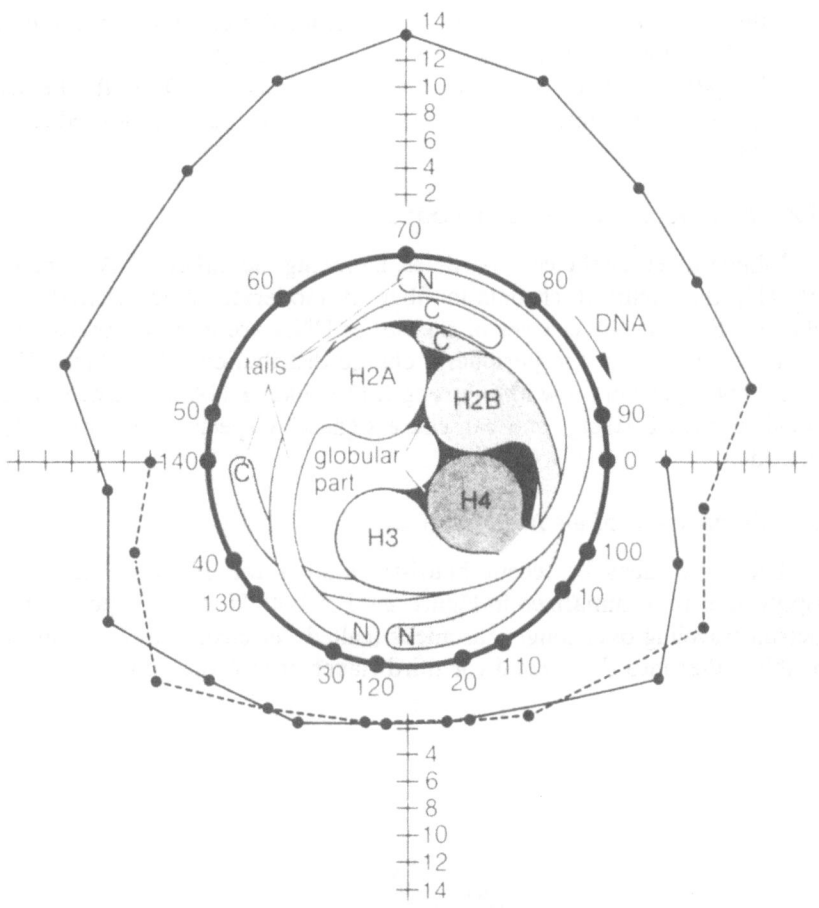

Figure 27. Structure and charge distribution of a simplified nucleosome core model according to Zinke[137]; relative basicity on the symmetry axis 1–14 (outside lines).

As a result of the strong electrostatic interaction in the inner rear region the f-value [cf. Eq. (14)] increases towards 1, because $z > 4$. This higher screening inside, in comparison to the outside value $f = 0.76$, causes a bending of DNA around the histone core. This model[137] explains the influence of ionic strength:

- low ionic strength increases the repulsion between both DNA strands in the front region ($\kappa^{-1} < 1$ nm → $\kappa^{-1} > 2$ nm) and hence an opening on this side is created.
- high ionic strength ($\kappa^{-1} < 0.5$ nm) results in competition from the weak binding of N-terminal regions of H2a and H2b, releasing these histones, not H_3–H_4.

as well as the reported polarographic results[135]:

- this DNA compaction causes no additional a.c. peaks, in agreement with π-state (cf. Sections 4.3.1.2.1 and 4.3.2.3.)
- in spite of the lower effective negative charge of DNA the heads of histones are slightly positive and hence may be better adsorbed (Figure 23).

5.3.2. The Chromatin Fiber Organization

Joining together the core particles, including the linker DNA complexed with H1, the compact chromatin fiber at moderate ionic strength[137] is obtained. Because the charges of the linker DNA are more compensated by H1, an asymmetry in the peripheral charge distribution takes place (Figure 28). At higher or very low ionic strength the linker DNA is free or stretched (so-called beaded string) and can cause signals by specific adsorption at the electrode.[135]

5.3.3. Comparison of electron pathways

The basic ideas of recent heuristic models for electron exchange of biopolymers are summarized in Figure 29. For both kinds of biopolymers the electron traveling over longer distances (called "electron hopping") must be possible either into the second (or third) layer or into the double helix.

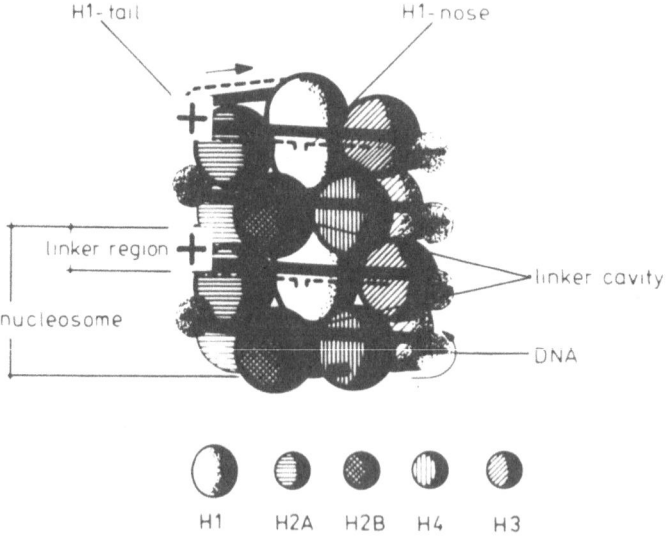

Figure 28. Two nucleosomes[137] forming a chromatin thread (moderate ionic strength). Head of histone H1 inside the pocket. The outside charge distribution is not symmetrical, due to the charges in the linear DNA region.

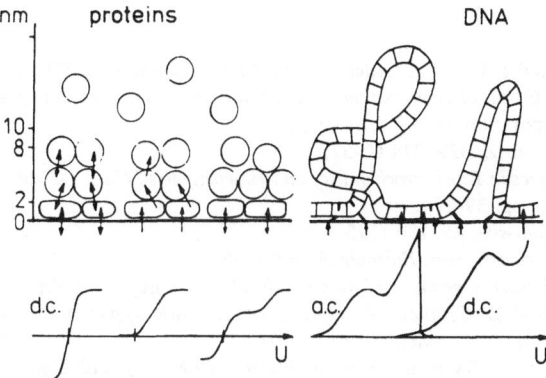

Figure 29. Pathways of electron exchange for proteins and nucleic acids (according to F. Scheller and H. Berg). Left part: adsorption layers of globular proteins and their polarograms: reversible cytochrome c_3 in polylayer; irreversible cytochrome c or metmyoglobin in polylayer; reversible lysozyme only in monolayer. Right part: segment adsorption of the double helix with irreversible reduction of external bases and entrapped bases.

6. Outlook

At the present time, the electrochemical behavior of proteins and nucleic acids is measurable and understandable in general. For the elucidation of details of the electrode mechanisms and of the conformational changes during adsorption it seems necessary to introduce and combine surface and spectroscopic techniques, because sometimes polarographic data seem to be ambiguous.

Further research trends[125] are:

- the recording of degradation and compaction processes brought about by the influence of chemical and physical agents,
- the study of interactions with inorganic and organic ligands,
- the characterization of nucleoprotein superstructures, phages, viruses, and organelles,
- the use of special electrodes for measurements in biological fluids,
- the generation of photoelectrons for biopolymers as scavengers,[120,121,138]
- analysis of fermentation broth and even single cells.

In most cases electrochemical methods are in competition with sophisticated spectroscopic techniques in this research.

References

1. J. Heyrovský and J. Babicka, *Collect. Czech. Chem. Commun.* **2**, 370 (1930).
2. E. Cohn and J. Edsal, *Proteins, Amino Acids, and Peptides as Ions and Dipolar Ions*, Reinhold Publishing Corporation, New York (1942).
3. H. Berg, *Biochem. Z.* **329**, 274 (1957).
4. H. Berg, in *Topics in Bioelectrochemistry and Bioenergetics*, Vol 1, G. Milazzo, ed., J. Wiley, London (1976), pp. 39–104.
5. E. Paleček, *Naturwiss.* **45**, 186 (1958).
6. E. Paleček, *Bioelectrochem. Bioenerg.* **8**, 469 (1981).
7. C. Tanford, *Physical Chemistry of Macromolecules*, J. Wiley, New York (1961).
8. D. Eisenberg and D. Crothers, *Physical Chemistry with Applications to the Life Sciences*, Benjamin Publishing Co., Menlo Park (1979).
9. B. Feinberg and M. Ryan, in *Topics in Bioelectrochemistry and Bioenergetics* Vol. 4, G. Milazzo, ed., J. Wiley, New York (1981), pp. 225–270.
10. J. Heyrovský and J. Kuta, *Grundlagen der Polarographie*, Akademie V., Berlin (1965).
11. B. Breyer and H. Bauer, *Alternating Current Polarography and Tensammetry*, Interscience, New York (1963).
12. M. Brezina and P. Zuman, *Polarography in Medicine, Biochemistry and Pharmacy*, Interscience, New York (1958).
13. W. Norde, in *Adhesion and Adsorption of Polymers*, Part B, Lieng-Huang Lee, ed., Plenum Publishing Corporation, New York (1980), pp. 801–825.
14. H. Jehring, *Elektrosorptionsanalyse mit der Wechselstrompolarographie*, Akademie V., Berlin (1974).
15. F. Scheller, H.-J. Prümke, H. Schmidt, and P. Mohr, *Bioelectrochem. Bioenerg.* **3**, 328 (1976).
16. A. Archakow, A. Kusnetsov, M. Izotov, and I. Karuzina, *Biofizika* **26**, 351 (1981).
17. T. Ikeda, K. Toriyama, and M. Senda, *Bull. Chem. Soc. Japan* **52**, 1937 (1979).
18. K. Santhanam, N. Jespersen, and A. Bard, *J. Am. Chem. Soc.* **99**, 274 (1977).
19. T. M. Cotton, S. G. Schultz, and R. Van Duyne, *J. Am. Chem. Soc.* **102**, 7960 (1980).
20. V. Brabec and I. Schindlerova, *Bioelectrochem. Bioenerg.* **8**, 451 (1981).
21. J. Haladjian, M. Pilard, P. Bianco, and P. Serre, *Bioelectrochem. Bioenerg.* **9**, 91–101 (1982).
22. J. Schellman, *Biopolymers* **14**, 999 (1975).
23. T. Record, C. Anderson, and T. Lohman, *Quart. Rev. Biophys.* **11**, 103 (1978).
24. P. von Hippel and K. Wong, *J. Biol. Chem.* **240**, 3909 (1965).
25. Y. Ilan, A. Shafferman, B. Feinberg, and Y. Lau, *Biochim. Biophys. Acta* **548**, 565 (1979).
26. K. E. Reinert, *Bioelectrochem. Bioenerg.* **8**, 301 (1981).
27. M. Stankovich and A. Bard, *J. Electroanal. Chem.* **85**, 173 (1977).
28. M. Stankovich and A. Bard, *J. Electroanal. Chem.* **86**, 189 (1978).
29. M. Fontaine, C. Rivat, C. Hamet, and C. Caullet, *Bioelectrochem. Bioenerg.* **4**, 242 (1977).
30. S. Kwee, *Bioelectrochem. Bioenerg.* **3**, 264 (1976).
31. N. Imai, Y. Umezawa, Y. Arata, and S. Fujiwara, *Biochim. Biophys. Acta* **626**, 501 (1980).
32. F. Scheller, G. Strnad, B. Neumann, M. Kühn and W. Ostrowski, *Bioelectrochem. Bioenerg.* **6**, 117 (1979).
33. T. Ikeda, S. Ando, and M. Senda, *Bull. Chem. Soc. Japan* **54**, 2189 (1981).
34. E. Suponeva, B. Kisselev, and L. Chekulaeva, *Bioelectrochem. Bioenerg.* **8**, 251 (1981).
35. F. Scheller and H. Prümke, *Stud. Biophys.* **60**, 137 (1976).
36. Chih-Ho Su and W. Heineman, *Anal. Chem.* **53**, 594 (1981).
37. K. Niki, T. Yagi, H. Inokuchi, and K. Kimura, *J. Electrochem. Soc.* **124**, 1889 (1977).
38. P. Bianco and J. Haladjian, *Bioelectrochem. Bioenerg.* **8**, 239 (1981).
39. B. Kuznetsov, N. Mestechkina, M. Izolov, I. Karuzina, A. Karyakin, and A. Archakov, *Biofizika* **44**, 1569 (1979).
40. F. Scheller, M. Jähnchen, G. Etzold, and H. Will, *Bioelectrochem. Bioenerg.* **1**, 478 (1974).

41. H. Berg and K. Kramarczyk, *Biochim. Biophys. Acta* **131**, 141 (1967).
42. M. Kuik and K. Krassowski, *Bioelectrochem. Bioenerg.* **9**, 419–425 (1982).
43. I. Ivanov and E. Rachlejewa, in *I. Jena Symposium (1962): Die Polarographie in Chemotherapie, Biochemie und Biologie*, H. Berg, ed., Akademie Verlag, Berlin (1964), pp. 207–211.
44. G. Ruttkay-Nedecky and B. Bezuch, in *Biological Aspects of Electrochemistry*, G. Milazzo, ed., Birkhäuser V., Basel (1971), pp. 553–562.
45. H. Berg, in *III. Jena Symposium (1965): Elektrochemische Methoden und Prinzipien in der Molekularbiologie*, H. Berg, ed., Akademie Verlag, Berlin (1966), pp. 479–484.
46. B. Kuznetzov, in *Biological Aspects of Electrochemistry*, G. Milazzo, ed., *Experientia, Suppl.* **18**, 381 (1971).
47. S. Mairanovsky, in *III. Jena Symposium (1965): Elektrochemische Methoden und Prinzipien in der Molekularbiologie*, H. Berg, ed., Akademie Verlag, Berlin (1966), pp. 473–478.
48. I. Kolthoff and S. Kihara, *J. Electroanal. Chem.* **96**, 95 (1979).
49. B. Feinberg and Y. Lau, *Bioelectrochem. Bioenerg.* **7**, 187 (1980).
50. J. Chmelik, J. Kadlecek, and V. Kalous, *J. Electroanal. Chem.* **99**, 245 (1979).
51. H. Berg, in *I. Jena Symposium (1962): Die Polarographie in der Chemotherapie, Biochemie und Biologie*, H. Berg, ed., Akademie Verlag, Berlin (1962), pp. 128–151.
52. V. Brabec and I. Schindlerova, *Bioelectrochem. Bioenerg.* **8**, 451 (1981).
53. J. Reynaud, B. Malfoy and A. Bere, *Bioelectrochem. Bioenerg.* **7**, 595 (1980).
54. J. Kulys and G. Svirmickas, *Anal. Chim. Acta* **117**, 115 (1980).
55. J. Kulys, M. Pesliakiene and A. Samalius, The development of bienzyme glucose electrodes, *Bioelectrochem. Bioenerg.* **8**, 81 (1981).
56. H. Berg and F. A. Gollmick, in *III. Jena Symposium (1965): Elektrochemische Methoden und Prinzipien in der Molekularbiologie*, H. Berg, ed., Akademie Verlag, Berlin (1966), pp. 533–544.
57. H. Berg, in *II. Jena Symposium (1963): Physikalische Chemie biogener Makromoleküle*, H. Berg, ed., Akademie Verlag, Berlin (1964), pp. 213–229.
58. B. Breyer, in *III. Jena Symposium (1965): Elektrochemische Methoden und Prinzipien in der Molekularbiologie*, H. Berg, ed., Akademie Verlag, Berlin (1966), pp. 575–577.
59. W. Heinemann, C. Anderson, and H. Halsall, *Science* **204**, 865 (1979).
60. M. Senda, T. Ikeda, T. Kakutani, K. Kano, and H. Kinoshita, *Bioelectrochem. Bioenerg.* **8**, 151 (1981).
61. H. Berg and D. Tresselt, *Pharmazie* **16**, 74 (1961).
62. H. Lücke and H. Berg, in *I. Jena Symposium (1962): Polarographie in Chemotherapie, Biochemie und Biologie*, H. Berg, ed., Akademie Verlag, Berlin (1964), pp. 275–277.
63. H. Berg. K. Granath, B. Nygard, J. Strassburger, and P. Weist, *J. Electroanal. Chem.* **36**, 167 (1972).
64. H. Berg, *Pharmazie* **11**, 239 (1956).
65. O. Fischer, J. Totusek, L. Trnkova, and E. Fischerova, submitted for publication.
66. Y. Ogawa, K. Kobayashi, T. Yagi, and K. Niki, Electrode reaction of cytochrome c_3 in the presence of various proteins, *Rev. Polarography (Japan)* **27**, 95 (1981).
67. M. Shinagawa and S. Kakumoto, Studies on semiconductive nature of Brdicka current, *Coll. Czech., Chem. Commun.* **47**, 2724–2734 (1982).
68. Sh. Gohtani and M. Takagi, Polarographic studies on Takaamylase A, *Rev. Polarography (Japan)* **27**, 89 (1981).
69. R. Wells, T. Goodman, W. Hillen, G. Horn, R. Klein, J. Larson, U. Müller, S. Neuendorf, N. Panayotatos, and S. Stirdivant, DNA-structure and gene regulation, *Progr. in Nucl. Acid Res. and Mol. Biol.* **24**, 167–267 (1980).
70. S. Englander, N. Kallenbach, A. Heeger, J. Krumhansl, and S. Litwin, Nature of the open state in long polynucleotide double helices: possibility of soliton excitation, *Proc. Natl. Acad. Sci. U.S.A.* **77**, 7222 (1980).

71. H. Fritzsche, Lou-Sing Kan, and P. Ts'o, Exchange behavior of the Watson-Crick NH protons of a RNA Miniduplex, *Biochemistry* **22**, 277–280 (1983).

72. P. Elving, Nitrogen heterocyclic compounds: electrochemical information concerning energetics, dynamics and mechanisms, in *Topics in Bioelectrochemistry and Bioenergetics*, Vol. 1, G. Milazzo, ed., J. Wiley, London (1976), pp. 179–286.

73. G. Dryhurst, *Electrochemistry of Biological Molecules*, Academic Press, New York (1977).

74. F. Oosawa, *Polyelectrolytes*, Marcel Dekker, New York, (1971).

75. G. Manning, The molecular theory of polyelectrolyte solutions with applications to the electrostatic properties of polynucleotides, *Quart. Rev. Biophys.* **11**, 179 (1978).

76. G. Manning, *Biopolymers* **20**, 2337 (1981).

77. K. Weller, H. Schütz, and I. Petri, *Stud. Biophys.* **87**, 187–188 (1982).

78. E. Paleček, in *I. Jena Symposium (1962): Die Polarographie in der Chemotherapie, Biochemie und Biologie*, H. Berg, ed., Akademie Verlag, Berlin (1964), pp. 270–274.

79. H. Berg and H. Bär, *Mber. Dtsch. Akad. Wiss. Berlin* **7**, 210 (1965).

80. H. Berg, H. Bär, and F. A. Gollmick, *Biopolymers* **5**, 61 (1967).

81. J. Flemming and H. Berg, *Mber. Dtsch. Akad. Wiss. Berlin*, **7**, 206 (1965).

82. J. Flemming and H. Berg, in *III. Jena Symposium (1965): Elektrochemische Methoden und Prinzipien in der Molekularbiologie*, H. Berg, ed., Akademie Verlag, Berlin (1966), pp. 559–563.

83. E. Paleček, in *Methods in Enzymology*, Vol. 21, L. Grossman and K. Moldave, eds., Academic Press, New York (1971), pp. 3–23.

84. P. Valenta and H. W. Nürnberg, *J. Electroanal. Chem.* **49**, 55 (1974).

85. I. Miller, *J. Mol. Biol.* **3**, 229 (1961).

86. J. Flemming and L. Pospisil, *Stud. Biophys.* **57**, 83 (1976).

87. G. Barker and D. McKeown, *Bioelectrochem. Bioenerg.* **3**, 373 (1976).

88. J. Flemming, *Biopolymers* **6**, 1697 (1968).

89. H. Berg, D. Tresselt, J. Flemming, H. Bär, and G. Horn, *J. Electroanal. Chem.* **21**, 181 (1969).

90. H. Berg, in *Polarographie, Probleme und Perspektiven*, (Allunionskonferenze Riga) Isdatelstvo Riga (1977), p. 308.

91. H. Berg and H. Bär, *Stud. Biophys.* **3**, 133 (1967).

92. H. Berg, J. Flemming, E. Paleček, and V. Vetterl, *Stud. Biophys.* **33**, 81 (1972).

93. J. Flemming and H. Berg, *Bioelectrochem. Bioenerg.* **1**, 459 (1974).

94. J. Flemming, *Bioelectrochem. Bioenerg.* **2**, 79 (1975).

95. H. Berg, G. Horn and J. Flemming, *Stud. Biophys.* **57**, 87 (1976).

96. H. Berg, G. Horn, J. Flemming, and V. Glezers, *Bioelectrochem. Bioenerg.* **4**, 404 (1977).

97. H. Berg, *Rev. Polarography (Japan)* **24**, Suppl. 1–2 (1979).

98. H. Berg, *Stud. Biophys.* **75**, 209 (1979).

99. H. Berg and G. Horn, *Bioelectrochem. Bioenerg.* **8**, 167 (1981).

100. H. Berg, *Abh. d. Sächs. Akad. Wiss. (Leipzig)* **54**, 1 (1981).

101. H. Berg, J. Evdokimov, H. Bär, and J. Warschawsky, *Molekularnaja Biologija* **2**, 830 (1968).

102. J. Reynaud, *Bioelectrochem. Bioenerg.* **7**, 267 (1980).

103. V. Brabec and E. Paleček, *Biopolymers* **11**, 2577 (1972).

104. M. Voriškova, E. Lukašova, F. Jelen, and E. Paleček, *Bioelectrochem. Bioenerg.* **8**, 487 (1981).

105. P. Valenta and P. Grahmann, *Electroanal. Chem.* **49**, 41–53 (1974).

106. E. Paleček and F. Jelen, *Bioelectrochem. Bioenerg.* **7**, 317 (1980).

107. J. Flemming and H. Berg, *Bioelectrochem. Bioenerg.* **3**, 241 (1976).

108. J. Flemming, *Bioelectrochem. Bioenerg.* **4**, 476 (1977).

109. A. Brajter-Toth, R. Goyal, M. Wrona, T. Lacava, N. Nguyen, and G. Dryhurst, *Bioelectrochem. Bioenerg.* **8**, 413 (1981).

110. V. Brabec, *Bioelectrochem. Bioenerg.* **8**, 437 (1981).

111. H. Berg, J. Flemming, and G. Horn, *Bioelectrochem. Bioenerg.* **2**, 287 (1975).

112. H. Berg, *Bioelectrochem. Bioenerg.* **4**, 522 (1977).

113. A. Silberberg, in *Ions in Macromolecular and Biological Systems*, 29th Colston Symp., D. Everett and B. Vincent, eds., Bristol (1978), pp. 1–10.
114. E. Neumann, in *Ions in Macromolecular and Biological Systems*, 29th Colston Symp., D. Everett and B. Vincent, eds., Bristol (1978), pp. 170–191.
115. I. Michaeli, in *Ions in Macromolecular and Biological Systems*, 29th Colston Symp., D. Everett and B. Vincent, eds., Bristol (1978), pp. 121–131.
116. Yu. Yevdokimov, V. Salyanov, and H. Berg, *Nucl. Ac. Red.* **9**, 743 (1981).
117. W. Norde and J. Lyklema, in *Ions in Macromolecular and Biological Systems*, 29th Colston Symp., D. Everett and B. Vincent, eds., Bristol (1978), pp. 11–40.
118. C. Simionescu, S. Dumitrescu, and V. Percec, in *Topics in Bioelectrochemistry and Bioenergetics*, Vol. 2, G. Milazzo, ed., J. Wiley, Chichester, New York (1978), pp. 151–204.
119. H. Berg, H. Schweiss, E. Stutter, and K. Weller, *J. Electroanal. Chem.* **15**, 415 (1967).
120. G. Barker, Lecture presented at the Heyrovsky Memorial Congress on Polarography, Prague (1980).
121. H. Berg, Lecture presented at the Heyrovsky Memorial Congress on Polarography, Prague (1980).
122. Y. Matsuzawa, T. Oki, T. Takeuchi, and H. Umezawa, *J. Antibiotics* **34**, 1596 (1981).
123. J.-M. Sequaris, E. Koglin, P. Valenta, and H. Nürnberg, *Ber. Bunsenges. Phys. Chem.* **85**, 512 (1981).
124. J. Sequaris, P. Valenta, H. Nürnberg, and B. Malfoy, in *Ions in Macromolecular and Biological Systems*, 29th Colston Symp., D. Everett and B. Vincent, eds., Bristol (1978), pp. 201–229.
125. G. Dryhurst, K. Kadish, F. Scheller, and R. Renneberg, in *Biological Electrochemistry*, Academic Press, New York (1982), pp. 398–519.
126. H. Berg, *Ergebn. Experim. Medizin* **24**, 275 (1977).
127. E. Bauer, H. Berg, and H. Schütz, *Faserforschung und Textiltechnik* **24**, 57 (1973).
128. H. Berg, G. Horn, and U. Luthardt, *Bioelectrochem. Bioenerg.* **8**, 537 (1981).
129. H. Berg, F. A. Gollmick, H. Triebel, E. Bauer, G. Horn, J. Flemming, and L. Kittler, *Bioelectrochem. Bioenerg.* **5**, 335 (1978).
130. H. Berg, *Bioelectrochem. Bioenerg.* **5**, 347 (1978).
131. T. Cummings and P. Elving, Electrochemical reduction of thymine in dimethylsulfoxide, *J. Electroanal. Chem. Interfacial Electrochem.* **102**, 237–248 (1979).
132. E. Paleček, E. Lukasova, F. Jelen, and M. Vojtiskova, *Bioelectrochem. Bioenerg.* **8**, 497 (1981).
133. H. Schütz, F. A. Gollmick, and E. Stutter, *Stud. Biophys.* **75**, 147 (1979).
134. H. Berg, U. Fiedler, J. Flemming, and G. Horn, "Corresponding electrode processes of DNA and proteins", *Bioelectrochem. Bioenerg.* (in press).
135. B. Malfoy, G. Sabeur, and J. Reynaud, *Biochem. Biophys. Res. Commun.* **94**, 996 (1980).
136. H. Berg, *Bioelectrochem. Bioenerg.* **3**, 359 (1976).
137. M. Zinke, *Bioelectrochem. Bioenerg.* **8**, 189 (1981).
138. K. Yamashita and H. Imai, Photocurrents observed with proteins. *Bioelectrochem. Bioenerg.* **5**, 650 (1978).
139. F. Scheller and G. Strnad, Electrode reactions of protein prosthetic groups, *Adv. Chem. Ser.* **210**, 219–236 (1982).
140. V. Brabec, V. Glezers, and V. Kadysh, Adsorption of desoxyribonucleic acid at mercury electrodes from alkaline solutions of higher ionic strength, *Coll. Czech. Chem. Commun.* **48**, 1257–1271 (1983).

4

Bioelectrocatalysis

M. R. TARASEVICH

Abstract

The structure and physicochemical properties of the enzymes which have been used to date to promote electrochemical reactions are briefly outlined. Methods of their immobilization are described. The status of research on redox transformations of proteins and enzymes at the electrode–electrolyte interface is discussed. Current concepts on the ways of conjugation of enzymatic and electrochemical reactions are summarized. Examples of bioelectrocatalysis in some electrochemical reactions are described. Electrocatalysis by enzymes under conditions of direct mediatorless transport of electrons between the electrode and the enzyme active center is considered in detail. Lastly, an analysis of the status of work pertaining to the field of sensors with enzymatic electrodes and to biofuel cells is provided.

1. Introduction

The term "bioelectrocatalysis" covers a group of phenomena associated with the acceleration of electrochemical reactions in the presence of biological catalysts—enzymes. It is a new trend in bioelectrochemical research which has sprung up at the junction of bioengineering and electrocatalysis.

M. R. TARASEVICH • Institute of Electrochemistry, USSR Academy of Sciences, Moscow, USSR.

Biological catalysts, i.e., enzymes, are unique in their properties as accelerators of numerous chemical processes occurring both in the living cell and outside the organism. In the first place, we would like to note the high rate of reactions occurring in the presence of enzymes: for some of them the number of elementary acts occurring at the active center of the enzyme per unit time amounts to up to $10^6 \, s^{-1}$. Not a single catalyst being used at present in industry permits such an acceleration of chemical processes to be achieved. A second, not less important, property of enzymes is their specificity. Numerous examples are known where an enzyme acts as a catalyst in the transformation of some single substance or a narrow group of substances of the same type. Furthermore, enzymes promote reactions at ordinary temperatures and in the region of neutral pH.[1]

As a rule, the region of ordinary temperatures and the use of heterogeneously fixed catalysts are considered to be the optimal conditions for carrying out electrochemical reactions. These requirements place serious constraints on the possibility of introducing catalysts already in use in chemical technological processes into electrochemical technology. One of the ways to solve this problem is the use of natural catalysts, i.e., enzymes, to accelerate electrochemical reactions, thus opening up prospects for development of radically new technological processes and devices.

Although the unique properties of enzymes have long been known, the history of bioelectrocatalysis is rather short: the first research into the use of enzymes as catalysts dates back to the 1970s.[2,3] Until quite recently enzymes had not been used in electrocatalysis for a number of reasons: the absence of pure enzyme preparations, instability of enzymes, impossibility of multiple application. It has become possible to overcome these difficulties due to improvements in methods for isolation and purification of enzymes, as well as preparation of immobilized enzymes. By way of immobilization it is possible to raise the stability of protein macromolecules, converting an enzyme into a heterogeneous state while its activity is either fully preserved or reduced insignificantly.[4]

The specifics of biocatalysts enable us to outline the following trends in their application in electrochemical systems:

1. Development of catalysts for carrying out electrochemical reactions in neutral media for which no sufficiently effective inorganic catalysts are available at present.
2. Development of bioelectrochemical systems permitting various inexpensive substances of organic nature to be used as fuel.
3. Using the specificity of enzymic activity to develop highly selective electrochemical sensors.

The most important problem of bioelectrocatalysis is the study of the mechanism of electron transfer between the active center of the enzyme and the electrode, and realization of effective pathways for conjugation of enzymatic and electrochemical reactions. There exist two possibilities of conjugation:

1. In the course of an enzymatic reaction there arises an intermediate which is subject to electrochemical conversion into an end product at lower overvoltage compared to the initial substrate.
2. In the course of an enzymatic reaction electron transfer takes place between the electrode and the active center either with the aid of low-molecular weight, easily diffused carriers (mediators) or by way of direct electrochemical oxidation or reduction of an active site in the enzymes, i.e., by direct electron exchange between the enzyme and the electrode.

Another aspect of bioelectrocatalysis is the application of electrochemical methods to study aspects of the mechanism of the enzymatic effect and, in particular, of the relationship between conformational transformations of proteins, the redox potential value at the active center, and the electron transfer rate. Understanding the specificities of biocatalysts opens up a possibility for developing synthetic models of enzymes on the basis of complex compounds of a nonprotein nature.

In the present chapter we shall briefly discuss the structure and physicochemical properties of those enzymes that have already been used to accelerate electrochemical reactions; the methods of their immobilization are described; the present status of research in redox transformations of proteins and enzymes at the electrode–electrolyte interface is dealt with; the existing concepts of the ways in which conjugation of enzymatic and electrochemical reactions takes place are summarized; examples of bioelectrocatalysis for a number of electrochemical reactions are described. In conclusion we give a brief summary of the status of research in the field of sensors with enzymatic electrodes and with respect to biofuel elements.

2. Physicochemical Properties of Enzymes

2.1. Classification of Enzymes and Their Structure

Enzymes are specific proteins performing the role of catalyst in living organisms. All enzymes can be subdivided into six classes.[5] Until now the enzymes of two classes, viz. oxidoreductases catalyzing oxidation–reduction reactions and hydrolases promoting hydrolysis reactions, have been used in electrochemical systems.

Oxidation–reduction reactions in the presence of oxidoreductases occur through the transfer of either electrons or hydrogen atoms. Table 1 shows certain enzymes in the oxidoreductase class, which are finding application in electrochemical research.

Table 1
Enzymes in the Oxidoreductase Class

Name	Prosthetic group	Molecular weight	Substrates
Catalase	4 hemes	225,000	H_2O_2
Peroxidase	heme	40,000	polyphenols, H_2O_2
Cytochrome oxidase	heme, Cu	200,000	O_2
Laccase	4 Cu	60,000	phenols, quinones, O_2
Hydrogenase	Fe—S	60,000	hydrogen
Glucose oxidase	2 FAD	150,000	glucose, O_2
Xanthine oxidase	FAD, Mo, Fe	280,000	xanthine
Aldehyde oxidase	FAD, Mo, Fe	300,000	aldehydes, O_2
Alcohol dehydrogenase	Mn	150,000	alcohols

Hydrolases catalyze the hydrolytic splitting of covalent bonds: ester, peptide, acid anhydride, etc. Table 2 contains data on hydrolases which have been utilized in research on bioelectrocatalysis.

An enzyme is a high-molecular weight protein with an active center where the binding of the substrate and its chemical transformation take place. The protein part of the enzyme consists of α-L-amino acids linearly interconnected by peptide bonds. The lower boundary of the molecular weight of proteins is arbitrarily assumed to be 5000. The upper boundary amounts to some hundred thousand or even millions. The structural diversity of proteins is due to a wide choice of amino acids and the presence of several levels of structural organization.

The primary structure of enzymes is determined by peptide covalent bonds joining each amino acid to the neighboring one, as well as by disulfide bonds joining sulfur atoms in two cysteine radicals.

The secondary structure is determined by the arrangement of the polypeptide chain sections in space. It is usually an L-spiral stabilized by hydrogen bonds.

Table 2
Enzymes in the Hydrolase Class

Name	Prosthetic group	Molecular weight	Substrates
Alkaline phosphatase	4 Zn	80,000	ethers of phosphoric acid
Acidic phosphatase			O-substituted ethers of thiophosphoric acid

The tertiary structure is determined by the convolution of the entire polypeptide chain as a whole into a compact three-dimensional system, for instance, a spherical globule. The principal forces determining the three-dimensional structure are hydrophobic interactions: the polypeptide chains convolute in such a manner as to prevent contact between hydrophobic parts and the aqueous medium. The hydrophobic nucleus of the globule is practically inaccessible to water molecules (Figure 1).

The binding of the substrate and its further transformation take place in the active center of the biocatalyst. The active center is formed from the enzymes of the polypeptide chain playing the role of a voluminal structure where amino acid radicals, viz. histidine imidazole, serine or tyrosine oxy groups, thiol groups of cysteine, ionized and nonionized carboxyl groups, etc., are specifically fixed in space. Numerous enzymes contain a nonprotein prosthetic group (coenzyme) in the macromolecule. The coenzyme is joined to the protein part (apoenzyme) either covalently or with the participation of hydrogen bonds, or hydrophobic or ion–ion interactions. The role of prosthetic groups is often played by such bioinorganic compounds as heme, nicotineamide adenine dinucleotide (NAD), and nicotineamide adenine dinucleotide phosphate (NADP), etc. As part of coenzymes, an important role in the catalytic activity of enzymes is played by different metals—iron, copper, zinc, molybdenum. Figure 2 shows by way of illustration the structure of the iron–sulfur cluster, an active center of hydrogenase, and of heme, an active center of peroxidase.

The structural specificities of proteins enable a great many radicals endowed with different functions to be concentrated at a single active center; this is the basis of the high catalytic activity and specific action of enzymes.

2.2. The Mechanism of Enzymatic Catalysis

The mechanism of reactions catalyzed by enzymes has a complex character and has been conclusively established by no means in every case. However, in spite of their complexity, enzymatic reactions proceed in conformity with the general laws of conventional chemical transformations.[1] Qualitative factors responsible for the catalytic effect of an enzyme are summarized in Reference (6):

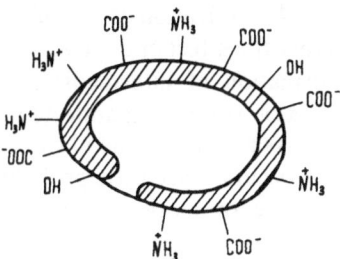

Figure 1. Schematic model of the protein globule.

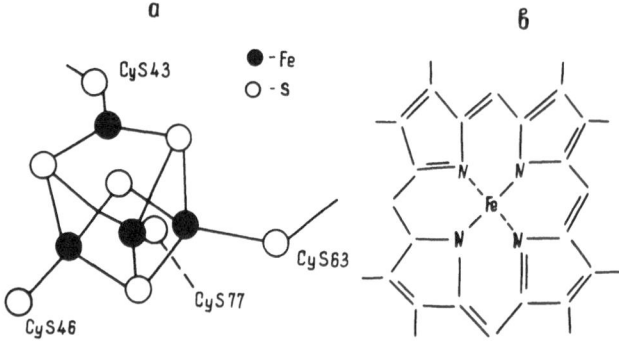

Figure 2. The structure of (a) the iron–sulfur cluster and (b) heme.

1. Great affinity between enzyme and substrate, i.e., a high probability for the formation of an enzyme–substrate complex, which is equivalent to a sharp increase in reagent concentrations under conventional conditions (proximity effect). Actually the acceleration mechanism in this case involves a stabilization of the activated complex due to hydrophobic or electrostatic interactions and, in certain cases, even the formation of hydrogen bonds. The kinetic role of stabilization of the activated state in enzymatic catalysis has been most adequately dealt with in Reference (7).

2. A rigorous reciprocal orientation of reagents, cofactors, and the active center (orientation effect). The induced character of correspondence between enzyme and substrate has been examined in Reference (8). Within the framework of this theory a more active and diversified role is attributed to the substrate. If the substrate fails to induce a proper arrangement at the active center, its settling on the protein does not lead to a reaction.

3. The impact of nucleophilic and electrophilic groups of the active center on the substrate at the contact area in the enzyme–substrate complex (the effect of synchronous intramolecular catalysis). The polyfunctional catalysis involves a great many processes: "push–pull" mechanisms, processes involving a relay charge transfer, as well as a general acid–base catalysis. Presumably, the enzyme in the initial state of the enzymatic reaction already contains structural elements of the transition state and in this case the reaction must be thermodynamically more advantageous.

4. Activation of the substrate by way of redistribution of electron density under the effect of the electroactive groups in the enzyme (polarization effect). Formation of intermediate compounds during fixation of reagents at the reaction center is manifested in covalent catalysis.

As a result of the above effects, a sharp increase in the absolute reaction rate occurs, since little probable higher-order reactions requiring the creation of three or more molecules are substituted by highly effective first-order reactions, i.e., reactions of intramolecular polyfunctional catalysis. All of the effects considered above make their contributions to enzymatic activity; however, no quantitative concepts of a single mechanism of enzymatic activity are yet available. It is universally assumed[6-11] that conformational effects play a determining role in enzymatic activity. A protein globule represents a dynamic system and a plausible physical idea favors the assumption that the energy of thermal motion or the energy acquired by the globule in the sorption of the substrate is converted into the energy of the enzyme–substrate complex, the result being an effective reduction of activation energy. Some general concepts on the mechanism of the elementary act in enzymatic catalysis are elaborated in References (9–11). Essentially they imply that the chemical electron function of the enzyme in transforming the substrate into a product is realized through conformational restructuring of the molecule. According to this hypothesis, the profile of the conformational free energy of the protein along the reaction coordinate is complementary to the energy profile of the chemical free energy. As a result, the total free energy is characterized by a reduced activation barrier (Figure 3). In this approach the property of the enzyme as a reaction medium is emphasized. This enables us, in estimating the respective contribution to enzymatic activity, to make use of the methods for calculating the rates of oxidation–reduction reactions in solutions by taking into account the dynamic role of the medium, which was first proposed by Marcus.[12]

2.3. The Structure and Functions of Proteins—Electron Carriers and Enzymes

In this section we shall briefly discuss the structure and functional specificities of proteins and enzymes for which data are available on their electrochemical behavior and on their acceleration of electrochemical reactions.

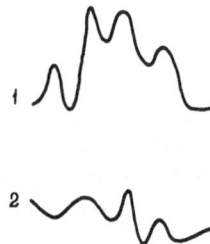

Figure 3. Complementary effect between conformational and chemical free energy. (1) Profile of chemical free energy, (2) profile of conformational free energy, (3) summary profile.

2.3.1. Heme-Containing Proteins and Enzymes

These compounds contain as a prosthetic group the porphyrin complex of iron(II)—heme—or iron(III)—hemin. These proteins perform different functions. They are capable of either reversibly combining oxygen for transport (hemoglobin) or storing it in a combined form (myoglobin), take part in electron transfer processes (cytochrome *b* and *c*), catalyze the reduction of oxygen to water (cytochrome *c* oxidase) and the oxidation of different functional groups by oxygen (cytochrome P450), as well as catalyze the decomposition of peroxides (catalase and peroxidase).

The schematic structure of heme is shown in Figure 2. Porphyrin acts as a ligand of Fe(II) or Fe(III). It is the variations in the type of protein substitutes or the method of securing coordination between heme and the protein portion that are the source of the above-listed functional diversity of heme-containing proteins.

2.3.1.1. Hemoglobin and Myoglobin

These proteins are the subject of voluminous literature partly summarized in Reference (13). Fe(II) is coordinated with protoporphyrin IX and the nitrogen atom in the imidazole ring of histidine which is present in the polypeptide chain. Myoglobin contains a polypeptide chain and a heme; hemoglobin consists of four polypeptide chains (two α and two β) and four hemes, each subunit being in contact with the other three. The hemoglobin molecule has a molecular weight of about 67,000 and its size is $50 \times 55 \times 69$ Å. Oxygen coordination is due to easy electron transfer from the t_{2g}-orbitals of iron to oxygen orbitals. According to the Pauling model,[14] the oxygen molecule is bound at a 120° angle to the porphyrin plane:

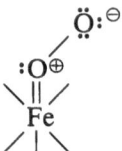

The Fe—O double bond in this structure arises from donation of a lone pair of electrons from the oxygen atom to one of the antibonding orbitals localized on the Fe atom and donation of Fe *d*-electrons from the d_{xz} and d_{yz} orbitals to the oxygen. Fe(II)-heme unbonded with protein is unable to reversibly bind oxygen in aqueous solutions. In hemoglobin, heme is placed in a "ligand pocket" creating a hydrophobic environment by a medium with a low dielectric permeability. An important role in the reversible bonding of oxygen is played by conformational changes in the protein globule and, among other things, by the interaction among subunits.

2.3.1.2. Cytochrome c

Cytochrome c with a molecular mass of about 12,000 contains a heme group and a polypeptide chain. The prosthetic group is combined with the protein through thioether bonds and through the coordination of the iron atom in the heme with the protein chains of amino acid radicals. As a result, the heme is immersed in the protein molecule and lies in a crevice-like fold formed by polypeptide chains.

The mechanism of electron transport has not yet been explicitly clarified; however, some specificities of this process have already been established.[15-17] Electron transfer was shown to be accompanied by significant conformational restructuring. Heme iron changes its valency, so that ferricytochrome passes over from the low-spin state to high-spin ferrocytochrome.[18] Presumably, electron transfer takes place along hydrophobic spaces in the macromolecule. In each half of the molecule, regions can be recorded from the heme to the surface, in which we can find only hydrophobic groups of side amino acid chains. Wherever hydrophobic regions come to the molecular surface they abut on the clusters of positively charged lysine radicals. There is proof that precisely these charged regions secure linkage with the other components of the electron-transport systems.

2.3.1.3. Cytochrome b₂

Cytochrome b_2 functions in yeasts both as an electron transport system and as lactate dehydrogenase, combining the functions of two different proteins—L-lactate dehydrogenase and cytochrome b—into a single catalytic unit. The molecular mass of the enzyme is approximately 235,000. It consists of four subunits, each containing a group of flavin mononucleotide (FMN) and heme. FMN can be reversibly separated from the protein. The absorption spectrum of b_2 corresponds to a low-spin complex. It is assumed that the transfer of electrons from lactate proceeds through flavin to the heme.[13]

2.3.1.4. Cytochrome oxidase

Cytochrome oxidase is a terminal respiratory enzyme catalyzing the oxidation of cytochrome by oxygen, involving the formation of water according to the reaction:

$$4c^{2+} + O_2 + 4H^+ \rightarrow 4c^{3+} + 2H_2O$$

Purified preparations of cytochrome oxidase are unstable and researchers have to deal, as a rule, with submitochondrial particles including, together with cytochrome oxidase, a part of the lipid membrane. The enzyme contains heme and copper in equimolar quantities as prosthetic groups. It apparently reacts with cytochrome c due to electrostatic interaction; interaction with oxygen is limited by the latter's rate of diffusion.

An established viewpoint today is that cytochrome oxidase consists of two subunits—cytochrome a and a_3. The molecular weight of such a dimer amounts to 200,000. There is one heme and one copper atom in each subunit with a molecular weight of 100,000.[19] Differences between cytochromes can be due either to different forms of bonding of copper atoms or heme with protein or to conformational changes in the protein affecting the spin state of iron in the heme. Available data enable us to conceive the electron transfer chain as follows: cytochrome $c \rightarrow$ cytochrome $a \rightarrow$ copper $\rightarrow$ cytochrome $a_3 \rightarrow$ O_2.

2.3.1.5. Cytochrome P450

Cytochrome P450[20] is an enzyme which catalyzes oxidase reactions of a mixed type:

$$RH + DH + H^+ + O_2 \rightarrow ROH + D^+ + H_2O$$

where DH is an electron donor. Heme in these proteins with a molecular mass of about 40,000 is localized in a hydrophobic environment and in an oxidized, low-spin state is combined with the thiol group of cysteine. Reduced cytochrome P450 exists in a high-spin state. Under the effect of inhibitors (urea, detergents, etc.) and other unfavorable factors, cytochrome P450 easily changes over into a non-active low-spin complex P420.

2.3.1.6. Peroxidase

Peroxidase is a monomeric globular protein of 50 Å diameter and about 40,000 molecular mass (horseradish peroxidase).[21] Peroxidase displays oxidoreductase activity in the oxidation of a wide range of substances, among them phenols and amines:

$$DH + H_2O_2 \rightarrow DOH + H_2O$$

The prosthetic group of peroxidase is a protoporphyrin complex of trivalent iron, in which the Fe $3d$-orbitals are occupied by five unpaired electrons. The peroxidase apoenzyme is a polypeptide chain of about 300 amino acids, among them methionine and cysteine. Part of the S—S and S—H groups in these amino acids lie on the outer surface of the protein globule. The porphyrin ring of hemin, the active center of peroxidase, is highly aromatic and sufficiently hydrophobic; therefore, in all probability hemin is linked with protein due to hydrophobic interactions and coordinative with the imidazole group in histidine. Furthermore, hemin is bound with protein also by two ionic bonds formed by the negatively charged propionic acid groups of hemin and the positively charged groups of the protein.[22-24] An analysis of literature data pertaining to the formation mechanism of the structure of the native protein reveals that the strength of the enzymic complex is not great and that the principal stabilizing effect is produced by the formation of the coordination bond between the protein ligand and heme iron.[25]

Peroxidase forms a number of complexes with hydrogen peroxide. The activation mechanism of substrate molecules in the active center of peroxidase involves the formation of a ternary complex: heme–H_2O–donor of electrons. The role of the iron ion consists of transferring electrons from the donor to the hydrogen peroxide molecule. Peroxidase is endowed, moreover, with the oxidase function which increases in the presence of divalent manganese ions.

2.3.1.7. Catalase

Both in structure and in functions, catalase is akin to peroxidase. Catalase is the most effective among all known catalysts for the decomposition of hydrogen peroxide ($k = 10^7 s^{-7}$) and it displays, moreover, a peroxidase activity. The catalase molecule with a molecular mass of 225,000 is comprised of four subunits, each of which contains a hematin group. Hydrogen peroxide forms three compounds with catalase. Decomposition of H_2O_2 or ROOH occurs as a result of interaction between compound I of the catalase and the second substrate molecule.[26] Not only hydroperoxides but other hydrogen donors, for instance, ethanol, are capable of reacting with compound I.

2.3.2. Proteins and Enzymes Containing Non-heme Iron

Iron-containing proteins[13] are those in which iron atoms are bound with sulfur-containing ligands. The number of labile sulfur atoms is usually equal to that of iron atoms which varies between 1 and 18. The molecular mass of these proteins is in the range of 6000 to 750,000 units. Such proteins effect electron transfer in photosynthesis, nitrogen fixation, and respiration in mitochondria. They are capable of transferring electrons under a potential close to that of a reversible hydrogen electrode. Their oxidation–reduction potential is between +0.35 and −0.49 V.

2.3.2.1. Ferredoxin

Ferredoxin is an electron carrier. Iron and labile sulfur are present in its active center. Depending on the material from which the protein is isolated, its molecular weight varies from 6000 to 10,000. The protein molecule contains one or two iron–sulfur clusters (Figure 2a). Such a cluster respresents a distorted cube in the apex of which are arranged, alternating by fours, sulfur and iron atoms, while the cysteine radicals are coordinated in such a manner that at each iron atom a tetrahedral environment is set up. Presumably, electron transfer takes place only at one end of the cube. If two clusters are present, they are interconnected through a polypeptide chain. Conformational changes in the protein matrix significantly modify the donor–acceptor and catalytic properties of the cluster.

2.3.2.2. Hydrogenase

The principal function of hydrogenase or hydrogen-ferredoxin-oxidoreductase is to secure the progress of metabolic processes through hydrogen oxidation. Under specific conditions the enzyme is capable of isolating hydrogen from water.[27,28] According to available literature data,[29,30] the hydrogenase active center is comprised of one or two iron–sulfur clusters interconnected through the polypeptide chain of the enzyme. In the reduction of the enzyme by hydrogen, the iron–sulfur (cysteine) bond breaks, with the formation of the hydride form of the enzyme. The bonding of iron ions with the sulfur ion in the structures under discussion is characterized, apparently, by an ionic or strongly polar character. It can be presumed precisely that the polar ionic Fe—S bond is the active center which is responsible for the heterolytic cleavage of the hydrogen molecule.

The molecular weight of different hydrogenases varies from 60,000 to 200,000 units. Most of the investigated hydrogenases are labile. The principal inactivating factors are thermal denaturing and oxidation by oxygen. The hydrogenase from *Thiocapsa roseopersicina* was found to be the most stable. The reaction rate constant for the majority of hydrogenases is about $\sim 10^2 \, sec^{-1}$.

The activation of molecular hydrogen by hydrogenases involves heterolytic cleavage of the hydrogen molecule with the transfer of two electrons onto the active center[29]:

$$E + H_2 \rightarrow E \vdots H^- + H^+$$

Energetically this reaction is advantageous (free energy, 33 kcal/mol) compared to the homolytic cleavage of the molecule (enthalpy, 104.5 kcal/mol). The enzyme–hydride complex is an intermediate compound for all substrates taking part in reaction with hydrogenases.

2.3.3. Copper-Containing Oxidases

The biological role of copper-containing proteins is associated with the transfer of electrons, and of oxygen, and with oxidation catalysis.[31] Oxidases have also been used in electrochemical research. They can be subdivided into four main groups: "blue" proteins—electron carriers (azurin, stellacyanin, plastocyanin) containing 1–2 copper atoms in the molecule; "blue" oxidases (laccase, ascorbatoxidase, ceruloplasmin) containing at least four copper atoms and catalyzing the reduction of O_2 to H_2O with a simultaneous oxidation of the substrate; "non-blue" oxidases (urease, cytochrome c oxidase) containing one or two copper atoms and catalyzing the reduction of O_2 to H_2O_2; and oxygenases (tyrosinase).

2.3.3.1. Laccase

n-Diphenol oxidase catalyzes the oxidation of phenols and *n*-diphenylenediamines. The terminal acceptor of electrons is oxygen. Laccase with a molecular weight of 60,000, isolated from fungi, has been the most thoroughly investigated to date. The laccase isolated from the lac tree, with a molecular weight of 140,000, consists, apparently, of two subunits.[32] The active center of the molecule contains four copper atoms. Apart from copper atoms, SH groups and S—S bridges were found to be present in laccases. The structure of the active center has not yet been elucidated.

It is known at present that copper in oxidases can have three different forms: Form I—Cu^{2+} ion imparting an intensive blue coloring to the protein; form II—Cu^{2+} with low absorption in the visible region; form III—copper not detected by EPR. The unusual spectral properties of form I are due to the geometry of the coordination sphere of Cu^{2+} in the form of a distorted tetrahedron, with cysteine sulfur and the nitrogen of two histidines as ligands. The copper of form II is similar in its spectral and paramagnetic properties to conventional copper(II) complexes in which water or OH^- ion is directly coordinated. Copper of type II is distinguished by its high affinity for anions for instance, F^-. Copper of form III, which is not detected by EPR, makes up 50% of the total copper content and is present in the protein in the form of a dimer Cu^{2+}—Cu^{2+} with paired electron spins. These data give evidence, apparently of a nonsymmetrical character of the active center of the enzyme.

Laccase catalyzes the reaction of four electrons of the substrate–reducer and four protons of the solvent with an oxygen molecule involving the formation of two water molecules. The most difficult problem is the conjugation of the one-electron stage of substrate oxidation and four-electron reduction of the oxygen. Summarizing the available literature, we can classify the function of the different types of copper ions in laccase in the following manner. Copper of form I apparently plays a significant role in substrate oxidation, since this form of copper exhibits the highest oxidation–reduction potential. Cu^{2+} of type II takes part in electron transfer from the primary acceptor Cu^{2+} of type I to the two-electron acceptor, Cu^{2+} of the third form. The highly cooperative pair of copper ions of type III makes possible the two-electron reduction of the oxygen molecule directly to peroxide. Cu^{2+} of the second type plays a significant role in the stabilization of intermediate compounds. Cu^{2+}(II) coordinates water and is the binding site of H_2O_2. The active center of O_2 reduction lies, obviously, in a space which may be either open or closed depending on the state of different acceptors in the protein. This property may be highly important for the stabilization of reactive intermediate compounds of oxygen.

2.3.3.2. Tyrosinase (Neurospora)[33]

Tyrosinase is an oxygenase with a molecular weight of 33,000. In contrast to laccase, there are only one or two copper atoms per molecule. Tyrosinase catalyzes the oxidation of catechins to the respective quinones with the reduction of O_2 to H_2O. Since one of the cooper ions is presnet in the form not detected by EPR, it is assumed that it is a diamagnetic Cu^{1+} ion. The essentially different structure of the active center of tyrosinase and of laccase points also to a different mechanism of their operation. Apparently, in the case of tyrosinase a progressive transport of electrons takes place from the substrate through Cu^{1+} to a coordinated oxygen molecule. In this case, the negative charges which arise are compensated by a redistribution of protons in the solvent.

2.3.4. Flavoproteins

Biological activity of flavin enzymes is determined by the capacity of their prosthetic groups—flavin adenine dinucleotide (FAD) or flavin mononucleotide (FMN)—to take part in the reversible process of oxidation–reduction. They perform the role of electron carriers from the substrate to either other carriers or to oxygen. The function of flavin involves two main aspects: activation of CH bonds and uncoupling of electrons.[34] Figure 4 shows a flavin redox system. In enzymes, flavins are often coordinated with Mo, Fe, or Co ions.

2.3.4.1. Glucose Oxidase

Glucose oxidase oxidizes a nonphosphorylated glucose molecule, i.e., brings about a direct oxidation of free glucose. Its molecular weight is about 150,000. It contains two FAD molecules as the prosthetic group. Glucose oxidation by means of glucose oxidase can be represented by the following equation:

$$C_6H_{12}O_6 + H_2O + O_2 \rightarrow C_6H_{12}O_7 + H_2O_2,$$

the primary oxidation product being δ-lactone which thereupon nonenzymatically converts into gluconic acid.[35]

2.3.4.2. Xanthine oxidase

The study of highly purified preparations of the enzyme has revealed that there are two FAD molecules, two molybdenum atoms, and eight iron atoms per protein molecule. Molybdenum is combined with the histidine and cysteine radicals of the protein molecule. In addition, a bond with FAD, of unknown

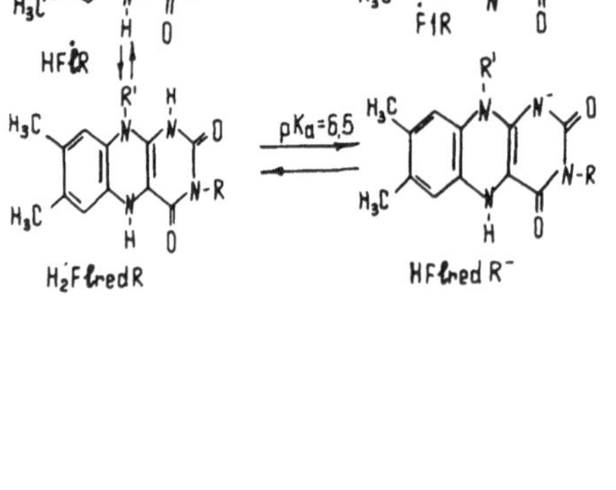

Figure 4. Flavin redox system.

type, is presumed to exist. Iron can be separated during dialysis against a mixture of dithionite and dipyridyl. The operation mechanism of xanthine oxidase has recently attracted a great deal of attention. It will be important to throw light on the way the electron is transferred within the enzyme molecule. It is assumed[35] that the electron transport is realized along the chain:

$$\text{substrate} \to Mo \to FAD \to Fe \to O_2$$

However, there also exists a different viewpoint postulating that enzymatic activity is not associated with the presence of iron in the molecule. Xanthine oxidase catalyzes the oxidation of the hydrated form of hypoxanthine and xanthine into xanthine and uric acid, respectively. The reaction in the presence of xanthine oxidase can be written down as follows:

$$\text{xanthine} + H_2O + O_2 \to \text{urate} + H_2O_2$$

Two examples of flavin oxidases differing in the structure of the active center show significant differences in the catalysis mechanism as well. While glucose oxidase is a narrowly specific enzyme, xanthine oxidase has a wide range of substrates. But for the majority of flavoproteins the characteristic feature is a rigorous specificity. The specificity of an enzyme is known to be determined by its protein part. It has been suggested that the protein component of flavin enzymes not only directs the operation of the prosthetic group but is directly engaged in electron (hydrogen) transfer with the participation of sulfhydryl groups.

2.3.5. Alkaline Phosphatase

In contrast to all of the enzymes listed above, alkaline phosphatase[36] belongs to the class of hydrolases which effect the cleavage of a covalent bond while simultaneously activating water:

$$AB + H_2O \rightleftarrows AH + BOH$$

Alkaline phosphatase from the bacterium *Escherichia coli* is the only well-characterized metalloenzyme among those carrying phosphate. The dimer molecule of the enzyme with a molecular mass of 80,000 contains from two to four zinc atoms. Alkaline phosphatase is a nonspecific phosphomono-esterase, and it also catalyzes the hydrolysis of compounds with a phosphoan-hydride bond (pyrophosphate and ATP) and of fluorophosphate. The substrate molecule in the active center directly interacts with a zinc ion. In this case we can presume that the site of phosphate addition lies in the contact region between two subunits. There are several possible explanations for the activating role of metal ions. Cations of divalent metals regain their charge in a neutral solution and can remove the negative charge of oxygen atoms in phosphate, thus facilitating the addition of the substrate to the enzyme. Furthermore, a metal ion is likely to cause the polarization of the respective P—O bonds. In drawing off electrons from the oxygen atom, the metal ion weakens the P—O bond and increases the positive charge on the phosphorus atom. This must facilitate both the nucleophilic attack and the rupture of the P—O bond. Another possible interpretation of the role played by a metal ion is the activation of water. The strength of O—H bonds in a coordinated water molecule significantly declines and the ionization of the molecule becomes possible.

3. Methods of Immobilization of Enzymes

The obstacles for the use of enzymes in electrocatalysis are their instability and the impossibility of multiple application of an enzyme which is in a homogeneous phase. Overcoming these difficulties is possible through the utilization of so-called immobilized enzymes, i.e., enzymes bound with the carrier. The most promising methods of immobilization—immobilization by way of adsorption on carriers; inclusion into the space lattice of gels; chemical methods of immobilization[37]—for use in bioelectrocatalysis will be presented in the following subsections.

3.1. Immobilization by Adsorption

The adsorption method is the simplest one and is often used in bioelectrocatalysis research. Essentially it involves the incubation of protein in the carrier suspension with the subsequent washing of the nonadsorbed protein. Adsorption of proteins on different types of surfaces is effected due to electrostatic, hydrophobic, and dispersion interactions. The most popular carriers are carbon, soot, clays, aluminum oxide, silica gel, and glass. The optimal inert carrier is glass. It has recently been shown that porous glass with calibrated pore size can be used for immobilization of enzymes by adsorption. An interesting method of immobilization by adsorption has been proposed[38,39] in which lipid is first adsorbed on carbon or silica gel and then the enzyme is adsorbed on the so-called "soft" surface of the lipid.

As a rule, the quantity of adsorbed enzyme increases with carrier surface, but only to a certain limit, i.e., until the dimensions of pores in the carrier become commensurable with the size of the protein globule. The catalytic activity of adsorbed enzymes is determined above all by the nature of the enzyme itself and the degree of surface filling. The majority of enzymes are characterized by declining activity with greater filling, which is possibly associated with the appearance of strong intermolecular interactions. A decrease of specific activity has been observed in the case of catalase adsorbed on graphite. In come cases with increasing adsorption a growth of catalytic activity is observed[37] or the activity is found to be independent of the extent of adsorption.

An exceptional simplicity of the adsorption method of immobilization renders its application very attractive; however, it is by no means acceptable in all cases because of possible desorption of the enzyme from the carrier surface and a partial denaturing of the enzyme due to the unfolding of the protein globule.

3.2. Immobilization by Inclusion into the Space Lattice of Gels

Immobilization through inclusion into a gel, similarly to immobilization by adsorption, is a physical method of protein fixation. Advantages of such a method are its simplicity and the absence of chemical modification of the enzyme molecules. Polyacrylamide, polyethylene glycol methacrylate, polysaccharide, polyionite, as well as various inorganic gels are used.

Polyacrylamide gel is formed in the reaction of acrylamide with N,N'-methylenebisacrylamide. It has the structure of a three-dimensional lattice encompassing the enzyme. By varying the relationship of the components, it is possible to regulate the properties of the gel: strength, elasticity, viscosity. When carrying out the immobilization of an enzyme in polyacrylamide gel, the yield is as high as 100%. As to its chemical nature, polyacrylamide gel is weakly hydrophobic and rather inert. It swells in an aqueous medium. The diffusion rate of low-molecular weight compounds in nonconcentrated gels is close to that in water. The lower activity of an enzyme immobilized into a gel is more often associated either with its reduced concentration or with the diffusion rate of the reaction components.

Polyethylene glycol methacrylate gel is formed during copolymerization of methacrylate ethylene glycol and dimethacrylate ethylene glycol. The gels obtained are similar to the polyacrylamide ones but have better mechanical properties and transparency, and are more hydrophobic.

Polysaccharide and polyionite gels are different in that their spatial structure is governed by hydrogen and ionic bonds rather than by covalent bonds. The formation of polyionite gels is effected with a reduction of the ionic strength of the solution containing polycations and polyanions. An amplification of electrostatic interaction between polymeric chains takes place with the formation of a gel, in the structure of which the enzyme is incorporated.

The methods of gel formation listed above are fairly mild and can be employed for immobilization of labile enzymes. A shortcoming of the immobilization in gels is the washing out of enzymes during the process of operation. Moreover, gels are, as a rule, nonelectroconductive, thus making it difficult to use immobilized enzymes in electrocatalysis.

3.3. Chemical Methods of Immobilization

Chemical methods of immobilization are being currently developed with high intensity. Their main advantage lies in the fact that even in the case of very prolonged utilization the enzyme does not go into the solution. However, the formation of additional chemical bonds often accounts for a complete or partial inactivation of the enzyme. In the case of chemical immobilization two approaches are possible: cross-linking of protein macromolecules and their covalent attachment to carriers.

3.3.1. Cross-linking with Bifunctional Reagents

On adding a reagent enzyme with two or more reactive groups to the solution, the protein globules are cross-linked with the formation of a space lattice. The compounds most often used for this purpose are glutaraldehyde (**1**), hexamethylene diisocyanate (**2**), adipimidate (**3**), diazobenzidine-3,3-dianisidine (**4**), 4,4-diisothiocyanatediphenyl-2,2-disulfonic acid (**5**), and *sym*-trichlortriazine (**6**):

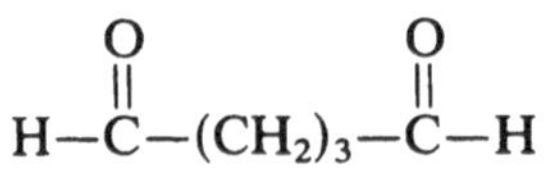

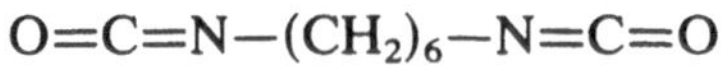

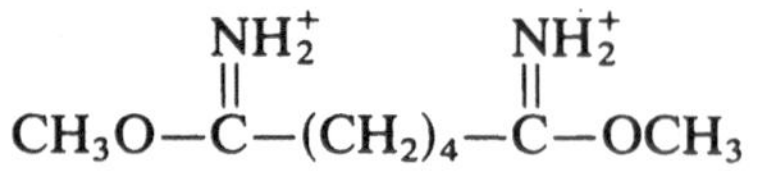

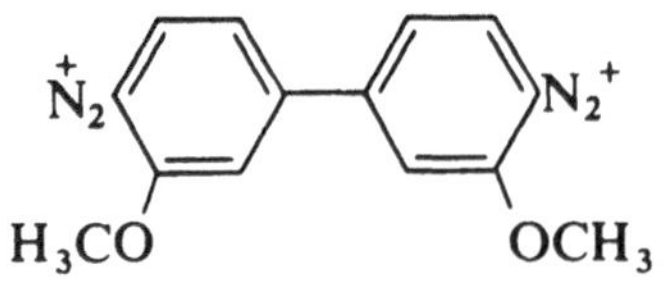

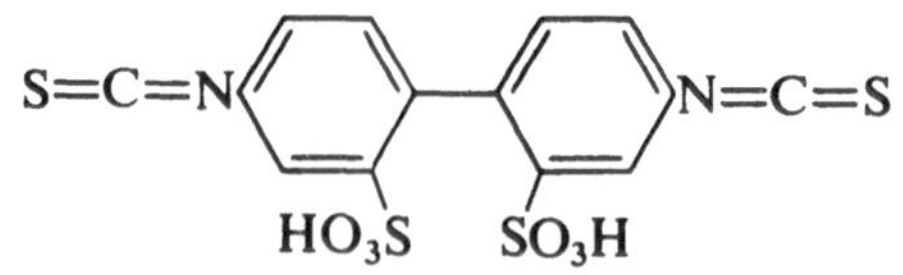

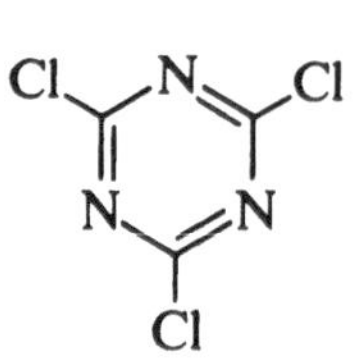

Immobilization of enzymes by cross-linking with bifunctional reagents is carried out by different methods: direct cross-linking of enzymes; cross-linking of enzymes with inert proteins; adsorption of enzymes on a water-insoluble carrier with subsequent processing using a bifunctional reagent. Albumin is often employed as an inert protein; cellophane, collagen and ion-exchange membranes are used as water insoluble carriers.

3.3.2. Covalent Linking to Carriers

This method of immobilization is particularly interesting when using enzymes as electrocatalysts, since it provides high stability and the possibility of using a wide range of carriers, including electroconductive ones. The principal reagents of the protein globule are the NH_2 group of lysine, COOH groups of aspartic and glutamic acids, histidine, cysteine, and tyrosine. Different types of carriers contain different groups which can enter into reaction with the structural elements of the protein globule. Organic polymers obtained through the copolymerization of maleic anhydride with ethylene or acrylic acid contain active anhydride groups. Dextran, saccharose, and cellulose have OH groups. In polyamide polymers (capron, nylon, silk) a peptide bond enters into reaction. In the case of polyacrylonitrile fibers, immobilization is effected through imidoethers. Linking of enzymes to inorganic carriers (porous glasses, metals, etc.) is realized through hydrated metal oxides. Linking of enzymes to various carbon materials (pyrographite, glassy carbon, soot, activated charcoal) is carried out with the use of functional groups ($\geqslant$C—OH, —COOH, $>$C=O, etc.) at their surface.

The activity of an enzyme immobilized by one of the chemical methods is determined by numerous factors: the properties of the enzyme, the properties of the carrier, the method of formation of the covalent bond with the carrier, etc. Therefore, it is difficult to predict the properties of the enzyme in every given case. The methods of immobilization, as well as specific examples of enzyme fixation, are discussed elsewhere.[37]

3.4. Properties of Immobilized Enzymes

Although general pathways of enzyme stabilization are not yet clear, numerous examples are available[37] which reveal the significant increase in the stability of immobilized enzymes with respect to unfolding (i.e., denaturing) of the protein globule under the effect of temperature, extreme pH values, denaturing compounds, etc. It has been shown that for each enzyme there is a possibility of selecting a carrier on which the enzyme, if stored under 4°C, will retain its activity for months.

Immobilization is, as a rule, accompanied by a declining activity of the enzymes. However, at present there are numerous examples of successful immobilization with the preservation of enzyme specificity and activity within 10 to 90% of its activity in the native state. This permits us to think that the problem of obtaining enzymic preparations which are close in physicochemical properties to heterogeneous catalysts and electrocatalysts, particularly as regards the preservation of enzymatic activity and specificity, can be successfully solved in each specific case.

4. Electrochemical Properties of Protein Macromolecules and Their Active Groups

The history of the use of electrochemical methods to study the properties of proteins, enzymes, and their component structural units is several decades old. Until quite recently, these studies were primarily associated with the use of polarography to solve analytical problems.[40,41] The polarographic method of analysis has become very popular for determination of proteins, enzymes, and nucleic acids.[42–44] It has been found that proteins irreversibly adsorb on mercury and in the presence of cobalt salts promote the evolution of hydrogen. In this case wave height is proportional to the amount of adsorbed protein. These studies have been summarized in some reviews and monographs.[40–42]

Much progress in the study of electrochemical properties of protein macromolecules has been achieved recently using another method, namely, the study of redox transformations of proteins, the carriers of electrons and enzymes, and their active groups at the electrode–electrolyte interface. This approach is intimately related to the use of enzymes to promote electrochemical reactions and pursues the purpose of elucidation of the mechanism of electron transport and the structural features of enzymes.

4.1. Electrochemical Properties of Active Groups

4.1.1. Heme

Heme and its derivatives are representatives of metalloporphyrins which are being intensively investigated both in catalysis and in electrocatalysis.[45] Both the ligand and the central atom in the metal can be subjected to electrochemical transformations. The ligand can add and yield two electrons. The potential difference necessary for the formation of a univalent anion and univalent cation amounts to about 1.4 V. These processes are observed, as a rule, only in nonaqueous solutions. Redox reaction potentials of central metal atoms lie within this interval. Redox transformations of the central metal ion are observed in aqueous solutions during adsorption of hemin on an inert electrode, for instance, an amalgamated gold electrode (Figure 5).[46] The rate of the process is controlled by adsorption.[47] In the case of maximum coverage of the surface, the area occupied by a molecule amounts to about 220 Å^2. It is close to the geometrical area of the molecule, 238 Å^2, and is evidence for a planar orientation of adsorbed molecules.

In the case of polarographic reduction of deuterohemin in 0.1–KOH[48] two waves are observed with $E_{H_2,1/2}^0 = -0.36$ V and -1.16 V, which correspond to the reduction of the central Fe^{3+} ion and the ligand, respectively. Half-wave potentials for the reduction of Fe^{3+} are observed to decline in the series: proto- > hemato- > meso- > deuterohemin. This shift reflects the variation of electron density at the central atom.

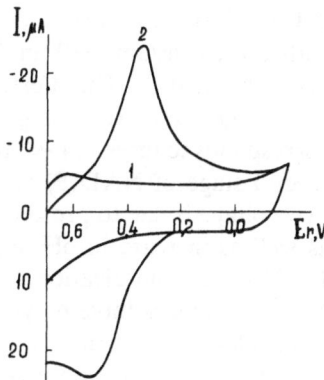

Figure 5. Potentiodynamic curves (2) on the amalgamated gold electrode in hemin solution (10^{-5} M, pH = 7.2). (1) Background curve.

These data, similarly to the results pertaining to other metalloporphyrins,[45] provide evidence for the reversible character of the redox transformations on electrodes.

4.1.2. NAD, NADP

Electrochemical transformations of nicotinamide adenine dinucleotide (NAD) and nicotinamide adenine dinucleotide phosphate (NADP) have been dealt with in several publications, since they are the most important hydrogen carriers. There are reviews on this topic.[49–51] The standard redox potential of the NAD(P)H/NAD(P) couple has been found to be -0.32 V.[52]

Kinetic and coulometric studies have been carried out on various electrode materials: mercury, gold, graphite, platinum,[53–55] both in aqueous and nonaqueous solutions, on natural and model compounds.[53–56] Electrochemical transformations of NAD(P) are schematically shown in Figure 6. The transfer of the first electron is reversible and is independent of

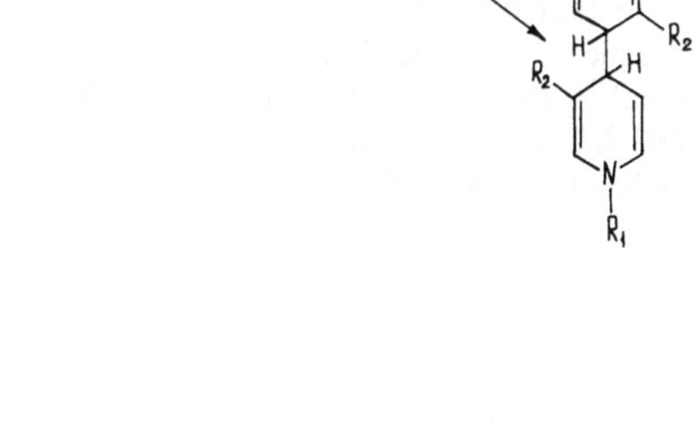

Figure 6. Electrochemical transformations of NAD.

pH. The NAD(P)$^{\cdot}$ thus formed undergoes a rapid and irreversible dimerization with a rate constant $\sim 10^6 \, m^{-1} \, s^{-1}$.[57] Higher values of dimerization rates have been obtained.[58] The dimer is completely inactive in reactions with several dehydrogenases. According to a recent publication[59] 90% of the product corresponds to three 4,4'-stereoisomers, 10% to three 4,6'-stereoisomers. The second stage of NAD(P) reduction is irreversible and depends on pH. The overvoltage is due to proton transfer. Potential shifts towards negative values, as well as increase in pH, lead to a higher yield of the enzymatically active 1,4-NADH. Dimerization can be significantly slowed down by cross-linking NAD to water-soluble polymers. An alternative method is the use of mercury electrodes coated with cholesterol oleate, on which the first reduction wave is suppressed, reduction proceeding at more negative potentials.[60]

An important role of adsorption during the reduction of NAD$^+$ has been emphasized.[61] Reduction of the pyridine ring in NAD$^+$ on a gold electrode takes place at more positive potentials compared to those in the case of nicotinamide mononucleotide (NMN$^+$) and nicotinamide. This is due to strong adsorption of the adenine part of NAD$^+$ on the electrode surface.

The results of electrooxidation of NADH on electrodes made of platinum and carbon materials[62-65] reveal that the process is irreversible: the slope of the polarization curve is between $(2.3)(3RT)/2F$ and $(2.3)(3RT)/F$. The half-wave potential at pH = 7 is about 0.9 V on platinum and about 0.6 V on glassy carbon. Adsorption plays a significant role in the oxidation process. On a pure platinum electrode NADH is oxidized to NAD$^+$, the latter slowly adsorbing and passivating the electrode.[61] NADH fails to adsorb on glassy carbon. NAD$^+$ obtained in the case of two-electron transfer oxidation of NADH rapidly adsorbs in a planar orientation, presumably, through the adenine group.[65] Thereupon a slow reorientation into a vertical position takes place, this involving interaction between parallel adenine and pyridine rings.

In order to accelerate electrooxidation of NADH, attempts are being made to develop chemically modified electrodes. Positive effects have been observed in the case of glassy carbon modified with 3,4-dihydroxybenzyl-amine[66] and dopamine.[67]

4.1.3. FAD, FMN

Flavin nucleotides—flavin adenine dinucleotide (FAD) and flavin mononucleotide (riboflavin-5'-phosphoric acid) (FMN)—as prosthetic groups can either comprise a metal or serve as cofactors without a metal. Electrochemical transformations of FAD and FMN are characterized by strong adsorption.[49] The area occupied by a FAD molecule on a mercury electrode is 280 Å^2, i.e., close to the geometrical dimensions corresponding to the molecular model. Reduction of FAD and FMN proceeds in two reversible crossing one-electron steps.[49] Reduction of FMN occurs in acid to the $FMNH:FMNH_2$ complex which is not further reduced.

The method of impulse polarization has been used[69] to show that FAD and FMN are reversibly reduced on the mercury electrode, the maximum height for $E_{H_2^0} = -0.20--0.19$ V for pH ~ 7 being directly proportional to the concentration of the prosthetic group.

4.2. Redox Transformations of Proteins and Enzymes on Electrodes

One of the problems arising in electrochemical studies of biopolymers is the preservation of native properties of protein macromolecules during their adsorption on the electrode. Investigations of conformational changes in both neutral and enzyme-active proteins on the mercury electrode by means of electrochemical methods have been carried out.[70–71] The quantity of adsorbed protein and its structure were estimated from the magnitude of the catalytic current of hydrogen evolution and the capacity of the double layer. An analysis of the data obtained makes it clear that for a great number of proteins at concentrations below 3×10^{-4} g/mL rapid ($k > 0.5 \, s^{-1}$) irreversible adsorption occurs, accompanied by surface denaturing. The protein is flattened and the protein globule unfolded to 5–10 Å thickness. This process is due to the high surface tension at the mercury–aqueous solution boundary (400 mN/m) and a relatively low surface pressure during protein adsorption from dilute solutions. Monolayer, or close to monolayer, coating of the electrode surface with denatured protein molecules leads to a sharp reduction of the surface tension to a value (10–20 mN/M) which is of the same order as that at the water–polar organic compound boundary. This layer prevents the denaturing of the next layers of protein. Therefore, it has been proposed[70] that in the pores of the first layer and in the subsequent layers the native molecules of protein (enzyme) are adsorbed. Surface denaturing must also become difficult in the case of increasing surface pressure due to a higher concentration of protein adsorbent. This accounts for the fact that it is possible to carry out investigations of electrochemical properties of proteins, the carriers of electrons and enzymes in oxidation–reduction reactions, in the native state. Similar ideas have recently been suggested.[72]

Thus the conformational behavior of proteins at the electrode–electrolyte interface is presumably determined by the properties of this boundary: surface tension, charge, hydrophobicity, and chemical structure. It is to be remembered, however, that the above picture presents only a qualitative character, because results of direct structural (for instance, spectral) investigations into protein conformation at the interphase boundary are not available.

Some specific examples of the study of electrochemical behavior of proteins and enzymes are given below.

4.2.1. Cytochrome c and Other Heme-Containing Proteins

The study of electrochemical properties of cytochrome c in the adsorbed state is dealt with in a number of publications.[46,60-62] The electrochemical transformation of cytochrome c on the mercury electrode was demonstrated for the first time in Reference (73). The reduction current for cytochrome c on a dropping mercury electrode has been found to be proportional to concentration within a wide range of values[75] (Figure 7).

The possibility of electrochemical reduction and oxidation of cytochrome c on a gold electrode[46,47] was demonstrated by means of spectroelectrochemical measurements. The measurements were made in a cell with an optically transparent (grid) electrode, thus permitting the spectra to be recorded directly during electrochemical measurements. Figure 8 shows the spectra of the electrochemically reduced (curve 1) and oxidized (curve 2) cytochrome c. The spectra of ferro- and ferricytochrome c obtained electrochemically are identical with those of the chemically reduced and oxidized forms of cytochrome c. Investigations have also shown that after electrochemical reduction of the protein its enzymatic activity is preserved.[73]

During variation of the degree of electrochemical oxidation of cytochrome, the reduction of heme iron, which is present at the active center of cytochrome c takes place. In the potentiodynamic curves measured on the gold amalgamated electrode, only one peak at the cathode branch of the curve is observed, with $E = 0.300$ V. In the anodic part of the i-E curve, no oxidation peak of cytochrome c, reduced during the cathodic portion, could be observed. Taking into account that the redox potential of cytochrome is 0.68 V, it may be concluded that electrochemical transformation of cytochrome c proceeds with significant overvoltage. However, it has been shown[76] that for the concentrations in the solution ranging from 10 mM to 30 mM, cytochrome c is reversibly oxidized and reduced on the mercury electrode. The authors are of the opinion that with an increasing concentration of cytochrome c the rate of the electrode process is limited by charge transfer across several protein layers. These results are in contradiction with the results obtained by the same authors[77] when a radiochemical method was used to determine the concentration of cytochrome c on mercury and it was shown that the surface area per protein molecule

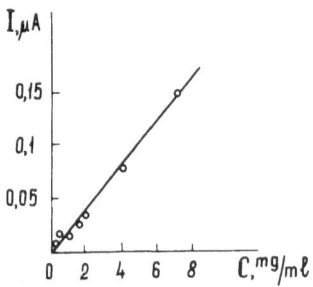

Figure 7. The cytochrome c reduction current versus concentration.

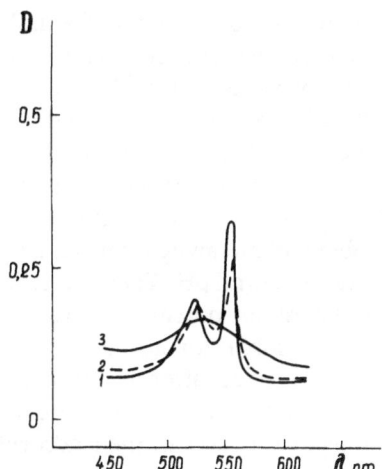

Figure 8. Spectra of cyctochrome c measured in a cell with an optically transparent electrode. Curve 1—cytochrome c reduced at $E_r < 0.3$ V; curve 2—cytochrome c oxidized at $E_r > 1.2$ V; curve 3—the initial solution of cytochrome c.

amounts to 2200 Å^2 in the plateau region of the adsorption isotherm. In the crystalline state, the molecule occupies about 1200 Å^2.

It appears more likely that significant overvoltage for the electroreduction and electrooxidation of cytochrome in the first monolayer is associated with conformational changes in the protein globule of cytochrome c during adsorption. Investigation of the spectra of cytochrome c adsorbed on silochrome has shown[78] that during adsorption cytochrome goes over into a high-spin state; the concentration of the high-spin form increases with decreasing coverage and such a high-spin hemoprotein is not reduced by ascorbate.[79] It is most likely that at low coverage cytochrome is adsorbed in a high-spin form and is reduced with much overvoltage. With a higher concentration of cytochrome c in solution and, consequently, on the electrode surface, the subsequent molecules are capable of adsorbing onto the first layer without undergoing great conformational variations. Irreversibility of the electrode behavior of cytochrome c in this case is associated with the difficulties involved in transferring electrons to the second adsorption layer, since the thickness of the adsorbed protein film is of the order of 10 Å.

Despite certain differences of opinion regarding the mechanism of the electrode reaction of cytochrome, highly significant is the fact that there is a direct exchange of electrons between the protein molecule's active center and the electrode.

It has been reported[80] that a reversible one-electron transformation of cytochrome c occurs on an indium oxide electrode. Cytochrome c yields quasi-reversible maxima[81] at $E_{H_2^0} \sim 0.25$ V on a gold electrode in the presence of 4,4'-bipyridine.

The electrode behavior of certain other cytochromes has also been investigated. It has been shown[82,83] that cytochrome c_3 yields a reversible wave on the mercury electrode, the reaction involving four successive transitions of electrons to four hemes in the molecule. The rate of reactions with cytochrome c at mercury electrodes is controlled by diffusion and is, apparently, uninhibited by adsorbed particles.[84] Two close-lying reduction peaks ($E_{H_2} \sim -0.17$ and -0.22 V have been observed). These values are fairly close to the redox potentials of three hemes in the cytochrome c molecule in the native state.

Apart from cytochromes, certain other heme-containing proteins have been studied: methemoglobin[75,85] and metmyoglobin.[86] Both methemoglobin and metmyoglobin can be quantitatively reduced on the mercury electrode at neutral pH. Their reductions require very high overvoltages compared to the redox potential corresponding to the native state. In the case of metmyoglobin, for instance, overvoltage amounts to about 1.0 V. However, the end product after oxidation by oxygen is native oxymyoglobin. These data reveal that, as in the case of cytochrome c, adsorbed protein molecules are subject to reduction, their adsorption not being accompanied by irreversible conformational changes.

High activity in the electrochemical reactions of myoglobin is displayed at a gold electrode modified by methyl or benzyl viologen.[87] For $E_{H_2}^0 = -0.725$ and $+0.25$ V there correspondingly occurs a complete reduction and oxidation of myoglobin.

4.2.2. Peroxidase

Electrochemical properties of this heme-containing enzyme have been studied.[88] The potentiodynamic curves measured on the gold amalgamated electrode in the presence of peroxidase in the electrolyte show two maxima—anodic and cathodic (Figure 9), near the potential $E \sim 0.10$ V. Replacement of the test electrolyte with electrolyte containing no peroxidase did not lead to a change in the variation of charging curves. This points to the irreversible

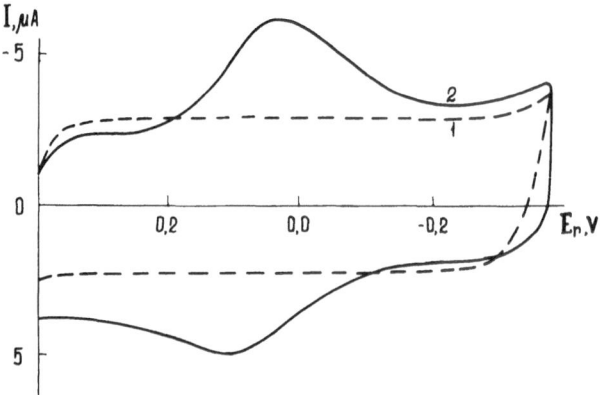

Figure 9. *i–E* curves measured on the gold amalgamated electrode in the presence of peroxidase solution (2). (1) Background curve, pH = 6.9.

character of peroxidase adsorption and the occurrence of electrochemical reaction in the adsorbed state.

Investigations in a cell with an optically transparent electrode were carried out to elucidate the nature of the observed oxidation–reduction reaction of adsorbed peroxidase. As is known, reduction and oxidation of hemin in the active center of peroxidase is accompanied by significant spectral changes. Figure 10 shows the spectra of native peroxidase, of peroxidase reduced by sodium dithionite, and of peroxidase during its electrochemical reduction when the potential of the working electrode corresponds to that of the maximum for cathodic peroxidase reduction on potentiodynamic curves. As follows from the figure, in the case of electrochemical reduction the peroxidase spectrum fails to undergo any apparent changes. Thus, spectroelectrochemical measurements made in a peroxidase solution indicate that the redox process observable on the i–E-curves is unrelated to reversible oxidation–reduction of the enzyme active center. Further research made on electrodes from carbon materials and in solutions with different pH suggests that the redox process observed on mercury proceeds with the participation of disulfide groups of the enzyme according to the reaction:

$$\text{R—S—S—R} + 2\text{Hg} \rightarrow 2\text{RSHg} + 2\text{H}^+ \underset{-2e}{\overset{+2e}{\rightleftarrows}} 2\text{RSH} + 2\text{Hg}$$

As has been shown,[89] this kind of reaction does take place on the mercury electrode in the presence of cystine. The i–E curves measured on an amalgamated electrode in the presence of peroxidase apoenzyme in the solution also have anodic and cathodic maxima, similar to those obtained in the peroxidase solution. Thus, in the potential region investigated, the S—S groups of the protein globule of peroxidase are electrochemically active. Heme iron in the active center does not take part in the observed redox process.

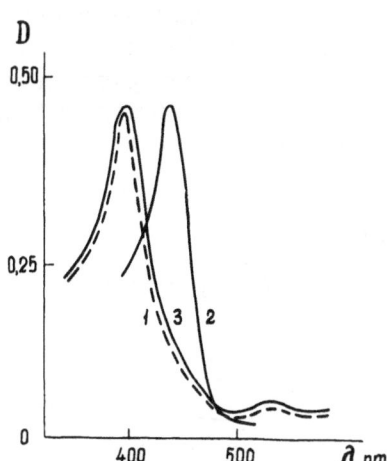

Figure 10. The spectra of peroxidase in a cell with an optically transparent electrode. (1) Native peroxidase; (2) peroxidase reduced by sodium dithionite; (3) peroxidase reduced at $E_r = +0.05$ V.

4.2.3. Ferredoxin

Electrochemical properties of ferredoxin have been investigated.[46,90–92] An easily discernible wave proportional to ferredoxin concentration is observed at the mercury electrode. Potentiodynamic i–E-curves demonstrate that the redox process is close to a reversible one. The data obtained favor the assumption that the ferredoxin active center is reduced on the electrode while preserving its native structure. However, such a conclusion cannot be deemed absolutely valid, since it does not provide for an increased possibility of reduction for the sulfur-containing groups available in ferredoxin. The half-wave potential of ferredoxin electroreduction amounts to 0.1 V, i.e., close to the reduction potential of sulfide groups in sulfur-containing proteins.[93]

A reversible reduction and oxidation of ferredoxin is also observed on a gold electrode modified by methyl viologen.[94]

4.2.4. Flavoproteins

From polarographic studies of proteins, some authors[95–97] have arrived at the conclusion that flavoproteins cannot be reduced on electrodes. Recently more thorough polarographic investigations were undertaken on the reduction of glucose oxidase,[69] flavin-containing protein,[69] and xanthine oxidase.[98]

On the impulse polarograms of glucose oxidase (GOD) a well-defined maximum at $E_{H_2} = 0.1$ V, as compared to $E_{H_2} = -0.2$ V for pure FAD, is observed for the reduction of one prosthetic group (FAD) (Figure 11, curve 1 and 2). In enzyme cleavage into an apoenzyme and a prosthetic group the maximum potential (curve 3) coincides with that corresponding to a pure FAD. This has enabled the authors to draw the conclusion that FAD bound with protein is subject to reduction. The magnitude of electric current for the reduction is proportional to the concentration of glucose oxidase. For an increase of the pH by unity, the curve is displaced towards the cathode side by 0.06 V, i.e., the transition of each electron onto the prosthetic group is accompanied by that of a proton. In contrast to this, the reduction potential of riboflavin-containing protein coincides with the reduction potential of free riboflavin which indicates that dissociation of the molecule occurs during adsorption.

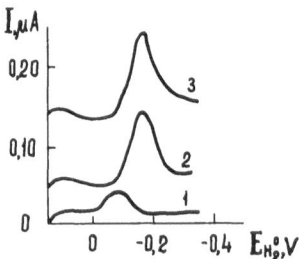

Figure 11. Differential impulse polarograms in 0.1 M phosphate buffer. (1) 8 μM GOD; (2) 10 μM FAD; (3) 8 μM (apoenzyme + FAD).

During the reduction of xanthine oxidase the impulse polarographic wave displays two maxima at potentials 0.08 and −0.01 V.[98] The position of the first maximum coincides with the enzyme redox potential. In a solution of denatured xanthine oxidase, the first maximum is absent. Accordingly, the view of the authors is that xanthine oxidase in this potential region is reduced in the native state. As for the other maximum, it is possibly associated with the reduction of sulfur-containing groups.

Thus, it can be seen that the study of the electrochemical behavior of protein macromolecules in the adsorbed state can provide highly important information on their redox transformations. It is necessary, however, to take into account the possibility of significant conformational restructuring of proteins in the adsorbed state. To elucidate the role of conformational changes it is necessary to undertake investigations combining electrochemical and spectral methods.

5. Methods of Conjugation of Electrochemical and Enzymic Reactions

Several methods of using biocatalysts to accelerate electrochemical reactions have already been developed.[99–100] Their schemes are shown in Figure 12 for the example of oxidation reactions.

The first method (Figure 12a, b) involves obtaining, in the course of an enzymatic reaction, intermediates (S*) whose electrochemical transformation on the electrode from a given material into the end product is realized at a lower overvoltage as compared to the initial substrate (S^{red}). The enzyme is solubilized or fixed on the electrode. In this case the overall process is likely to be accelerated; however, it is obvious that not all of the free energy of the reaction

$$S^{red} \rightarrow S + ne$$

is used electrochemically. Nonetheless, this scheme has certain development prospects, since electrochemical transformations of the majority of organic substances proceed at high overvoltage.

Another, more general, method involves electron exchange between the electrode and the active center of the enzyme where the substrate is activated and reacts (Figure 12c, d). In this caser the enzyme plays the typical role of an electrocatalyst and its effect corresponds to the notion of "bioelectrocatalysis". The central problem is effecting electron transport between the enzyme active center and the electrode. The solution of this problem can be approached by two radically different methods.

In the first case (Figure 12c, e), electron transfer from the active center onto the electrode is effected by low-molecular weight carriers—mediators.

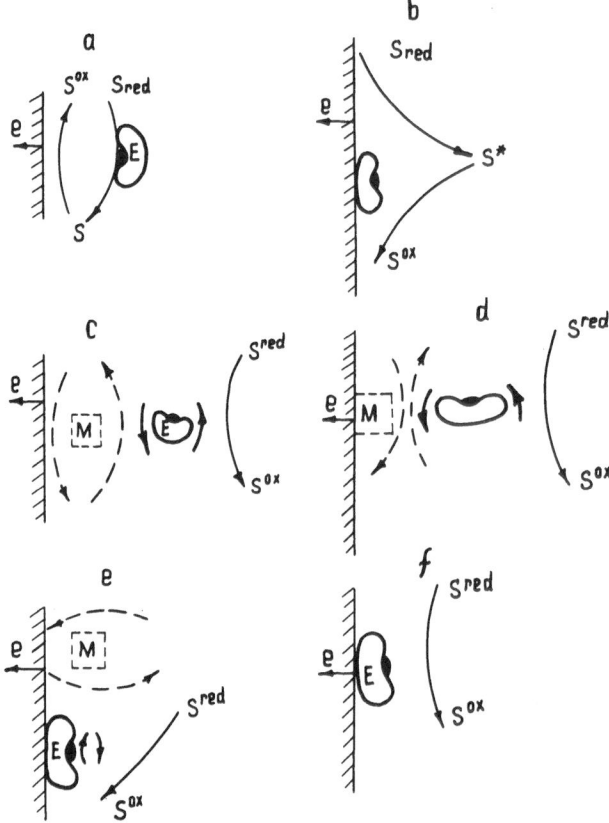

Figure 12. Schemes showing the application of enzymes to promote electrochemical reactions.

According to the "c" scheme, all reaction components (substrate, mediator, and enzyme) are present in a solubilized state. The reaction takes place in the bulk solution, while the electron transport is effected by a low-molecular weight carrier. According to schemes "d" and "e," one of the components, either the mediator or enzyme, is immobilized, but electron transfer is brought about by the mediator.

In order that a compound be used as a mediator, it must satisfy quite a number of requirements: 1) the interaction stage between the mediator and the enzyme active center must be fast (the mediator must be a "specific" substrate of the enzyme); 2) the normal oxidation–reduction potential of the mediator must be close to that of the reaction concerned; 3) the mediator should be subject to electrochemical oxidation (or reduction) on the electrode made from a given material under conditions close to reversible ones. By no means are all of the known mediators able to meet the above requirements. In Table 3 the characteristics of certain mediator compounds which have been used in bioelectrocatalysis are given.

Table 3
Electrochemical Properties of Certain Mediators

Nos.	Name	Enzyme	E^0, V	Electrode material	Reference
1	Methyl viologen	Hydrogenase	0.011	Carbon materials	(101)
2	Quinone	Peroxidase	0.71	Graphite, gold	(92)
3	TMPD	Cytochrome c oxidase	+0.72	Graphite	(93)
4	Pyrocatechol	Tyrosinase	+0.7	Carbon materials	(96)
5	Phenazine methosulfate	Cytochrome b_2	+0.54	Platinum	(100)

The second, or more accurately, the third method (Figure 12f) represents a radically different organization of electron transport—direct electron exchange between the enzyme active center and the electrode. This variant appears to be most alluring, since it enables us to eliminate the intermediate link associated with the mediator transfer of electrons from the enzyme onto the electrode, and to develop a heterogeneous system not different from conventional, electrocatalytically active electrodes. However, by no means in every case, as will be demonstrated further, is it possible to realize a reaction without the participation of a mediator. Therefore, the search for optimal mediators continues to be an important task.

6. The Use of Enzymes to Accelerate Electrochemical Reactions

In the course of work on bioelectrocatalysis all three ways of using enzymes to accelerate electrochemical reactions have been taken into account. These research areas cover the following electrochemical reactions: hydrogen, oxygen, and oxidation reactions of certain organic compounds. In each case, different methods were employed to achieve the conjugation of enzymatic and electrochemical reactions.

6.1. Hydrogen Reaction

The hydrogen reaction involves the processes of ionization and evolution of molecular hydrogen. To achieve activation of these processes, hydrogenases from fluorotrophic bacteria *Thiocapsa roseopersicina*[101–107] and *Chloropseudomas ethylica*[108] have been used.

6.1.1. Mediator Mechanism of Electron Transport

The mediator transport of electrons from the active center of the enzyme onto the electrode has been used to carry out the reaction of anodic oxidation of molecular hydrogen on carbon electrodes.[101]–[104] Methyl viologen (MV) was chosen as mediator. The scheme of reactions proceeding in such a system can be described as follows:

$$ e \begin{array}{c}\longleftarrow\end{array} \quad \binom{MV^+}{MV^{2+}} \quad \binom{E}{E'} \quad \binom{\frac{1}{2}H_2}{H^+} $$

where E and E' are the reduced and oxidized forms of the hydrogenase active center. MV is a well-known substrate of hydrogenase which is usually employed to determine the enzyme activity.[107] Figure 13 is an illustration of the reaction kinetics in the system hydrogen–hydrogenase–MV. Hydrogenase catalyzes both the formation of hydrogen from the reduced form of MV (Figure 13a) and the reverse reaction, the reduction of MV by molecular hydrogen. A typical form of the kinetic curve for MV reduction by hydrogen is shown in Figure 13b. A detailed analysis of the reaction kinetics of hydrogenases is given in Reference (109). What is important here is that MV is a sufficiently active substrate of hydrogenase. This makes it possible to reach rapidly an equilibrium in the hydrogen–MV system, in a neutral solution:

$$ MV^{2+} + \tfrac{1}{2}H_2 \rightleftarrows MV^+ + H^+ $$

Methyl viologen, MV^{2+}, is easily reducible, yielding a relatively stable cation–radical MV^+. The reversible potential of the MV^{2+}/MV^+ redox couple is +0.011 V, i.e., rather close to the equilibrium potential of the hydrogen electrode in the same solution. The polarographic behavior of MV on the mercury dropping electrode has been studied.[110] The results show that MV

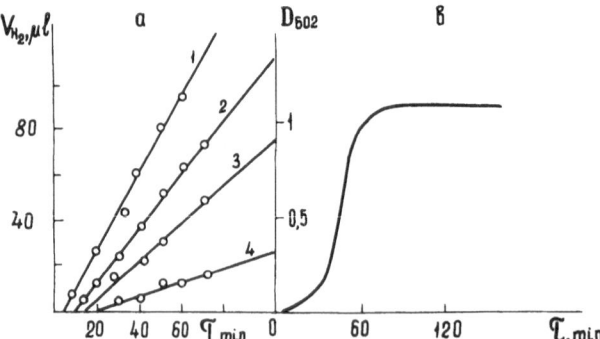

Figure 13. Reaction kinetics of MV with hydrogen in the presence of hydrogenase. (a) Evolution of molecular hydrogen from the solution MV^+(4 mM) + sodium dithionite (10 mM) at pH = 7.8, 21°C and protein concentration (mg/mL): (1) 0.625; (2) 0.5, (3) 0.315, (4) 0.125. (b) Reduction of the mediator by molecular hydrogen in the solution MV^{2+}(2.5 mM) + 0.1 M KCl + 0.06 M Na_3PO_4 at pH = 7.2, 25°C and protein concentration 0.3 mg/mL.

is capable of accepting either one or two electrons from the mercury electrode:

$$MV^{2+} + e \rightarrow MV^{+}$$

$$MV^{+} + e \rightarrow MV^{0}$$

Similar data were obtained[101,102] on the amalgamated gold and pyrographite electrodes. The half-wave potentials are equal to: $E_{1/2}^{I} = -0.03$ V and $E_{1/2}^{II} = -0.37$ V and are practically independent of the nature of the electrode. The anode–cathode polarization curves obtained in the presence of a mixture of the oxidized and reduced forms of MV are given in Figure 14. An analysis of the kinetics of mediator oxidation and reduction at the electrode reveals that the process proceeds on the carbon electrode under close to reversible conditions and is controlled by concentration polarization. Thus, MV fully satisfies the above-formulated requirements of mediators for electron transport in electrochemical systems with the participation of enzymes.

It has been shown[101–103] that in the system hydrogenase–methyl viologen–molecular hydrogen there exists the possibility of the latter's oxidation on the pyrographite electrode in the region of potentials close to the equilibrium potential of the hydrogen electrode reaction. At $E_r = 0.1$–0.2 V, the rate of the overall process is controlled by diffusion constraints involved in transporting reduced methyl viologen to the electrode; the current density on the rotating disc pyrographite electrode is linearly dependent on $\omega^{1/2}$. The rate of the process can be significantly increased by raising the concentration of the enzyme (Figure 15a) and the mediator (Figure 15b). Attainment of the maximum current value in the first case will be possible when the rate of enzymatic oxidation of methyl viologen becomes equal to that of its enzymatic reduction. Under high mediator concentrations, a significant departure of the maximum current from linear dependence is also observed. This effect is explained by the formation of dimer forms of reduced MV:

$$2MV^{+} \rightleftarrows (MV^{+})_2$$

(the constant K of this equilibrium being equal to 1.5×10^{-3} mol/l).

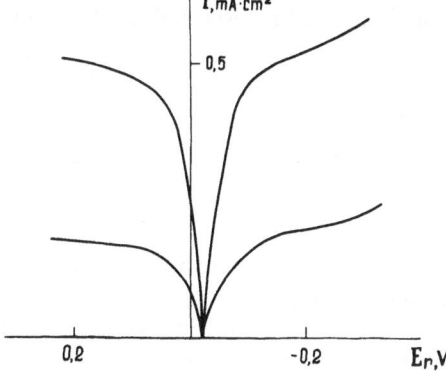

Figure 14. Anodic and cathodic polarization curves of MV chloride on the pyrographite electrode at concentrations 1 mM (curves 1 and 1′) and 5 mM (curves 2 and 2′) in the solution 0.1 M KCl + 0.06 M Na$_3$PO$_4$ at pH = 7.2, 25°C; electrode rotation velocity: 600 rev/min.

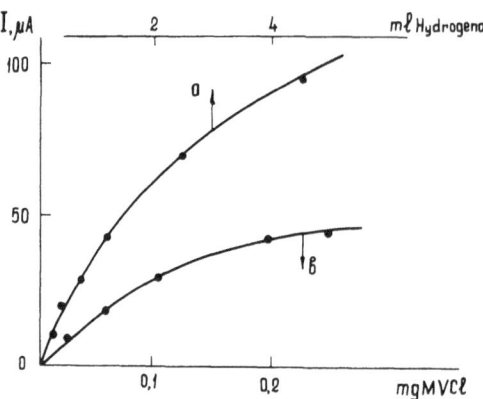

Figure 15. Maximum anodic currents in the system hydrogen–hydrogenase–MV–pyrographite electrode (a) versus the enzyme concentration in the solution containing 8 mM MV$^+$ and (b) versus mediator concentration. Electrode rotation velocity: 600 rev/min.

In this case, the dependence of the concentration of the monomeric form on total concentration of MV$^+$ takes the form

$$[MV^+] = \left\{ \left(\frac{K}{2} \right)^2 + K[MV^+]_0 \right\}^{1/2} - \frac{K}{2} \tag{1}$$

Hence, it follows that for $K \gg [MV^+]_0$ a linear dependence of the concentration of the mediator monomeric form on the total concentration of the reduced form must be observed. For $K \ll [MV^+]_0$ Eq. (1) will take the form

$$[MV^+] = (K[MV^+]_0)^{1/2} - \frac{K}{2} \tag{2}$$

i.e., under high mediator concentrations a linear dependence of maximum anodic currents of hydrogen oxidation on the square root of the concentration of the reduced mediator form must be observed. This provides an explanation for the data shown in Figure 15b.

As is known, carbon electrodes are inert for the electrochemical oxidation of hydrogen. The study of the system hydrogen–hydrogenase–MV–carbon electrode has revealed that it is possible to realize electrooxidation of molecular hydrogen under close to reversible conditions. Considering the data on the mechanism of hydrogen activation by hydrogenases, the general scheme of reactions can be described by the following equations:

$$H_2 + E \rightleftharpoons EH_2$$

$$EH_2 \rightleftharpoons EH^- + H^+$$

$$EH^- + MV^{2+} \rightleftharpoons MV^{\cdot+} + E$$

$$MV^{\cdot+} \rightarrow MV^{2+} + e$$

The next step for the improvement of the kinetics of this reaction is made possible through adsorptional immobilization of hydrogenase. In the case of adsorption on soot of hydrogenase from *T. roseopersicina*, up to 90% of the activity of the native enzyme is preserved. Figure 16 shows the enzymatic and

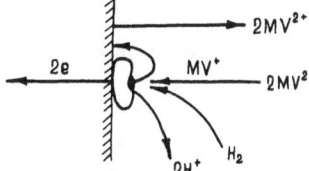

Figure 16. The processes involved in hydrogen activation with immobilized hydrogenase.

electrochemical processes proceeding on the electrode with hydrogenase immobilized on soot. As a result of the enzymatic reaction proceeding in close proximity to the electrode surface, there arises the reduced form of methyl viologen which is capable of oxidation or diffusion in the bulk solution. On the basis of studies of the i–E and i–l dependences (Figure 17) on the sectional semi-immersed electrode[110] for different mediator concentrations it was shown that the overall electrode reaction takes place over a wide range of potentials in the region of mixed kinetics with respect to hydrogen. The activity of native enzyme and on the degree of coverage of the carrier (soot) surface with hydrogenase (Table 4).

The activity of immobilized hydrogenase is higher, the greater the amount of enzyme adsorbed on soot. We can explain this kind of dependence on the basis of model concepts concerning protein concentration at the electrode surface, which have been described above. At low coverage, the adsorption of protein macromolecules takes place at the most active sites on the carrier surface. As a result of strong interaction with the carrier, there is a possibility of either partial denaturing of the enzyme or its unfavorable orientation. With increasing coverage, the interaction forces between the enzyme and the substrate decrease, the molecules are oriented in the most favorable manner, and the activity of the enzyme approaches that of the native form. Experimental results shown in the table permit us to compare the rate of electrochemical oxidation of hydrogen in the presence of immobilized enzyme with that of the

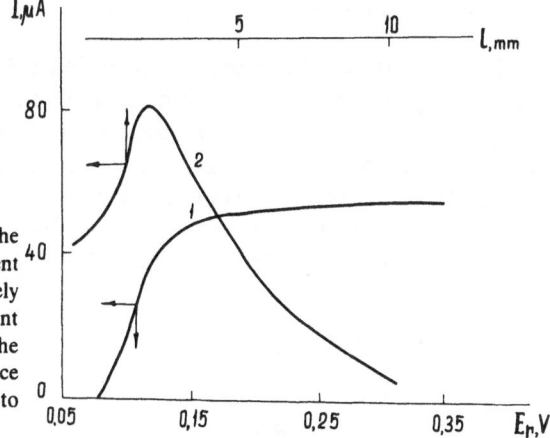

Figure 17. Measurements on the sectional electrode. Curve 1: current versus potential on the completely immersed electrode; curve 2: current versus the degree of immersion of the electrode, E_r = +0.15 V, l = distance from the top edge of the electrode to the electrolyte surface.

Table 4

Activity of Immobilized Hydrogenase Versus the Degree of Coverage of Soot Surface

Nos.	$\theta, \dfrac{\text{mg protein}}{\text{mg soot}}$	$I, \dfrac{\mu A^a}{\text{mg protein}}$	$V \cdot 10^{10} \dfrac{\text{mol } H_2{}^b}{\text{min} \cdot \text{mg protein}}$
1	0.025	146	3.4
2	0.050	207	3.6
3	0.075	270	3.9
4	0.100	259	3.8
5	0.125	348	3.95
6	0.150	340	4.2

^{a} On the basis of data obtained by the method of the revolving disc electrode for $E_r = 0.15$ V.
^{b} On the basis of gaseous chromatographic analysis with respect to hydrogen isolation.

enzymatic reaction of hydrogen evolution. Calculations reveal that the rate constant for the evolution of hydrogen by the enzyme is about 2.5 times as high as that of electrochemical oxidation of hydrogen in the presence of immobilized hydrogenase.

Temperature exerts a perceptible influence on the rate of the process concerned. The maximum current value is observed at 40–50°C. A further increase in temperature causes a sharp decrease in electrode activity.

In the presence of immobilized hydrogenase, it is possible to realize both the electrochemical oxidation of hydrogen and its evolution. To carry out the latter reaction, it is necessary to generate the reduced form of methyl viologen in the bulk of the solution.

Thus, in the system described a heterogeneous catalyst is used, though the mediator is present in the bulk of the solution.

6.1.2. Mediatorless Electron Transport

For the case of the hydrogen evolution reaction accelerated by hydrogenase, a system involving the mediatorless transport of electrons between the hydrogenase active center and the electroconductive matrix, i.e. electrode, has been developed.[105-107] Taking into account the fact that methyl viologen is a specific mediator of hydrogenase, a new method has been worked out for the immobilization of hydrogenase in an electroconductive polymer, the structural units of which are similar to methyl viologen:

$$(-CH{=}\underset{\underset{CH_2{-}\overset{+}{N}}{|}}{C}{-})_n \quad Br^- \qquad\qquad n = 8\text{–}10$$

$(-CH=C-)_n$
|
$CH_2-\overset{+}{N}$⟨benzene⟩—⟨benzene⟩$\overset{+}{N}-CH_3$ $n = 8\text{--}10$
Br^- Br^-

$(-CH_3-\overset{|}{CH})_n$—⟨benzene⟩$\overset{+}{N}-CH$—⟨benzene⟩ molec. mass $= 40{,}000$
Br

The proposed method for obtaining immobilized enzymes involves coprecipitation of polycations with enzymes under the influence of the half-reduced lithium salt of tetracyanodiquinone dimethane. The enzyme is immobilized in water-permeable polymers with high electrical conductivity (up to 10^{-2} ohm^{-1} cm^{-1}). The product is a black powder with an enzyme content as high as 500 mg per 1 g of the carrier.

The electrocatalytic activity of the immobilized enzyme was studied by the suspension electrode method for the hydrogen evolution reaction. Curve 1 in Figure 18 characterizes hydrogen evolution at $E_r = -0.25$ V as compared to the pure polymer (curve 2). Since the system contains no mediator, a direct electron exchange between the matrix and the active center of the enzyme must take place.

It is interesting to note that hydrogenase immobilized in a semiconductor exhibits a higher specific activity. In addition, the immobilization of the carrier significantly stabilizes the enzyme (Figure 19).

6.2. Oxygen Reaction

Oxygen reactions, i.e., processes involved in evolution and electroreduction of molecular oxygen, are most important in electrochemistry.[111] The

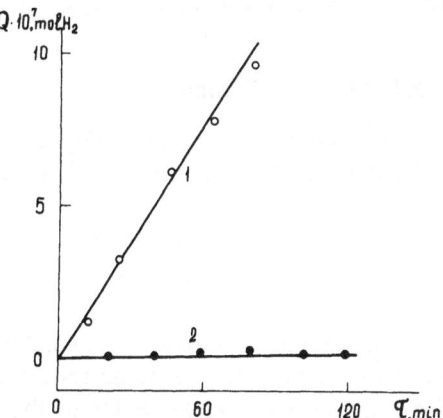

Figure 18. Kinetic curves for the electrochemical isolation of hydrogen on the polymeric suspension electrode. (1) Hydrogenase immobilized into polymer; (2) polymer suspension.

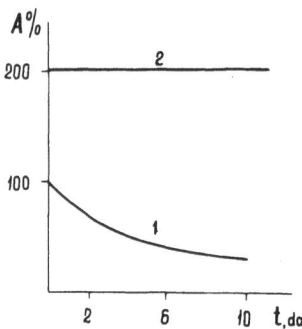

Figure 19. Stabilization of hydrogenase *T. roseopersicina* in open air by immobilization in a semiconductor complex: (1) native enzyme, (2) immobilized enzyme.

anodic reaction proves to be the basic one in quite a number of electrolytic productions while the cathodic reaction is important in electrochemical energy conversion. Research carried out in the past 10–15 years has shown that the problem of developing a reversible oxygen electrode can hardly be solved on the basis of metallic electrocatalysts. One of the new promising systems may be enzymes, since some of them accelerate the process of oxygen reduction directly to water without forming hydrogen peroxide; others activate the water molecule with the subsequent hydroxylation of oxidation of organic compounds. It is to be noted that enzymes function only at neutral pH's and ordinary temperatures. It is precisely for these conditions that electrocatalysts for the oxygen reactions are generally unavailable.

In investigations[112–118] of the use of enzymes to promote oxygen reaction, different methods of electron transport from the enzyme active center to the electrode have been employed. In the research performed to date it has been possible to accelerate only the cathodic reduction of oxygen and hydrogen peroxide. Carbon has been used, as a rule, as the electrode material, at which the electroreduction of molecular oxygen in neutral electrolytes proceeds with significant overvoltage. In addition, carbon electrodes happen to be good substrates for immobilization, since they are easily modified by various groups that can be used for the binding of protein macromolecules.

6.2.1. Mediator Transport

The simplest bioelectrocatalytic method to activate the oxygen reaction is by the use of enzymes with oxidase activity and a mediator:

$$\xrightarrow{2e} \Big\| \quad \begin{pmatrix} Q \searrow \\ \nwarrow HQ \end{pmatrix} \quad \begin{pmatrix} 2E \searrow \\ \nwarrow 2E^{\cdot} \end{pmatrix} \quad \begin{pmatrix} \tfrac{1}{2}O_2 + 2H^+ \\ \searrow \quad H_2O \end{pmatrix}$$

An attempt was made[114] to make use of the oxidase function of peroxidase to promote the cathodic reduction of oxygen. The electron transport from the enzyme active center to the electrode occurred with the aid of a mediator, the quinone–hydroquinone pair. Quinones are specific substrates of peroxidase and are practically reversibly oxidized and reduced on the pyrographite electrode. However, the redox potential of the quinone–hydroquinone pair is 0.70 V, i.e., much below the reversible potential of the reaction $O_2 \to H_2O$, which is 1.23 V. The investigations were carried out in neutral solutions on the pyrographite rotating electrode. Under these conditions, electroreduction of molecular oxygen proceeds with a significant overvoltage. Figure 20 shows the curves obtained with a rotating disc electrode, made from pyrographite, in the presence of peroxidase and a mediator in the solution. As follows from the data obtained, the half-wave potential of oxygen electroreduction is displaced to the positive side by about 0.30 V.

An investigation[115] into the possibility of activating cathodic oxygen reduction by means of cytochrome c oxidase, or, more accurately, submitochondrial particles (SMP) isolated from mitochondria of the bull's heart and enriched with the enzyme has been reported.[115] Investigations were carried out on the pyrographite electrode, using the TMPD-dihydrochloride–WB (perchlorate) couple as mediator. The choice of such a carrier was due to the fact that TMPD is the most active substrate of cytochrome c oxidase. The redox potential of the TMPD–WB couple is 0.72 V and redox transformations on the pyrographite electrode are practically reversible.

Cytochrome c oxidase catalyzes the oxidation of TMPD by oxygen to WB which thereupon is reduced on the electrode. Current versus operation time in such a system is shown in Figure 21. The reaction rate constant calculated on the basis of current measurements is $8.3 \times 10^2 \, s^{-1}$. But this calculation refers only to the initial period of operating time. The enzyme preparation retained its activity during 10 days, while the mediator proved to be rather unstable, since the reaction of TMPD oxidation to WB was continuing further with the formation of products which were isolated from the reaction. It appears feasible to undertake a further improvement of this system, because cytochrome c oxidase is one of the few enzymes capable of high-rate reduction

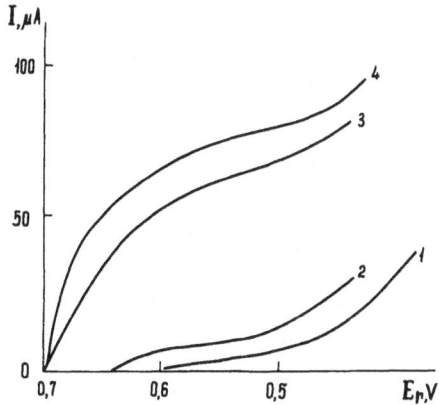

Figure 20. Cathodic reduction of oxygen on the disc pyrographite electrode at pH = 7.2. (1) Background curve; (2) polarization curve with the presence in the solution of the quinone–hydroquinone couple, (3) curve obtained in the presence of peroxidase (0.5 mg/mL) and the quinone–hydroquinone couple, (4) curve measured in the presence of peroxidase (0.5 mg/mL), the quinone–hydroquinone couple, and divalent manganese ions (0.1 mM).

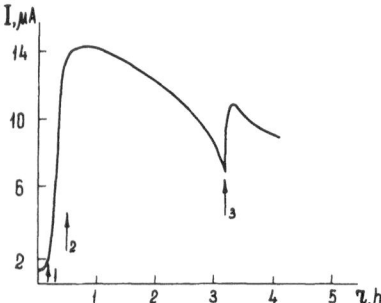

Figure 21. Current variations of the short-circuited element versus time under different conditions. Arrows show the moments of introduction of components in the following order: (1) 0.5 mM TMPD, (2) 0.6 mg/mL SMP, (3) buffer solution containing O_2 (260 μM). At zero time the solution was saturated with O_2 (260 μM). pH = 7.0, 25°C.

of oxygen to water, without the formation of hydrogen peroxide. This, however, requires an enzyme with a higher degree of purification.

In the case of laccase[117] and tyrosinase,[118] use was made of their specific substrates as mediators, for which the electrochemical transformations on carbon electrodes are sufficiently reversible.

Laccase immobilized by adsorption on soot promotes the process of hydroquinone oxidation by molecular oxygen. Quinone is reduced cathodically, thus accounting for a decrease in overvoltage over that for molecular oxygen reduction by about 0.3 V.

Figure 22 illustrates the data obtained for tyrosinase in the presence of oxygen and various mediators in the solution. It is seen that the highest stationary potential ($\sim$0.9 V) was achieved in the case of caffeic acid. The highest stability is observed in the system with a pyrocatechin-o-quinone, which preserved its activity for 50 hours.

The further improvement of the mediator technique for electron transport cells for the use of redox couples with a higher equilibrium potential approaching that of the oxygen reaction. However, according to available data, redox

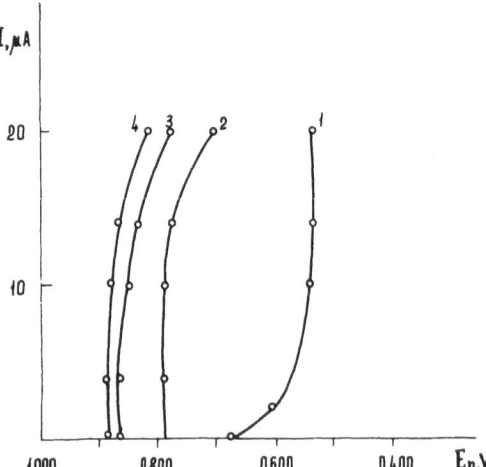

Figure 22. Polarization curves of oxygen reduction on the soot electrode in a solution of pH 6.4 containing 1.5 mg/mL tyrosinase and different mediators ($10^{-3}\,M$): (1) without mediator or enzyme; (2) pyrocatechol; (3) chlorogenic acid; (4) caffeic acid.

couples with a higher potential are distinguished by low solubility in aqueous solutions and their oxidized form is unstable.

6.2.2. Mediatorless Activation of Electroreduction of Molecular Oxygen

Direct and sufficiently valid proof of the acceleration of electroreduction of molecular oxygen by enzymes under conditions of direct mediatorless electron exchange between electrode and active center was first presented in References (116–118). However, these studies were preceded by certain investigations with protein carriers which suggested the existence of such a mechanism. The acceleration of oxygen electroreduction by cytochrome *c* was first evidenced[112] by the observation that it produces an exaltation of oxygen reduction current on a mercury electrode. The activation of the pyrographite electrode was achieved by using cytochrome *c* and hemoglobin.[13] These proteins are not enzymes, but they can be looked upon as simpler models of complex biological catalysts. Investigations using the rotating ring-disc electrode method gave proof that in the presence of adsorbed proteins the rate of electroreduction of oxygen increases. Furthermore, whereas on pyrographite molecular oxygen is reducible only to hydrogen peroxide, on iron-containing complexes a parallel reaction of direct oxygen reduction to water takes place. Figure 23 shows polarization curves on a pyrographite disc in the presence of the aforementioned proteins.

The results on the activation of electroreduction of oxygen by cytochrome *c*, as well as data on the investigation of the electrochemical properties of cytochrome, permit us to conclude that in this case there is a direct electron exchange between the active center and the electrode.

However, the most interesting results on the activation of electroreduction have been obtained with laccase and tyrosinase. The introduction of either laccase or tyrosinase into a neutral (pH ~ 6.0) solution saturated with oxygen has been shown to result in a sharp displacement of the stationary potential of the soot electrode towards positive values and an acceleration of electroreduction of oxygen within the potential range 0.6–1.2 V (Figure 24). In

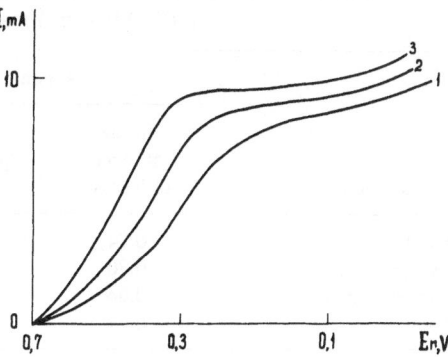

Figure 23. Polarization curves of oxygen reduction (1) on pyrographite, (2) on pyrographite covered with hemoglobin, and (3) with cytochrome *c.* ω = 1300 rev/min, pH 6.86.

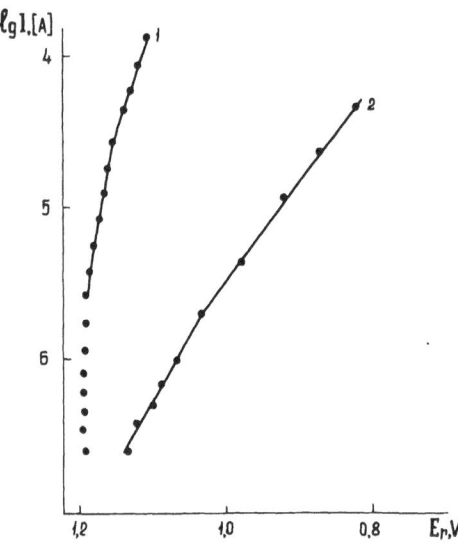

Figure 24. Polarization curves of oxygen reduction on the soot electrode with (1) adsorbed laccase and (2) tyrosinase.

the cases of both laccase and tyrosinase, the investigations using the rotating ring-disc electrode method revealed that for $E_r > 0.6$ V no hydrogen peroxide is formed.

The most reproducible data have been obtained with laccase. Table 5 gives the values of stationary potentials which arise in the presence of laccase on different electrode materials. The highest potential value, $+1.207$ V, which is close to the equilibrium potential of the oxygen electrode, is observed on soot electrodes which had been previously kept in a laccase solution $(10^{-5}\ M)$ for twenty-four hours. The enzyme adsorption on soot electrodes is practically irreversible, and after immobilization the enzyme retains catalytic properties in the absence of laccase in the solution.

Table 5

Stationary Potentials (V) on Different Electrode Materials in the Presence $(10^{-7}\ M)$ and Absence of Laccase

Material	Argon		Oxygen	
	In the absence of laccase	In the presence of laccase	In the absence of laccase	In the presence of laccase
Gold	0.74	0.77	0.80	0.98
Pyrographite	0.70	0.80	0.80	1.12
Glassy carbon	0.66	0.73	0.75	1.13
Soot	0.71	0.91	0.75	1.16

The enzymatic nature of electrocatalysis was proven by the specific inhibition of electrocatalytic effects by fluoride and azide ions and also by inactivation through heating. Laccase promotes molecular oxygen reduction only within the pH region where the enzyme retains its native structure (Figure 25).

Investigations of the dependence of the stationary potential on the soot electrode with adsorbed laccase on the solution pH and on oxygen pressure have revealed that the values of the parameters $\partial E/\partial(pH)$ and $\partial E/\partial \log P_{O_2}$ correspond to the values of the coefficients in the Nernst equation for the O_2/H_2O couple. An insignificant departure of the equilibrium potential from the theoretical value is due to a local pH rise in the adsorbed protein layer as compared to the pH in the bulk of the solution.

A typical polarization curve for oxygen reduction on the soot electrode with adsorptionally immobilized laccase is shown in Figure 24. The potential–current density plots, obtained by either increasing or decreasing the potentials, are practically coincident in the potential range 1.2–0.6 V. In the case of potentials below 0.6 V the enzyme is partly inactivated, presumably due to the effect of hydrogen peroxide arising in this potential region on free sites of the soot. The anodic curves measured on the soot electrode with immobilized laccase reveal that, in contrast to the cathodic process, laccase virtually fails to catalyze anodic oxidation of water, except in the potential region close to 1.2 V. This can be related to the oxidation of copper atoms in the laccase active center under high anodic potentials and to a disruption of the structure of the active center.

The experimentally measured kinetic parameters $\partial E/\partial \log I$, $\partial E/\partial \log P_{O_2}$ and $\partial E/\partial(pH)$ (Table 6) were used to propose a scheme for the oxygen

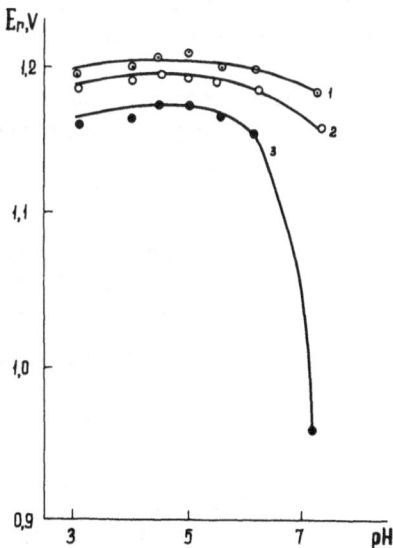

Figure 25. Potential at the electrode with immobilized laccase versus solution pH at currents (μA): (1) 0.0; (2) 1.1; (3) 17.5. Investigations were conducted in acetate and phosphate–acetate buffer solutions.

reduction reaction with immobilized laccase (here E stands for laccase):

$$E + O_2 \rightleftarrows EO_2$$

$$EO_2 + e \rightleftarrows EO_2^-$$

$$EO_2 + H^+ \rightarrowtail EO_2H \qquad (\Delta E = 1.15\text{--}1.10 \text{ V})$$

$$EO_2H + 2e + H^+ \rightarrow EO^- + H_2O \qquad (\Delta E = 1.20\text{--}1.15 \text{ V})$$

$$EO^- + e + H^+ \rightleftarrows LOH^-$$

$$LOH^- + H^+ \rightleftarrows L + H_2O$$

This scheme describes well the experimental dependences of the above parameters, if we assume that $\alpha = 0.5$. The value of the stoichiometric number, $\nu \sim 1.1$, is also in good agreement with this scheme. The presence of a slow two-electron step can be explained by the presence in laccase of a two-electron acceptor (copper of type 3) from which a synchronous transfer of two electrons to the oxygen molecule takes place. The type I copper is coordinated with ionogenic groups that can secure the protonation of the coordinated oxygen molecule. The first electron transfer in the scheme is a rapid one. It can contribute to the strengthening of the complex and facilitate its protonation. An increase of the slope of the curve to 0.06 V for $E_r < 1.15$ V in this scheme can be attributed to a retardation of the protonation stage.

The observed experimental kinetic parameters can also be interpreted within the framework of slow one-electron steps, if we take into account the possibility of a barrierless and an activationless mechanism. In this case the increase of the slope from 0.03 V to 0.06 V corresponds to a variation of the limiting stage from the barrierless ($\alpha = 1$) transfer of the second electron to the activationless ($\alpha = 0$) transfer of the third electron. Experimental data presently available do not permit us to make a choice between the two mechanisms.

The results obtained make it clear that laccase in the immobilized state is an effective catalyst for the oxygen reduction reaction in neutral solutions.

Table 6
Kinetic Parameters of Oxygen Electroreduction in
the Presence of Adsorbed Laccase for $E_r > 1.15$ V[a]

Kinetic parameter	$\times \dfrac{2.3RT}{F}$
$\partial E / \partial \log I$	-0.5
$\partial E / \partial (\text{pH})$	-1.0
$\partial E / \partial \log P_{O_2}$	$+0.5$

[a] For $E_r < 1.15$ V the value of $\partial E / \partial \log I$ increases to 0.06 V.

Comparison of the exchange currents calculated per laccase molecule ($i_0 = 3 \times 10^{-25}$ A/laccase molecule) and per surface platinum atom ($i_0 = 8 \times 10^{-27}$ A/platinum atom) reveal that the activity of the enzyme is significantly higher compared to the most active catalyst for cathodic reduction of oxygen. However, the activity of immobilized laccase is about 30 times lower than that of the solubilized preparation. Therefore, it can be expected that the improvements in the method of laccase immobilization will permit its activity in the electrochemical system to be further increased.

The adsorptional method of immobilization used in this work significantly adds to the stability of the enzyme. An electrode with laccase immobilized on soot showed stable performance for 50 hours, the polarization current being $87.5\ \mu A$.

The above result is, in fact, the first sufficiently reliable example of the acceleration of an electrochemical reaction due to an enzyme involving direct electron exchange between the enzyme active center and the electrode.

6.2.3. Mediatorless Activation of Electroreduction of Hydrogen Peroxide

Hydrogen peroxide is in certain cases a stable intermediate product of oxygen electroreduction.[119] The equilibrium potential for the H_2O_2/H_2O couple is 1.76 V. However, even on the most active platinum electrode the stationary potential value in the presence of hydrogen peroxide is not greater than $E_r \simeq 1.0$ V. The possibility of a mediatorless activation of hydrogen peroxide electroreduction was reported in Reference (120). In the presence of peroxidase in hydrogen peroxide solution the stationary potential set up on gold, pyrographite, and soot ranges from 1.05 to 1.24 V. The potential value increases with increase in the specific surface of the electrode, its maximum value corresponding to the soot electrode.

Figure 26 shows the polarization curve of H_2O_2 reduction of a soot electrode with immobilized peroxidase. In other work it was shown that, in the absence of a mediator, peroxidase fails to accelerate the process of oxygen

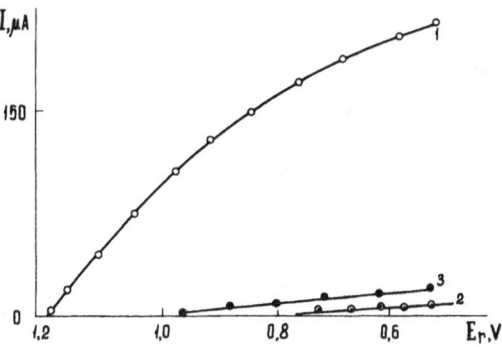

Figure 26. Polarization curves for hydrogen peroxide reduction on a soot electrode: (1) in the presence of immobilized peroxidase; (2) in the absence of peroxidase; (3) in the presence of peroxidase and phenyl-hydrazine.

reduction,[114] while its active center, hemin, accelerates O_2 and H_2O_2 reduction rather insignificantly.[113] This favors the assumption that the observed cathodic process is due to the electrochemical reduction of hydrogen peroxide; peroxidase, which preserves its molecular integrity, plays the role of the catalyst.

Phenylhydrazine,[121] a specific inhibitor of peroxidase, completely inactivates the electrode (Figure 26, curve 3). Plots of peroxidase activities versus hydrogen peroxide concentrations in solution and in the electrochemical system are similar and pass through a maximum at $C_{H_2O_2} \sim 5 \times 10^{-4}$–$2 \times 10^{-3}$ mol/l. Above these concentrations, the peroxidase activity abruptly falls off, the reason being, apparently, the inhibiting effect of H_2O_2. As seen from Figure 27, the pH dependences for the electrochemical and enzymatic systems are similar. These data provide additional evidence for the participation of peroxidase as a molecular entity in hydrogen peroxide electroreduction.

Organic metals[122] have also been used as the electrode material for the adsorption of peroxidase.[123] The electrochemical properties of these organic metals, which are complexes of N-methylphenazine (NMP$^+$) or N-methylacridine (NMA$^+$) with the anion radical of tetracyano-n-quinodimethane (TCNQ$^-$) with composition NMP$^+$TCNQ$^-$, NMP$^+$(TCNQ$^-$)$_2$, or NMA$^+$TCNQ$^-$, have been reported.[124] Hydrogen peroxide electroreduction on the organic metals takes place in the presence of peroxidase for $E_r < 0.2$ V (Figure 28). In the case of the enzyme adsorbed on NMP$^+$TCNQ$^-$, the reaction starts at $E_r > 0.8$ V. In the opinion of the authors,[122] these data point to the presence of a direct electron exchange between the electrode and peroxidase. There are, however, certain apprehensions that, because of partial solubility, the organic metals may act as mediators.

6.2.4. On the Optimal Catalyst for Oxygen Reaction

The realization of a direct electron exchange between the electrode and the enzyme opens up a radically new approach for the application of enzymes to promote electrochemical reactions. On the other hand, the data obtained

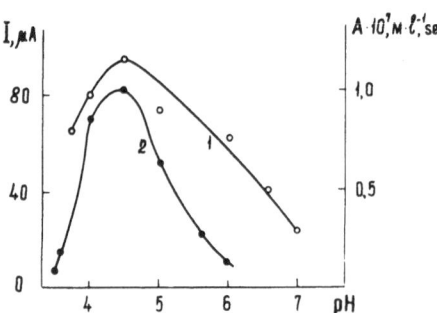

Figure 27. The activity of immobilized peroxidase versus pH: (1) enzymatic oxidation, o-dianisidine; (2) electrochemical system ($E_r = 1.05$ V); $C_{H_2O_2} = 1.8 \times 10^{-3}$ M.

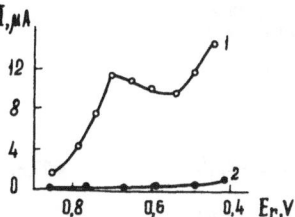

Figure 28. Polarization curves of hydrogen peroxide reduction ($4 \times 10^{-3} M$) (1) in the presence of peroxidase adsorbed on NMP^+TCNQ^- electrode and (2) in the absence of peroxidase.

for the oxygen reaction permit us to formulate concepts concerning the structure of an optimal synthetic catalyst to accelerate the oxygen reaction.[125]

A summary of the available data[126] points to two main causes for the irreversibility of this reaction:

1. The need for the rupture (formation) of the double bond in the oxygen molecule. Only the pairs of reactions $O_2 \rightleftarrows H_2O$ and $H_2O_2 \rightleftarrows H_2O$ in which the O—O bond breaks are irreversible.
2. The progress of the reaction through an intermediate formation of hydrogen peroxide. Although $E^0(H_2O_2/H_2O) > E^0(O_2/H_2O)$, hydrogen peroxide is observed to be reduced at significantly more negative potential than molecular oxygen.

With an optimal catalyst, the activation of oxygen molecules and of water (OH ions) and an increase of their energy up to the required level, apparently close to the first singlet state of O_2 (Figure 29), is achieved in the course of their adsorptional linking or coordination. Therefore, the reaction center must effectively link up and easily yield both these species, or there must be two reaction centers which can exchange electrons rapidly. Oxygen in the ground state has two unpaired electrons with parallel spins. Therefore, it can be presumed that the optimal course of the reaction (Figure 30) must first involve the transfer of the first electron (probably, retarded) and thereupon a synchronous transfer of two electrons inside the reaction complex. The breaking (formation) of the O—O bond and the addition (release) of protons also takes place inside the reaction complex, either in the course of the synchronous electron transfer or immediately after it. This excludes an intermediate formation of the kinetically stable hydrogen peroxide.

The author is of the opinion that progress in this direction will permit us to develop synthetic catalysts not inferior in activity to natural enzymes. First steps along this path have already been made.[127]

6.3. Oxidation of Organic Compounds

This field of enzyme application can be of the greatest interest, because, on the one hand, electroorganic reactions require high overvoltages and, on the other, the use of enzymes is bound to open the way for carrying out reactions which do not proceed on conventional electrocatalysts.

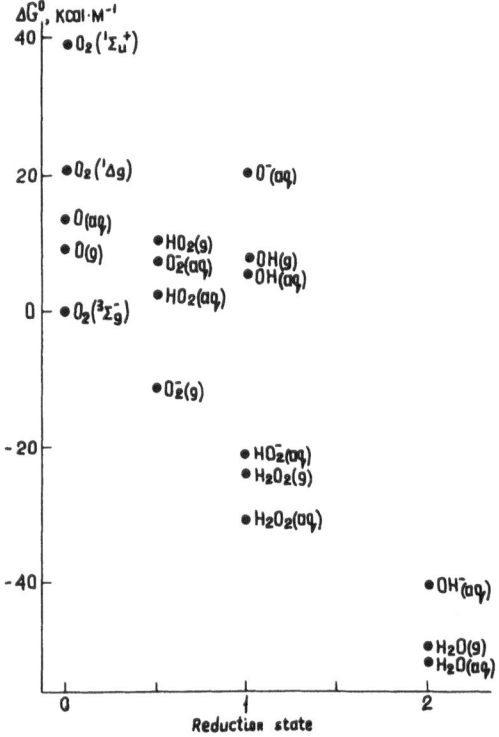

Figure 29. Energy diagram for oxygen-containing species in the system O_2/H_2O.

6.3.1. Oxidation of Glucose

The reaction of D-glucose oxidation to δ-glucolactone is accelerated by glucose oxidase. The transport of electrons to the electrode[2,3,128,129] was realized with the aid of mediators; such studies[2,3] are historically the first examples of the successful use of the mediator approach in bioelectrocatalysis.

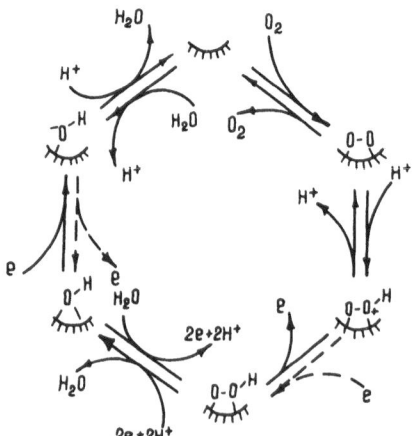

Figure 30. The scheme for the optimal route of oxygen reaction.

Glucose oxidase is immobilized in a polyacrylamide matrix. Oxygen and quinone which serve as mediators are reduced to hydrogen peroxide and hydroquinone, respectively. Since their oxidation starts at $E_r > 0.7$–0.9 V, it is possible to displace the initial potential for the oxidation of glucose by 0.6–0.8 V towards positive values compared to its stationary electrochemical oxidation to gluconic acid on platinum.[130]

The mediatorless electrooxidation of glucose has been realized in the presence of glucose oxidase adsorbed on an electrode made of organic metal.[122] Oxidation occurs at $E_r > 0.5$ V and the electrode preserved its activity for more than 100 days.

6.3.2. Oxidation of DL-Lactate

To accelerate the electrooxidation of lactic acid, NAD$^+$-dependent lactate dehydrogenase has been used with FMN as a mediator.[131] For the same purpose (as well as for the oxidation of alcohols)[132] NAD-dependent dehydrogenases with two conjugated pairs of mediators have been proposed:

$$
2e + 2H^+ \underset{\nwarrow}{\overset{\nearrow}{\left(\begin{array}{c} PMS - H \\ +H+ \\ PMS^+ \end{array}\right)}} \underset{+H^+}{\left(\begin{array}{c} NAD^+ \\ NADH \end{array}\right)} E \left(\begin{array}{c} C-H_2 \\ C \end{array}\right)
$$

where PMS is phenazine methosulfate. Even though we can accelerate the electrooxidation of the organic substrates in these systems, they are too complex to be used either for practical purposes or in fundamental research.

The use of cytochrome b_2[133,134] as an enzyme permits us to simplify the system, since this introduces the possibility of using the electrochemically reversible couples ferro/ferricyanide and PMS/PMSH$_2$ as mediators.

The mediatorless transport of electrons from the active center of cytochrome b_2 to organic metal during lactate oxidation has been reported.[122,134] The enzyme was adsorptionally immobilized on the finely-crushed organic metal[122] and this mixture was deposited on a platinum electrode. As evidenced by the data in Figure 31, lactate is oxidized in such a system in the region $E_r > 0.5$ V. Other organic substrates namely, glucose and saccharose, are not oxidized. Figure 32 gives a comparison between the bioelectrocatalytic activity of cytochrome b_2 in the adsorbed state and its enzymatic activity in the native state at different pH's. The similarity of the two curves indicates, that the rate of the overall catalytic process is limited by the enzymatic stage, rather than by electron transport.

6.3.3. Oxidation of Alcohols

There are a number of investigations[131,132,135] in which an attempt was made to use enzymes to promote the electrooxidation of alcohols. The aim

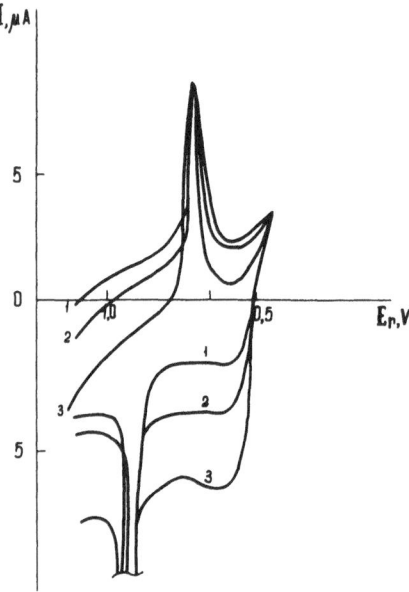

Figure 31. I–E curves on NMF$^+$ TCHM$^-$ with cytochrome b_2 in phosphate buffer at lactate concentrations (mM): (1) 0; (2) 0.59; (3) 5.9. Potential scanning velocity: 0.2 V/min.

pursued in these studies[135] was to achieve the bioelectrocatalytic oxidation of methanol. The general scheme of the process of methanol oxidation can be represented as follows:

$$CH_3OH \underset{E_1}{\rightleftharpoons} HCHO \xrightarrow[E_2]{} HCOOH \xrightarrow[E_3]{} CO_2$$

with the use of three enzymes: alcohol dehydrogenase—E_1, aldehyde dehydrogenase—E_2, and formate dehydrogenase—E_3. The overall reaction is the oxidation of CH_3OH to CO_2. The main difficulty associated with this reaction is that NADH is electrochemically inactive, presumably due to the formation of dimeric forms in the process of electrolysis. It was suggested in this work that methyl viologen, in the presence of NADH dehydrogenase, be used as a mediator for the transport of electrons from NADH to the electrode. The

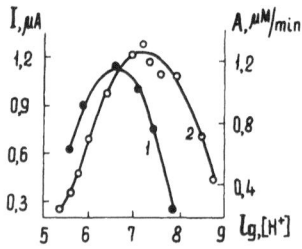

Figure 32. Dependences of the current of lactate electrooxidation by cytochrome b_2 on the organic metal (1) and on enzymatic activity in the native state (2).

scheme of formate oxidation in the presence of two mediators can be represented as follows:

$$\xleftarrow{2e} \left|\left(\substack{2MV^+ \\ \\ 2MV^{2+}}\right) E_0 \left(\substack{NADH \\ \\ NAD}\right) E_3 \left(\substack{HCOOH \\ \\ CO_2}\right)\right.$$

where E_0 is NADH dehydrogenase. The investigations were carried out on a rotating disc pyrographite electrode. The maximum values of diffusion current were attained at a potential of 0.3 V (Figure 33). The reaction rate was limited by diffusion of the reduced form of MV^+ to the electrode. The observed current values were constant over a two-week period. Comparison of the results of this study on the graphite electrode with literature data on direct oxidation of formate on platinum electrodes[136] reveals that in the presence of a mediator and enzyme the electrooxidation of formate proceeds at more positive potentials.

6.3.4. Hydroxylation of Aniline and Other Substrates

An interesting trend[137] is associated with the use of cytochrome P450 in bioelectrocatalysis. This is due to its ability to promote the hydroxylation of a wide range of substrates, thus permitting effective methods for decontamination to be developed. The existing difficulties are due to the low stability of the enzyme and to the irreversible character of electrochemical regeneration of NADH. Therefore, attempts have been made[137] to carry out the hydroxylation reaction in the absence of NADH. Aniline was employed as a model substrate, since the quantity of hydroxylated product can be easily determined by anodic oxidation. It was shown[137] that, under conditions of electroreduction of O_2 to H_2O_2 in the presence of P450 in the solution, the demethylation of benzphetamine, aminopyrine, and n-nitroanisole, and the hydroxylation of aniline and the steroids progesterone and testosterone can occur. Hydroxylation of aniline by hydrogen peroxide in the presence of P450 was experimentally proved.

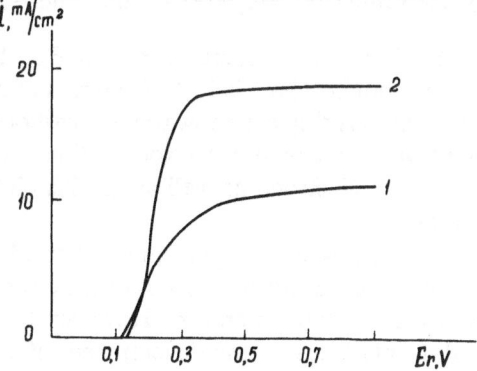

Figure 33. $i-E$ dependences obtained on the rotating disc pyrographite electrode in the system NADH–methyl viologen–NADH dehydrogenase–HCOOH. Rotation velocity: (1) 200 rev/min; (2) 600 rev/min.

The mechanism of the reactions in such a system is not clear at present; the authors, however, presume that the oxidation of the P450 substrate occurs via intermediates (probably of a peroxide nature) arising during oxygen electroreduction.

6.3.5. Hydrolysis of the Ether Bond

It has been shown[138,139] that there exists the possibility of accelerating electrochemical reactions by means of enzymes catalyzing the hydrolytic cleavage of covalent, for instance, ether, bonds. If a compound RX undergoes an electrochemical transformation at high overvoltage

$$RX + H_2O \rightarrow R'O + HX + nH^+ + ne,$$

then, by using a biocatalyst which facilitates the cleavage of the RX bond, yielding an electrochemically active product ROH:

$$RX + H_2O \rightarrow ROH + HX$$

$$ROH \rightarrow R'O + nH^+ + ne$$

it is possible to accelerate the overall process. The investigated enzymes were alkaline phosphatase, β-glucosidase, and cholinesterase which were used to promote the electrooxidation of monophosphates of pyrocatechol and n-aminophenol, β-D-glucoside n-aminophenol, and indoxyl acetate, respectively.

In the case of alkaline phosphatase, the reaction proceeds in the following manner:

pyrocatechol monophosphate $\xrightarrow[\text{H}_2\text{O} - 2e]{\text{electrooxidation}}$ o-quinone + phosphate + 2H$^+$

pyrocatechol monophosphate $\xrightarrow[\text{H}_2\text{O}]{\substack{\text{alkaline} \\ \text{phosphatase}}}$ pyrocatechol + phosphate

pyrocatechol $\xrightarrow[-2e]{\text{electrooxidation}}$ o-quinone + 2H$^+$

As the half-wave potential, Ph = 8 $E_{r_{1/2}}$ for the oxidation of pyrocatechol monoorthophosphate is 1.16 V and that of pyrocatechol is 0.86 V, the polarization curve is displaced towards the cathode side. This is illustrated in Figure 34 showing data on the electrooxidation of pyrocatechol monoorthophosphate on a paste electrode as well as on this electrode modified by alkaline phosphatase.

In conclusion, it is to be noted that other publications on the acceleration of electrode transformations of organic compunds in the presence of enzymes are available in the literature. However, these are of a rather empirical nature, and mainly focus on the development of electrochemical sensors.

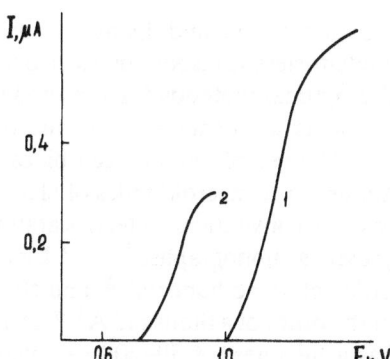

Figure 34. Polarization curves in a 1 mM solution of pyrocatechol monoorthophosphate (pH = 8) on the rotating (333 rev/min) paste electrode (1) and the rotating paste electrode modified by alkaline phosphatase (2).

7. Mechanism of Bioelectrocatalysis

Investigations on the mechanism of bioelectrocatalysis are still in the initial stages and thus this section will be devoted to the formulation of problems that must be solved to elaborate the theory of bioelectrocatalysis.

Whenever an enzyme is solubilized and the exchange of electrons with the electrode is realized by means of a mediator, the processes in the enzyme active center must be described by the laws of enzymatic catalysis and those on the electrode by the laws of electrochemical kinetics.

In the case of immobilization (adsorption) of the mediator or enzyme, the electrode surface represents an environment essentially different from the native conditions of the protein macromolecule. Therefore fundamental investigations on the conformational transformations at the interface are needed for the development of methods to optimize the components of the bioelectrochemical system at the interface. The problem of the possible effect of the field of the double electric layer, in whch the enzyme finds itself, on the operation of the latter's active center has not been examined so far.

A complex and radically new situation evolves in the case of a direct, mediatorless, transport between the enzyme active center and the electrode. Apart from the problems mentioned above, some new fundamental questions arise, which have not been encountered either in electrochemistry or enzymology. In the case of preservation of the molecular integrity of the immobilized enzyme, electrochemical transformations of the substrate in this system take place at large (some 10-Å) distances from the conductive phase. Therefore, it is necessary to investigate the mechanism of electron transfer and of the distribution of the potential jump (the structure of the electric double layer) in the electrode–enzyme–electrolyte system. The electrode becomes the donor or acceptor of electrons when the reaction proceeds at the enzyme active center. This implies a change in the functioning mechanism of the enzyme as compared to the native conditions. The chemical and electronic structure of the electrode surface must play an extremely important role in the orientation and fixing of the enzyme and in the organization of the mediatorless transport of electrons. This calls for elaboration of concepts on the optimal methods for the modification of the electrode surface according to the nature of the enzyme and the type of electrochemical reaction.

The use of immobilized biocatalysts is likely to bring about certain variations in the macrokinetics of electrochemical reactions, as compared with the case of conventional electrocatalysts which have been analyzed in detail in previous monographs.[140,141] Linear dimensions of the enzymes are of the order of some hundred Å and the area occupied by them at the interface is of the order of a thousand Å^2. The matrix in which the enzymes are immobilized has a thickness of 50–200 μm, which is commensurable with the parameters of the diffusion layer. Furthermore, the specificities of the enzymatic kinetics may exert an effect on the macrokinetic characteristics of the system as a whole.

Fundamental research on the above problems will make it possible to work out concepts on the mechanism of bioelectrocatalysis and to elaborate the scientific basis for developing optimal bioelectrocatalysts.

7.1. Maximum Rates of Bioelectrocatalytic Reactions

The studies of the kinetics of bioelectrocatalytic transformations show that in some systems (for instance, adsorbed laccase[117]) the kinetic parameters correspond to the phenomenology of electrochemical kinetics, while in other systems (for instance, lactate oxidation[122]) they fit the phenomenology of enzymatic catalysis. In the latter case,[122] we observe a hyperbolic dependence of anode current on the substrate concentration, as expected from the Michaelis–Menten equation.[1] The absence of a general theory of bioelectrocatalysis does not permit us to examine the kinetics of electrochemical reactions in the presence of enzymes under different conditions. At present we can only try to estimate the scope of possible accelerations of electrochemical reactions by making some simple assumptions.

Let us consider what reaction rates could be expected in reactions catalyzed by an enzyme in the form of a monolayer adsorbed on an electrode. The size of an average enzyme is about 20×50 Å and the area occupied by it on the electrode, ~ 1000 Å^2. Let us assume, moreover, that all the macromolecules are functioning effectively and the electron transport is not a slow step, nor is substrate diffusion. Then the current in the system will be determined by the catalytic activity of the enzyme itself. Figure 35 shows the distribution of enzymes with respect to reaction rate constant.[109] It is seen most typically $k \sim 10^2$ s^{-1} while for many enzymes $k \sim 10^3$ s^{-1}. If we take the latter value as the maximum estimate, then in the case of substrate concentration of about 10^{-3} M the current density in the kinetic regime will be about 1 mA/cm^2. This is close to the exchange current density for the hydrogen electrode reaction and is many orders higher than the exchange current density for the oxygen reaction, and even higher relative to that for electrooxidation

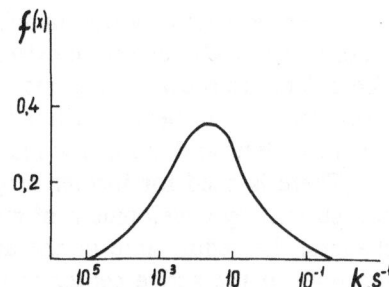

Figure 35. The distribution curve of enzymes with respect to the reaction rate constant (first-order) in enzymatic catalysis.

reactions of organic compounds.[140] Therefore, if we succeed in solving the electron transport and macrokinetic problems, the prospects for the activation of electrode reactions using enzymes are favorable.

7.2. The State of Adsorptionally Immobilized Enzymes

The results obtained on the activation of electroreduction of oxygen by adsorbed laccase,[117] of hydrogen peroxide by adsorbed peroxidase,[120] and of lactate by adsorbed cytochrome b_2[122] confirm the possibility of producing heterogeneous bioelectrocatalysts with activities significantly superior to those of known electrocatalysts.

On the basis of all the available data, we can assert that enzymes immobilized on an electrode which catalyze electrochemical transformations preserve their molecular integrity and principal bioelectrocatalytic characteristics. The proof of this is found in the following three groups of experimental facts.

1. The results of investigations of enzymatic activity in the immobilized state. When laccase, peroxidase, and other enzymes are adsorbed on dispersive electroconductive carriers, their enzymatic activity perceptibly declines, thus indicating a denaturing of part of the macromolecules. In this case, however, there is a sufficient quantity of denatured enzyme molecules strongly bonded with the carrier, which display a specific biocatalytic activity in model reactions of the substrates. It is these molecules of the enzyme, as will be seen further, that are responsible also for the electrocatalytic activity of the system.

2. Parallelism in the specific inhibition of electrocatalytic and enzymatic activity. The specific inhibitors of a particular enzyme are observed also to suppress its electrocatalytic activity in the adsorbed state. Experimental data demonstrate that α,α'-dipyridyl completely suppresses the reaction of hydrogen evolution by immobilized hydrogenase[106]; fluorine ions inactivate laccase in the reaction of oxygen electroreduction[117]; and diphenylhydrazine has the same effect on peroxidase in the reaction of hydrogen peroxide electroreduction.[120] A complete parallelism is also observed in the inactivating effect of hydrogen peroxide on peroxidase in the electrochemical reaction and enzymatic oxidation of o-dianisidine.

3. Preservation by the immobilized enzymes under the electrocatalysis conditions of a characteristic extreme dependence of reaction rate upon pH. These data are shown in Figures 25, 27, and 32 for laccase, peroxidase, and cytochrome b_2, respectively. In all cases, a parallelism is observable between electrocatalytic and enzymatic reactions.

There is need for further progress by more fundamental research and, particularly, by development of methods for examination of conformational changes, the redox state of the active center, and the coordination of the substrate in the active center of the enzymes immobilized at the interface boundary. The use of combined spectral–electrochemical methods appears to be the most fruitful approach in such investigations.

7.3. Electron Transfer in the Mediatorless Method of Bioelectrocatalysis

The observations listed above enable us to conclude that under the conditions of mediatorless bioelectrocatalysis, the adsorbed enzymes are functioning in a state close to the native one. Therefore, there is no basis for the assumption that the orbitals of the metal atom or of any other groups in the active center overlap with those of the electroconductive substrate. In the presence of such a direct overlap we would expect a significant change in the functioning mechanism of the active center. Moreover, the active center of the enzymes lies, as a rule, rather deep in the hydrophobic space of the protein globule.[1,7]

Therefore, it appears reasonable to assume the existence of the tunneling mechanism of electron transfer between the electrode and the enzyme active center. The possibility of a sub-barrier electron transfer at large distances (10–20 Å) has been shown, at least theoretically, for electrochemical systems[142] and is widely used to discuss electron transport in biological systems.[143]

For different models[144,145] of the tunneling mechanism of electron transfer the rate of this process is given by the expression

$$W(r) = \xi \exp\left(-\frac{r}{A}\right) \tag{3}$$

where r is the distance over which the electron is transferred, ξ is a constant, and A is determined by the form of the potential barrier. Consequently, the efficiency of the electron transfer must be dependent on the distance of the active center from the electrode. For $r < r_{cr}$, the rate of electron transfer is faster than that of the reaction at the active center. For $r > r_{cr}$ the rate of tunneling is significantly below that of the catalytic reaction. In this case, the efficiency of electron transfer must exponentially decrease with increasing distance.

The effect of the distance between the active center and the electrode on the reaction rate has been studied[146] using as an example the electrocatalysis of the oxygen reduction reaction by laccase adsorbed on soot. Variation in the distance between the active center and the electroconductive substrate was achieved by inserting an intermediate monolayer of lipid molecules: flatly and vertically oriented cholesterol molecules and vertically oriented lecithin molecules (scheme in Figure 36). In this case, the conditions of obtaining compact lipid monolayers were fulfilled. The subsequent setting of laccase did not lead to their desorption.

In this comparison, the value of electrocatalytic activity for all electrodes was normalized with respect to the rate of the model enzymatic reaction hydroquinone oxidation. The data obtained are shown in Figure 37. It is seen that a sharp reduction in the reaction rate is observed within a narrow range of distances, and r_{cr} amounts to about 20 Å. The shaded area corresponds to calculated data for a parabolic barrier height of 4–5 eV and other simple but by no means explicit assumptions.

Nonetheless, the concepts of tunneling for electron transfer are quite promising for the interpretation of certain effects of bioelectrocatalysis.

8. Prospects for Practical Utilization of Bioelectrocatalysis

The theory of bioelectrocatalysis is only making its initial progress; however, in the field of practical application we can already point out two areas in which the use of biocatalysts is likely to yield essentially new results.

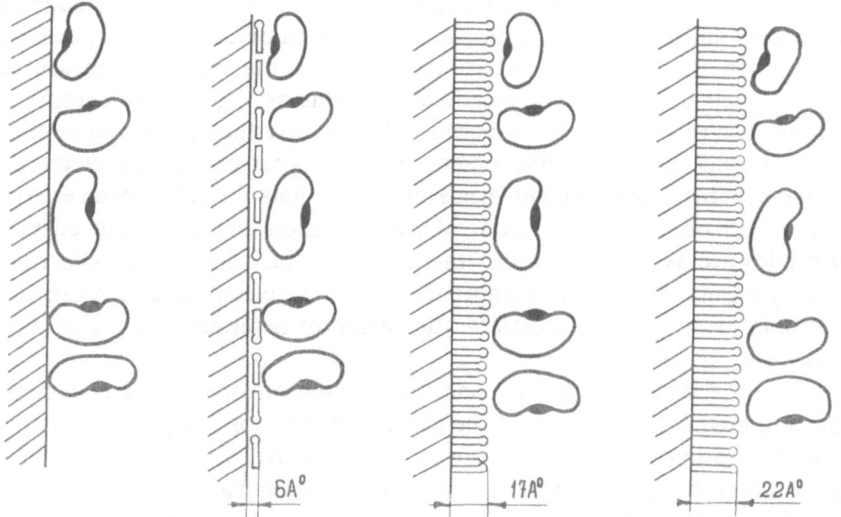

Figure 36. The scheme of enzyme adsorption on the electrode with different lipid interlayers.

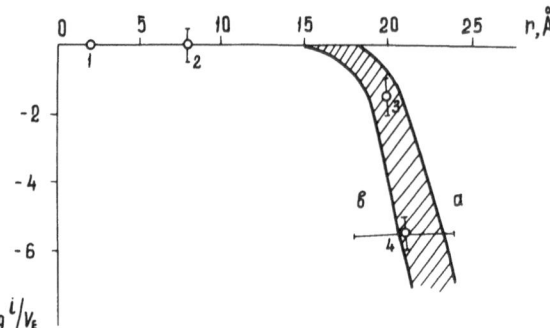

Figure 37. The relative oxygen reduction rate versus the distance between the electrode and the enzyme: for laccase adsorbed (1) on soot; (2) on cholesterol with planar orientation; (3) on vertically oriented cholesterol; (4) on lecithin. The curves a and b are calculated for barrier heights of 4 and 5 eV, respectively.

1. The high specificity of enzymes permits us to develop enzymic electrodes for electrochemical sensors needed for environment and medico–biological applications.
2. Application of biocatalysts, above all mediatorless types, in electrochemical energetics.

    The most complete reviews of enzymic electrodes, their design and characteristics are given in references (100) and (147). Enzymic electrodes consist of an immobilized enzyme, in the layer of which the given enzymatic reaction is realized, and an electrochemical element sensitive to variations in the concentration of the substrate or reaction product. In amperometric sensors, an element of this type is an electrode where the electrochemical transformation occurs and the current is measured; in potentiometric sensors, conventional or ion-selective electrodes are employed and the potential is measured. Here we shall briefly discuss the general operating principles of amperometric enzymic electrodes and sensors, because they involve the most direct application of bioelectrocatalytic phenomena. Certain types of enzymic electrodes and their characteristics are shown in Table 7. The immobilization of an enzyme on the electrode is realized, as a rule, by inclusion into a gel, but enzymic electrodes are available in which the enzyme is held in the near-electrode layer by means of a dialysis membrane.[134,153] The electrode design can vary according to the type of substrate, the region of determined concentration, service life, etc.

    For the example of glucose concentration sensors, we can see how diversified are methods for control of the content of the substrate. Figure 38 gives as an illustration the scheme of the enzymic glucose electrode[100] in which the reaction of glucose oxidation by oxygen takes place. The measure of glucose concentration is either the loss of oxygen in the system or the quantity of hydrogen peroxide produced. On the other hand, sensors containing two

Table 7

Enzymic Electrodes and Their Characteristics

Analyzed substance	Electrode	Stability (months)	Determination time (min)	Working concentration interval (M)	Reference
1. Glucose	Pt (O_2)	4	1	10^{-1}–10^{-5}	(148)
2. Alcohols	Pt (H_2O_2)	14	1	2×10^{-2}–10^{-4}	(149)
	Pt (O_2)	4	0.5	0.5–100 mg/100 mL	(150)
3. L-Amino acids	Pt	4–6	0.2	10^{-3}–10^{-5}	(151)
4. Urea	Pt (O_2)	4	0.5	10^{-2}–10^{-4}	(152)
5. DL-Lactate	Pt (PMS)		3	3×10^{-5}–10^{-6}	(134)
6. Glucose	Pt (quinone)		1	10^{-6}–10^{-3}	(153)
7. Lactic acid	Pt[Fe(CN)$_6$]$^{3+}$		1		(153)

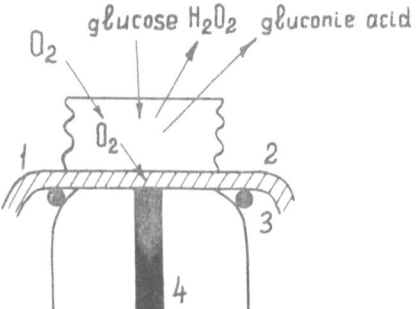

Figure 38. A schematic illustration of the enzymic glucose electrode based on the oxygen membrane sensor: (1) a layer of immobilized glucose oxidase; (2) semipermeable Teflon membrane; (3) reference electrode; (4) cathode.

enzymes have been proposed: glucose oxidase and catalase for decomposition of hydrogen peroxide. In this case the oxygen concentration is determined. Determinations of glucose[153] are made using quinone as an oxidizer; its decrease in concentration serves as a measure of glucose concentration.

Sensors using the principle of acceleration of electrochemical reactions by enzymes have been developed for determination of more than 30 different compounds[100] and their number is continuously increasing. Some of them are commercially manufactured and find application in clinical practice.[154]

The overwhelming majority of amperometric sensors described in the literature employ the mediator method of electron transfer. An exception is the proposal of a mediatorless sensor of lactate concentration.[134] It can be expected that by expanding the range of enzymes for which the mediatorless method of electron exchange with the electrode occurs, some significant simplifications can be made for a great many of the already available sensors.

The study of bioelectrocatalysis for processes and devices involving energy transformation proves to be a problem which is far more difficult, but at the same time of much greater scope, than the development of sensors. Repeated attempts have been made to create fuel cells using the products of enzymatic reactions[128,155] including implanted biofuel cells.[156] However, these studies have not made any significant progress. Only the discovery of a direct, mediatorless method for using enzymes to promote electrochemical reactions can enable us to undertake the development of heterogeneous electrocatalysts which will be superior to known electrocatalysts, due, in part, to their applicability in neutral electrolytes at room temperature. It is precisely for these conditions that good electrocatalysts are lacking at present.

The fact that there are fairly broad and specific areas of practical applications of bioelectrocatalysis indicates that further vigorous pursuit of this field of research is worthwhile. In conclusion, it is to be noted that the progress of bioelectrocatalysis follows the same trend as the general development of catalysis—integration in a single catalyst of the advantages of heterogeneous, homogeneous, and enzymatic catalysis.

References

1. W. P. Jencks, *Catalysis in Chemistry and Enzymology*, McGraw-Hill Book Company, New York (1969).
2. S. J. Updike and G. P. Hicks, *Nature* **214**, 986 (1967).
3. L. B. Wingard, C. C. Liu, and N. L. Nagda, *Biotechnol. Bioeng.* **13**, 629 (1971).
4. O. R. Zaborsky, *Immobilized Enzymes*, CRC Press, Cleveland, Ohio (1974).
5. *Enzyme Nomenclature Recommendation (1972) of the International Union of Pure and Applied Chemistry and the International Union of Biochemistry*, Elsevier, Amsterdam (1972).
6. A. E. Braunshtein, *Journ. D. I. Mendeleev All-Union Chemical Society* **8**, 81 (1963). In Russian.
7. I. V. Berezin and K. Martinek, *Principles of Physical Chemistry of Enzymatic Catalysis*, Vysshaya Shkola Publishers, Moscow (1977). In Russian.
8. D. Koshland, *Proc. Natl. Acad. Sci. U.S.A.* **44**, 98 (1958).
9. M. V. Vol'kenshtein, *Journ. D. I. Mendeleev All-Union Chemical Society* **17**, 293 (1972). In Russian.
10. B. Havsteen, *Conformational Changes in Enzymes Accompanying Catalysis*, University Press, Aarhus (1968).
11. R. Lamry and R. Biltonen, in *Structure and Stability of Biological Macromolecules*, S. N. Timashefe and J. D. Fasman, eds., Marcel Dekker, Inc., New York (1969).
12. R. Marcus, *J. Chem. Phys.* **24**, 966 (1956).
13. J. M. Rifkind, in *Inorganic Biochemistry*, Vol. 2, G. B. Eichhorn, ed., Elsevier Scientific Publishing Company, Amsterdam (1973, 1975).
14. L. Pauling, *Nature* **203**, 182 (1964).
15. G. R. Moore and R. J. P. Williams, *Coord. Chem. Rev.* **18**, 125 (1976).
16. E. C. Slater, B. F. Van Gelder and K. Minnaert, in *Oxidases and Related Redox Systems*, T. E. King, ed., John Wiley, New York (1965).
17. N. Sutin and J. K. Yandell, *J. Biol. Chem.* **147**, 6932 (1972).
18. D. Dickerson, T. Takano, and E. Margoliash, *J. Biol. Chem.* **246**, 1511 (1971).
19. O. Hayaishi, ed., *Molecular Oxygen in Biology: Topics in Molecular Oxygen Research*, North-Holland Publishing Company, Amsterdam (1974).
20. A. I. Archakov, *Microsomal Oxidation*, Nauka Publishers, Moscow (1975), p. 270. In Russian.
21. T. Asakura and T. Yonetani, *J. Biol. Chem.* **244**, 537 (1969).
22. J. Y. H. C. Shih, L. L. Shannon, E. Kay, and I. J. Lew, *J. Biol. Chem.* **246**, 4546 (1971).
23. M. Tamura, T. Asakura, and T. Yonetani, *Biochem. Biophys. Acta* **268**, 292 (1972).
24. R. Banerjee, *Biochem. Biophys. Acta* **64**, 368 (1962).
25. A. P. Savitskii, N. N. Ugarova, and I. V. Berezin, *Biologicheskaya Khimiya*, **3**, 1242 (1977). In Russian.
26. B. Chance, in *The Enzymes*, Vol. 2. Part 1, J. B. Sumner and K. Myrback, eds., Academic Press, New York (1951), p. 446.
27. E. N. Kondrat'eva and I. N. Gogotov I. N., *Mikrobiologiya*, **38**, 938 (1969). In Russian.
28. I. N. Gogotov, N. A. Zorin, and V. M. Ushakov, *Mikrobiologiya* **42**, 21 (1973). In Russian.
29. H. Gitlitz and A. I. Krasna, *Biochemistry* **14**, 2561 (1975).
30. T. H. Moss, A. I. Bearden, R. G. Bartsch, M. A. Cusanovich, and A. S. Pietro, *Biochemistry* **7**, 1591 (1968).
31. R. Malkin and B. G. Malmström, *Adv. Enzymol.* **33**, 177 (1970).
32. B. Reinhammar, *Biochim. Biophys. Acta* **205**, 35 (1970).
33. M. Fling, N. H. Horowitz, and S. F. Heinemann, *J. Biol. Chem.* **238**, 2045 (1963).
34. V. Ya. Yakovlev, ed., *Coenzymes*, Meditsina Publishers, Moscow (1973). In Russian.
35. T. Nakazawa, S. Yamamoto, T. Maki, H. Takeda, H. Kajita, M. Nozaki, and O. Hayaishi, *Flavins and Flavoproteins*, Academic Press, New York (1968).

36. G. L. Eichhorn, ed., *Inorganic Biochemistry*, Vol. 1, Elsevier Scientific Publishing Company, Amsterdam (1973, 1975).
37. I. V. Berezin, V. K. Antonov, and K. Martinek, eds., *Immobilized Enzymes*, MGU Publishers, Moscow (1976). In Russian.
38. E. S. Vorob'eva and O. M. Poltorak, *Vestnik Mosk. Universiteta, Chem. Ser.* 17 (1966). In Russian.
39. O. M. Poltorak and E. S. Vorob'eva, *J. Phys. Chem.* **40**, 1665 (1966). In Russian.
40. I. D. Ivanov, *Polarography of Proteins, Enzymes, and Amino Acids*, USSR Academy of Sciences Publishing House, Moscow (1961). In Russian.
41. J. Heyrovsky and J. Kuta, *Zaklady Polarografie*, Cesk. Akad., Prague (1962).
42. N. L. Weinberg and H. R. Weinberg, *Chem. Rev.* **68**, 449 (1968).
43. S. R. Betso, M. N. Klapper, and Z. Andreassen, *Experientia, Suppl.* **18**, 157 (1971).
44. B. A. Kuznetsov, *Experientia, Suppl.* **18**, 381 (1971).
45. M. R. Tarasevich and K. A. Radyushkina, *Catalysis and Electrocatalysis by Metalloporphyrins*, Nauka Publishers, Moscow (1982).
46. M. R. Tarasevich and V. A. Bogdanovskaya, *Bioelectrochem. Bioenerg.* **3**, 589 (1976).
47. C. F. Kolpin and H. S. Swofford, *Anal. Chem.* **50**, 916 (1978).
48. F. Scheller, H. J. Prümke, P. Mohr, J. Seyer, and A. Makower, *Stud. Biophys.* **62**, 223 (1977).
49. B. Janik and P. Elving, *Chem. Rev.* **68**, 295 (1968).
50. P. J. Elving, in *Topics in Bioelectrochemistry and Bioenergetics*, Vol. 1, G. Milazzo, ed., Wiley, New York (1976), p. 179.
51. A. L. Underwood and J. N. Burnett, in *Electroanalytical Chemistry*, Vol. 6, A. J. Bard, ed., Marcel Dekker, New York (1973).
52. P. Leduc, D. Thevenot, and R. Bunet, *Bioelectrochem. Bioenerg.* **3**, 491 (1976).
53. P. Leduc and D. Thevenot, *Bioelectrochem. Bioenerg.* **1**, 96 (1974).
54. J. Moiroux and P. J. Elving, *J. Am. Chem. Soc.* **102**, 6533 (1980).
55. J. Moiroux and P. Elving, *Anal. Chem.* **50**, 1056 (1978).
56. K. Sasaki and A. Kitani, *J. Electroanal. Chem.* **94**, 201 (1978).
57. C. O. Schmakel, K. S. Santkanam, and P. J. Elving, *J. Am. Chem. Soc.* **97**, 5083 (1975).
58. H. Hanschmann and H. Berg. *Bioelectrochem. Bioenerg.* **8**, 71 (1981).
59. H. Jaegfeldt, *Bioelectrochem. Bioenerg.* **8**, 355 (1981).
60. M. Aizawa, S. Suzuki, and M. Kubo, *Biochem. Biophys. Acta* **444**, 886 (1976).
61. K. Takamura, A. Mori, and F. Kusu, *Bioelectrochem. Bioenerg.* **8**, 229 (1981).
62. H. Jaegfeldt, *J. Electroanal. Chem.* **110**, 295 (1980).
63. W. J. Blaedel and R. A. Jenkins, *Anal. Chem.* **47**, 1337 (1975).
64. R. D. Braun, K. S. V. Santhanam, and P. J. Elving, *J. Am. Chem. Soc.* **97**, 2591 (1975).
65. J. Moiroux and P. J. Elving, *J. Electroanal. Chem.* **102**, 93 (1979).
66. D. Chi-Sing Tse and T. Kuwana, *Anal. Chem.* **50**, 1315 (1978).
67. C. Degrand and L. L. Miller, *J. Am. Chem. Soc.* **102**, 5728 (1980).
68. W. Friedrich, L. Müller, and M. Moritz, *J. Electroanal. Chem.* **69**, 361 (1976).
69. F. Scheller, G. Strnad, B. Neumann, M. Kühn, and W. Ostrowski, *J. Electroanal. Chem.* **104**, 117 (1979).
70. B. A. Kuznetsov and G. P. Shumakovich, *Bioelectrochem. Bioenerg.* **2**, 35 (1975).
71. G. P. Shumakovich and B. A. Kuznetsov, *Biofizika* **24**, 777 (1979). In Russian.
72. F. Scheller and G. Strnad, in press.
73. S. Betso, M. Klapper, and L. Anderson, *J. Am. Chem. Soc.* **94**, 8197 (1972).
74. M. R. Tarasevich and V. A. Bogdanovskaya, in *IFIAS Workshop on Physico-Chemical Aspects of Electron Transfer Processes in Enzymes*, IFIAS, Stockholm (1977), pp. 119–129.
75. B. A. Kuznetsov, G. P. Shumakovich, and N. M. Mestechkina, *Bioelectrochem. Bioenerg.* **4**, 512 (1977).
76. F. Scheller and H. Prümke, *Stud. Biophys.* **60**, 137 (1976).
77. F. Scheller, H. Prümke, H. E. Schmidt, and P. Mohr, *Bioelectrochem. Bioenerg.* **3**, 328 (1976).

78. M. V. Genkin, R. M. Davydov, O. V. Krylov, and L. A. Blyumenfeld, *Biofizika* **22**, 158 (1977). In Russian.
79. M. V. Genkin, R. M. Davydov, O. V. Krylov, and L. A. Blyumenfeld, *Dokl. Akad. Nauk. SSSR* **232**, 367 (1977). In Russian.
80. P. Yeh and T. Kuwana, *Chem. Lett.* 1145 (1977).
81. M. J. Eddowes and H. A. O. Hill, *J. Am. Chem. Soc.* **101**, 4461 (1979).
82. K. Niki, T. Yagi, H. Jnokuchi, and K. Kimura, *J. Electrochem. Soc.* **124**, 1889 (1977).
83. K. Niki, T. Yagi, H. Jnokuchi, and K. Kimura, *J. Am. Chem. Soc.* **101**, 3335 (1979).
84. P. Bianco and J. Haladjian, *Bioelectrochem. Bioenerg.* **8**, 239 (1981).
85. M. Janchen, F. Scheller, H. J. Prumke, P. Mohr, and G. Etzold, *Stud. Biophys.* **39**, 1 (1973).
86. F. Scheller and M. Janchen, *Stud. Biophys.* **46**, 153 (1974).
87. J. E. Stargardt and F. M. Houkridge, *Anal. Chem.* **50**, 930 (1978).
88. A. I. Yaropolov, M. R. Tarasevich, and S. D. Varfolomeev, *Bioelectrochem. Bioenerg.* **5**, 18 (1978).
89. I. M. Kolthoff, W. Stricks, and N. Tanaka, *J. Am. Chem. Soc.* **77**, 4739 (1955).
90. B. A. Kiselev, A. A. Kazakova, V. B. Yevstigneev, V. K. Gins, and E. I. Mukhin, *Biofizika* **21**, 35 (1976).
91. C. L. Hill, J. Renaud, R. H. Holm, and L. E. Mortenson, *J. Am. Chem. Soc.* **99**, 2549 (1977).
92. P. Bianko and J. Haladjian, *Biochem. Biophys. Res. Commun.* **78**, 323 (1977).
93. M. Senda, T. Ikeda, T. Kakutani, and K. Kano, *Bioelectrochem. Bioenerg.* **8**, 151 (1981).
94. H. L. Lendrum, R. T. Salmon, and F. M. Haukridge, *J. Am. Chem. Soc.* **99**, 3154 (1977).
95. S. Kwee and H. Lund, *Bioelectrochem. Bioenerg.* **2**, 231 (1975).
96. F. Duke, K. Kunst, and L. King, *J. Electrochem. Soc.* **116**, 32 (1969).
97. F. Knobloch, *Methods Enzymol.* **18B**, 305 (1971).
98. B. A. Kuznetsov, N. M. Mestechkina, and G. P. Shumakovich, *Bioelectrochem. Bioenerg.* **4**, 1 (1977).
99. M. R. Tarasevich, *J. Electroanal. Chem.* **104**, 587 (1979).
100. Yu. Yu. Kulis, *Analytical Systems on the Basis of Immobilized Enzymes*, Moklas Publishers, Vilnius (1981). In Russian.
101. M. R. Tarasevich, V. A. Bogdanovskaya, V. S. Bagotskii, S. D. Varfolomeev, A. I. Yaropolov, and I. V. Berezin, *Elektrokhimiya* **13**, 892 (1977). In Russian.
102. S. D. Yarfolomeev, A. I. Yaropolov, I. V. Berezin, M. R. Tarasevich, and V. A. Bogdanovskaya, *Bioelectrochem. Bioenerg.* **4**, 314 (1977).
103. I. V. Berezin, S. D. Varfolomeev, A. I. Yaropolov, V. A. Bogdanovskaya, and M. R. Tarasevich, *Dokl. Akad. Nauk SSSR* **225**, 105 (1975). In Russian.
104. V. A. Bogdanovskaya, S. D. Varfolomeev, M. R. Tarasevich, and A. I. Yaropolov, *Elektrokhimiya* **16**, 763 (1980). In Russian.
105. S. D. Varfolomeev, S. O. Bachurin, I. V. Osipov, K. A. Aliev, I. V. Berezin, and V. A. Kabanov, *Dokl. Akad. Nauk SSSR* **239**, 348 (1978). In Russian.
106. S. O. Bachurin, S. D. Varfolomeev, I. V. Tysyachnaya, V. E. Davydov, G. V. Mavrenkova, and I. V. Berezin, *Dokl. Akad. Nauk SSSR* **253**, 370 (1980).
107. S. D. Varfolomeev, S. O. Bachurin, and Ali Nagui, *J. Mol. Catalysis No. 9*, 223 (1980).
108. Ch. D. Toai, S. D. Varfolomeev, I. N. Gogotov, and I. V. Berezin, *Molek. Biologiya* **10**, 452 (1976). In Russian.
109. S. D. Varfolomeev, *Energy Conversion by Biocatalytic Systems*, MGU Publishers, Moscow (1981). In Russian.
110. S. F. Chernyshov, R. Kh. Burshtein, M. R. Tarasevich, and A. V. Yalyshev, *Elektrokhimiya* **6**, 949 (1970). In Russian.
111. M. R. Tarasevich, A. Sadkovsky, and E. Yeager, in *Comprehensive Treatise of Electrochemistry*, Vol. 4, Plenum Press, New York, 1981.
112. S. R. Betso, M. H. Klapper, and L. B. Andreasson, *Biological Aspects of Electrochemistry, Proc. of the 1st Int. Symp.*, Rome (1971), pp. 157–162.

113. M. R. Tarasevich and V. A. Bogdanovskaya, *Bioelectrochem. Bioenerg.* **2**, 69 (1975).
114. A. I. Yaropolov, S. D. Varfolomeev, I. V. Berezin, V. A. Bogdanovskaya, and M. R. Tarasevich, *FEBS Lett.* **71**, 306 (1976).
115. I. V. Berezin, A. S. Pobochin, V. V. Kupriyanov, and V. N. Luzikov, *Bioorganicheskaya Khimiya* **3**, 989 (1977). In Russian.
116. S. D. Varfolomeev, V. A. Bogdanovskaya, A. I. Yaropolov, M. R. Tarasevich, and I. V. Berezin, *Dokl. Akad. Nauk SSSR*, **240**, 615 (1978). In Russian.
117. M. R. Tarasevich, A. I. Yaropolov, V. A. Bogdanovskaya, and S. D. Varfolomeev, *J. Electroanal. Chem.* **104**, 393 (1979).
118. V. A. Bogdanovskaya, E. F. Gavrilova, M. R. Tarasevich, and D. I. Stom, *Elektrokhimiya* **16**, 1596 (1980). In Russian.
119. M. R. Tarasevich and V. S. Bagotskii, *Elektrokhimiya* **14**, 1340 (1978). In Russian.
120. A. I. Yaropolov, V. Malovik, S. D. Varfolomeev, and I. V. Berezin, *Dokl. Akad. Nauk SSSR* **249**, 1399 (1979). In Russian.
121. H. Hidaka and S. Udenfriend, *Arch. Biochem. Biophys.* **140**, 174 (1970).
122. M. L. Khidekel and E. I. Zhilyaeva, *Journ. D. I. Mendeleev All-Union Chem. Soc.*, **23**, 506 (1978). In Russian.
123. Yu. Yu. Kulis, A. S. Samalius, and G.-J. S. Svirmickas, *FEBS Lett.* **114**, 7 (1980).
124. C. D. Jaeger and A. J. Bard, *J. Am. Chem. Soc.* **101**, 1690 (1979).
125. M. R. Tarasevich, *J. Res. Inst. Catalysis, Hokkaido University* **28**, 355 (1980).
126. M. R. Tarasevich and E. I. Khrushcheva, in *Kinetics of Complex Electrochemical Reactions*, V. E. Kazarinov, ed., Nauka Publishers, Moscow (1981). In Russian.
127. M. R. Tarasevich, M. E. Fol'pin, V. A. Bogdanovskaya, S. B. Orlov, G. N. Novodarova, and E. M. Kolosov, *Elektrokhimiya* **18**, 340 (1982). In Russian.
128. E. J. Lahoda and C. C. Liu, *Biotechnol. Bioeng.* **17**, 413 (1975).
129. F. Scheller, D. Pfeiffer, M. Kühn, J. Hundertmark, A. Quade, M. Jänchen, G. Lande, H. Holesch, and H. Dittmer, *Acta Biol. Med. Ger.* **39**, 671 (1980).
130. I. A. Avrutskaya and M. Ya. Fioshin, *J. Prikl. Khim.* **42**, 2994 (1969). In Russian.
131. S. Suzuki, F. Takahashi, I. Satch, and N. Sonobe, *Bull. Chem. Soc. Japan* **48**, 3246 (1975).
132. A. Malinauskas and Yu. Yu. Kulis, *Anal. Chim. Acta* **98**, 31 (1978).
133. Yu. Yu. Kulis and K. V. Kadzyauskene, *Dokl. Akad. Nauk SSSR* **239**, 636 (1978). In Russian.
134. Yu. Yu. Kulis and G.-J. S. Svirmickas, *Anal. Chim. Acta* **117**, 115 (1980).
135. R. P. Petukhova, A. M. Yegorov, A. A. Sukhno, B. I. Podlovchenko, Yu. V. Rodionov, and I. V. Berezin, *Theses for the 2nd Conference on Electrocatalysis*, Moscow (1978), p. 54. In Russian.
136. T. Loucka and J. Weber, *J. Electroanal. Chem.* **21**, 329 (1969).
137. F. Scheller, R. Renneberg, W. Schwarze, G. Strnadt, K. Rommerening, H.-J. Prümke, and P. Mohr, *Acta Biol. Med. Ger.* **38**, 503 (1979).
138. Yu. Yu. Kulis, V. I. Razumas, and A. A. Malinauskas, *Dokl. Akad. Nauk SSSR* **245**, 394 (1979). In Russian.
139. Yu. Yu. Kulis, V. Razumas, and A. Malinauskas, *J. Electroanal. Chem.* **116**, 11 (1980).
140. Yu. A. Chizmadzhev, V. S. Markin, M. R. Tarasevich, and Yu. G. Chirkov, *Macrokinetics of Processes in Porous Media*, Nauka Publishers, Moscow (1971). In Russian.
141. I. G. Gurevich, Yu. M. Vol'fkovich, and V. S. Bagotskii, *Liquid Porous Electrodes*, "Nauka i Tekhnika" Publishers, Minsk (1974). In Russian.
142. R. R. Dogonadse, A. M. Kuznetsov, and L. Ulstrup, *J. Electroanal. Chem.* **79**, 267 (1977).
143. L. A. Blyumenfeld, *Problems of Biological Physics*, Nauka Publishers, Moscow (1977). In Russian.
144. E. A. Moelwyn-Hughes, *Physical Chemistry*, Vol. 1, Pergamon Press, London–New York–Paris (1961).
145. K. I. Zamaraev and R. F. Khairutdinov, *Usp. Khim.* **47**, 992 (1978). In Russian.

146. A. I. Yaropolov, T. K. Sukhomlin, A. L. Karyakin, S. D. Varfolomeev, and I. V. Berezin, *Dokl. Akad. Nauk SSSR* **962**, 112 (1982). In Russian.
147. G. G. Guilbault, *Handbook, of Enzymatic Analysis,* Marcel Dekker, New York (1976).
148. H. Hilsson, A. C. Akerlung, and K. Mosbach, *Biochem. Biophys. Acta,* **320**, 529 (1973).
149. G. G. Guilbault and G. J. Lubrano, *Anal. Chim. Acta* **60**, 254 (1972).
150. L. C. Clark, *Enzyme Engineering,* Interscience, New York (1972).
151. L. C. Clark, *Conf. Dynamics Septic Shock Man.* **11**, 13 (1968).
152. G. G. Guilbault and J. G. Montalvo, *J. Am. Chem. Soc.* **92**, 2533 (1970).
153. D. L. Williams, A. R. Doing, and A. Korosi, *Anal. Chem.* **42**, 118 (1970).
154. D. N. Gray, M. N. Keyes, and B. Watson, *Anal. Chem.* **49**, 1067A (1977).
155. H. A. Videla and A. J. Arvia, *Biological Aspects of Electrochemistry, Proc. of the 1st Int. Symp.,* Rome (1971), pp. 667–674.
156. T. Mori, T. Sato, Y. Natuo, T. Tosa, and I. Chibata, *Biotechnol. Bioeng.* **14**, 663 (1972).

5

Electrochemical Aspects
of Bioenergetics

EDMOND F. BOWDEN, FRED M. HAWKRIDGE,
and HENRY N. BLOUNT

1. Introduction

The application of electrochemical techniques to the study of biological systems finds its roots in work done more than fifty years ago.[1] This early work, based principally on potentiometric studies, provided extensive thermodynamic information for a number of biological molecules. More recently there has been a remarkable expansion in the application of electrochemical techniques to problems in biological systems. A number of scientific disciplines that focus on the thermodynamics, kinetics, and mechanisms of biological electron transfer are responsible for this enhanced activity. The basic processes by which energy is transduced between chemical and electrical domains in biological systems are being intensely investigated.

The advent of new electrochemical methodologies as well as the use of established electrochemical techniques in novel ways has contributed to the realization that bioelectrochemistry can provide new insights into bioenergetics. In concert with these advances in techniques and procedures, there has been a shift in the emphasis of this field from thermodynamic studies to kinetic and mechanistic investigations. The success of early potentiometric studies of

EDMOND F. BOWDEN and FRED M. HAWKRIDGE • Department of Chemistry, Virginia Commonwealth University, Richmond, Virginia 23284. *HENRY N. BLOUNT* • Center for Catalytic Science and Technology and Brown Chemical Laboratory, The University of Delaware, Newark, Delaware 19711.

biological molecules derived primarily from the use of small redox molecules that served to couple the redox state of the biological system with the potential-indicating electrode. In the absence of these redox coupling molecules, called mediators, it was not possible to obtain accurate potentiometric data on a reasonable laboratory time scale during chemical redox titrations of biological molecules. Mediated potentiometric redox titrations have been widely used to acquire Nernstian responses for biological redox molecules and have provided reliable stoichiometric and thermodynamic data for these systems. These data have afforded a foundation for understanding the mechanisms of electron transfer in many diverse processes such as photosynthesis, oxidative phosphorylation, and nitrogen fixation.

Mediators have also found use in voltammetric studies which seek to characterize the kinetics of the electron transfer reactions between the electrochemically generated mediator reactant and the biological molecule.[2] Mediated voltammetric studies of biological molecules employing a number of techniques have since been widely applied to problems in bioenergetics.

An outgrowth of mediated potentiometric redox titrations has been the evolution of various models which attempt to explain why biological molecules do not readily communicate their redox states directly to potential-indicating electrodes. Insulation of the redox site from the electrode by the protein portion of the molecule, the lack of the requisite enzyme character on an electrode surface, and adsorption of the biological molecule on the electrode surface are among the features of models which have been proposed.

Nevertheless, voltammetric techniques have long been applied to the study of direct electron transfer processes of biological molecules.[3] Such early voltammetric studies revealed a high degree of electrochemical irreversibility in the direct heterogeneous electron transfer reactions between electrodes and biological molecules. The irreversible nature of direct electron transfer observed in these early studies precluded the accurate and precise characterization of the electron transfer stoichiometry and thermodynamics of biological molecules. The models alluded to above were often used to account for the irreversible electron transfer kinetics observed in these studies.

The past decade has been marked by significant advances in the application of electrochemistry to the study of biological molecules which react both directly at electrodes and indirectly through electrochemically generated mediators.[4-8] The increasing availability of well-characterized biological samples together with advances in electrochemical methodology have served to stimulate work in bioenergetics. There is increasing interest in using electrochemistry to study the molecular events associated with biological molecule electron transfer kinetics and mechanisms. This interest has been stimulated by the extensive literature which focuses on the stoichiometry and thermodynamics of biological electron transfer systems. The increasing exchange between the biochemical and electrochemical communities portends continued and accelerated activity in bioelectrochemistry.

The aim of this chapter is to critically evaluate the advances in bioelectrochemistry with emphasis on the implications of the more recent kinetic and mechanistic studies. The electrochemical techniques which have been used in these studies and the procedures for extracting bioenergetic parameters are described. No attempt is made to cover the vast literature on bioenergetics derived from potentiometric measurements. The biological molecules discussed in this chapter are principally those involved in biological electron transfer reactions and those which contain one or more metal redox sites embedded within the protein. Other chapters in this volume consider the bioelectrochemistry of small (Chapter 2) and very large (Chapter 3) biological molecules.

2. Interfacial Properties of Electrodes and Biological Membranes

In this section, a brief survey of some important interfacial properties of electrodes and biological membranes in contact with aqueous solutions is presented. This survey is intended to serve two purposes. First, it is clear that the depth of understanding of interfacial bioelectrochemical data depends heavily upon the degree to which the electrode interfacial region in the absence of the biological reaction is understood. Thus, the survey points out many, but certainly not all, of the interfacial features and properties with which bioelectrochemical experimentalists may have to be concerned. In-depth descriptions and extensive bibliographies for the individual topics may be found by consulting the references given. A second purpose served by the interfacial survey which follows is to establish some comparative basis for evaluating electrode reactions of biological redox molecules in terms of their physiological reactions at membrane interfaces. It is readily apparent that electrode and membrane interfaces exhibit some significantly different features. Nevertheless, there are some key similarities which are likely to exert similar influences upon heterogeneous redox reactions. For example, as described below, both electrode and membrane interfaces are electrically charged, and furthermore structured water layers exist at both. The extent to which such comparisons can be drawn while still retaining their validity is a point deserving additional attention.

2.1. The Metal Electrode/Aqueous Solution Interface

The interfacial properties of metal electrodes in contact with aqueous solutions, usually containing electrolytes, have been the focus of considerable attention. The majority of this work has used mercury for the electrode phase [9-11] but recent attention has been increasingly focused on solid electrodes.[12-15] Many views of these interfaces have been proposed. The Gouy–Chapman–Stern model, proposed in the early part of this century, provides a description of the electrode/aqueous electrolyte interface that remains central to current views. This model pictures the electrical double layer that exists at electrode/aqueous solution interfaces as shown in Figure 1.

When a potential is imposed on a metal electrode in contact with an aqueous electrolyte solution, any excess charge on the electrode resides on its surface. If the electrode is negatively charged, for example, there will be an excess of electrons in a very thin layer at the metal surface. Ions, solvent molecules and other species in the solution phase will interact with this charged surface to form a structured interfacial region generally referred to as the electrical double layer. The net charge on the solution side of the double layer is opposite in sign to the excess charge on the metal. Ions residing in the double layer region are usually classified as being specifically adsorbed or nonspecifically adsorbed.

Specifically adsorbed ions[11,14] are those which directly contact the electrode surface. As indicated in Figure 1, specifically adsorbed ions are considered to be desolvated, and they displace solvent molecules adjacent to the electrode surface. Iodide, which is weakly solvated by water, is a good example of an ion which tends to specifically adsorb on electrode surfaces. The nature of specific adsorption is a function of both electrostatic and chemical interactions between the electrode and the ion. Specific adsorption can significantly alter the interfacial potential profiles as well as the kinetics of interfacial reactions. The thin solution layer closest to the electrode surface which contains specifically adsorbed ions as well as solvent molecules is often called the inner layer or the Helmholtz layer. The inner Helmholtz plane (IHP) is considered to pass through the centers of specifically adsorbed ions (see Figure 1).

Nonspecifically adsorbed ions are those ions which retain their primary solvation shells and which are concentrated adjacent to the electrode surface due to electrostatic forces only. However, because of thermal motion, the nonspecifically adsorbed ions are actually distributed in a layer extending some distance from the electrode surface. This layer is called the diffuse layer, and

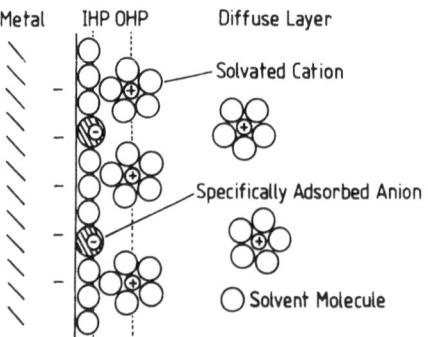

Figure 1. Schematic representation of metal electrode-aqueous electrolyte double layer structure. Adapted from Reference (9) with permission.

its thickness depends on the ionic strength. In dilute electrolytes, the diffuse layer can extend more than 100 angstroms into the solution phase.[16] The nature of the diffuse layer can have an important bearing on electron transfer rates since the actual potential felt by a reactant close to the electrode surface is dependent upon it. For reactants which do not penetrate the inner layer and actually contact the electrode surface, the outer Helmholtz plane (OHP) is usually considered to be the approximate site for electron transfer. The OHP contains the centers of nonspecifically adsorbed ions at their position of closest approach to the surface, as shown in Figure 1.

The potential profile across a metal/aqueous electrolyte interface is represented in Figure 2. This figure is not intended to be quantitative, but rather is intended to demonstrate the effect of a change in potential. As shown, a potential change for a metal electrode is manifested almost entirely as a change on the solution side of the double layer. In reasonably concentrated electrolyte (above *ca.* 0.1 M), most of the potential drop occurs over a distance of only several angstroms next to the electrode surface. Field strengths in this region can reach as high as 10^7 V/cm.[13]

Another very important property of a given electrode/solution interface with regard to its structure and functioning is its potential of zero charge (pzc).[11,15] This is the unique value of electrode potential at which the excess charge on the electrode is zero. Acquisition of pzc data for solid electrodes is dependent on the chemical nature of the electrode surface, and wide variations in pzc values for electrodes of the same metal in identical solutions have been reported.[9]

Brief consideration is now given to the solvent structure at metal/aqueous electrolyte interfaces.[11,13,17] Several molecular models have been proposed which treat a single layer of water molecules at the metal surface. Within the layer, the individual water molecules (or clusters of molecules) are allowed to have certain orientations. In the earliest and simplest molecular model, an inner-layer water molecule is oriented as a result of its dipole interaction with the charge on the metal electrode. Orientation is limited to either of the two positions in which the molecular dipole is perpendicular to the electrode surface. More realistic treatments have since been described which variously

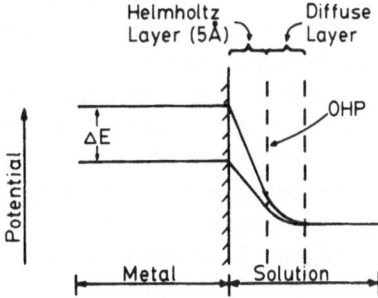

Figure 2. Potential–distance profiles for metal electrode/solution interfaces showing the effect of changing the bias voltage by ΔE.

consider contributions from lateral dipole interactions between water molecules, dipole image forces, hydrogen bonding, water–metal chemical interactions, and water quadrupole effects.[11,13,17] Although the exact nature of interfacial water structure remains an unsettled issue, it is clear that such water is structurally distinct from bulk water, being in a more ordered state. Another important feature of interfacial water structure, namely, its extent from the electrode surface into solution, has not been treated in the proposed molecular models.[17] Conway[13] has concluded from available experimental data that the significant interfacial water layer at metal electrodes encompasses only one to two molecular layers, i.e., 4–5 Å.

The interfacial water structure at a metal electrode has also been considered in terms of the hydrophobic/hydrophilic nature of the metal[13,18] as commonly determined through contact angle measurements. Hydrophilic electrodes that are readily wetted by water result in water structuring at the electrode surface that is believed to be different from that which occurs at more hydrophobic electrodes. The history of an electrode can result in divergent results with respect to the hydrophilic properties of its surface. Gold has recently been shown to be hydrophilic when the surface is clean.[19,20] Exposure to ambient laboratory atmosphere alters this property and this may account for conflicting reports which have addressed this question.

2.2. The Semiconductor/Aqueous Solution Interface[16,21,24–26]

Semiconductor electrodes are finding increasing use in the study of biological electron transfer reactions by electrochemical techniques. While some of these materials can exhibit metal-like behavior as electrodes, important differences exist in comparison to metals. Typical charge carrier densities (n) for metals and intrinsic semiconductors are about 10^{22} and $10^{15}/cm^3$, respectively.[22] Intrinsic semiconductors are materials which owe their conductivity to thermal excitation of electrons from the valence band to the conduction band. The doping of intrinsic semiconductors with electron-donating or -accepting impurities results in n-type and p-type semiconductors, respectively. Such materials exhibit higher charge carrier densities and are thus more conductive than the parent materials.

Certain oxides can be doped to yield semiconductor materials. Optically transparent semiconductor electrodes of doped tin oxide and indium oxide have been used successfully in the study of biological electron transfer reactions (*vide infra*). Although doping over several orders of magnitude of charge carrier density is possible with these oxides, highly doped specimens ($n = 10^{20}–10^{21}/cm^{3}$ [23]) have seen the most widespread use as electrodes. With respect to electrical conductivity, these materials exhibit a considerable degree of metallic character and have been classed as degenerate semiconductors.[16]

Figure 3 illustrates the band model usually employed in discussions of the electronic structure of semiconductors. Electrons can reside at energy

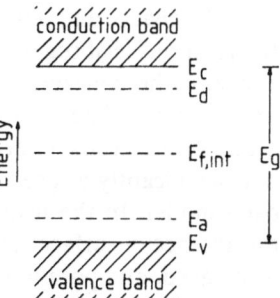

Figure 3. Band model for semiconductors. Energy levels shown are: E_c, the conduction band edge; E_v, the valence band edge; E_g, the band gap; $E_{f,int}$, the Fermi energy for an intrinsic semiconductor; E_d, electron donors in n-type semiconductors; E_a, electron acceptors in p-type semiconductors.

levels within the conduction or valence bands, but not at energy levels within the band gap (E_g) which separates them. For an intrinsic semiconductor, thermal fluctuations will partially populate the otherwise vacant conduction band with electrons from the otherwise completely filled valence band. Equal numbers of conduction band electrons and the valence band holes are the current carriers in this type of material. The Fermi level, defined as the electronic energy level which has a one-half probability of being occupied, is located at the middle of the band gap, as shown in Figure 3. Also shown in that figure are typical energy levels for the electron donor and acceptor impurities which result respectively in n- and p-type semiconductors.

As with metal electrodes, the interpretation of interfacial bioelectrochemical data obtained with semiconductor electrodes will depend heavily upon an understanding of interfacial structure and properties. Characterization of semiconductor interfaces is often more difficult than with metals because, in part, the charge and potential distribution are more complex. Unlike metals, the electric field arising from excess charge near the semiconductor surface can extend deep into the semiconductor bulk, giving rise to a space charge region. The thickness of the space charge region is a function of carrier concentration, becoming thinner as the carrier concentration increases. The similarities between the semiconductor space charge region and the solution diffuse layer discussed earlier have been noted often.

Because of the space charge region in semiconductors, changes in applied potential are not necessarily restricted to the solution side of the double layer, as was the case for metal electrodes. Instead, it is possible for a change in applied potential to be taken up in the semiconductor through the phenomenon of band bending. Band bending is important in the treatment of semiconductor electrode kinetics because of the requirement that electrons involved in the reaction must originate from near the Fermi level of the semiconductor.[16] With band bending, the Fermi level near the surface can appear in a region of charge carriers rather than in the band gap where charge carriers are generally absent. One aspect unique to semiconductor electrode kinetics is the possibility of reaction limitation due to depletion of carriers.

The charge and potential distribution in the semiconductor/aqueous electrolyte interfacial region can depend on several factors. One of these, as noted above, is the concentration of carriers. Another, often dominating factor, is surface states. The presence of a high concentration of surface states having available electronic energy levels different from the band energies can result in a significantly altered interfacial energetic picture compared to the simple band model. In the usual picture of an intrinsic semiconductor electrode in the absence of surface states, the solution side of the double layer is considered to be fixed, and potential changes are confined to the semiconductor interior, as discussed above. However, with increasing carrier concentration and/or surface state concentration, potential changes can influence the solution side of the double layer as well. Degenerate semiconductors are extreme examples of this behavior in which metal-like behavior is exhibited.

Metal oxide semiconductor electrodes also differ from bare metal electrodes with respect to interactions with water. Interfacial region in which water properties differ significantly from those found in the bulk phase is generally more extensive than for metal electrodes. Significant interfacial water structure can extend to several molecular layers from oxide surfaces.[13] Also, the inner monolayer of water can be rotationally immobile due to hydrogen bonding, a feature that is absent at pure metal surfaces.

Furthermore, metal oxide electrode surfaces are sites for acid/base equilibria.[21,26] This situation leads to a strong solution pH dependence of the surface charges and, therefore, the potential drop in the double layer. In the absence of specific ion adsorption other than H^+ and OH^-, there will exist a particular solution pH for which the excess surface charge on the metal oxide is zero. This value of pH is called the zero point of charge (not to be confused with the pzc). If specific adsorption of ions other than H^+ and OH^- occurs, then the solution pH at which no excess surface charge exists is called the isoelectronic point or the point of zero zeta potential.[16]

From the brief discussion above, it is apparent that the structure and potential profile at the semiconductor/electrolyte interface can be quite complex. Nevertheless, it seems apparent that sustained attention with regard to this problem will yield substantial benefits for bioelectrochemical research.

2.3. Biological Membranes

Biological membranes are considerably more complex than the models discussed above for electrode/aqueous electrolyte interfaces and there might appear to be few lines of comparison to be drawn between the two cases. Biomembranes possess two-dimensional structure and have a thickness of approximately 100 angstroms.[27,28] They are formed primarily from amphiphilic phospholipids, which impart the two-dimensional structure to the membrane, and from proteins. In aqueous solutions, the long hydrophobic tails of the lipids are found in the interior of the membrane while the polar head

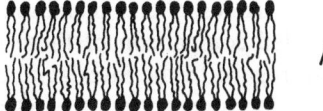

A

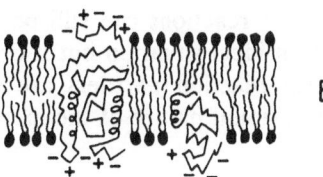

B

Figure 4. Membrane models. A. Black, or bimolecular, lipid membrane. Adapted from Reference (28) with permission. B. Lipid–globular protein mosaic model. Adapted from Reference (27) with permission.

groups are found on the surfaces of the membrane as illustrated in Figure 4A. This is a model for the artificial "black lipid membrane."[28]

Real biomembranes are more complex and include various protein and nonprotein components. Figure 4B shows a common mosaic model for a biomembrane that includes protein in its fabric. Proteins associated with biomembranes are classified into two primary types, integral and peripheral. Peripheral proteins can be extracted from the biomembrane and are pictured as being weakly associated with the membrane. Integral proteins, illustrated in Figure 4B, are embedded in the membrane and may have protruding ionic regions. Integral proteins are difficult to extract from the biomembrane and disruption of the membrane itself is usually required in order to remove them.

The structure of the biomembrane illustrated in Figure 4A places the polar head groups at the surface of the membrane and gives rise to an electrical double layer in aqueous solution.[28-31] Because of the charged surface, oppositely charged electrolyte ions will reside near the surface of the membrane in a fashion analogous to the Helmholtz region at an electrode/aqueous electrolyte interface. At this interface not only a large potential drop likely occurs but also a diffuse layer of electrolyte ions extends into solution.

Another similarity between the interfaces of electrodes and biomembranes in contact with aqueous solutions is the existence of an ordered water layer at their respective surfaces. Considerable attention has been given to understanding the structure of water at biomembrane surfaces.[13,32-34] Studies of phosphatidylcholine bilayers indicate that an average of ca. 24 water molecules are associated with each lipid head group.[34] These water molecules exhibit properties different from those corresponding to bulk phase water. Assuming a planar bilayer surface with an area of 50 Å^2/head group, an effective hydration water thickness of ca. 4–5 molecular layers can be calculated. Evidence has also been given for the innermost water molecules of hydration being tightly bound.[34] These features are qualitatively similar to those given above in the description of interfacial water structure at metal oxide surfaces. The general concept of a layer of structured water is thus common to interfaces of electrodes[13] and membranes[34] in contact with aqueous phases. A more detailed comparison of these related situations is a worthwhile undertaking with regard to the interfacial behavior of biological molecules.

Electrode/aqueous electrolyte interfaces lack the molecular specificity associated with biomembranes in controlling the enzymatic character of biological redox reactions. However, the interfacial role in biological electron transfer reactions can still be probed by electrochemical techniques. This is not an ideal situation but it does offer some advantages over experiments in which reaction partners are studied under conditions in which the two-dimensional nature of the membrane is entirely lost.

3. Thermodynamic Studies of Biological Molecules

The determination of the thermodynamics of an electron transfer reaction relies on the experimental constraints that were discussed in the Introduction section. The reliable determination of the formal potential of a redox couple is dependent upon the observation of responses which are Nernstian. For the redox process given by

$$O + ne \underset{k_{b,h}}{\overset{k_{f,h}}{\rightleftharpoons}} R \tag{1}$$

where $k_{f,h}$ and $k_{b,h}$ are the forward and back heterogeneous electron transfer rate constants, respectively, the Nernst equation

$$E = E^{0\prime} - \frac{RT}{nF} \ln \frac{[R]}{[O]} \tag{2}$$

where $E^{0\prime}$ is the formal potential of the redox couple and all other terms have their usual definitions, will be obeyed if the rate of heterogeneous electron transfer is rapid relative to the time scale of the measurement. In practical terms, this means that when an increment of reducing or oxidizing equivalents is added to a sample containing O and R the redox concentration ratio given in Eq. (2) must be stoichiometrically shifted and rapidly result in an invariant potentiometric response that is in agreement with the Nernst equation. At 25°C a plot of E versus $\log [R]/[O]$ must be linear and have a slope of -0.0591 V/n and an intercept equal to the formal potential of the redox couple. The same results should be obtained for both reductive and oxidative titrations of the sample.

Biological samples alone do not often exhibit Nernstian responses when chemical titrations are potentiometrically monitored. It has long been recognized that chemical redox titrations of biological molecules will exhibit potentiometric responses which are Nernstian if organic redox molecules are included in the biological sample.[1] These organic molecules, called mediators, are selected so that their formal potentials will be close to the formal potential of the biological molecule. A series of mediators is often used with formal potentials which are *ca.* 70 mV apart so that a wide range of potentiometric measurements is feasible. This approach is of particular importance when a series of biological molecules are being studied in the same sample. A concern in this type of experiment is the agreement between successive reductive and oxidative titrations of the same biological sample. During a chemical titration, the volume of the sample changes and spent titrant accumulates in the sample. Successive reductive and oxidative titrations of a biological sample can give rise to hysteresis in the potentiometric responses that may be due to these effects.

An alternative to relying on the potentiometric measurements alone during a chemical redox titration of a biological molecule involves optically monitoring the change in the redox state of the biological molecule during the titration. The redox state of the biological molecule is determined from the optical absorption data together with the measured potentiometric response at each point in the titration. Agreement between these two sets of data expressed in the form of the Nernst equation provides more confidence in the measured thermodynamic parameters of a biological molecule, and this procedure has found wide use.[35-37]

Certain advantages arise in the use of electrochemical techniques compared to chemical redox titrations in the determination of biological molecule thermodynamic parameters. Electrochemical techniques which couple the principle of mediation between electrodes and biological molecules to overcome irreversible electrode reactions with optical monitoring of the redox state of the sample have recently been developed. The following two sections address the principles and applications of these techniques in the study of biological molecule thermodynamics.

3.1. Optically Transparent Thin-Layer Electrochemistry

The technique of optically transparent thin-layer electrochemistry (OTTLE) was first applied to the characterization of the stoichiometry and thermodynamics of horse heart cytochrome *c* by Heineman *et al.*[38]. This report showed how the special features of OTTLE can be combined with optical monitoring of a mediated biological electrode response to provide a simple, accurate, and precise means of characterizing the stoichiometry and thermodynamics of a biological molecule. The mechanism of mediation is described by the following equations:

$$\text{electrode} \qquad M_O + ne \rightleftarrows M_R \qquad (3)$$

$$\text{solution} \qquad M_R + B_O \rightleftarrows M_O + B_R \qquad (4)$$

where M_O/M_R and B_O/B_R are the oxidized and reduced forms of the mediator and the biological molecule, respectively. In the work by Heineman *et al.*[38] the mediator was 2,6-dichlorophenolindophenol and as mentioned above the biological molecule was cytochrome *c*. The formal potential of the mediator is 0.226 V vs. NHE and the formal potential of cytochrome c is 0.260 V vs. NHE.[38]

The unique advantages of OTTLE arise both from the very short distances over which molecules must diffuse to undergo electron transfer at the electrode surface and from the facility for acquiring optical measurements to monitor the redox state of solution resident species which is provided by an optically transparent electrode (OTE).[39] The time (t, seconds) required to attain redox equilibrium in an OTTLE cell may be estimated by

$$t = l^2/(2D_O) \tag{5}$$

where l is the longest distance (cm) that a reacting species must diffuse to reach the electrode (often approximated by the pathlength of the OTTLE cell) and D_0 is the diffusion coefficient (cm^2/s) of the electroactive species. For a typical OTTLE pathlength of *ca.* 0.2 mm and a diffusion coefficient of 5×10^{-6} cm^2/s it is evident that the time required to achieve redox equilibrium is about 1 min.

The major liability in using OTTLE is the requirement that the optically monitored species exhibits a large difference in its absorption spectra between the oxidized and reduced forms, i.e., the sample must have a large difference molar absorptivity. Fortunately, many biological molecules have difference molar absorptivities in excess of 10^4 M^{-1} cm^{-1} and are therefore amenable to study by OTTLE.

Figure 5 shows the first OTTLE results reported for a biomolecule (i.e., cytochrome c[38]). These data were acquired by applying potentials to a gold minigrid electrode near the formal potential of cytochrome c and recording each spectrum after achieving redox equilibrium between the electrode, the

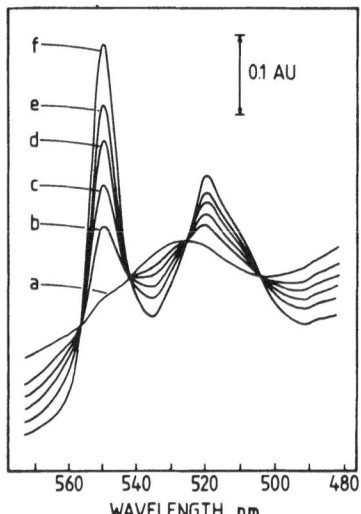

Figure 5. Spectra of cytochrome c at different values of $E_{applied}$, mV vs. SCE. Cell pathlength = 0.22 cm, 0.5 mM cytochrome c, 0.01 mM 2,6-dichlorophenolindophenol. (a) 250.0, (b) 50.0, (c) 30.0, (d) 10.0, (e) −10.0, and (f) −250.0. Adapted from Reference (38) with permission.

mediator, and cytochrome *c*. The concentration ratio of [O]/[R] can be calculated for each applied potential using the equation

$$[O]/[R] = (A_R - A)/(A - A_O) \qquad (6)$$

where, at the monitored wavelength, A is the absorbance at a given applied potential and A_R and A_O are the absorbance values at the same wavelength for the totally reduced and oxidized samples, respectively. From these calculated values, a Nernst plot of log ([O]/[R]) versus applied potential is constructed as shown in Figure 6 for the data of Figure 5. The *n* value of Eq. (1) and the formal potential of cytochrome *c* are determined from the slope and intercept, respectively.

The results shown in Figure 5 can be obtained by applying potentials in random order. This procedure provides very accurate and precise values for n and $E^{0\prime}$ while establishing the overall reversibility of the redox reaction of a biological molecule. All of these advantages are realized in mediated OTTLE studies of biological molecules with speed, simplicity, and use of small quantities of the biological sample. Figure 7 shows an OTTLE cell which is easily fabricated, assembled, and used.[40]

From the foregoing discussion, it is apparent that a study of a given biological molecule by mediated OTTLE will require access to a mediator having the desired formal potential as well as optical properties in both the oxidized and reduced forms which do not interfere with the optical response of the biological sample. Many mediators suitable for use in OTTLE studies of biological molecules have been described.[1,7,41–45]

Since the first report of the application of mediated OTTLE to the study of the redox properties of biological molecules,[38] the technique has been widely used for this purpose. Table 1 summarizes this body of work. A number of novel applications of OTTLE described in this table have served to provide new insights into the thermodynamics of biological molecule redox reactions. One of these applications is described in some detail in the following discussion.

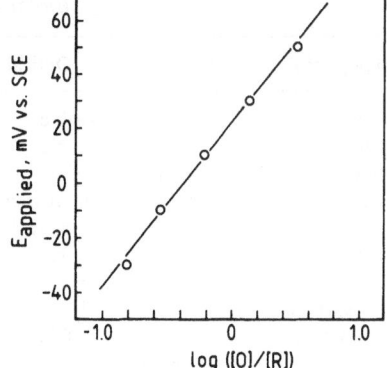

Figure 6. Typical plot of $E_{applied}$ vs. log ([O]/[R]) for the results shown in Figure 5. $E^{0\prime} =$ 262 (± 0.8) mV vs. NHE. Adapted from Reference (38) with permission.

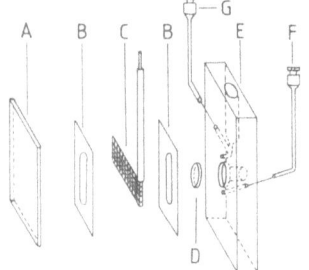

Figure 7. Small-volume optically transparent thin layer electrochemical cell. (A) Quartz cover plate, (B) Teflon spacer, (C) gold minigrid optically transparent electrode, (D) quartz disc, (E) plastic body, (F) inlet syringe port, (G) Pt syringe needle for auxiliary electrode. Adapted from Reference (40) with permission.

The temperature dependence of the formal potential of cytochrome c was evaluated in the presence of various electrolyte ions in water and D_2O.[48–50] The formal potential of cytochrome c exhibited biphasic dependence on temperature in water in the presence of chloride. When these experiments were repeated under conditions in which the counterion for chloride, sodium, was replaced by potassium the same biphasic behavior was observed. Even when the effect of specific binding of the chloride ion to cytochrome c was taken into account, biphasic behavior persisted. However, when water was replaced by D_2O a linear dependence of the formal potential of cytochrome c on the temperature was obtained. Figure 8 shows the results obtained for the temperature dependence of the formal potential of cytochrome c in water in the presence and absence of halide ions.

These results were ascribed to higher-order phase changes in water solutions containing chloride. The inflection in the temperature dependence shown in Figure 8 occurs at 42°C. It was postulated that the higher-order phase change for water in the presence of chloride at this temperature may be related to the inability of advanced forms of life to survive temperatures above this value.

Table 1
Optically Transparent Thin-Layer Electrochemical Studies of
Biological Molecules

Biological redox molecule	Optical probe	Experimental aim	References
Cytochrome c	Visible	n value/$E^{0\prime}$	38, 40, 46, 47
Cytochrome c	Visible	Temperature study	48, 49, 50
Photosystem I subchloroplasts	Visible	Cryogenic redox titration	51
Photosystem II subchloroplasts	Fluorescence	n value/$E^{0\prime}$	51
Ferredoxin	Circular dichroism	n value/$E^{0\prime}$	51
Myoglobin	Visible	n value/$E^{0\prime}$	52
Blue copper proteins	Visible	n value/$E^{0\prime}$	53
Cytochromes c_{551}, c_2, HiPIP	Visible	n values/$E^{0\prime}$	47
Vitamin B_{12}	Visible	n value/$E^{0\prime}$	54–58

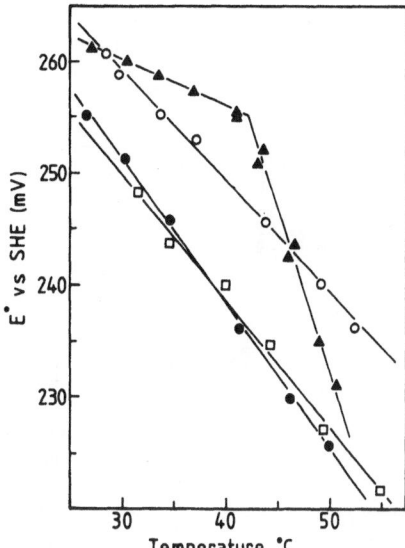

Figure 8. Temperature dependence of the formal potential of horse heart cytochrome c in $0.10\,M$ sodium halide solutions in H_2O. Solutions contained $0.10\,M$ sodium phosphate pH 7.00 buffer. (○) F^-; (▲) Cl^-; (●) Br^-; (□) I^-. Adapted from Reference (49) with permission.

3.2. Indirect Coulometric Titrations at Optically Transparent Electrodes

Indirect coulometric titrations (ICTs) of biological molecules with OTEs are also based on the mechanism illustrated in Eqs. (3) and (4). Here again the use of a mediator provides a means of coupling a biological molecule electron transfer reaction to an electrode in order to overcome irreversible heterogeneous electron transfer kinetics. An advantage of applying ICTs at OTEs to biological molecules is the ability to directly measure the n value of the sample. Just as in the case of mediated OTTLE, the sample volume remains constant during repetitive titrations. In contrast to mediated OTTLE the formal potential of the mediator in an ICT should be removed from that of the biological molecule so that the equilibrium constant for Eq. (4) is large. By measuring the charge associated with the electrolysis of the mediator, Eq. (3), under the same conditions, the charge associated with Eq. (4) is indirectly known. The same principles which have been described for ICTs in the determination of nonbiological molecules must be applied.[59] The current efficiency of the reaction mechanism given in Eqs. (3) and (4) must first be established in control experiments.

An excellent example of the application of ICTs at OTEs to the study of a mixture of cytochrome c and cytochrome c oxidase is shown in Figure 9. In this experiment, each titration increment is 1.04×10^{-6} equivalents and methyl viologen was the reductive mediator. In addition to evaluating the n value of cytochrome c and the n values of each heme of cytochrome c oxidase, the data of Figure 9 can be plotted in log–log form for the ratios of the redox states of the two sample components[35] as shown in Figure 10. Knowing the

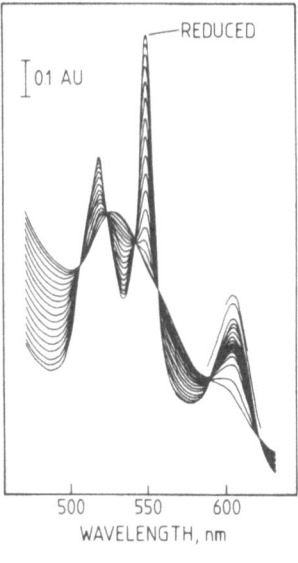

Figure 9. Spectra of cytochrome c (17.5 μM)–cytochrome c oxidase (12.5 μM heme a) mixture during reductive indirect coulometric titration using methyl viologen. Titrant increments of 1.04×10^{-6} equivalents; spectra recorded after each addition of reductive charge. Final two spectra at 605 nm were taken after excess methyl viologen cation radical was present. Adapted from Reference (60) with permission.

n value for each component from the ICT permits calculation of a best-fit response to determine the formal potentials of both cytochrome c oxidase hemes as illustrated by the fit of the calculated solid line to the data.[60]

The applications of ICTs at OTEs to the study of biological molecule thermodynamics and stoichiometry are summarized in Table 2. A number of

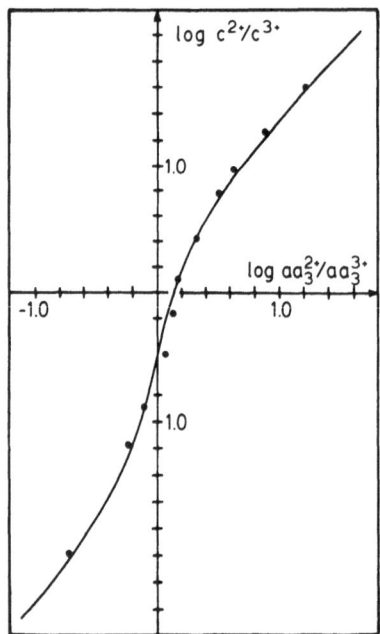

Figure 10. Plot of log [cytochrome c (reduced)/cytochrome c(oxidized)]versus log [cytochrome c oxidase (reduced)/cytochrome c oxidase (oxidized)] for reductive indirect coulometric titration of cytochrome c–cytochrome c oxidase mixture. Data points taken from Figure 9 and solid line is computer calculation for the cytochrome c formal potential of 250 mV, the cytochrome c oxidase heme a formal potential of 210 mV, and the cytochrome c oxidase heme a_3 formal potential of 350 mV. Adapted from reference (60) with permission.

Table 2
Indirect Coulometric Titrations of Biological Molecules at
Optically Transparent Electrodes

Biological redox molecule	Reference
Cytochrome c	61, 62
Cytochrome c/cytochrome c oxidase mixture	60
Cytochrome c oxidase	63, 64, 65
Cytochrome c oxidase/CO complex	66
Complex IV of mammalian oxidative phosphorylation	67
Vesicle-resident cytochromes	68
Photosystem II	69
Flavoproteins	70

reviews have described the technique and enumerated its advantages compared to alternative approaches.[7,41,42,71-74]

4. Kinetics and Mechanisms of Biological Electron Transfer Reactions

The preceding section described bioelectrochemical studies which focused on the determination of the stoichiometry and thermodynamics of biological molecule electron transfer reactions. Central to these studies was the use of a mediator which coupled the biological molecule to the electrode so that Nernstian responses could be obtained. Following the initial report of such work,[2] there were a number of studies which sought to characterize the *direct* electron transfer reactions of biological molecules at electrodes. In this section, the evolution of techniques that have been applied to the characterization of the direct electron transfer kinetics and mechanisms of biological molecules at electrodes will be traced. The models which have been proposed will be presented and evaluated in light of recent developments in this field of work.

The goal of much of this work has been to elucidate the physiological implications of biological electron transfer reactions. Kinetics provide indirect evidence for reaction mechanisms and results must be interpreted with caution. However, these data, together with the thermodynamic results obtained in related studies, provide guidance in understanding physiological electron transfer processes.

The material of this section will focus on results of homogeneous electron transfer kinetic studies involving reactions between electrochemically generated redox reactants and on direct heterogeneous electron transfer kinetic studies. An overview of the work in this area concludes this chapter.

4.1. Homogeneous Electron Transfer Kinetic Studies

The first determination of the homogeneous electron transfer kinetics for the reaction of an electrochemically generated species with a biological molecule was reported by Ito and Kuwana.[2] In that study, the reaction of electrochemically generated methyl viologen cation radical with ferredoxin–NADP reductase was evaluated using single potential step chronoabsorptometry and chronocoulometry. Studies of this reaction in the presence of NADP established that the indirect generation of reduced ferredoxin–NADP reductase according to Eqs. (3) and (4) resulted in the catalytic formation of enzymatically native NADPH. Direct reduction of NADP leads to the formation of an enzymatically inactive dimer product. This work used spectroelectrochemical[75,76] and chronocoulometric[77] techniques which had been developed and applied to the catalytic regeneration mechanism.

A chronoamperometric method for evaluating the kinetics of homogeneous electron transfer between an electrochemically generated mediator/reactant and biological molecules was subsequently reported by Ryan et al.[78]. The effects of solution ionic strength and the charge of the electrochemically generated reactant on the electron transfer kinetics with cytochrome c and cytochrome c_2 were reported in that work.

Several reports have evaluated the homogeneous electron transfer kinetics of cytochrome c using potential step spectroelectrochemistry. These reports together with other studies of biological homogeneous electron transfer reaction kinetics are summarized in Table 3. Evaluation of the kinetics of these reactions requires caution in that the small diffusion coefficients of the biological molecules studied relative to those of the electrochemically generated reactants mandates consideration of these parameters in data analysis.[81,83]

Table 3
Homogeneous Electron Transfer Kinetic Studies of Biological Molecules by Electrochemical Techniques

Biological molecule	Electrochemically generated reactant[a]	Reference
Ferredoxin–NADP reductase	MV^+	3
Cytochrome c_2	ferrous-EDTA	78
Cytochrome c	MV^+	79, 81, 82–84
Cytochrome c	BV^+	80
Cytochrome c	Diquat$^+$	80
Cytochrome c oxidase	MV^+	82
Ferredoxin	V^+_{670}	85

[a] MV^+, cation radical of 1,1'-dimethyl-4,4'-bipyridinium dichloride; BV^+, cation radical of 1,1'-dibenzyl-4,4'-bipyridinium dibromide; Diquat$^+$, cation radical of 1,1'-ethylene-2,2'-bipyridinium dibromide; V^+_{670}, cation radical of 1,1'-propylene-2,2'-bipyridinium dibromide.

4.2. Heterogeneous Electron Transfer Kinetic and Mechanistic Studies

Following the initial report of Kôno and Nakamura[3] of the direct electrochemical reduction of cytochrome c at a platinum electrode, the study of heterogeneous reactions of biological redox molecules has been slow to develop. Griggio and Pinamonti[86] reported in 1965 that cytochrome c underwent direct reaction at the dropping mercury electrode at an overpotential of ca. −0.36 V indicating a highly irreversible electrode reaction (see Table 4). The decade of the 1970's marked the beginning of direct electrochemical studies of biological molecules with an emphasis on the kinetics and mechanisms of these reactions. Whereas mercury electrodes were principally employed during this period, the advent of solid electrodes at which biological molecules undergo heterogeneous electron transfer reactions has led to an increasing interest in the use of bioelectrochemical techniques during the latter part of this decade. At present it is clear that protein structure, electrode material, and solution conditions are important determinants of heterogeneous electron transfer kinetics of biological systems. Cytochrome c, the most widely studied biological redox molecule, is considered first, followed by results which have been reported for other biological molecules. A selected summary of literature references to direct electrochemical studies of heme proteins is given in Table 4.

4.2.1. Cytochrome c at Mercury Electrodes

The first detailed polarographic study of cytochrome c was reported by Betso, Klapper, and Anderson.[87] Since that report the reaction of cytochrome c at mercury electrodes has been extensively studied, principally by the groups of Scheller,[88–94] Kuznetsov,[95–99] and Haladjian.[100–102] Several models for the electron transfer reaction of cytochrome c at mercury have been presented but a consensus on the mechanism does not presently exist. A unifying point in these models is the significant adsorption of cytochrome c which occurs at mercury electrodes and subsequently affects electron transfer with freely diffusing molecules.[87,93,97,100,103] Adsorption is evidently the basis of the concentration dependence of the electron transfer kinetics of cytochrome c, which ranges from polarographically reversible at low concentrations to quasireversible and irreversible at higher concentrations.[87,89,97,102] The anomalously small diffusion coefficients calculated from diffusion-controlled cytochrome c polarographic currents[87,89,97] is a second striking feature which stems from adsorption phenomena.

Three mechanisms for the reduction of cytochrome c were originally proposed by Betso et al.[87]: hole conduction through the protein fabric of adsorbed molecules, electron transfer between the mercury surface and the exposed heme edge of adsorbed molecules followed by desorption, and electron transfer between the mercury surface and the exposed heme edge of adsorbed

Table 4
Selected Summary of Electrochemical Studies of Proteins[a]

Protein studied	Electrode used in study	References
Cytochrome c	Hg	86–*88*, 89–*100*, 101–103, 106
	Hg/Au	109
	Ag	114
	Au	38, 109–113
	Ni	99
	Pt	2, 87, 100, 109, 113
	Si	113
	In_2O_3	115, *120*, *126*
	SnO_2	115, *118*, *119*, *120*, 126
	Au/bip[b]	110–112, 127–*129*, *130*
	Pt/bip	134
	MVMG[c]	*119*, *120*, *164*
	BIM[d]	113
Cytochrome c_3	Hg	148, *149*–154
	Au	153, 157
	Au/bip	158
Cytochrome c_7	Hg	159
Cytochrome b_5	Hg	99
Cytochrome P450	Hg	99, 143
	C	143
Myoglobin	Hg	88–92, 94, 97, 160, 161
	MVMG	136, *163*, *164*, 165
Hemoglobin	Hg	88–92, 94, 97, 98, 160, 162
Ferredoxin	Hg	109, 171, 172
	MVMG	135, *173*, *174*
Peroxidase	Hg	97, 109, 168, 169
Catalase	Hg	97, 169

[a] References in italics cite heterogeneous electron transfer kinetic parameters.
[b] Gold/4,4'-bipyridine and gold/1,2-bis (4-pyridyl) ethylene adsorbed electrodes,
c Methyl viologen-modified gold electrodes.
d Bipyridinium-immobilized mediator electrodes.

molecules followed by rotational diffusion of the adsorbed molecules with subsequent electron transfer to freely diffusing molecules.

The two models for the reduction of cytochrome c at mercury which have received the most attention are those of Kuznetsov and Scheller. In Kuznetsov's model,[96,97,99] a multilayer structure of adsorbed cytochrome c molecules was proposed. An irreversibly adsorbed layer of unfolded cytochrome c molecules forms first at the mercury electrode surface and covers all but about 20% of the electrode surface. A layer of weakly adsorbed native cytochrome c molecules then begins to form on top of the first layer and electron transfer occurs between the mercury electrode and weakly adsorbed molecules which are found in the pores of the primary layer. Electron transfer can then proceed intermolecularly both between adsorbed cytochrome c molecules and between adsorbed and diffusing molecules at the protein/solution interface, resulting in net reduction current. Figure 11 shows the direct current polarograms of oxidized cytochrome c at several concentrations.[97] The limiting currents observed in these experiments were only about 23% of the expected diffusion-limited responses. However, the limiting currents were linearly dependent on the square root of the mercury column height indicating a mass transfer-limited electrode reaction. Similar behavior had been previously reported.[87]

Scheller's mechanism[88,89] also involves an irreversibly adsorbed layer of cytochrome c molecules but with a smaller degree of unfolding of the molecule. It has been recently suggested, moreover, that the first adsorbed layer may involve only minor deformation of the cytochrome c molecules.[94,104] Electron transfer was initially proposed to occur via a hopping mechanism through the adsorbed layer of denatured molecules to native molecules in a weakly adsorbed second layer.[89,91] The presence of pores in the first adsorbed

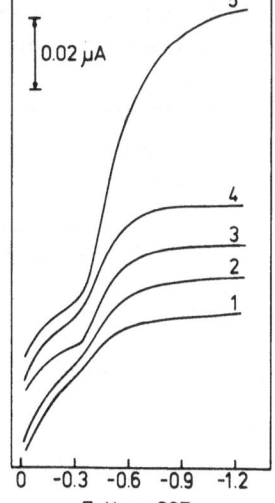

Figure 11. Direct current polarograms of ferricytochrome c at various concentrations. Sample contained Tris-H_2SO_4 buffer, pH 7.2, drop time = 5.6 s, mercury flow rate = 0.813 mg/s. Concentrations of cytochrome c are (1) 40 μM, (2) 80 μM, (3) 120 μM, (4) 160 μM, (5) 320 μM. Adapted from Reference (98) with permission.

layer[96,97] was subsequently incorporated into this model for cytochrome c reduction at mercury surfaces.[93,94] Electron transfer to native molecules is then pictured as occurring simultaneously through the adsorbed molecules of the first layer and directly at the mercury in the pores of this layer. However, electron transfer through the primary absorbed layer of structurally deformed cytochrome c molecules remains as a difference between the models of Scheller and Kuznetsov.

A related model, put forth by Haladjian *et al.*,[100] involves an adsorbed layer of cytochrome c molecules which are unfolded and flattened so that direct reduction at the mercury surface can occur. Electron transfer then occurs through this primary flattened layer of adsorbed molecules to native molecules which weakly adsorb on top of the first layer. Evidence for the adsorbed layer of cytochrome c being denatured was provided by film transfer experiments after the procedure of Kuznetsov.[97] Figure 12 shows the cyclic voltammetry observed at a hanging mercury drop in electrolyte alone and after exposure of the mercury drop to a 108 μM solution of cytochrome c, washing of the drop, and reimmersion in the electrolyte.[100] No oxidative response is evident and repeated potential cycling of the electrode results in the loss of the reductive peak. If the electrode is removed from solution and exposed to air, the adsorbed layer is oxidized as evidence by cyclic voltammetry on reimmersing the electrode in electrolyte. Since native cytochrome c is not oxidized by air, this result indicates that the adsorbed layer is formed irreversibly and involves denatured cytochrome c molecules. This model[100] appears similar to Scheller's original model.[89,91]

It is apparent that the cytochrome c electron transfer reaction at mercury electrodes is complex and dependent on a number of parameters. The adsorption of cytochrome c at mercury is in and of itself a complicated process. The formation of the first monolayer is rapid and chemically irreversible.[89,90,95,100] The concentration dependence of the double layer capacitance due to cytochrome c adsorption on mercury is shown in Figure 13.[93] By measuring the

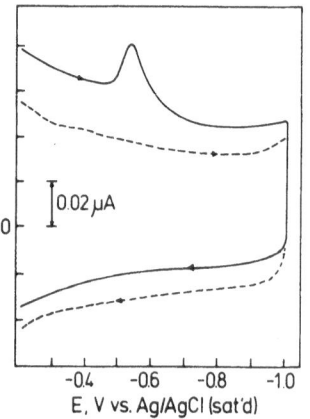

$0.02 \mu A$

0 —

-04   -06   -08   -1.0

E, V vs. Ag/AgCl (sat'd)

Figure 12. Adsorption study of cytochrome c at the hanging mercury drop electrode by cyclic voltammetry. Solution contained 0.01 M Tris-HCl buffer, pH 7.6, electrode area = 0.017 cm². (– – –) Supporting electrolyte alone. (——) Adsorbed cytochrome c after immersion of hanging mercury drop in 108 μM cytochrome c solution for 1 min, washing for *ca.* 30 s, and immersion in supporting electrolyte alone. Scan rate = 50 mV/s. Adapted from Reference (100) with permission.

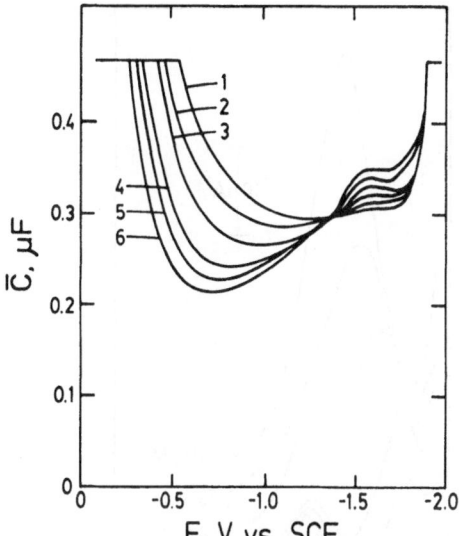

Figure 13. Differential capacitance of cytochrome c solution. Solution contained 0.1 M KCl. Cytochrome c concentrations are (1) 0, (2) 2 μM, (3) 4 μM, (4) 6 μM, (5) 8 μM, and (6) 12 μM. Adapted from Reference (93) with permission.

relative decrease in differential capacitance near the pzc as a function of time, Scheller *et al.* further demonstrated that the rate of this initial adsorption process is diffusion limited.[93] These results are shown in Figure 14, and at a concentration of 3.10 μM it is clear that the formation of this first layer is complete for the times studied.

Under experimental conditions similar to those of Scheller and co-workers,[93] Serre *et al.*[102] also found that the mercury drop surface was saturated with adsorbed cytochrome c at a concentration above *ca.* 5 μM for a 5-s drop time using differential pulse polarography. Figure 15 shows that an adsorption postpeak is concentration-dependent only below this concentration.

Although rapid adsorption of cytochrome c at mercury electrodes is widely accepted based on diverse experimental results,[87–103] the structure of this primary adsorption layer remains a point of controversy. If the cytochrome c

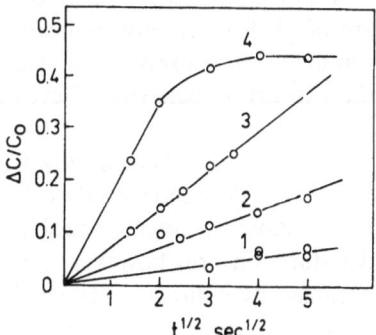

Figure 14. Relative decrease in differential capacitance of cytochrome c solutions at the dropping mercury electrode as a function of time. Differential capacitance measurements obtained near the point of zero charge, -700 mV vs. SCE. Solutions contained 0.1 M KCl. Cytochrome c concentrations are (1) 0.38 μM, (2) 0.76 μM, (3) 1.54 μM, and (4) 3.10 μM. Adapted from reference (93) with permission.

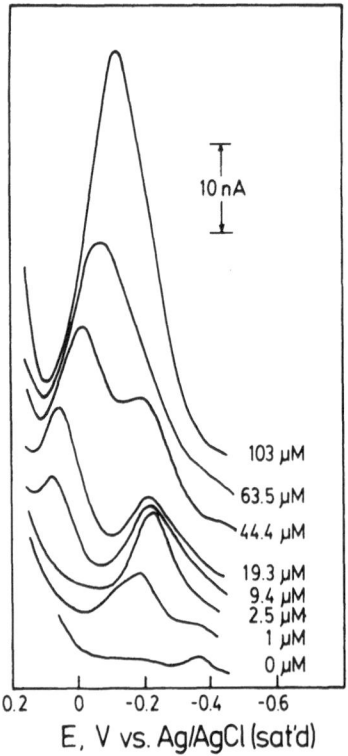

10 nA

103 μM

63.5 μM

44.4 μM

19.3 μM
9.4 μM
2.5 μM
1 μM
0 μM

0.2   0   -0.2   -0.4   -0.6

E, V vs. Ag/AgCl (sat'd)

Figure 15. Differential pulse polarograms of cytochrome c at various concentrations in 0.10 M Triscacodylate buffer, pH 6.05, and 0.10 M sodium perchlorate. Adapted from Reference (102) with permission.

molecules are adsorbed, to what extent do structural changes occur on adsorption and what is the time dependence of denaturation?

Kuznetsov[95] studied Brdicka catalytic waves for several nonheme proteins and proposed that complete unfolding accompanied adsorption of these biological molecules. Cytochrome c gave rise to weak Brdicka currents,[95] probably as a consequence of having only three hidden sulfur atoms per molecule. Senda *et al.*[103,105] have recently stated that cytochrome c Brdicka currents result from the heme group and the methionine-80 sulfur atom, which is the native ligand for the sixth coordination site. This proposal requires at least partial unfolding of cytochrome c upon adsorption. Brdicka solutions are of pH 9 to 10 and Brdicka currents are measured at *ca.* -1.2 V; hence caution must be exercised in using these results to draw conclusions regarding the adsorption behavior of cytochrome c at neutral pH and more moderate potentials.

The aforementioned work from the laboratories of Scheller[94] and Haladjian[100,102] does indicate that the mercury-adsorbed cytochrome c molecules are denatured under neutral pH conditions. Details of the extent of this denaturation are, however, not clear. The effective electrode surface area occupied by a directly adsorbed cytochrome c molecule has been calculated from both differential capacitance and radiochemical data. The value of

2000 Å^2,[93] returned by both calculations, is nearly double that expected for a native molecule. Such data suggest that the molecule must be significantly unfolded at the mercury surface in order to exhibit this enlarged cross-section. As discussed by Scheller,[94] though, such a conclusion is not unambiguous since other factors may be contributing to this effect. For example, if directly adsorbed cytochrome c molecules are assumed to exist in nearly native conformations and also to retain their hydration sheaths, one would also expect to find an effective electrode surface area per molecule which is larger than would be predicted from the crystallographic structure.

The differential pulse polarographic (DPP) results which have been given[101,102] may provide a basis for better understanding the adsorption-induced denaturation of cytochrome c at mercury. An interesting feature of the DP polarograms is the adsorption peak observed at the lowest cytochrome c concentrations, e.g., as shown in Figure 15. The negative displacement of this peak on the potential axis with respect to the $E^{0\prime}$ of cytochrome c, i.e., ca. -0.25 V, may be a direct reflection of denaturation. It has been noted that changes in conformation of cytochrome c which result in its decreased activity in enzyme systems are generally accompanied by a redox potential lowering of approximately the same value.[106] In the present situation, if the DPP adsorption peak is reasonably assumed to arise from cytochrome c molecules directly adsorbed on the mercury surface, than it would seem that a large degree of denaturation is not necessarily involved. The actual denaturation may be comparable in extent to the pH-induced state III to state IV conformational change (see below), in which methionine-80 is displaced as an iron ligand and the heme crevice is opened.

To summarize the foregoing paragraphs on primary adsorption structure, it appears certain that cytochrome c adsorbed directly on mercury is denatured to some degree. At the DME in neutral pH solution, redox potential shifts of ca. -0.25 V upon adsorption suggests heme crevice opening but not complete unfolding. At the HMDE, however, as shown in Figure 12, a more extreme negative shift of ca. -0.6 V is associated with adsorbed cytochrome c molecules and this fact suggests that more denaturation occurs in this case, perhaps involving significant chain unfolding. These observations indicate that adsorption structure is time dependent on the order of seconds to minutes and may differ between DME and HMDE experiments. The work of Kuznetsov *et al.*[96] further indicates, as noted previously, that the primary adsorption layer is a porous structure, ostensibly because of the inability of these large molecules to pack closely together. Processes which occur following formation of the primary monolayer of adsorbed cytochrome c are less clear.

The effect of adsorption on the polarographic electron transfer rates of cytochrome c is particularly evident with regard to the dependence of reaction overpotential on cytochrome c concentration. As long as the concentration of cytochrome c is kept below ca. 10–30 μM in pH 6–7 electrolytes, its reduction is polarographically reversible.[87,92,94,96,102] However, on raising solution concentration above these levels, the polarographic half-wave potential or peak potential shifts to increasingly negative values. Betso *et al.*[87] and subsequently other workers[94,97,102] observed that the magnitude of the reduction overpotential is proportional to log $(1/C)$, where C is the cytochrome c solution concentration. Figure 16 shows this behavior for a series of buffers. Although all five electrolyte systems give rise to the relationship just described, clear differences are apparent with regard to their positioning on the potential axis. Such effects, which are not well understood, are considered below. For the pH 6.05 Tris-cacodylate/perchlorate systems, the region of kinetic reversibility referred to above is visible at concentrations under *ca.* 10 μM.

To account for the linear dependence of overpotential on log $(1/C)$, Betso *et al.*[87] proposed that the reduction of diffusing cytochrome c molecules was being inhibited by adsorption. It was noted that this same relationship was already known for DME reaction inhibition by electroinactive adsorbates. Kuznetsov[99] has recently challenged this view on the grounds that the electrochemical and interfacial properties associated with the cytochrome c reaction are not comparable to the original systems used to deduce the relationship between overpotential and concentration. In his view, the pore structure of the primary adsorption layer, which has been presented above, must be taken into account. Cytochrome c molecules are proposed to adsorb in the pores in a reversible Langmuirian manner. Increased adsorption due to higher solution concentration causes increased two-dimensional surface pres-

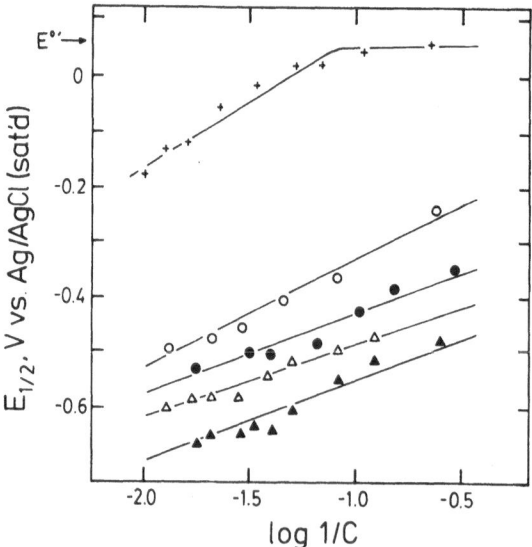

Figure 16. Plot of the half-wave potential of cytochrome c as a function of log l/[cytochrome c (μM)] for different media: (○) Tris-perchlorate, (△), Tris-acetate, (▲), Tris-cacodylate (0.01 M, pH 7.6), (+) [0.10 M Tris-cadocylate + 0.10 M sodium perchlorate], and (●) Tris-chloride buffer (pH 6.05). Adapted from Reference (102) with permission.

sure, which in turn tends to force out molecules adsorbed in the pores. Such a process will increase the distance between the electrode surface and the reducible molecules in the pores, which will result in the increased reduction overpotential. This thought-provoking model also leads to a linear dependence of overpotential on $\log (1/C)$ at sufficiently high concentrations. The original paper[99] should be consulted for details of the theoretical development.

The distinctive polarographic kinetic behavior of cytochrome c has also given rise to the concept of adsorption, or concentration, regions.[92] In essence, the transitions from one kinetic region to another are delineated in terms of the solution cytochrome c concentrations associated with them. These concentrations are then physically associated with particular modes of adsorption behavior at the DME surface. The use of solution concentration in this manner is a convenience since the drop growth rate and drop life, as well as solution conditions, will also impact the extent to which adsorption has proceeded at a particular time. In the lowest concentration region, below about 5 to 10 μM, formation of the primary adsorption layer is described as occurring during the drop life of the electrode. In the DP polarograms of Figure 15, the adsorption peak apparent at low concentrations is evidently a manifestation of this particular kinetic regime. In the second concentration region, which extends from about 5–20 μM, it has been proposed that formation of the primary monolayer is complete and the second layer of reversibly adsorbed molecules begins to form during the drop life. The second region corresponds to polarographic reversibility. Above this second concentration range, the secondary adsorption layer is evidently completed before the end of the drop life and increasing irreversibility occurs. It should be pointed out that the transitions between these regions may not necessarily be sharp, especially the transition between the second and third regions.

Turning to the subject of the diffusion limited behavior of cytochrome c at the DME it is clear that, in terms of the linearity of i_L vs. C plots, this reaction is well-behaved over a wide concentration range.[87,89,97,98] However, as noted previously, the diffusion coefficient calculated from such plots is consistently low. Several explanations have been proposed for this effect. Betso et al.[87] felt that inhibited diffusion through the adsorbed primary layer accounted for the low diffusion coefficient. Berg[104] ascribed the low value to the blocking of the electrode surface which inhibited electron transfer to freely diffusing molecules. In his recent paper, Kuznetsov[99] has considered this question in terms of a chemically modified electrode concept in which three electron transfer reactions are occurring. These transfers occur between the electrode and cytochrome c molecules in pores, between pairs of cytochrome c molecules within the adsorption layer structure, and between adsorbed and diffusing molecules at the protein/solution interface. This latter reaction, which is analogous to the homogeneous cytochrome c self-exchange reaction, is unaffected by the electrode potential and is considered to be the rate-limiting step leading to the anomalously small diffusion currents. In effect, all of the diffusing oxidized molecules which reach the protein/solution interface do not react, even though the molecules within the adsorption layer are fully reduced.

Solution pH significantly affects the magnitudes of both the polarographically measured diffusion coefficient and the overpotential for cytochrome c reduction. The overpotential dependence on pH was first described by Griggio and Pinamonti[86] and later by others.[87,89,102] The former authors found that the overpotential was independent of pH from 6.0 to 8.5 but shifted negatively by about 75 mV per pH unit between 8.5 and 9.6. Qualitative agreement with these results is found in the latter work referred to above.[87,89,102]

Cytochrome c exists in several pH-dependent conformational states.[107] Native ferricytochrome c, known as State III, undergoes a conformational change to State IV as solution pH is raised from neutral values. This conformational change has a pK of 9.35.[107] Serre et al.[102] used differential pulse polarography to demonstrate the effect of the conformational change on reduction overpotential.[102] Figure 17 shows the shift in the differential pulse polarographic peak potential with increasing pH and the simultaneous loss of the 695 nm absorbance maximum of State III.[102] The pK of this conformational transition was determined to be *ca.* 1.5 pH units below the 9.35 value mentioned above. This difference was ascribed to the perchlorate ion and this effect will be discussed below.

Solution pH also influences the magnitude of the diffusion-limited current. It is apparent that the cytochrome c State III/State IV conformation change accounts for the decrease in the diffusion-limited current which is observed

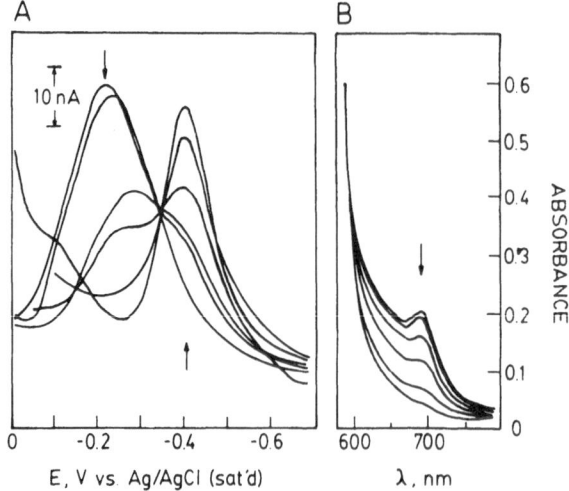

Figure 17. Effect of pH variation on (A) the differential pulse polarograms and (B) on the 695 nm absorbance band of 114 μM cytochrome c. Solution contained mixed 0.01 M Tris-cacodylate and 0.10 M sodium perchlorate buffer/electrolyte. For a series of solution pH values, 6.4, 7.0, 7.6, 8.2, and 9.5, the arrows indicate the direction of change in the responses with increasing pH. Adapted from Reference (102) with permission.

at basic pH values. [87,89,102] Both Griggio and Pinamonti[86] and Betso *et al.*[87] reported a second more negative direct current polarographic response as the solution becomes more alkaline. The pH dependence of this second wave coincides with the results of Serre *et al.*[102]

The importance of the electrolyte composition in the reduction of cytochrome *c* has been carefully studied by Haladjian's group.[100–102] The disagreement in the pK of the State III/State IV conformational change determined by differential pulse polarography in the presence of perchlorate compared to previous work[107] was ascribed to perchlorate binding to cytochrome *c*. This explanation is consistent with recent NMR, EPR, and 695 nm optical measurements which demonstrated that perchlorate binds more favorably to the State IV conformation.[108] In the presence of perchlorate, a pK of 8.0 was determined by ^{35}Cl NMR, in excellent agreement with the polarographic results. Halajian's group had demonstrated the importance of perchlorate binding to cytochrome *c* earlier.[101] Differential pulse polarographic peak potentials and morphologies were dependent on perchlorate below 0.1 M, again in excellent agreement with the NMR results.[108]

The differences between the results of Griggio and Pinamonti[86] and of Betso *et al.*[87] for the pH dependence of polarographic half-wave potentials is now apparent. The former workers did not have perchlorate present in their samples while the latter workers had 0.1 M $NaClO_4$ present in their samples.

Specific ion effects are apparent in the work of Serre *et al.*[102] as previously shown in Figure 16 for a series of anions. The effect of ionic strength has also been shown to be important with regard to the polarographic response of cytochrome *c*.[101] These results point to involvement of anions in the electrode reaction of cytochrome *c* at mercury but the nature of this effect is not clear at this time.

4.2.2. Cytochrome c at Solid Electrodes

Solid voltammetric electrodes which have been used to study the electron transfer reactions of cytochrome *c* can be divided into three classes: metals, metal oxide semiconductors, and chemically modified metals and semiconductors. Literature on the latter two electrode classes has been published only since 1977. Literature concerned with bare metal electrodes dates at least to the work of Kôno and Nakamura[3] in 1958. In that paper, the reduction of ferricytochrome *c* at pH 7.0 was found to be electrochemically irreversible at platinum. Overpotentials between −0.5 to −1.2 V were employed to observe measurable rates of reduction. It was suggested that the reduction of cytochrome *c* might be occurring via electrogenerated hydrogen. Tarasevich and Bogdanovskaya[109] have since demonstrated that cytochrome *c* can be reduced at platinum electrodes which have been previously exposed to hydrogen evolution. The direct reduction of cytochrome *c* at platinum electrodes poised at potentials positive of hydrogen evolution has been described as proceeding at very slow rates.[87,109] Slightly faster rates have been observed at gold.[109] The reoxidation of cytochrome *c* at platinum and gold was found to proceed even more slowly than the reduction.[109] Some investigators have failed to detect a reaction corresponding to cytochrome *c* reduction at gold[38,110–113] and platinum electrodes.[113] Similarly slow direct electron transfer to cytochrome *c* has been observed at nickel[99] and silver[114] electrodes.

Adsorption of cytochrome *c* on bare metal electrodes has been addressed by several authors. Cotton *et al.*[114] used surface-enhanced resonance Raman spectroscopy to demonstrate that cytochrome *c* adsorbs on silver electrodes. These results indicated that cytochrome *c* is not extensively denatured in the adsorbed state due to the close correlation between the surface adsorbed and bulk solution spectra. However, spectral correlation also indicated that adsorbed cytochrome *c* existed in the oxidized state at an applied potential of +0.04 V. Since this potential is more than 0.2 V negative of the formal potential of cytochrome *c*, it would be expected that adsorbed cytochrome *c* should be reduced. If, however, there is limited denaturation at this applied potential, the formal potential may be shifted to more negative values. In the previous discussion of the adsorption of cytochrome *c* on mercury[102] loss of the methionine-80 ligand bond, some opening of the heme crevice, and increased heme exposure to solvent result in a negative shift in the formal potential of cytochrome *c* (see Figure 17).

Haladjian *et al.*[100] studied the adsorption of cytochrome *c* on platinum. When a platinum electrode was first exposed to a cytochrome *c* solution and then washed with water the electrochemical response of Fe(II)/Fe(III) was inhibited. This result indicates that cytochrome *c* adsorbs strongly enough to resist removal on washing with water.

No adsorption of cytochrome *c* was evident at gold in a.c. impedance measurements reported by Eddowes *et al.*[112]. No detectable change in the electrode double layer capacitance was found upon addition of cytochrome *c* to an electrolyte solution.

Yeh and Kuwana[115] were the first to report on the electrochemistry of cytochrome *c* at doped metal oxide semiconductor electrodes. A nearly reversible electrode reaction was indicated by the cyclic voltammetry and differential pulse voltammetry of cytochrome *c* at tin-doped indium oxide electrodes. Except for the calculated diffusion coefficient, all of the characteristics of the electrochemistry of cytochrome *c* at this electrode indicated that the electrode reaction was well-behaved. A value of 0.5×10^{-6} cm^2/s was determined for the diffusion coefficient which, like previously determined values at mercury,[87,89] is lower than the value obtained by nonelectrochemical methods (i.e., 1.1×10^{-6} cm^2/s[116,117]). The electrochemical response of cytochrome *c* at tin oxide semiconductor electrodes was reported to be quasi-reversible, although no details were given.[115]

The heterogeneous electron transfer kinetics of cytochrome *c* at tin-doped indium oxide and fluoride-doped tin oxide optically transparent electrodes

(OTEs) were determined by spectroelectrochemical techniques.[118-120] Using the new technique of derivative cyclic voltabsorptometry (DCVA)[121] the heterogeneous electron transfer kinetics for the reduction of ferricytochrome *c* were found to be quasi-reversible while the oxidative electrode reaction exhibited irreversible kinetics.[118] The reductive reaction exhibited a formal heterogeneous electron transfer rate constant of 2.2×10^{-6} cm/s and an electrochemical transfer coefficient of 0.31.[118] However, the oxidative electrode reaction did not agree with these kinetic values indicating that the electrode reaction did not behave in accordance with Butler–Volmer theory.[122,123] The same results were obtained for the reaction of cytochrome *c* at tin oxide and indium oxide OTEs using single potential step chronoabsorptometry (SPS/CA)[124,125] and asymmetric double potential step chronoabsorptometry (ADPS/CA)[118] again indicating a lack of agreement with Butler–Volmer theory.[120]

In a recent report, it was shown that the reaction of cytochrome *c* at indium oxide OTEs exhibits an electrochemical response consistent with Butler–Volmer theory when high-quality commercial samples of cytochrome *c* are further purified.[126] The electrode reaction is quasi-reversible and the CV and DCVA results of these experiments are shown in Figure 18. These data

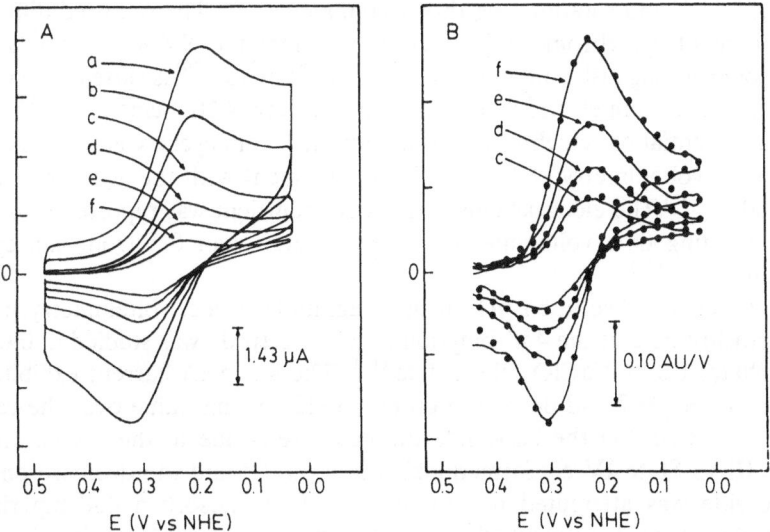

Figure 18. (A) Cyclic voltammetry of purified cytochrome *c* at doped indium oxide optically transparent electrodes. Solution contained 73 μM cytochrome *c*, 0.21 *M* Tris, 0.24 *M* cacodylic acid, pH 7.0, 0.20 *M* ionic strength. Electrode area = 0.71 cm². Potential scan rates in mV/s are: (a) 100; (b) 50; (c) 20; (d) 10; (e) 5.0; (f) 2.0. (B) Derivative cyclic voltabsorptometry of purified cytochrome *c* at a tin-doped indium oxide optically transparent electrode. Same conditions as described above. Circles are calculated derivative cyclic voltabsorptometric responses for 73 μM cytochrome *c*, formal potential = 0.260 V, *n* = 1.0, diffusion coefficient of oxidized and reduced cytochrome $c = 1.2 \times 10^{-6}$ cm²/s, difference molar absorptivity at 416 nm = 57,000 M^{-1} cm⁻¹, formal heterogeneous electron transfer rate constant = 1.0×10^{-3} cm/s, and electrochemical transfer coefficient = 0.5. Adapted from Reference (126) with permission.

afforded a formal heterogeneous electron transfer rate constant of 1×10^{-3} cm/s and an electrochemical transfer coefficient of 0.5 and the calculated diffusion coefficient for cytochrome c agreed with nonelectrochemically determined values. These results strongly suggest that the correct interpretation of cytochrome c electron transfer kinetics and mechanisms from electrochemical results requires the use of highly purified samples.

Chemical modification of electrode surfaces has been demonstrated to enhance the rates of electron transfer with cytochrome c compared to bare electrodes. Hill's group has been highly successful with gold electrodes on which 4,4'-bipyridine is adsorbed.[110–112,127–129] Adsorption of 1,2-bis(4-pyridyl)ethylene on gold results in equally facile electrochemical responses[110,111,130] for cytochrome c. The following discussion is restricted to the use of 4,4-bipyridine since this case has been more extensively studied, but may apply as well to the 1,2-bis(4-pyridyl)ethylene/gold electrode. The modified surface is formed by immersing a polished gold electrode in aqueous electrolyte containing 5 to 10 μM 4,4'-bipyridine together with cytochrome c. A formal heterogeneous electron transfer rate constant of 1.5×10^{-2} cm/s has been determined for the reaction of cytochrome c at this surface using a.c. impedance and rotating disk electrode methods.[129,130] This electrode reaction has also been studied by cyclic voltammetry,[110,111] a.c. cyclic voltammetry,[110,111] and rotating ring-disk voltammetry.[129] The reported diffusion coefficient of cytochrome c from cyclic voltammetry, 9.4×10^{-7} cm^2/s,[111] and from rotating disk voltammetry, 1.1×10^{-6} cm^2/s,[129] is close to the value determined by nonelectrochemical methods, 1.16×10^{-6} cm^2/s.[116,117] The formal potential of cytochrome c as determined from cyclic voltammetry and a.c. cyclic voltammetry, +0.255 V, is in agreement with the accepted value, +0.260 V.[131] The electrochemical transfer coefficient was reported to be 0.5 from rotating disk voltammetry[129] but greater than 0.5 from a.c. cyclic voltammetry.[111]

The effect of solution pH on the magnitude of a.c. voltammetry peaks for cytochrome c at the 4,4'-bipyridine/gold electrode was studied in 0.02 M phosphate, 1.0 M NaClO$_4$ electrolyte.[111] The a.c. peak current exhibited a maximum at pH 7 and decreased on either side of this value over the range pH 5 to pH 10. On the basic side the decrease is due to the cytochrome c State III to State IV conformational transition[131] and the decrease on the acidic side was attributed to protonation of the adsorbed 4,4'-bipyridine molecules. A pK for the alkaline transition was given as 9.1 and it was noted to be close to the previously reported value of 9.3.[131] However, this value is different from the values of 7.8 to 8.0 previously reported for perchlorate-containing solutions[102,108] and the reason for this difference is not clear.

The effect of the 4,4'-bipyridine/gold interface on the electron transfer reactions of cytochrome c is of considerable interest. The current view pictures 4,4'-bipyridine molecules adsorbed on the gold electrode in a perpendicular, end-on orientation, forming a monolayer. Cytochrome c molecules reversibly bind to the adsorbed layer via hydrogen bonding of surface lysine residues to nitrogen lone pairs of the 4,4'-bipyridine molecules. This interaction would then orient the cytochrome c molecules so that their heme edges point towards the electrode enhancing electron transfer.[128,129]

A detailed study by Albery *et al.*[129] of the reaction of cytochrome c at the 4,4'-bipyridine/gold electrode using rotating disc voltammetry and a.c. modulated rotating ring-disc voltammetry led to the proposal of the following mechanism

$$A_\infty \underset{k_D}{\overset{k_D}{\rightleftharpoons}} A_0 \tag{7}$$

$$A_0 \underset{\Gamma_L k_{des,A}}{\overset{k_{ads,A}}{\rightleftharpoons}} A_{ads} \tag{8}$$

$$A_{ads} \underset{\Gamma_L k_{-e}}{\overset{\Gamma_L k_e}{\rightleftharpoons}} B_{ads} + e \tag{9}$$

$$B_{ads} \underset{k_{ads,B}}{\overset{\Gamma_L k_{des,B}}{\rightleftharpoons}} B_0 \tag{10}$$

$$B_0 \underset{k_D}{\overset{k_D}{\rightleftharpoons}} B_\infty \tag{11}$$

where Γ_L (mol/m^2) is the surface concentration of cytochrome c adsorption sites on the modified electrode, k_D (m/s) is the mass transport rate constant at the rotating disc electrode, $k_{ads,A}$ and $k_{ads,B}$ (cm/s) are adsorption rate constants, $k_{des,A}$ and $k_{des,B}$ (s^{-1}) are desorption rate constants, and k_e and k_{-e} (s^{-1}) are potential-dependent rate constants for the forward and backward electron transfer reactions between adsorbed molecules and the electrode. Adsorption of both redox forms of cytochrome c exhibited fair agreement with Langmuir behavior. Using transition state theory a free energy profile corresponding to Eqs. (7) to (11) was devised. The free energy of activation for adsorption of either redox form of cytochrome c was given as 31 kJ/mol, and for desorption and electron transfer of cytochrome c a value of 63 kJ/mol was cited. The profile is symmetrical with equal adsorption energies for both redox forms of cytochrome c. Albery *et al.*[129] have suggested that the mechanism given by Eqs. (7) to (11) is general for electron transfer reactions of metalloproteins at electrodes and that considerable binding energy for Eq. (8) is necessary to overcome the activation energy.

Although the above mechanism is consistent with the cytochrome c reaction at 4,4'-bipyridine/gold electrodes it is not yet possible to judge the general nature of its applicability due to the unavailability of detailed studies with different proteins and electrodes. Electron transfer proteins such as cytochrome c_3[132,133] appear likely to have low energies of activation, and reversible binding to a surface would, in that case, play a lesser role than with a molecule such as cytochrome c, whose heme is extremely well protected.

Nevertheless, the proposal that reversible binding plays a major role in many electrode reactions of biological molecules is an important contribution to the understanding of these reactions.

In a recent report, it was demonstrated that adsorption of 4,4'-bipyridine on platinum led to quasi-reversible rates of electron transfer with cytochrome c as evidenced by cyclic voltammetry.[134] However, the concentration of 4,4'-bipyridine required to produce this electrochemical response was five times that which is required at gold electrodes. This difference was ascribed to the difference in the tendency of 4,4'-bipyridine to adsorb at gold and platinum electrodes. These results[134] indicate that the use of 4,4'-bipyridine may be applicable to other solid electrodes as well for the study of cytochrome c electron transfer reactions.

Another type of chemically modified electrode which has been used to study the heterogeneous electron transfer reactions of cytochrome c was described by Landrum et al.[135] This electrode surface is prepared by applying a potential of −0.73 V to a gold electrode in contact with a solution of 1,1'-dimethyl-4,4'-bipyridinium dichloride (methyl viologen) at neutral pH. A reaction occurs which results in the irreversible formation of a layer on the gold with complete loss of the original redox properties of the reactant. The molecular structure of the surface layer and the reaction mechanism leading to its formation are not known. The surface layer has been studied by scanning electron microscopy,[135,136] secondary ion mass spectrometry and electron spectroscopy for chemical analysis (ESCA),[137] and spectroelectrochemistry.[138] ESCA revealed considerable amounts of carbon and oxygen in the surface layer but no nitrogen[137] indicating loss of nitrogen from the pyridinium rings of the reactant. It appears that atomic hydrogen, surface adsorbed on the gold, is involved in the electrochemically driven reaction which leads to the formation of the surface layer.[138]

This chemically modified electrode exhibits quasi-reversible rates of heterogeneous electron transfer for the reduction of cytochrome c as determined by single potential step chronoabsorptometry.[120] However, reductive and oxidative experiments did not yield kinetic results in agreement with Butler–Volmer theory.

One other type of chemically modified electrode has been used in the study of cytochrome c, as described by Lewis and Wrighton.[113] In this work platinum, gold, and p-type silicon substrates were modified with a polymeric overlayer which contained bipyridinium-type one-electron redox centers. These immobilized mediators transfer electrons between the electrode surface and diffusing ferricytochrome c at the potential of the mediators which is ca. 0.59 V more negative than the formal potential of cytochrome c. Oxidation of ferrocytochrome c is precluded because of the difference in the formal potentials of cytochrome c and the immobilized mediator. This same group has recently immobilized 2,3,4,5-tetramethyl-1-(dichlorosilylmethyl)-[2]-ferrocenophane on platinum.[139] The juxtaposition of the formal potentials of cytochrome c and this immobilized mediator, 0.260 V and 0.286 V vs. NHE, respectively, permits the reduction and oxidation of cytochrome c at this electrode surface.

4.2.3. Other Cytochromes

The heterogeneous electron transfer characteristics of other cytochromes have been investigated. Cytochrome c_3 is the most widely studied while some results are available for cytochrome c_7, cytochrome b_5, and cytochrome P450. The latter two cytochromes will be discussed first.

Cytochrome b_5 is a membrane-bound electron transfer protein with a formal potential of 0.020 V vs. NHE.[140,141] It functions in the desaturation of fatty acids in mammalian cells, especially in the liver, and it has been associated with cytochrome P450 systems, erythrocytic methemoglobin reduction, and mitochondria. One end of the cytochrome b_5 molecule is highly hydrophobic and anchors it to the membrane while the other end is hydrophilic and encases a single protoheme. The heme has two histidine axial ligands and is highly isolated from solvent. There has been one report of the direct reduction of cytochrome b_5.[98] This work utilized cytochrome b_5 samples prepared both by detergent solubilization, which leads to the native molecule, and by trypsin solubilization, which removes part of the hydrophobic end of the molecule. The identity of the sample preparation was not always indicated and the sample purity was cited as *ca.* 50–60%.

Cytochrome b_5 was found to be reducible at the DME at an overpotential of *ca.* −0.26 V for 30 μM concentration in 0.05 M phosphate buffer, pH 7.0. From 70 to 300 μM the diffusion current was linear with concentration but the magnitude was much lower than predicted. At a vibrating nickel electrode trypsin-isolated cytochrome b_5 was reduced at an overpotential of *ca.* −0.78 V while the detergent-isolated sample required *ca.* −1.08 V. Under the same experimental conditions, cytochrome c was reduced at an overpotential of *ca.* −0.62 V.

Cytochrome P450 is also membrane bound and is known to function in the metabolism of drugs in mammalian liver microsomes.[142] It has a molecular weight of from 44,000 to 55,000, one protoheme group in a single polypeptide chain, and is difficult to isolate because of its membrane association. Scheller *et al.* reported 10% reduction of partially purified cytochrome P450 at a mercury pool electrode at −0.76 V.[143] A polarographic wave at $E_{1/2} \simeq$ −0.34 V was tentatively assigned to the reduction of cytochrome P450. However, Kuznetsov *et al.* found no polarographic wave for a sample of 90–95% purity.[98] No reduction of cytochrome P450 was reported at nickel[98] or carbon.[143] From these limited data it is clear that cytochrome P450 is directly reduced at a much slower rate than other cytochromes studied to date.

Cytochrome c_3 and c_7 undergo direct electrochemical reduction at extremely fast rates compared to other cytochromes. Cytochromes c_3 has four heme groups imbedded in a single polypeptide chain with a molecular weight of ca. 12,000.[144] The hemes in cytochrome c_3 exhibit considerable solvent exposure compared to other cytochromes and the X-ray crystal structures of two cytochromes c_3 molecules were recently reported.[145,146] Cytochrome c_3 molecules are found in sulfate-respiring bacteria and have formal potentials of approximately -0.20 to -0.30 V vs. NHE.

Yagi et al.[147] were the first to describe the direct reduction of cytochrome c_3 and a glassy carbon electrode was employed in that work. Niki's group[148-151] and Haladjian's group[152-154] have subsequently studied cytochrome c_3 and their results are in good agreement with each other. The direct reduction of cytochrome c_3 at mercury is very fast as determined by normal pulse polarography,[148,149] differential pulse polarography,[149,152] scan reversal pulse polarography,[149] and cyclic voltammetry.[148,149,153] Niki et al.[149] have estimated the lower limit for the formal heterogeneous electron transfer rate constant to be 0.1 cm/s for cytochrome c_3 from Desulfovibrio vulgaris, strain Miyazaki, and the diffusion coefficient of ferricytochrome c_3 was determined from normal pulse polarography to be 0.94×10^{-6} cm^2/s.

Sokol et al.[150] showed that the voltammetric response of cytochrome c_3 arises from four relatively independent hemes with closely spaced formal potentials. Digital simulation results fitted to experimental data indicated four independent heme redox centers having formal potentials of -0.226, -0.278, -0.298, and -0.339 V. Each heme is apparently chemically unique and does not interact significantly with the other hemes in electron transfer.

Bianco and Haladjian[154] reported similar results for cytochrome c_3 from Delsulfovibrio desulfuricans, strain Norway. They resolved four theoretical one-electron processes from the experimental responses obtained by differential pulse polarography and linear sweep voltammetry. These results are shown in Figure 19. The four formal potentials were given as -0.168, -0.306, -0.364, and -0.398 V. The first one-electron reduction step appears as a separate peak in the differential pulse polarogram and as a shoulder in the linear sweep voltammogram as shown in Figure 19. These authors used Stellwagen's[155] relationship between heme solvent exposure and redox potential, since the X-ray crystal structure for this cytochrome c_3 is known,[145] and assigned formal potentials to each individual heme. It will be very interesting to see if this prediction is supported by future work using this approach for other systems.

There is considerable evidence that cytochrome c_3 strongly adsorbs on mercury. Mercury drops from a DME formed in solutions containing cytochrome c_3 do not coalesce for 30 minutes[148] and electrocapillary curves are lowered.[152] At the hanging mercury drop electrode cyclic voltammetry of cytochrome c_3 from Desulfovibrio vulgaris, strain Hildenborough, and also from Desulfovibrio desulfuricans, strain Norway, gave further evidence of adsorption.[153] Each sample exhibited a peak at ca. 0 V characteristic of adsorption, i.e., current was directly proportional to scan rate. Film transfer

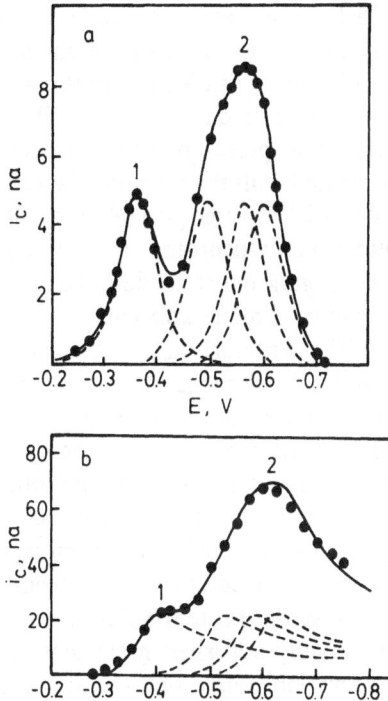

Figure 19. Voltammetric responses of cytochrome c_3 from *D. desulfuricans*, strain Norway. Solution contained 18.4 μM cytochrome c_3, 0.01 M *Tris*-HCl buffer pH 7.6. (a) Differential pulse polarogram, (——) experimental result, (●) calculated points, (---) theoretical differential pulse polarograms for four monoelectronic reversible electron transfer reactions. (b) Linear sweep voltammogram, same conditions as above. Adapted from Reference (154) with permission.

experiments employing a hanging mercury drop electrode which was exposed to a cytochrome c_3 solution, rinsed, and then immersed in buffer showed a peak at *ca.* 0 V and a second more negative peak.[153] These experiments indicate strong adsorption by cytochrome c_3 and the peak potentials suggest that structural changes occur upon adsorption.

Haladjian's group[152] studied these same cytochrome c_3 molecules by differential pulse polarography. An adsorption prepeak was linear with increasing concentration up to a concentration of 4 μM and 7 μM for *D. vulgaris* and *D. desulfuricans*, respectively, at an 0.5-s drop time. Above these concentrations the amplitude of the prepeak leveled off, indicating that a primary adsorption monolayer had been formed.

Differential capacitance measurements by Niki *et al.*[151] for cytochrome c_3 from *D. vulgaris*, strain Miyazaki, were consistent with irreversible, diffusion-limited adsorption for 4-s drop times above a concentration of 10 μM. The surface excess of cytochrome c_3 was calculated to be 0.92 × 10^{-11} mole/cm^2. Niki *et al.*[151] also investigated the a.c. polarographic behavior of cytochrome c_3 at the reversible half-wave potential. The capacitive peak height was frequency independent while the resistive peak height decreased with increasing frequency to a value of zero above 2000 Hz. These results were fit to a Laitinen–Randles[156] equivalent circuit yielding an *n* value of

0.75 for the electrode reaction of adsorbed cytochrome c_3. This low value may be due to the assumption that concentrations of oxidized and reduced cytochrome c_3 molecules in the adsorbed layer are equal since the measurements were made at the reversible half-wave potential. Previously described differential pulse polarographic results indicated that the potentials of peaks associated with the adsorbed layer are removed from the reversible half-wave potential[152] which possibly accounts for this discrepancy. Having four separate hemes in cytochrome c_3 may also add complications. Nevertheless, this encouraging effort to describe protein adsorption in terms of an equivalent circuit is a unique approach.

The structure of cytochrome c_3 adsorbed on mercury is unknown. Niki *et al.*[151] have suggested that only a small degree of denaturation is involved, but there is a lack of unambiguous, direct evidence concerning this point. The role that adsorbed cytochrome c_3 plays in the reduction of diffusing molecules at mercury electrodes is undoubtedly a central one. Niki *et al.*[151] have proposed that the reduction of bulk ferricytochrome c_3 occurs at sites occupied by adsorbed cytochrome c_3 molecules. In other words, electron transfer to diffusing molecules does not occur at the mercury surface in this model, but rather at the protein/solution interface via electron movement through the irreversibly adsorbed molecules. This proposal is supported by at least two observations. First, polarographically evaluated diffusion coefficients for cytochrome c_3 exhibit reasonable values, in contrast to the low values previously discussed for cytochrome *c*. And second, if mixed primary adsorption layers are formed from cytochrome c_3 and cytochrome *c*, the cytochrome c_3 diffusion-limited reduction current decreases as the fraction of cytochrome c_3 in the adsorption layer decreases.[151] Thus it appears that the model proposed by Niki *et al.*[151] is likely to be correct for cytochrome c_3 at mercury, although such mechanistic proposals must be viewed with caution at this point.

A striking feature of the electrochemistry of cytochrome c_3 is its extremely fast electron transfer rate at gold electrodes. Bianco *et al.*[153] demonstrated that the cyclic voltammetric behavior of *D. vulgaris* and *D. desulfuricans* cytochrome c_3 at gold disc electrodes was reversible up to the highest scan rate employed, 200 mV/s. One difference between the results obtained at gold compared to those obtained at mercury is the absence of an adsorption prepeak at gold. This difference has not been studied in detail to date. Singleton *et al.*[157] also reported facile electron transfer for cytochrome c_3 from *Desulfovibrio africanis*, strain Benghazi, at gold electrodes.

Eddowes *et al.*[158] reported nearly reversible electron transfer kinetics for the reaction of cytochrome c_3 from *Desulfovibrio desulfuricans*, strain Norway, at 4,4'-bipyridine/gold electrodes. The morphology of these cyclic voltammetric results was different from those reported for bare gold electrodes.[153] This difference may be due to use of different electrolytes, pH 7.6 Tris-HCl[153] compared to pH 7 phosphate/NaClO$_4$,[158] or to the different electrode surfaces.

Bianco and Haladjian[159] recently reported the polarographic behavior of cytochrome c_7, also denoted $c_{551.5}$, from *Desulfuromonas acetoxidans*. This molecule contains three hemes, has a molecular weight of *ca.* 9,800 daltons, and a formal potential of *ca.* −0.19 V. Their results indicated that the electrochemical responses of cytochrome c_3 and cytochrome c_7 are very similar.

4.2.4. Myoglobin and Hemoglobin

The electrochemical behavior of myoglobin and hemoglobin has been extensively studied, particularly at the mercury electrode. Both molecules serve to store and transport dioxygen when their heme irons are in the reduced state. Interest in the redox chemistry of the heme irons in these molecules has in part prompted electrochemical studies of myoglobin and hemoglobin.

Scheller's group[88–92,94,160,161], especially, and also Kuznetsov's[97,98] have provided the available polarographic information for hemoglobin and myoglobin. A major finding of these studies is that the polarographic electron transfer kinetics of hemoglobin and myoglobin are much more irreversible that those observed for cytochrome *c*. Hemoglobin and myoglobin are reduced at mercury with overpotentials of *ca.* −0.4 V[89,92,97] and −0.8 V[89,91,92,161], respectively. As in the case of cytochrome *c*, adsorption phenomena participate in these electrode reactions. The models described for cytochrome *c* which were discussed earlier were also developed to account for the behavior of hemoglobin and myoglobin.

The reduction of myoglobin at mercury is strongly influenced by pH. No reduction is evident at pH 7, a maximum response occurs at pH 5, and again no reduction is evident below pH 4.5.[89,161] The half-wave potential is nearly constant in the pH range of 4.5 to 5.5. The diffusion-limited current is dependent upon pH and myoglobin concentration. At pH 6 the diffusion-limited current is linearly dependent on myoglobin concentration between *ca.* 60 and 130 μM.[89,91,161] The maximum diffusion-limited current observed at pH 5 agrees with the accepted value for the diffusion coefficient for metmyoglobin.[89] This result differs from that which has been observed for the reduction of cytochrome *c* at mercury as discussed earlier (i.e., the experimental diffusion coefficient was low).

The polarographic reduction of methemoglobin exhibits a well-defined wave at an overpotential of *ca.* −0.4 V. The earliest report of methemoglobin reduction at mercury indicated partial reduction of the tetrahemic moiety.[162] However, later work by Scheller *et al.*[89] indicated complete reduction of methemoglobin at mercury. In contrast to myoglobin, both the half-wave potential and the diffusion-limited current for methemoglobin reduction were independent of pH over the range 5.5 to 8.[89] The half-wave potential shifted to more negative values with increasing methemoglobin concentration.[92] Diffusion-limited currents and differential pulse polarographic peak currents were linearly dependent on methemoglobin concentration between 10 and 100 μM.[88,89] The polarographically determined diffusion coefficient was smaller than the value determined by nonelectrochemical methods[89] as has been observed for cytochrome c.

Evidence for the adsorption of metmyoglobin and methemoglobin on mercury electrodes is consistent with irreversible, diffusion-controlled adsorption during the formation of the primary monolayer. The same general experimental strategy has been used to characterize these processes[81,89,91,160] as was used in the study of cytochrome c. Scheller et al.[88,90,94] have used differential capacitance data to compute the area occupied by single adsorbed molecules. The most recent values cited for metmyoglobin and methemoglobin are 3500 Å^2 and 11,400 Å^2, respectively.[94] These values are much larger than the maximum areas calculated from the crystal structures for the surface of a rectangular prism (i.e., 1600 Å^2 and 4500 Å^2 for myoglobin and hemoglobin, respectively) and can suggest protein unfolding. As noted earlier, Scheller[94] has cautioned against using such data as unambiguous evidence for protein unfolding in the adsorbed state since hydration of the adsorbed molecules is ignored. Kuznetsov et al.[98] concluded from film transfer studies that adsorbed methemoglobin forms a flattened primary layer of adsorbed molecules together with overlayers of reversibly adsorbed native molecules.

Few reasons have been given as to why the electron transfer behavior of metmyoglobin and methemoglobin differs significantly from that of cytochrome c. Scheller and Prümke[92] have suggested that these molecules exhibit much slower rates of electron transfer at mercury compared to cytochrome c due to the lack of exposure of the heme edges at the molecules' surfaces. It is not clear why myoglobin, but not hemoglobin, is polarographically reduced in a strongly pH-dependent manner.

Two solid electrodes have been used to study the heterogeneous electron transfer properties of myoglobin. Stargardt et al.[136] demonstrated that the methyl viologen-modified gold electrode[135] described earlier catalyzes the reduction and reoxidation of metmyoglobin. Cotton et al.[114] have studied the adsorption behavior of myoglobin on bare silver electrodes using surface-enhanced resonance Raman spectroscopy. No reports dealing with hemoglobin redox behavior at solid electrodes have been found.

At the methyl viologen-modified gold electrode the reduction and reoxidation of neutral pH metmyoglobin solutions occurs in 5 and 15 minutes, respectively.[136] Repetitive reduction/oxidation cycles were reproducible and no denaturation was evident in the 350 to 600 nm adsorption region. Using single potential step chronoabsorptometry, Bowden et al.[163] subsequently determined the heterogeneous electron transfer kinetic parameters for the reduction of metmyoglobin at this electrode. This reaction was found to be highly irreversible. The method of channel flow hydrodynamic voltammetry has also been applied to the characterization of the heterogeneous electron transfer kinetic parameters for the reduction and oxidation of myoglobin at the methyl viologen-modified gold electrode by Castner and Hawkridge.[164] This steady-state kinetic method yielded a formal heterogeneous electron transfer rate constant which was larger than that obtained by single potential step chronoabsorptometry[163] but the reaction was still irreversible. The difference between the kinetic parameters obtained for the reaction of myoglobin by these two techniques is probably due to the differences in the methods, i.e., transient versus steady state. Castner's film transfer experiments[165] and Stargardt's electron spectroscopy for chemical analysis experiments[137] indicated that myoglobin adsorbs on the methyl viologen-modified gold electrode surface.

Cotton *et al.*[114] employed surface-enhanced resonance Raman spectroscopy (*vida supra*) to unequivocally show that both myoglobin and cytochrome *c* adsorb on bare silver electrodes. However, in contrast to cytochrome *c*, the surface Raman spectrum of myoglobin showed no clear correlation with the solution Raman spectrum. It was suggested[114] that, upon adsorption, myoglobin undergoes a spin-state change from high to low and that the adsorbed molecules exist reduced at -0.36 V and oxidized at $+0.040$ V, close to its formal potential of $+0.046$ V.[1] These results indicate that myoglobin adsorption on silver leads to denaturation.

4.2.5. Peroxidase and Catalase

Peroxidase and catalase form another class of heme proteins[166,167] that function as catalysts in the decomposition of hydrogen peroxide to water through the oxidation of a second substrate. Various peroxidases utilize different substrates and exhibit differing degrees of structural variation.[166] All peroxidases contain one heme functionality which is not covalently bound to the polypeptide chain and have molecular weights which vary between 35,000 and 60,000. The axial sites of the heme iron are thought to be occupied by a histidine imidazole nitrogen and a water molecule, although this point is not settled.[166] Catalase is a tetramer with a molecular weight of 60,000 and axial heme coordination sites are thought to be similar to peroxidase[167] in their ligation.

There have been few direct electrochemical studies of peroxidase and catalase due to the highly irreversible nature of these electrode reactions. Horseradish peroxidase was found to be electroinactive at the dropping mercury electrode.[97] Tarasevich and co-workers[109,168] observed a cyclic voltammetric response for the electron transfer of horseradish peroxidase at an amalgamated gold electrode. However, this response was ascribed to the disulfide bonds of the protein at neutral pH and not to the heme group. No response was detected at pyrolytic graphite electrodes.[168]

In an earlier study Brown[169] reported that both horseradish peroxidase and beef liver catalase were electroinactive polarographically under anaerobic conditions. However, under conditions which gave rise to the primary hydrogen peroxide complexes of these proteins, a reductive response was observed. Brown attributed these responses to the direct reduction of the respective hydrogen peroxide complexes of peroxidase and catalase.[169] In one other report catalase was reported to be electroinactive at the dropping mercury electrode.[97]

4.2.6. Other Biological Molecules

Many other proteins have been studied electrochemically. This section is not exhaustive and seeks to focus on those biological molecules which have been more extensively characterized.

Ferredoxins function physiologically as electron transfer agents and contain iron–sulfur prosthetic groups.[170] They typically undergo one-electron redox reactions at negative formal potentials. Voltammetric studies of ferredoxins at mercury electrodes have been characterized by strong adsorption accompanied by major denaturation.[109,171,172] Sulfur atom interactions with mercury are known to be involved in these processes.[109]

Turning to solid electrodes, Landrum et al.[135] found that spinach ferredoxin could be reduced and reoxidized repetitively at quasi-reversible rates at the methyl viologen-modified gold minigrid electrode. The formal potential indicated by cyclic voltammetry was approximately the same as the formal potential of ferredoxin, -0.428 V, indicating that the electrode reaction involved native molecules. Crawley and Hawkridge[173] used single potential step chronoabsorptometry to determine the heterogeneous electron transfer kinetics for the reduction of ferredoxin at this electrode surface using irreversible electron transfer theory.[124] These data were reevaluated using the more recent theory[125] which applies to heterogeneous electrode reactions exhibiting any degree of reversibility.[174] The formal heterogeneous electron transfer rate constant was found to be 6.5×10^{-5} cm/s and the electrochemical transfer coefficient was 0.42.

Polarographic reduction of the flavin adenine dinucleotide (FAD) prosthetic group of certain proteins has received some attention. These proteins transfer hydrogen atoms from organic molecules to the FAD riboflavin segment, which is an organic moiety.[175] Several FAD-containing proteins exhibit reduction potentials of $ca.$ -1.0 V in pH 7 media including D-amino acid oxidase,[97] glucose oxidase,[176,177] and cholesterol oxidase.[177] Protein adsorption is evident[97,177] and Ikeda et al.[177] have proposed that the voltammetric peaks mentioned above arise from reduction of irreversibly adsorbed FAD groups. In the case of cholesterol oxidase there is a difference between the reduction potential for adsorbed and solution-resident molecules indicating that surface denaturation is occurring.[177] Scheller et al.[176] have suggested that the reduction potential for glucose oxidase adsorbed on the electrode and resident in solution are the same which indicates that upon adsorption no denaturation occurs. Xanthine oxidase appears to behave differently from the other FAD proteins discussed above. The reduction potential for this molecule is the same as that of free FAD, namely, -0.19 V.[97] This led Kuznetsov *et al.*[97] to conclude that upon adsorption xanthine oxidase releases FAD which is then observed polarographically in the free form.

The direct electrochemical reduction and oxidation of bacteriorhodopsin, a bacterial photoreceptor protein, have recently been described.[178,179] The redox site of this molecule is organic, being a conjugated double bond system and a Schiff base. As previously noted for other proteins, strong adsorption onto mercury is evident[178,179] with bacteriorhodopsin. The adsorbed molecules undergo reduction–oxidation reactions near -0.8 V[178,179] which appear to be polarographically reversible.[179] An oxidation wave is observed at platinum at $+0.80$ V and has been ascribed to the chromophore.[179]

There are four additional sections of literature, which are not concerned primarily with the electron transfer between electrodes and protein prosthetic groups, but which are potentially germane to such studies. These areas include Brdicka currents, disulfide/sulfhydryl reactions, protein oxidations, and nucleic acid electrochemistry. Brdicka hydrogen currents arise in suitably buffered cobalt-containing solutions at the dropping mercury electrode and are enhanced by certain proteins. These phenomena have been studied for nearly 50 years and they remain mechanistically challenging.[180] As described in an early review,[181] much work on Brdicka currents was directed at the development of a polarographic test specific for cancer but this thrust has been abandoned. The recent fundamental work in this area from the groups of Kuznetsov[95,96] and Senda[103,105,182] has provided considerable insight into protein adsorption processes at mercury electrodes.

The existence of polarographically reducible disulfide bonds in proteins appears to be intimately related to many Brdicka reactions.[95,96,180–182] The reduction of disulfides to sulfhydryls at mercury electrodes, and the reverse process, have also been studied extensively.[183] Insulin[184] and other disulfide-containing proteins give rise to polarographic currents characteristic of this reaction.

Individual amino acids in proteins undergo oxidation reactions at potentials on the positive side of *ca.* 0.7 V. Tryptophan and tyrosine residues are oxidizable at carbon,[185–188] platinum,[187,188] and gold[187,188] electrodes while histidine is oxidizable at carbon.[187,188] The sulfur-containing amino acids cysteine, cystine, and methionine have been shown to be oxidized at carbon.[186,188,189] Adsorption processes are important in these reactions.

An extensive literature[104,180] describes the polarography of nucleic acids. These studies have been basically concerned with the effects of the interfacial electric field on the conformational structure of nucleic acids and this subject is treated in Chapter 3 of this volume.

5. Conclusions

Understanding biological electron transfer reactions requires elucidation of the molecular events associated with these processes. The broad interdisciplinary studies which have addressed these problems have provided extensive information regarding these crucial reactions. However, from the preceding material it is clear that many aspects of biological electron transfer are not understood. The ambitious task of solving any of these problems will require scientific inquiry from a variety of quarters. The application of electrochemical techniques to problems in bioelectrochemistry holds promise for contributing unique information and thereby additional insights regarding these issues. The thermodynamics, kinetics, and mechanisms attendant to physiological electron transfer reactions are highly appropriate for attack by bioelectrochemical technqiues.

The recent advances in electrochemical techniques, particularly when coupled to complementary physical or chemical probes, point to a fertile furture for the field of bioelectrochemistry. Techniques which provide molecularly specific information with time resolution approaching that required to monitor physiological electron transfer events are becoming increasingly available in a wide range of disciplines.

The thermodynamics associated with a broad range of bioelectrochemical reactions have been established with a high degree of agreement between experimental approaches. These data have provided the framework for understanding the macroscopic events associated with these reactions. Attention is increasingly being focused on the microscopic events which provide the reaction specificity in physiological electron transfer processes. While fundamental information regarding these aspects of bioenergetics will not necessarily arise from the use of more demanding experimental techniques compared to more basic approaches, it is clear that progress will be best realized by contributions from both quarters.

Acknowledgements

We gratefully acknowledge support from the following agencies during the writing of this chapter and for support of the work done in these laboratories which is described herein: National Science Foundation (PCM 79-12348), National Institutes of Health (GM 27208-02), the University of Delaware Institute of Neuroscience (NIH Biomedical VII), and the North Atlantic Treaty Organization. FMH wishes to thank the Department of Chemistry, University of Delaware, for time and support in the form of a sabbatical during which his contribution to this chapter was written.

References

1. W. M. Clark, *Oxidation-Reduction Potentials of Organic Systems*, Williams and Wilkins Pub. Co., Baltimore (1960).
2. M. Ito and T. Kuwana, *J. Electroanal. Chem.* **32**, 415 (1971).
3. T. Kôno and S. Nakamura, *Bull. Agr. Chem. Soc. Japan* **22**, 399 (1958).
4. D. T. Sawyer, ed., *Electrochemical Studies of Biological Systems*, American Chemical Society Symposium Series 38, Washington, D.C. (1977).
5. G. Dryhurst, *Electrochemistry of Biological Molecules*, Academic Press, New York (1977).
6. M. Blank, ed., *Bioelectrochemistry: Ions, Surfaces, Membranes*, American Chemical Society Advances in Chemistry Series 188, Washington, D.C. (1980).
7. G. S. Wilson, in *Methods in Enzymology* Vol. 54, S. Fleisher and L. Packer, eds., Academic Press, New York (1978), pp. 396–410.
8. K. M. Kadish, ed., *Electrochemical and Spectrochemical Studies of Biological Redox Components*, American Chemical Society Advances in Chemistry Series 201, Washington, D.C. (1982).
9. A. J. Bard and L. R. Faulkner, *Electrochemical Methods*, John Wiley & Sons, New York (1980).
10. J. O'M. Bockris, B. E. Conway, and E. Yeager, eds., *Comprehensive Treatise of Electrochemistry*, Vol. 1, Plenum Press, New York (1980).
11. J. O'M. Bockris and A. K. N. Reddy, *Modern Electrochemistry*, Plenum Press, New York (1970), Chapter 7.
12. R. Parsons, *J. Electroanal. Chem.* **118**, 3 (1981).
13. B. E. Conway, *Adv. Colloid Interface Sci.* **8**, 91 (1977).
14. M. A. Habib and J. O'M. Bockris, in *Comprehensive Treatise of Electrochemistry*, Vol. 1, J. O'M. Bockris, B. E. Conway, and E. Yeager, eds., Plenum Press, New York (1980), pp. 135–219.
15. A. N. Frumkin, A. O. Petrii, and B. B. Damaskin, in *Comprehensive Treatise of Electrochemistry*, Vol. 1, J. O'M. Bockris, B. E. Conway, and E. Yeager, eds., Plenum Press, New York (1980), pp. 221–289.
16. S. R. Morrison, *Electrochemistry at Semiconductor and Oxidized Metal Electrodes*, Plenum Press, New York (1980).
17. R. Parsons and R. M. Reeves, *J. Electroanal. Chem.* **123**, 141 (1981).
18. F. Franks, in *Water—A Comprehensive Treatise*, Vol. 4, F. Franks, ed., Plenum Press, New York (1975), pp. 1–94.
19. T. Smith, *J. Colloid Interface Sci.* **75**, 51–55 (1980).
20. G. L. Gaines, *J. Colloid Interface Sci.* **79**, 295 (1981).
21. S. M. Ahmed, in *Oxides and Oxide Films*, Vol. 1, J. W. Diggle, ed., Marcel Dekker, New York (1972), pp. 319–517.
22. M. A. Omar, *Elementary Solid State Physics*, Addison-Wesley Pub. Co., Reading, Massachusetts (1975), pp. 145, 273.
23. N. R. Armstrong, A. W. C. Lin, M. Fugihira, and T. Kuwana, *Anal. Chem.* **48**, 741 (1976).
24. R. Memming, in *Electroanalytical Chemistry*, Vol. 2, A. J. Bard, ed., Marcel Dekker, New York (1979), pp. 1–84.
25. Yu. V. Pleskov, in *Comprehensive Treatise of Electrochemistry*, Vol. 1, J. O'M. Bockris, B. E. Conway and E. Yeager, eds., Plenum Press, New York (1980), pp. 291–328.
26. I. Uchida, H. Akahoshi, and S. Toshima, *J. Electroanal. Chem.* **88**, 79 (1978).
27. A. White, P. Handler, E. L. Smith, R. L. Hill, and I. R. Lehman, *Principles of Biochemistry*, 6th Edition, McGraw–Hill, New York (1978), p. 303.
28. M. K. Jain, *The Bimolecular Lipid Membrane*, Van Nostrand Reinhold, New York (1972), pp. 37–41.

29. D. E. Goldman, in *Perspectives in Membrane Biophysics*, D. P. Agin, ed., Gordon and Breach Science Pub., New York (1972), pp. 205–210.

30. A. A. Pilla, in *Bioelectrochemistry*, H. Keyzer and F. Gutman, eds., Plenum Press, New York (1980), pp. 353–396.

31. R. Pethig, *Dielectric and Electronic Properties of Biological Materials*, John Wiley & Sons, New York (1979), pp. 150–185.

32. M. K. Jain, *The Bimolecular Lipid Membrane*, Van Nostrand Reinhold, New York (1972), pp. 385–388.

33. W. Drost-Hansen and J. S. Clegg, eds., *Cell-Associated Water*, Academic Press, New York (1979).

34. H. Hauser, in *Water—A Comprehensive Treatise*, Vol. 4, F. Franks, ed., Plenum Press, New York (1975), pp. 209–303.

35. K. Minnaert, *Biochim. Biophys. Acta* **110**, 42 (1965).

36. P. L. Dutton, in *Methods in Enzymology*, Vol. 54, S. Fleischer and L. Packer, eds., Academic Press, New York (1978), pp. 411–435.

37. G. S. Wilson, J. C. M. Tsibris, and I. C. Gunsalus, *J. Biol. Chem.* **248**, 6059 (1973).

38. W. R. Heineman, B. J. Norris, and J. F. Goelz, *Anal. Chem.* **47**, 79 (1975).

39. R. W. Murray, W. R. Heineman, and G. W. O'Dom, *Anal. Chem.* **39**, 1666 (1967).

40. C. W. Anderson, H. B. Halsall, and W. R. Heineman, *Anal. Biochem.* **93**, 366 (1979).

41. W. R. Heineman, *Anal. Chem.* **50**, 390A (1978).

42. R. Szentrimay, P. Yeh, and T. Kuwana, in *Electrochemical Studies of Biological Systems*, D. T. Sawyer, ed., American Chemical Society Symposium Series 38, Washington, D. C. (1977).

43. M. L. Meckstroth, B. J. Norris, and W. R. Heineman, *Bioelectrochem. Bioenerg.* **8**, 63 (1981).

44. S. Kwee, Pteridine mediators in the electrolysis of biological macromolecules, *Chem. Biol. Pteridines, Proc. 5th. Int. Symp.*, W. Pfleiderer, ed., Berlin, 1975, pp. 671–680.

45. R. T. Salmon and F. M. Hawkridge, *J. Electroanal. Chem.* **112**, 253 (1980).

46. B. J. Norris, M. L. Meckstroth, and W. R. Heineman, *Anal. Chem.* **48**, 630 (1976).

47. V. T. Taniguchi, N. Sailasuta-Scott, F. C. Anson, and H. B. Gray, *Pure Appl. Chem.* **52**, 2275 (1980).

48. C. W. Anderson, H. B. Halsall, W. R. Heineman, and G. P. Kreishman, *Biochem. Biophys. Res. Commun.* **76**, 339 (1977).

49. G. P. Kreishman, C. W. Anderson, C.-H. Su, H. B. Halsall, and W. R. Heineman,. *Bioelectrochem. Bioenerg.* **5**, 196 (1978).

50. G. P. Kreishman, C.-H. Su, C. W. Anderson, H. B. Halsall, and W. R. Heineman, in *Bioelectrochemistry: Ions, Surfaces and Membranes*, M. Blank, ed., American Chemical Society Advances in Chemistry Series 188, Washington, D.C. (1980).

51. F. M. Hawkridge and B. Ke, *Anal. Biochem.* **78**, 76 (1977).

52. W. R. Heineman, M. L. Meckstroth, B. J. Norris, and C.-H. Su, *Bioelectrochem. Bioenerg.* **6**, 577 (1979).

53. N. Sailasuta, F. C. Anson, and H. B. Gray, *J. Am. Chem. Soc.* **101**, 455 (1979).

54. D. Lexa, J. M. Saveant, and J. Zickler, *J. Am. Chem. Soc.* **99**, 2786 (1977).

55. T. M. Kenyhercz, T. P. DeAngelis, B. J. Norris, W. R. Heineman, and H. B. Mark, Jr., *J. Am. Chem. Soc.* **98**, 2469 (1976).

56. T. M. Kenyhercz and H. B. Mark, Jr., *J. Electrochem. Soc.* **123**, 1656 (1976).

57. H. B. Mark, Jr., T. M. Kenyhercz, and P. T. Kissinger, in *Electrochemical Studies of Biological Systems* D. T. Sawyer, ed., American Chemical Society Symposium Series 38, Washington, D.C. (1977).

58. D. Lexa, J. M. Saveant, and J. Zickler, *J. Am. Chem. Soc.* **102**, 4851 (1980).

59. L. Szebellédy and Z. Somogyi, *Z. Anal. Chem.* **112**, 385 (1938).

60. W. R. Heineman, T. Kuwana, and C, R. Hartzell, *Biochem. Biophys. Res. Commun.* **50**, 892 (1973).

61. F. M. Hawkridge and T. Kuwana, *Anal. Chem.* **45**, 1021 (1973).

62. Y. Fugihira, T. Kuwana, and C. R. Hartzell, *Biochem. Biophys. Res. Commun.* **61**, 538 (1974).
63. W. R. Heineman, T. Kuwana, and C. R. Hartzell, *Biochem. Biophys. Res. Commun.* **49**, 1 (1972).
64. J. L. Anderson, *Anal. Chem.* **51**, 2312 (1979).
65. L. N. Mackey, T. Kuwana, and C. R. Hartzell, *FEBS Lett.* **36**, 326 (1973).
66. J. L. Anderson, T. Kuwana, and C. R. Hartzell, *Biochem.* **15**, 3847 (1976).
67. R. Szentrimay and T. Kuwana, *Anal. Chem.* **50**, 1879 (1978).
68. T. Kula, E. Stellwagen, R. Szentrimay, and T. Kuwana, *Biochim. Biophys. Acta* **634**, 279 (1981).
69. B. Ke, F. M. Hawkridge, and S. Sahu, *Proc. Nat. Acad. Sci. U.S.A.* **73**, 2211 (1976).
70. M. T. Stankovich, *Anal. Biochem.* **109**, 295 (1980).
71. T. Kuwana and W. R. Heineman, *Acc. Chem. Res.* **9**, 241 (1976).
72. R. L. McCreery, *CRC Crit. Rev. Anal. Chem.* **7**, 89 (1978).
73. W. R. Heineman, F. M. Hawkridge, and H. N. Blount, in *Electroanalytical Chemistry*, Vol. 13, A. J. Bard, ed., Marcel Dekker, New York (1982), pp. 1–113.
74. D. A. Yates, R. Szentrimay, and T. Kuwana, *Anal. Biochem.* **102**, 271 (1980).
75. N. Winograd, H. N. Blount, and T. Kuwana, *J. Phys. Chem.* **73**, 3456 (1969).
76. H. N. Blount, N. Winograd, and T. Kuwana, *J. Phys. Chem.* **74**, 3231 (1970).
77. J. H. Christie, *J. Electroanal. Chem.* **13**, 79 (1967).
78. M. D. Ryan, J.-F. Wei, B. A. Feinberg, and Y.-K Lau, *Anal. Biochem.* **96**, 326 (1979).
79. T. Kuwana and W. R. Heineman, *Bioelectrochem. Bioenerg.* **1**, 389 (1974).
80. E. Steckhan and T. Kuwana, *Ber. Bunsenges. Phys. Chem.* **78**, 253 (1974).
81. L. Mackey, E. Steckhan, and T. Kuwana, *Ber. Bunsenges. Phys. Chem.* **79**, 587 (1975).
82. L. N. Mackey and T. Kuwana, *Bioelectrochem. Bionerg.* **3**, 596 (1976).
83. M. D. Ryan and G. S. Wilson, *Anal. Chem.* **47**, 885 (1975).
84. F. R. Shu and G. S. Wilson, *Anal. Chem.* **48**, 1676 (1976).
85. L. H. Rickard, H. L. Landrum, and F. M. Hawkridge, *Bioelectrochem. Bioenerg.* **5**, 686 (1978).
86. L. Griggio and S. Pinamonti, *Atti Ist. Veneto Sci. Lett. Arti, Cl. Sci. Mat. Natur.* **124**, 15 (1965–66).
87. S. R. Betso, M. H. Klapper, and L. B. Anderson, *J. Am. Chem. Soc.* **94**, 8197 (1972).
88. F. Scheller, M. Jänchen, G. Etzold, and H. Will, *Bioelectrochem. Bioenerg.* **1**, 478 (1974).
89. F. Scheller, M. Jänchen, J. Lampe, H.-J. Prümke, J. Blanck, and E. Palecek, *Biochim. Biophys. Acta* **412**, 157 (1975).
90. F. Scheller, M. Jänchen, and H.-J. Prümke, *Biopolymers* **14**, 1553 (1975).
91. F. Scheller, H.-J. Prümke, H. E. Schmidt, and P. Mohr, *Bioelectrochem. Bioenerg.* **3**, 328 (1976).
92. F. Scheller and H.-J. Prümke, *Stud. Biophys.* **60**, 137 (1976).
93. F. Scheller, H.-J. Prümke, and H. E. Schmidt, *J. Electroanal. Chem.* **70**, 219 (1976).
94. F. Scheller, *Bioelectrochem. Bioenerg.* **4**, 490 (1977).
95. B. A. Kuznetsov, *Experientia, Suppl.* **18**, 381 (1971).
96. B. A. Kuznetsov, N. M. Mestechkina, and G. P. Shumakovich, *Bioelectrochem. Bioenerg.* **4**, 1 (1977).
97. B. A. Kuznetsov, G. P. Shumakovich, and N. M. Mestechkina, *Bioelectrochem. Bioenerg.* **4**, 512 (1977).
98. B. A. Kuznetsov, N. M. Mestechkina, M. V. Izotov, I. I. Karuzina, A. V. Karyakin, and A. I. Archakov, *Biochemistry-U.S.S.R.* **44**, 1234 (1979); translated from *Biokhimiya*, **44**, 1569 (1979).
99. B. A. Kuznetsov, *Bioelectrochem. Bioenerg.* **8**, 681 (1981).
100. J. Haladjian, P. Bianco, and P.-A. Serre, *Bioelectrochem. Bioenerg.* **6**, 555 (1979).
101. J. Haladjian, P. Bianco, and P.-A. Serre, *J. Electroanal. Chem.* **106**, 397 (1980).
102. P.-A. Serre, J. Haladjian, and P. Bianco, *J. Electroanal. Chem.* **122**, 327 (1981).
103. T. Ikeda, H. Kinoshita, Y. Yamane, and M. Senda, *Bull. Chem. Soc. Japan* **53**, 112 (1980).

104. H. Berg, *Bioelectrochem. Bioenerg.* **4**, 522 (1977).
105. T. Ikeda, Y. Yamane, H. Kinoshita, and M. Senda, *Bull. Chem. Soc. Japan* **53**, 589 (1980).
106. E. Margoliash and A. Schejter, *Adv. Protein Chem.* **21**, 113 (1966).
107. R. E. Dickerson and R. Timkovich, in *The Enzymes*, Vol. XI-A, P. D. Boyer, ed., Academic Press, New York (1975), pp. 397–547.
108. T. Andersson, J. Ångström, K.-E. Falk, and S. Forsén, *Eur. J. Biochem.* **110**, 363 (1980).
109. M. R. Tarasevich and V. A. Bogdanovskaya, *Bioelectrochem. Bioenerg.* **3**, 589 (1976).
110. M. J. Eddowes and H. A. O. Hill, *J. Chem. Soc., Chem. Commun.* 771 (1977).
111. M. J. Eddowes and H. A. O. Hill, *J. Am. Chem. Soc.* **101**, 4461 (1979).
112. M. J. Eddowes, H. A. O. Hill, and K. Uosaki, *Bioelectrochem. Bioenerg.* **7**, 527 (1980).
113. N. S. Lewis and M. S. Wrighton, *Science* **211**, 944 (1981).
114. T. M. Cotton, S. G. Schultz, and R. P. Van Duyne, *J. Am. Chem. Soc.* **102**, 7960 (1980).
115. P. Yeh and T. Kuwana, *Chem. Lett.* 1145 (1977).
116. H. Theorell, *Biochem. Z.* **285**, 207 (1936).
117. A. Ehrenberg and S. Paleus, *Acta Chim. Scand.* **9**, 538 (1955).
118. E. E. Bancroft, H. N. Blount, and F. M. Hawkridge, *Biochem. Biophys. Res. Commun.* **101**, 1331 (1981).
119. E. E. Bancroft, H. N. Blount, and F. M. Hawkridge, *Adv. Chem. Ser.* **201**, 23–49 (1982).
120. E. F. Bowden, F. M. Hawkridge, and H. N. Blount, *Adv. Chem. Ser.* **201**, 159–171 (1982).
121. E. E. Bancroft, J. S. Sidwell, and H. N. Blount, *Anal. Chem.* **53**, 1390 (1981).
122. J. A. V. Butler, *Trans. Faraday Soc.* **19**, 734 (1924).
123. T. Erdey-Gruz and M. Volmer, *Z. Phys. Chem., Abt. A* **150** 203 (1930).
124. D. E. Albertson, H. N. Blount, and F. M. Hawkridge, *Anal. Chem.* **51**, 556 (1979).
125. E. E. Bancroft, H. N. Blount, and F. M. Hawkridge, *Anal. Chem.* **53**, 1862 (1981).
126. E. F. Bowden, F. M. Hawkridge, J. C. Chlebowski, E. E. Bancroft, C. Thorpe, and H. N. Blount, *J. Am. Chem. Soc.* **104**, 7641 (1982).
127. A. E. G. Cass, M. J. Eddowes, H. A. O. Hill, K. Uosaki, R. C. Hammond, I. J. Higgins, and E. Plotkin, *Nature* **285**, 673 (1980).
128. K. Uosaki and H. A. O Hill, *J. Electroanal. Chem.* **122**, 321 (1981).
129. W. J. Albery, M. J. Eddowes, H. A. O. Hill, and A. R. Hillman, *J. Am. Chem. Soc.* **103**, 3904 (1981).
130. M. J. Eddowes, H. A. O. Hill, and K. Uosaki, *J. Am. Chem. Soc.* **101**, 7113 (1979).
131. R. E. Dickerson and R. Timkovich, in *The Enzymes*, Vol. XI-A, P. D. Boyer, ed., Academic Press, New York (1975), pp. 397–547.
132. K. Niki, T. Yagi, H. Inokuchi, and K. Kimura, *J. Am. Chem. Soc.* **101**, 3335 (1979).
133. P. Bianco, G. Faugue, and J. Haladjian, *Bioelectrochem. Bioenerg.* **6**, 385 (1979).
134. I. Taniguchi, T. Murakami, K. Toyosawa, H. Yamaguchi, and K. Yasukouchi, *J. Electroanal. Chem.* **131**, 397 (1982).
135. H. L. Landrum, R. T. Salmon, and F. M. Hawkridge, *J. Am. Chem. Soc.* **99**, 3154 (1977).
136. J. F. Stargardt, F. M. Hawkridge, and H. L. Landrum, *Anal. Chem.* **50**, 930 (1978).
137. J. F. Stargardt, M. S. Thesis, Virginia Commonwealth University, 1981.
138. E. F. Bowden and F. M. Hawkridge, *J. Electroanal. Chem.* **125**, 367 (1981).
139. S. Chao, J. L. Robbins, and M. S. Wrighton, *J. Am. Chem. Soc.* **105**, 181 (1984).
140. F. S. Mathews, E. W. Czerwinski, and P. Argos, in: *The Porphyrins*, Vol. VII-B, D. Dolphin, ed., Academic Press, New York (1979), pp. 108–147.
141. B. Hagihara, N. Sato, and T. Yamanaka, in *The Enzymes* Vol. XI-A, P. D. Boyer, ed., Academic Press, New York (1975), pp. 550–593.
142. B. W. Griffin, J. A. Peterson, and R. W. Estabrook, in *The Porphyrins*, Vol. VII-B, D. Dolphin, ed., Academic Press, New York (1979), pp. 333–375.
143. F. Scheller, R. Renneberg, G. Strnad, K. Pommerening, and P. Mohr, *Bioelectrochem. Bioenerg.* **4**, 500 (1977).
144. R. E. Dickerson and R. Timkovich, in *The Enzymes*, Vol. XI-A, P. D. Boyer, ed., Academic Press, New York (1975), pp. 397–547.

145. R. Haser, M. Pierrot, M. Frey, R. Payan, J. P. Astier, M. Bruschi, and J. LeGall, *Nature* **282**, 806 (1980).

146. Y. Higuchi, S. Bando, M. Kusunoki, Y. Matsuura, N. Yasuoka, M. Kakudo, T. Yamanaka, T. Yagi, and H. Inokuchi, *J. Biochem.* **89**, 1659 (1981).

147. T. Yagi, M. Goto, K. Nakano, K. Kimura, and H. Inokuchi, *J. Biochem.* **78**, 443 (1975).

148. K. Niki, T. Yagi, H. Inokuchi, and K. Kimura, *J. Electrochem. Soc.* **124**, 1889 (1977).

149. K. Niki, T. Yagi, H. Inokuchi, and K. Kimura, *J. Am. Chem. Soc.* **101**, 3335 (1979).

150. W. R. Sokol, D. H. Evans, and K. Niki, *J. Electroanal. Chem.* **108**, 107 (1979).

151. K. Niki, Y. Takizawa, K. Kumagai, R. Fujiwara, T. Yagi, and H. Inokuchi, *Biochim. Biophys. Acta* **636**, 136 (1981).

152. P. Bianco and J. Haladjian, *Biochim. Biophys. Acta* **545**, 86 (1979).

153. P. Bianco, G. Faugue, and J. Haladjian, *Bioelectrochem. Bioenerg.* **6**, 385 (1979).

154. P. Bianco and J. Haladjian, *Electrochim. Acta* **26**, 1001 (1981).

155. E. Stellwagen, *Nature* **275**, 73 (1978).

156. H. A. Laitinen and J. E. B. Randles, *Trans. Faraday Soc.* **51**, 54 (1955).

157. R. Singleton, Jr., L. L. Campbell, and F. M. Hawkridge, *J. Bacteriol.* **140**, 893 (1979).

158. M. J. Eddowes, H. Elzanowska, and H. A. O. Hill, *Biochem. Soc. Trans.* **7**, 735 (1979).

159. P. Bianco and J. Haladjian. *Bioelectrochem. Bioenerg.* **8**, 239 (1981).

160. M. Jänchen, F. Scheller, H.-J. Prümke, P. Mohr, and G. Etzold, *Stud. Biophys.* **39**, 1 (1973).

161. F. Scheller and M. Jänchen, *Stud. Biophys.* **46**, 153 (1974).

162. S. R. Betso and R. E. Cover, *J. Chem. Soc., Chem. Commun.* 621 (1972).

163. E. F. Bowden, F. M. Hawkridge, and H. N. Blount, *Bioelectrochem. Bioenerg.* **7**, 447 (1980).

164. J. F. Castner and F. M. Hawkridge, *J. Electroanal. Chem.* **143**, 217 (1983).

165. J. F. Castner, Ph. D. Dissertation, Virginia Commonwealth University, 1981.

166. W. D. Henson and L. P. Hager, in *The Porphyrins*, Vol. VII-B, D. Dolphin, ed., Academic Press, New York (1979), pp. 295–332.

167. G. R. Schonbaum and B. Chance, in *The Enzymes*, Vol. XIII-C, P. D. Boyer, ed., Academic Press, New York (1976), pp. 363–408.

168. A. I. Yaropolov, M. R. Tarasevich, and S. D. Varfolomeev, *Bioelectrochem. Bioenerg.* **5**, 18 (1978).

169. G. L. Brown, *Arch. Biochem. Biophys.* **49**, 303 (1954).

170. R. Malkin, in *Iron–Sulfur Proteins*, Vol. II, W. Lovenberg, ed., Academic Press, New York (1973), pp. 1–26.

171. M. Senda, T. Ikeda, T. Kakutani, K. Kano, and H. Kinoshita, *Bioelectrochen. Bioenerg.* **8**, 151 (1981).

172. S. D. Varfolomeev, A. I. Yaropolov, I. V. Berezin, M. R. Tarasevich, and V. A. Bogdanovskaya, *Bioelectrochem. Bioenerg.* **4**, 314 (1977).

173. C. D. Crawley and F. M. Hawkridge, *Biochem. Biophys. Res. Commun.* **99**, 516 (1981).

174. C. D. Crawley, M. S. Thesis, Virginia Commonwealth University (1982).

175. A. White, P. Handler, E. L. Smith, R. L. Hill, and I. R. Lehman, *Principles of Biochemistry*, 6th ed., McGraw–Hill, New York (1978), p. 325.

176. F. Scheller, G. Strnad, B. Neumann, M. Kühn, and W. Ostroski, *Bioelectrochem. Bioenerg.* **6**, 117 (1979).

177. T. Ikeda, S. Ando, and M. Senda, *Bull. Chem. Soc. Japan* **54**, 2189 (1981).

178. B. Czochralska, M. Szweykowska, N. A. Dencher, and D. Shugar, *Bioelectrochem. Bioenerg.* **5**, 713 (1978).

179. E. P. Suponeva, B. A. Kiseley, and L. N. Chekulaeva, *Bioelectrochem. Bioenerg.* **8**, 251 (1981).

180. H. Berg, *Top. Bioelectrochem. Bioenerg.* **1**, 39 (1976).

181. O. H. Müller, *Meth. Biochem. Anal.* **11**, 329 (1963).

182. M. Senda, T. Ikeda, and H. Kinoshita, *Bioelectrochem. Bioenerg.* **3**, 253 (1976).

183. M. R. Smyth and W. F. Smyth, *Analyst* **103**, 529 (1978).

184. M. T. Stankovich and A. J. Bard, *Electroanal. Chem.* **85**, 173 (1977).

185. V. Brabec and V. Mornstein, *Biochim. Biophys. Acta* **625**, 43 (1980).
186. V. Brabec and V. Mornstein, *Biophys. Chem.* **12**, 159 (1980).
187. B. Malfoy and J. A. Reynaud, *J. Electroanal. Chem.* **114**, 213 (1980).
188. J. A. Reynaud, B. Malfoy, and A. Bere, *Bioelectrochem. Bioenerg.* **7**, 595 (1980).
189. J. A. Reynaud, B. Malfoy, and P. Canesson, *J. Electroanal. Chem.* **114**, 195 (1980).

6

Electrochemical Aspects of Metabolism

MICHAEL N. BERRY, ANTHONY R. GRIVELL, and PATRICIA G. WALLACE

1. Introduction

1.1. Objectives

The opportunity to write a Chapter on this topic presents something of a challenge. Apart from the prescient suggestions of a few individuals,[1-3] the importance of electrochemical processes in intermediary metabolism has received little recognition until comparatively recently. Even now, the details of biological electrochemical mechanisms remain obscure and it is not feasible to write a comprehensive account without indulging in a substantial degree of speculation. We do not see this as necessarily undesirable. Abundant standard textbook and review articles are available for those who wish to familiarize themselves with conventional concepts concerning metabolic regulation. We see our task as providing an alternative and integrative point of view that, it is to be hoped, will lead to new experimental approaches. The time seems well overdue for a multidisciplinary approach to the understanding of metabolic regulation and we trust that this review may serve towards that end.

MICHAEL N. BERRY, ANTHONY R. GRIVELL, and PATRICIA G. WALLACE ● Department of Clinical Biochemistry, School of Medicine, Flinders University of South Australia, Bedford Park, 5042, South Australia, Australia.

1.2. Scope

In recent years there has been a gradual recognition that the principles of solution chemistry do not provide a satisfactory basis for explaining many phenomena associated with the living state. Concomitantly, the importance of vectorial electrochemical reactions in living cells has been increasingly emphasized.[4-7] In general, however, descriptions of electrochemical phenomena have been restricted to processes associated with cellular membranes recognized as possessing electrodic properties.[8,9] Events in the cytoplasmic compartment have been excluded from consideration on the assumption that reactions occurring in this milieu would be scalar in nature, involving soluble chemical components interacting only by diffusion and random collision. We cast doubt on the validity of this assumption in the first part of this Chapter (Section 2), in which new ideas concerning cellular structure and mechanisms of vectorial charge transfer are reviewed. Evidence is presented that electrochemical events occur within all compartments of the cell and are essential for the establishment and maintenance of what we recognize as the "living state."

In the second part of the article (Section 3) we examine the possible involvement of electrochemical processes in metabolic regulation. Current concepts based on the principles of solution chemistry are critically reviewed and the advantages of an alternative electrochemical approach are demonstrated. The question of what factors govern overall metabolic rate as manifested by O_2-uptake and heat production in the living cell or organism are explored from an electrochemical viewpoint.

Finally (Section 4) we consider the possible role of superoxide (O_2^-) as an essential intermediate for many biological electrochemical reactions, which relate to energy transduction and message transmission. We advance the suggestion that derangements of the normal metabolism of superoxide may be the basis for many of the major diseases afflicting mankind.

2. The Living Cell as an Electrochemical System

2.1. Structural Analogies between Living Cells and Electrochemical Devices

Most biochemists have long since discarded the concept that cells consist of a membranous sac containing a soup full of enzymes and substrates. The recognition that cells contain discrete organelles such as nuclei, mitochondria, peroxisomes, endoplasmic reticulum, Golgi apparatus, lysosomes, etc. is virtually universal. Nevertheless, in examining reports dealing with the intermediary metabolism of these units, the impression is often gained that the organelles themselves are still regarded as membranous sacs with soluble contents, and that outside the organelles, enzymes and substrates are thought to interact entirely by processes involving free diffusion and random collision.

How realistic is this view? A considerable amount of evidence has become available that a substantial quantity of the cellular water assumes a structured form.[10] Moreover, the penetrating studies of the electron microscopists have now revealed that surrounding all cellular organelles and pervading the whole cytoplasm is a delicate fine structure, the cytoskeleton, comprising microtubules, microfilaments, and a variety of other structural elements such as intermediate filaments.[11,12] In motile cells the cytoskeleton is particularly extensive and it seems likely that an architecture of similar nature exists in all cells.

Recently a further level of organization has been recognized. Morphological investigations have convincingly demonstrated a ubiquitous proteinaceous ultrastructure—a microtrabecular lattice—which pervades the cytoplasmic compartment of living cells.[13,14] It is linked to the cytoskeleton and embraces the cellular organelles. This interconnection of cellular organelles is readily demonstrated in damaged[15] or intact[16] isolated liver cells, where the mitochondria show no Brownian motion, even when displaced to one pole of the cell by high-speed centrifugation (100,000 g) of hepatocyte suspensions.† In K. R. Porter's view‡ all catalytic activity of the so-called "soluble" cytoplasm may be carried out by enzymes adsorbed to the microtrabecular lattice or to cellular membranes. In support of this it has been shown that the aqueous supernatant produced within various types of cell, stratified by high-speed centrifugation, is devoid of enzyme activity.[17,18] Moreover, fractionation experiments have demonstrated that many enzymes designated as "soluble" are, in fact, adsorbed to intracellular surfaces.[19–24]

Thus, the living cell is pictured as a two-phase system; a solid-state phase within and adjacent to which the enzyme-catalyzed reactions of intermediary metabolism take place, and a bulk aqueous phase containing low M_r organic solutes and ions. This bulk phase ramifies throughout the cell in intimate relationship with the components of the solid-state phase, and can be likened to an intracellular circulatory system. In other words, the cytoplasm can be regarded as having the structure and composition of a gel, and hence represents a system with a massive surface/volume ratio.[10,25,26] This design is highly appropriate for an electrochemical system—the solid-state phase representing a multi-electrode array, and the aqueous phase the electrolyte.[9] It is at the surface between the two phases that electrical processes would be initiated; the greater the surface area of the interface, the greater the possible current flow.

2.2. Examples of Electrochemical Phenomena in Living Systems

There is now general agreement that the synthesis of ATP in the transducing membranes of mitochondria, chloroplasts, and bacteria is an

---

† Unpublished observations.
‡ Personal communication.

electrochemical process,[5,7–9,27–30] although the exact molecular mechanisms remain to be elucidated. While some modification[8,28] may well be required of Mitchell's original "chemiosmotic hypothesis," which argues that a proton electrochemical gradient across a transducing membrane can promote the synthesis of ATP,[5,27] there has developed a consensus that electrical energy provides the primary driving force. A vectorial electric field enables the phosphorylation of ATP, a process which, in the absence of the energy-dependent charge separation induced by a redox reaction or by light, would be thermodynamically highly unfavorable. In fact, recent preliminary reports describe the synthesis of ATP by chloroplasts and by submitochondrial particles in response to an applied electric field,[31,32] before significant accumulation of protons can be detected, and it seems likely that the establishment of an intramembranous electric field precedes the creation of the proton gradient.

Electrochemical phenomena are by no means confined to reactions involving the synthesis of ATP within transducing membranes. In recent years electrical changes within cells have been observed in processes as diverse as differentiation[33] and exocrine secretion.[34] It is now appreciated that so-called "non-excitable" cells respond to certain stimuli, often hormonal in nature, with a transient alteration in the polarity of their plasma membrane.[35] This event invariably precedes any measurable chemical change. Indeed for both the catecholamines and for insulin a case has been made that an alteration of membrane polarity is an initial and essential step in their cellular action.[35]

In addition to these events at the plasma membrane, the "electrogenic" energy-dependent nature of the translocation of many types of ions or molecules across intracellular membranes has been widely recognized.[36,37] Moreover, there is now compelling evidence for cytoplasmic electrochemical processes. Recent studies indicate that intracellular ionic currents can steer protein molecules to their ultimate destination within *Cecropia* oocytes by a process of ultramicroelectrophoresis.[38] Moreover, steady cytoplasmic ion currents are found to play a role in the growth of fucoid eggs,[39] the regeneration of the stumps of amputated amphibian limbs,[33] and the healing of mammalian epidermal wounds.[33] They can also be readily demonstrated at several stages of the life cycle of the water mold *Blastocladiella emersonii.*[40]

2.3. Mechanisms of Electric Field Generation and Charge Conduction

2.3.1. Charge Transfer by Semiconduction

2.3.1.1. Conduction of Electrons

The very high mobility of electrons relative to that of protons or other ions implies that electron translocation, associated with redox reactions, is likely to be the primary mechanism whereby electric fields are generated within living cells. The electron affinity of O_2 or other acceptors overcomes the ionization potential of H atoms derived from a metabolic fuel, resulting in the separation of protons and electrons which pass down their electrochemical potential gradients via separate paths. Because electrons can move much faster than protons, the conducting system is polarized, leading to the establishment of an electric field.

The transfer of electrons between carriers embedded within a membrane is a well-recognized phenomenon. Nevertheless, there is still considerable uncertainty as to the mechanisms of electronic charge transfer. Although doubts have been raised concerning the feasibility of electron conduction within proteins, a large body of evidence has now accrued that such semiconduction does indeed take place.[41-44] We envisage that proteins in the solid-state phase of the cell exist as macromolecular complexes and that electron transfer between the individual moieties occurs by semiconduction, rather than by diffusion and random collision of dissociable prosthetic groups. Thus, although a bound prosthetic group within the redox protein may well be the primary site of interaction with substrate, electron flow from the prosthetic group through the body of the protein may subsequently occur.[6,43] Likewise, charge transfer between contiguous proteins may not involve prosthetic groups at the immediate site at which electron transfer between the macromolecules takes place.

At present there is little information available concerning these processes, which essentially can be regarded as mechanisms involving chemical radicals.[45] As electron flow occurs, a series of radical intermediates (compounds containing an unpaired electron) is created, the process being analogous to that known to occur, for example, in the mitochondrial, chloroplast, or endoplasmic reticulum electron transport systems.[46-48] Generally, free radicals are considered to be highly reactive species.[45] However, it seems likely that the radical product formed within the solid-state phase may move only a very short distance before donating its unpaired electron to the next component of the metabolic sequence. Moreover, the enzyme–coenzyme radical, created in the redox reaction, may well be relatively stable by virtue of the opportunity for charge delocalization within the protein molecule.[49]

While there may still be some argument as to the importance of electron semiconduction in nonmembranous areas of the cell, there is general agreement that electron conduction within membranes plays a fundamental role in the living state, providing the primary driving force for synthetic metabolic processes. Even so, details of these processes remain obscure, in particular the molecular mechanisms relating to energy transduction. In recent years the theories of Mitchell[5,27] have led to the now generally held view that electric fields, established as a consequence of electron flow within and between cellular macromolecules, generate a protonic electrochemical gradient enabling the flow of proton current ("proticity"[8,27]) within the cell.

2.3.1.2. Conduction of Protons

The conduction of protons within proteins has been experimentally established,[50-52] and a number of plausible mechanisms have been proposed. One attractive theory[53,54] envisages protons as flowing in an ordered and vectorial manner through a series of hydrogen-bonded chains within a protein, a process analogous to the flow of protons in ice[55] (Figure 1). It seems possible that several polypeptides could interact to form ah extended series of hydrogen-bonded chains permitting charge transfer between individual macromolecules.[53] The ice-like nature of the ordered hydration shell surrounding contiguous cytoplasmic proteins would not be anticipated to provide a barrier to proton conduction. Indeed, structured water[10] itself might well provide hydrogen-bonded chains for the vectorial conduction of protons along macromolecular or membranous surfaces.

It is likely, therefore, that living systems possess mechanisms for vectorial conduction of not only the electrons, but also the protons which originate from various metabolic reactions. Important sources and sinks for protons are the numerous redox reactions which occur within the living cell. These generate or consume protons, depending on the direction of the reaction, e.g.:

$$lactate + NAD^+ \rightleftharpoons pyruvate + NADH + H^+ \tag{1}$$

Other major sources of protons within living cells are the reactions involving cleavage of the phosphate anhydride bonds of ATP (or other "high-energy" compounds). Thus the phosphorylation of glucose generates a proton:

$$glucose + ATP^{4-} \rightleftharpoons glucose\text{-}6\text{-}phosphate^{2-} + ADP^{3-} + H^+ \tag{2}$$

as do other phosphorylation reactions involving the cleavage of ATP. Likewise,

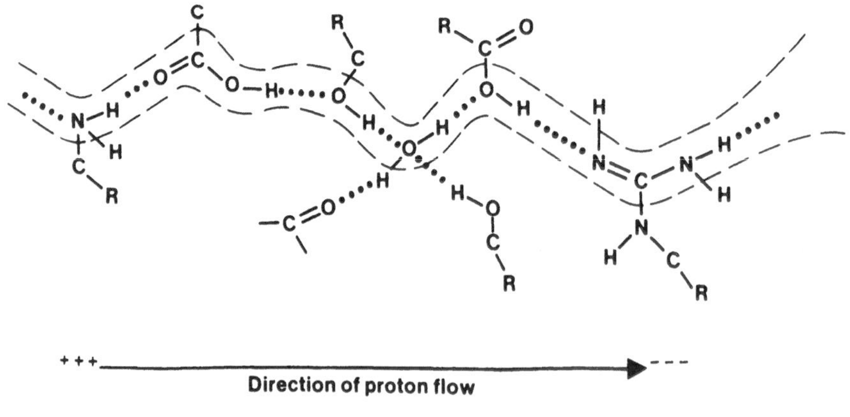

Figure 1. A series of H-bonds (····) serving as a proton-conducting pathway within a cytoplasmic or intramembrane protein.

the splitting of ATP, catalyzed by Na^+K^+-ATPase, generates a proton:

$$ATP^{4-} + H_2O \rightleftharpoons ADP^{3-} + Pi^{2-} + H^+ \tag{3}$$

Conversely, the central reaction of oxidative phosphorylation [reaction (3) in reverse] consumes a proton, as do other ATP-synthesizing reactions, e.g.:

$$\text{phosphoenolpyruvate}^{3-} + ADP^{3-} + H^+ \rightleftharpoons \text{pyruvate}^- + ATP^{4-} \tag{4}$$

Current can flow only if the generated protons are insultated from the bulk aqueous phase. Evidence from *in vitro* enzymological studies provides an indication of how this may be achieved. In kinase-catalyzed reactions, for which the mechanism of catalysis has been explored, it has been found that reactions take place deep within a cleft in the enzyme, a cleft from which bulk water is excluded.[56] Even *in vitro*, protons are not released immediately into the medium on ATP splitting.[57] Thus, it can be predicted that *in vivo* they do not instantaneously dissipate their energy as heat motion through hydration in the bulk aqueous phase, but rather flow vectorially for a finite period within the kinase and, possibly, contiguous enzymes of the associated metabolic pathway, before entering the bulk aqueous phase of the cell. Similar considerations should apply to proton release or consumption in NAD(P)-linked reactions.[57] In these types of process the enzyme participates not only as a catalyst of the chemical reaction but also as a mediator of charge transfer. In fact, vectorial charge relay is a not unusual feature of enzyme catalysis.[50,51]

It seems probable that proton current, generated by redox reactions and by ATP cleavage, flows through intracellular "circuits," specific for the various functions associated with the living state. Proton-conducting pathways, along the planes of energy-transducing membranes, have been proposed.[58,59] The idea of a "protoneural network" of proton-transfer proteins is particularly attractive. Insulation of the proton flow, along hydrogen-bonded chains from protein to protein, would be aided by the ambience of structured water at the surface of the particulates. Continuity would be ensured, though, by direct protein–protein interaction, involving quaternary-type contacts. As is well known, such interactions involve hydrogen-bonding across an interface which is highly hydrophobic in nature.[51,60] Such interconnection may be important in allowing proton flow to extend into the bulk aqueous phase, via the microtrabecular lattice. (A similar role of the lattice, in electron semiconduction was suggested by Lewis.[42])

Through the medium of this proticity, the cell accomplishes mechanical, osmotic, thermal, and electrical work.[61] In this view, the central role of ATP and similar compounds in these functions is related not merely to their possession of "high-energy" bonds, but to their stability in the aqueous phase, which allows them to act as reservoirs which can provide proton current on demand. The presence of equilibrating mechanisms such as the adenylate kinase system[62] ensures that local ATP depletion at any point in the cell is rapidly buffered.

2.3.2. Charge Transfer in the Bulk Aqueous Phase

The bulk aqueous phase represents much of the mass of the cell. The question arises as to whether *in vivo* redox reactions take place in this phase. *In vitro* the generally accepted mechanism for redox reactions in an aqueous medium involves "hydride ion" transfer.[63] This term is used to denote the passage of one proton and two electrons from substrate to coenzyme with an additional proton being released to the medium. The reduced coenzyme can then transfer the electrons and proton to an acceptor molecule, an additional proton being taken up from the medium.

The present state of our knowledge is too limited to come to firm conclusions about the importance of redox reactions in the bulk aqueous phase. We have recently obtained substantial evidence for the existence of specific channels conducting reducing-equivalents between cellular compartments,[64,65] and this would argue against the presence of freely diffusible coenzyme charge carriers. Nevertheless, there is a possibility that charge transfer by processes involving diffusion of soluble components does take place in localized regions of the cell. If this is the case, there must be mechanisms for the coordination of solid-state semiconduction with bulk aqueous phase charge transfer. This seems a profitable field for further investigation.

The bulk aqueous phase also acts as a store for key metabolites such as ATP. It may also serve as an energy store over and above its content of nucleotides. The cellular membranes enclose various compartments[66] and it seems likely that most (if not all) of these membranes can maintain ion and proton gradients across them.[36,37,67–69] These gradients thus represent free energy stores which can be utilized for cellular work. It should be emphasized, however, that these gradients are established and maintained as a consequence of redox reactions in the solid-state phase of the cell. They are produced as a result of energy flow and do not reflect thermodynamic equilibrium.

2.4. Coordination of Proton and Metabolic Flux

It is also necessary to address the question of how the rate of charge transfer is coordinated with metabolic flux so precisely that no significant accumulation of metabolic intermediates occurs. This could readily be achieved if the proton flow along a series of enzymes forming a metabolic pathway (the glycolytic sequence for example) were coordinated both in time and space with the movement of the anionic reaction products between the active centers of the corresponding enzymes (Figure 2). Under such conditions, the flow of protons through contiguous macromolecules might be anticipated to bring about conformational changes in the enzymes that would promote interaction of active centers, so that the reaction products were translocated without release into the bulk aqueous phase. The advantages of such a process for cellular economy and efficiency have been extensively discussed.[70,71]

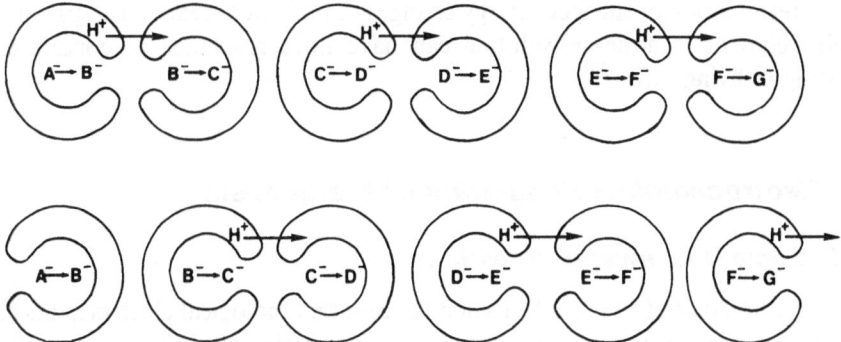

Figure 2. A schematic representation of coordination of proton and metabolic flux along a metabolic pathway. Proton flow between contiguous enzymes brings about conformational changes which facilitate passage of metabolic products between enzymes of the pathway.

Moreover, proton flux would serve to order an enzyme sequence in the appropriate catalytic configuration as long as substrate was available and metabolism was taking place (Figure 2). It is of interest that the microtrabecular lattice described by Porter is not a permanent structure, but can be dispersed by metabolic inhibitors or by cold.[14] The possibility must be seriously considered that the evanescent nature of the lattice represents the assembly and disaggregation of the enzymes of metabolic pathways in response to substrate availability, regulatory effectors, or changes in the strength of relevant electric fields.

The parallel flow of protons through the hydrogen-bonded chains of the protein during enzymic catalysis is exemplary of the principle of complementarity espoused by Lumry[72]—namely that the enzyme buffers the free energy changes associated with substrate–product transformation. The hydrogen-bonded chain is an energized system in that it contains an extra proton,[53] injected during the catalytic process. Thus, the protein during catalysis is in an "energized" state, relaxing only after passage of the proton (together with the product of the reaction) to the contiguous enzyme (or to the proton-conducting infrastructure). Accordingly, some of the free energy associated with substrate transformation is transduced throughout the enzyme sequence.[71,72] In a test tube, where generally no organizing electric field is present, proton (and energy) transfer between contiguous enzymes does not occur, so that any generated proton is released readily into the bulk medium; much of the free energy of the reaction would in these circumstances be measured as that associated with hydration of the proton.[73] In the living cell, this free energy is largely conserved in the cyclical conformational changes associated with proton flow between and within the protein enzymes that catalyze the sequential reactions of each metabolic pathway. It follows that the determination of the free energy changes of individual reaction steps from their behavior in aqueous solution can have little relevance to conditions existing in living cells. [73]

3. Electrochemical Regulation of Metabolism

3.1. Some Unanswered Questions

The first part of this article argued the case that biological electrochemical reactions represent the power source for all driven biological phenomena. We now examine the mechanisms that control the rate of power generation and utilization in living systems. Intimately related to this is the question of how the rate of heat production by living cells is regulated.

Classical thermodynamic reasoning identifies the ATP-splitting and ATP-generating reactions of the cell as the major sources of cellular heat production. Such reactions are thought to be irreversible processes, occurring far from thermodynamic equilibrium, with the free energy not converted to bond energy being necessarily lost as heat. [74–76] This concept is difficult to reconcile with the empirical observation that heat loss (or heat production under steady-state conditions of heat balance) reflects the "surface law" [77]; i.e., for any given temperature gradient and set of environmental conditions, heat loss is approximately proportional to body surface area. If animal heat production represented the sum of the heat generated by a multiplicity of scalar biochemical reactions, taking place far from equilibrium, it would be expected to be a function of body mass (representing the cellular content of enzymes and metabolites) rather than of surface area.

Another area of considerable conceptual difficulty relates to the observation that the adult animal exists in a more-or-less steady-state, well removed from equilibrium. When observed day-to-day there is generally little variation in body composition while the animal is on a consistent diet, the mass of protein, carbohydrate, and fat remaining approximately constant. This constancy is maintained despite the rapid turnover of the body's constituents. The living system is not static but dynamic, the steady-state being maintained because synthetic reactions exactly balance degradative processes. How this balance is achieved and what governs the overall rate of metabolic flux and heat production are the main topics of this section.

3.2. Current Concepts of Metabolic Regulation

The ability of animal organisms to perform reductive syntheses of complex macromolecules from simple precursors in the presence of a strongly oxidative and catabolism-promoting environment has been for some a source of great wonderment and speculation. [78] This, however, is not reflected in most biochemical textbooks in which the ability of living systems to synthesize macromolecules from low M_r precursors is attributed to the special properties of a class of molecules containing "high-energy" bonds, of which ATP is regarded as the most important. It is assumed that some of the free energy released during catabolism is conserved in the bond energy of ATP and that subsequent hydrolysis of this compound provides the necessary free energy for synthetic reactions. Thus, ATP is viewed as acting as a coupling agent between catabolic and anabolic processes. The belief that the great majority of cellular reactions utilizing "high-energy phosphate bonds" are scalar in nature, involving diffusion and random collision of metabolites, precludes any assumptions other than that the concentrations (activities) of the adenine nucleotides (and in some instances that of inorganic phosphate) must ultimately determine the direction and rate of metabolic flow. A major difficulty is that the steady-state cellular levels of ATP (as well as those of ADP and AMP) remain virtually constant with time over a wide range of metabolic flux.[74] Thus, no net free energy is made available through ATP turnover.[73] Hence it would seem on first principles that the free energy change over the span of any metabolic pathway would be the decisive factor determining the feasibility for metabolic flow through it. But if this were the case, gluconeogenesis or lipid synthesis from lactate could not be accomplished by living systems, in view of the unfavorable ΔG for these processes. Evidently, other forces must be at work, but current concepts fail to identify them.

Certainly, the multiplicity of redox reactions occurring within the living cell are not seen to have important regulatory effects, since the prevailing view is that the majority of these operate close to equilibrium.[75,79-81] This outlook is again a consequence of the notion that cytoplasmic reactions take place through molecular diffusion leading to random collision, the frequency of which is a function of metabolite and coenzyme concentration. The high activity of many redox enzymes in relation to overall metabolic flux and their apparent reversibility is thus considered to present *prima facie* evidence that cytoplasmic redox reactions must function close to chemical equilibrium.

It is noteworthy that direct measurement of the concentration of the components of cellular redox reactions is not in keeping with this conclusion. Indeed, measurements of actual $[NAD^+]/[NADH]$ ratios[82] give values at least one thousandfold too low[83-85] in relation to the anticipated near-equilibrium value. However, the worth of direct measurement is discounted (in this line of reasoning) on the basis that only freely soluble components could participate in the redox reactions under study. Accordingly, much of the NAD is assumed to be "bound" and unavailable for reaction.[79-81,83-85] Concentrations of "free" coenzymes can then be assumed that will, when substituted in the appropriate equations, support the conclusion that each redox reaction is operating near to equilibrium. This circular argument leads to the assumption that 99.9% of cytoplasmic NADH is "bound" and to a calculated value for "free" cytoplasmic NADH of less than 1 μM. This result seems hardly plausible since the redox enzymes of glycolysis (for example) can be present in concentrations several orders of magnitude higher than this.[86]

Our conclusion that current concepts fail to answer the fundamental question of what provides the driving force determining the directionality and rate of metabolic flow is best illustrated by consideration of the pathways of hepatic glycolysis and gluconeogenesis. Present views of the organization of these pathways are outlined in abbreviated form in Figure 3. Since the overall (negative) ΔG for the conversion of glucose to lactate is about 200,000 kJ per mole, it is not obvious why lactate is stoichiometrically converted to glucose in the starved state. Rather, it might reasonably be predicted from Figure 3 that the only consequence of addition of lactate (or lactate plus pyruvate) to the liver cell would be to bring about a compensatory change in the relative concentrations of glyceraldehyde 3-phosphate and 1,3-diphosphoglycerate. The tacit assumption that a net metabolic flow—rather than "futile cycling"[87–89]—is feasible in the scalar system outlined in Figure 3 has never been confirmed experimentally. Nor can the scheme explain why, in the fasting state, net metabolic flow is in the direction of glucose synthesis whereas glycolysis (and lipid synthesis) is favored in the livers of fed rats. As mentioned previously, the "phosphorylation potential" ([ATP]/[ADP][Pi]) shows no significant change between these two nutritional states.[74]

Since thermodynamic considerations lead to the conclusion that the system outlined in Figure 3 cannot function in the manner proposed, conventional views argue for some means of kinetic regulation.[74,76,90,91] It is suggested that certain reactions (marked with an asterisk in Figure 3) are rate-limiting steps. These steps are catalyzed by enzymes having a substantially lower V_{max} (maximal catalytic activity) than other enzymes of the pathways. Moreover, the activity of each regulatory enzyme is subject to the influence of various modulators, which may be either inhibitors or activators and which generally are themselves metabolic intermediates. In a manner as yet unexplained the concentrations (activities) of these putative modulators are continually altered so that they interact with the regulatory enzymes in a way which consistently produces optimal rates of metabolic flow through the pathways of the cell.

The complexity of such regulatory mechanisms seems somewhat daunting and to our knowledge no actual experiments have been designed which adequately test these proposals. Since directionality is determined by turning on appropriate switches, an extreme degree of coordination is required to ensure that rates of each potentially regulator step are equal. (In metabolically active tissues, glycolytic flux can change 100-fold without significant alteration in the levels of the intermediary metabolites of the pathway.[74]) The suggestion that (in the scheme of Figure 3) the overall flux in a particular direction could be regulated by allowing glycolysis and gluconeogenesis to proceed simultaneously (futile cycling[87–89]) adds a further complication. Since the regulatory steps are functioning far from equilibrium (according to present views),

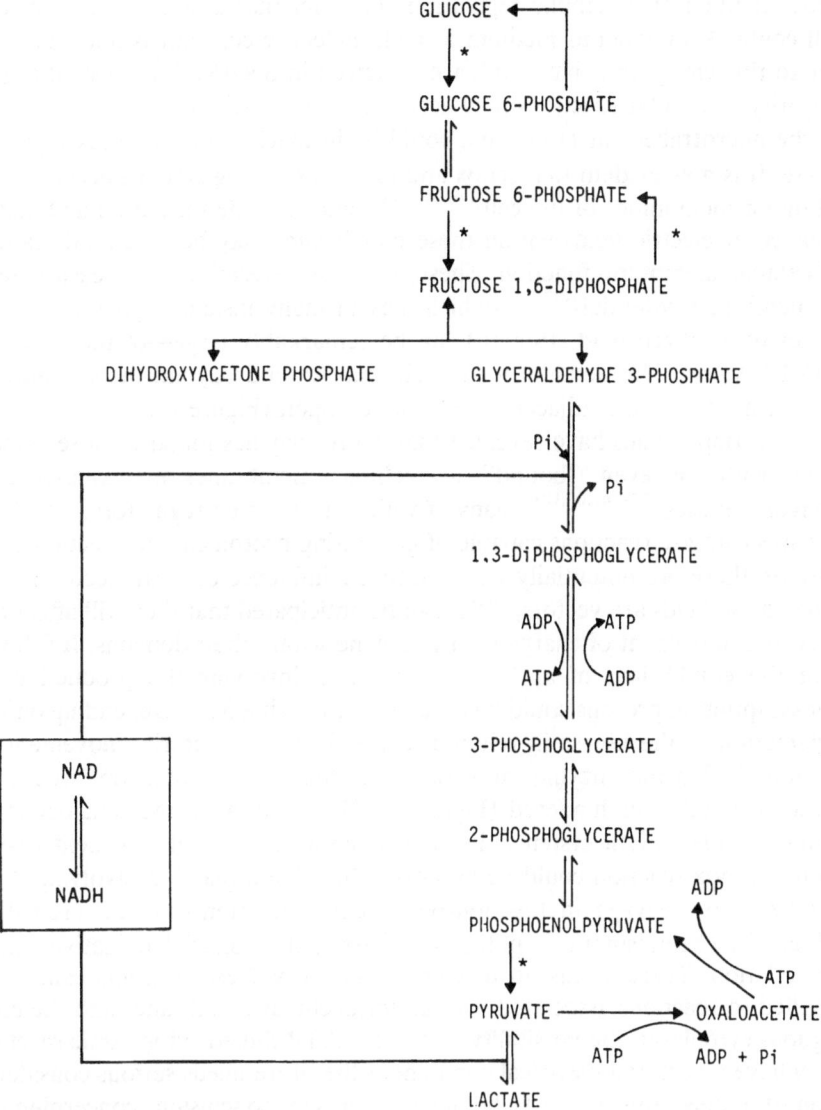

Figure 3. An abbreviated outline of current concepts of the glycolytic and gluconeogenic pathways. Putative regulatory steps are indicated with an asterisk.

the postulated model is seen as being not only extraordinarily complex, but highly inefficient in terms of energy conservation.

3.3. Influence of Cellular Electric Fields on Metabolic Processes

The likelihood that electric fields can be generated and sustained in living cells offers the opportunity for an alternative electrochemical approach to the understanding of metabolic regulation. The fact that a large portion of the cell contains an aqueous medium of high dielectric constant is not seen as a bar to this viewpoint, since evidence reviewed in Section 2 suggests that the majority of cellular reactions may occur in close association with membranes or the microtrabecular lattice and could be insulated from the bulk aqueous phase. It is now evident that redox and ion-translocatng systems occur within all major membranes of the cell,[36,37,92–96] and that the formation and maintenance of electric fields within these membranes may be a general feature of cellular membrane function. These fields, as determined by measurement of membrane potentials[97–99] (which may in many instances give underestimates of localized field strength) can be remarkably large—of the order of 10^5–10^6 V/cm. The domain of the field may extend beyond the confines of the membrane to the adjacent cytoplasmic region (Figure 4).

Fraction studies have revealed that many enzymes found in close association with, or even "bound" to, cellular membranes are kinases and dehydrogenases,[20–24,100,101] many of which are deemed regulatory.[74,74,90,91] Kinases catalyze reactions capable of generating proton charge (Section 2.3.) and are therefore potentially sensitive to the influence of local electric fields. Since these fields are vectors,[102] it can be anticipated that they will affect the vectorial movement of charge in any enzyme within their domains. It follows that the equilibrium of any cellular reaction involving the production or consumption of protons could be influenced by such a field. Depending on the orientation of the kinase in relation to the field, the vectorial movement of protons within the enzyme (an essential element of the reaction) would be either facilitated or hindered (Figure 4). Hence, in a manner analogous to inanimate electrolytic systems, the energy contained in an electric field arising from a redox reaction could be harnessed to drive a reaction involving ATP synthesis or cleavage in the nonspontaneous direction—as first argued in Mitchell's chemiosmotic hypothesis[5,27] for mitochondrial oxidative phosphorylation. There seems no *a priori* reason why these concepts cannot be applied to reactions associated with all intracellular membranes and the contiguous cytoplasm; the possibility that every ATP-linked cytoplasmic reaction may have a vectorial electrical component therefore needs serious consideration. If, indeed, this proves to be the case, present conclusions concerning the regulation of cytoplasmic metabolic pathways will need modification. The determination of whether or not electrochemical systems are poised far from equilibrium must take into account the electrical as well as the chemical potential.

For example, measurement of metabolite levels and enzyme V_{max} values in muscle or liver would appear to indicate that in these tissues the reaction catalyzed by phosphofructokinase:

$$\text{Fructose-6-phosphate}^{2-} + \text{ATP}^{4-} \rightleftharpoons \text{Fructose-1,6-diphosphate}^{4-} + \text{ADP}^{3-} + \text{H}^+$$

$$(5)$$

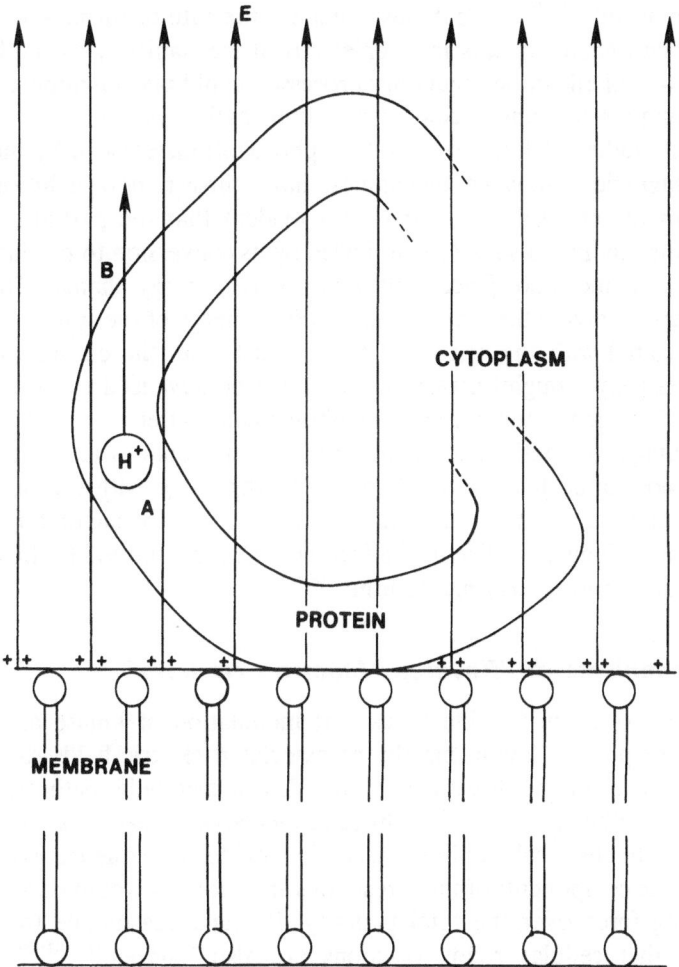

Figure 4. Schematic representation of a membrane-associated electric field extending into the adjacent cytoplasm. Proton flow through H-bonded chains with the contiguous protein is facilitated in the direction A → B, but impaired in the opposite direction.

lies far to the right.[74] However, if *in vivo* the kinase is located within the domain of an electric field so that the flow of protons from the site of catalysis is opposed, the system at steady-state may reflect a balance between the countervailing forces of chemical and electrical energies (the "static head" of irreversible thermodynamics[102]). Under this circumstance, the rate of the reaction could be influenced by the relative concentrations of fructose-6-phosphate and fructose-1,6-diphosphate. Collapse of the electric field, as in anoxia, would allow the reaction to proceed rapidly to the right under the now unrestrained influence of the chemical potential. Likewise, a heavy drain of proton current, during cellular work[61] (or through poisoning with an uncoupling agent[103-105]), would also enhance the rate of formation of fructose-1,6-diphosphate. In this manner electrical mechanisms, by modulating the expression of allosteric regulatory processes, could play an important part in controlling interactions of cellular metabolic pathways.

Recent studies of kinases (or rather phosphotransferases), by means of nuclear magnetic resonance techniques, have thrown new light on their mechanisms of catalysis.[106,107] It is now evident that the partial reactions involving the binding of substrate, followed by its conversion to product while still enzyme-bound, take place with minimal free energy change. The rate-limiting step, involving the major change in free energy of the system, appears to be associated with the release of the product from the enzyme. On this basis, a relatively straightforward proposal can be advanced to explain how an electric field generated within a membrane could alter the equilibrium of a kinase-catalyzed reaction, occurring within the domain of that electric field. It can be postulated that electrostatic forces associated with the binding of product to enzyme are opposed by the electric field, thereby facilitating release of the product (Figure 4). The field effects could be transmitted to the binding site through a proton relay mechanism.[50-52]

3.4. Mechanisms of "Reversed Electron Transfer"

It does not seem feasible to us that modulation of kinase activity by regulation of proton flow in membrane-associated electric fields can be the sole determinant of the directionality and flux of metabolic pathways. If, as suggested by others,[75,79-81] dehydrogenase-catalyzed reactions were to be poised close to chemical equilibrium, the inevitable consequence would be a pooling of the components of these reactions and proticity would be ineffective as a driving force over the total pathway. It seems reasonable to assume, therefore, that cellular redox reactions are also "driven."[64,108-110] The electrochemical interpretation of metabolism presented here provides a straightforward mechanism for explaining how this could come about.

In aerobic organisms it can be shown experimentally that an intact respiratory chain is an essential requirement for the occurrence of synthetic processes (at physiological rates). Gluconeogenesis from lactate or pyruvate, for example, can be blocked by respiratory inhibitors even though relatively high levels of cellular ATP are sustained.[111] We interpret this as an indication that some of the free energy released during the redox reactions of respiration is utilized to drive the reductive syntheses associated with gluconeogenesis.

Analogous processes occur in inanimate systems. From elementary chemical principles it is known that when two metallic "half-cells" are coupled by a salt bridge, the direction and extent of the reaction between them can be determined from knowledge of the concentrations of the various components and the value of their standard electrode potentials. The more negative the standard electrode potential, the greater the reducing power of the half-cell.

For example, since the standard electrode potential of the Cu/Cu^{2+} half-cell is $+0.34$ V and that of Zn/Zn^{2+} is -0.76 V, it can be predicted that under standard conditions metallic zinc will reduce Cu^{2+} to Cu, whereas metallic Cu cannot reduce Zn^{2+} without energy input. However, an applied voltage can be utilized to drive the reduction of Zn^{2+} by Cu.

Similar considerations apply to biological systems. The standard electrode potential of the 3-hydroxybutyrate/acetoacetate half-cell is -0.26 V and that of the lactate/pyruvate half-cell -0.19 V.[112] Hence in the presence of the respective dehydrogenases and the appropriate coenzyme (NAD), 3-hydroxybutyrate will reduce pyruvate to lactate with accumulation of acetoacetate.[113] In theory this reaction also could be driven backwards by the input of electrical energy, and this can be shown to occur experimentally under aerobic conditions.† Indeed, the facility to drive redox reactions in a thermodynamically unfavorable direction must be a fundamental requirement for maintenance of the steady-state of living systems. Otherwise, it would not be feasible for rates of reductive syntheses to match those of oxidative degradations.

The utilization of energy to drive electrons against an electrochemical gradient in biological systems has been termed "reversed electron transfer" and has been an intensely debated topic for more than twenty-five years. The concept was clearly expounded by Krebs and Kornberg in 1957.[114]

> The fact that the synthesis of, and the release of energy from, ATP is independent of the redox scale allows ATP to be used as an "energy currency" or "energy transmitter" in a wide range of situations. It can in particular act as an energy link between two oxido–reduction systems. This means that one oxido–reduction (e.g., the oxidation of ferrocytochrome by O_2), by producing ATP, can theoretically drive another similarly ATP-producing oxido–reduction backwards. Thus the system $NADH + Fp = NAD^+ + FpH_2$ could be driven from right to left to produce NADH, providing that the coupling of oxido-reduction and ATP synthesis in the system is reversible.

The analogy of a series of oxido–reduction cells was used, whereby a high electromotive force in one cell could drive a second cell backwards (see above). The situation regarding the action of ATP was seen as exactly analogous to the action of electrons in the oxido–reduction cell system,[114,115] in that ATP acted as an "energy currency," energetically linking systems of widely different potential.

These conjectures were soon supported by experimental evidence obtained with mitochondrial preparations, first by Chance and Hollunger[116] and later by many others,[117-120] who demonstrated that reversed electron transfer could involve not only flavin-coupled systems but also NAD-, NADP, and cytochrome-linked reactions. It was subsequently realized that ATP itself might not be directly involved in the energy transfer,[121] and the notion of a "high-energy intermediate" of oxidative phosphorylation was invoked

† Unpublished observations.

instead.[122] With the development of the "chemiosmotic hypothesis" by Mitchell[5,27] and suggestive evidence for the "reversibility" of oxidative phosphorylation,[123,124] the view has gradually developed that the phenomenon of reversed electron transfer represents another mode of membrane energy transduction driven by electrochemical processes, and depends on appropriate orientation of the transducing complexes within the inner mitochondrial membrane.[125,126]

If the mechanism of reversed electron transfer is indeed analogous to that of oxidative phosphorylation, it must likewise involve energy transduction rather than direct chemical coupling. Since redox reactions release or consume protons the appropriate orientation of a dehydrogenase in a membrane electric field could facilitate or hinder the related redox reaction, by promoting or depressing proton flow (cf. Figure 4). To mediate the proton flow associated with adenine nucleotide turnover we have postulated the existence of networks of hydrogen-bonded chains extending through and between proteins located within or closely associated with cellular membranes. It is conceivable that other such networks exist for the proton conduction associated with reversed electron flow, which must utilize several separate NAD(P) pools.

The actual mechanism of electron transfer appears to involve various "shuttles" whereby a reducing-equivalent carrier (e.g. malate) undergoes oxidation at one site and its oxidized form is reduced at another location.[127–130] Not infrequently, the two redox sites are in different cellular compartments. Although the prevailing view regards these shuttle components as readily diffusible, our own experimental work suggests that they remain enzyme- or carrier-bound throughout the redox cycle.[64,65]

We have now accrued considerable evidence that reversed electron transfer provides the driving force that poises the redox couples of living cells and enables reductive syntheses.[64,65,110] This work is still in progress, but already an understanding of the mechanisms by which energy flow drives proton current to poise both the phosphorylation state (by control of ATP synthesis) and the various cellular redox states (by control of reversed electron transfer) seems closer.

3.5. Regulation of Energy Flow and Heat Production

3.5.1. Cyclic Flows in Living Systems

In Figure 5 we present an alternative scheme for the glycolytic and gluconeogenic pathways, consistent with the ideas expressed in Section 3.4. The two pathways are combined into a single integrated system and we indicate the various sites where the cycle is driven, in a unidirectional manner, by proton current associated with ATP-linked or redox reactions. The diagram is greatly simplified in that, in the living cell, the system represents only one of a multiplicity of inextricably interlinked cycles, involving not only carbohy-

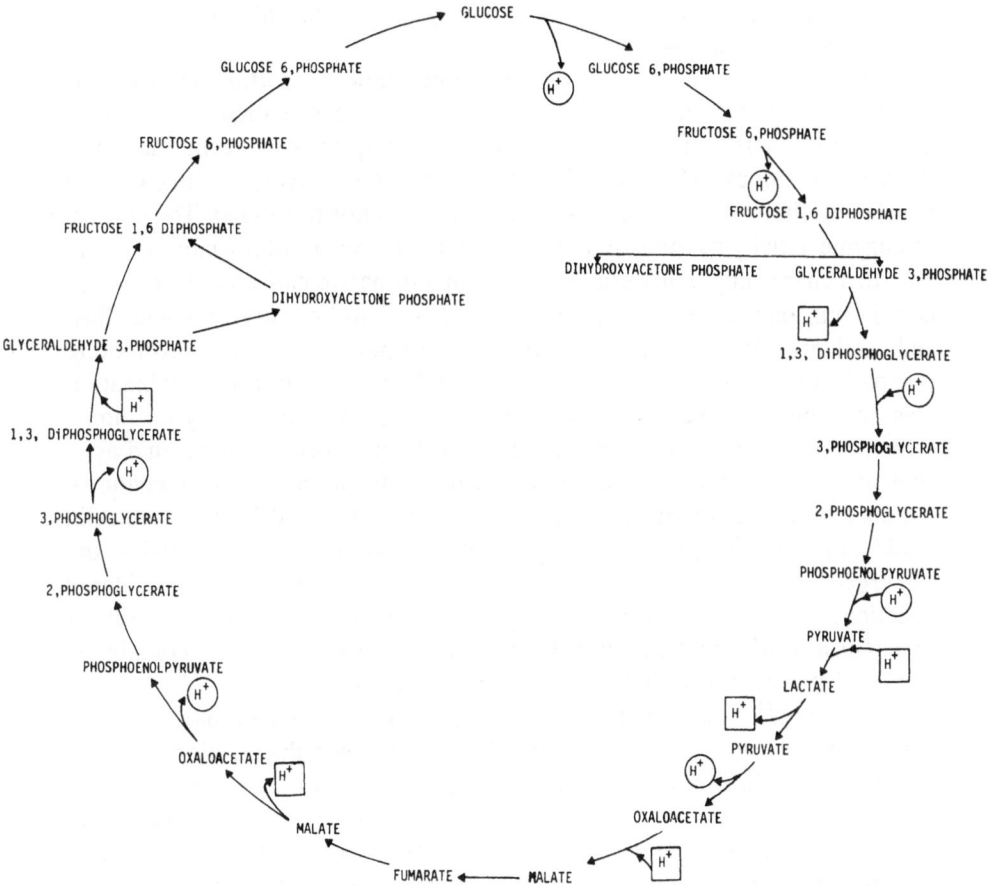

Figure 5. An alternative representation of cellular glycolysis and gluconeogenesis: $\boxed{H^+}$ represents proton flow associated with redox reactions; $\text{(H}^+\text{)}$ represents proton flow related to ATP turnover.

drate, but lipid, protein, and nucleic acid turnover. Nor does the scheme show the points where material flow between cycles occurs.

If we assume that enzyme activity is not rate-limiting and that the free energy change for any individual step of the cycle is small (cf. Section 3.3.), we can infer that net flux through any step will be a function of the ratio of substrate to product, the rate increasing the further the step is displaced from equilibrium.[131,132] At steady-state all steps will be operating at an equal rate (including the material flows in and out, which are not shown in the scheme). The system can now be perturbed in one of two ways: either by a change in the rate of material flow into or out of the system or, alternatively, there may be a change in the rate of energy flow, represented by an increase or decrease in proton current flux. In each case the steady-state will be perturbed and if the change is maintained, then new values of concentrations and flows will emerge. If the perturbation is only temporary the system will gradually return to the previous steady-state.

The system as described presents homeostatic properties, in that it is self-regulatory and does not require switches, being essentially under thermodynamic control. Hence, it is very much simpler in concept than that described previously (Section 3.2.). Of course, the possibility of a fine control overlaying the basic thermodynamic regulation is not precluded. The control of enzyme activity by induction or repression, or even allosteric effects, would lead to a channeling of material flow into an alternative pathway, since cycling could occur only at the rate of the slowest step. This fine control would allow cells and organisms to utilize resources in an optimal manner by promoting one pathway over another when a particular substrate or substrate combination was available. Alternatively, control of flux could be achieved by hormonal switching of critical proton current circuits. In this way it can be envisaged that in the starved state glucagon could facilitate the hepatic conversion of lactate to glucose. On the other hand, in the liver of the well-fed animal insulin could stimulate lipogenesis by enhancing the proton current driving the "energy-dependent transhydrogenase."[133] We have previously considered (Section 3.3.) how anoxia or a heavy drain of proton current as a result of mechanical work could alter flux. Of course, if the energy source is completely removed the system will decay to chemical equilibrium (death).[73]

Morowitz[134] has emphasized how energy flow through a model system will lead to the cyclic passage of material around loops in the reaction network. If the system is in contact with a large isothermal reservoir, a steady-state will be maintained as the material moves around the cycles and the input energy flows out as heat—"Material cycling is seen to be a very general feature of chemical networks kept at a far-from-equilibrium steady-state by the constant influx of energy in a form capable of electronic excitation." Morowitz sees such cycling as not uniquely biological: rather the biological example is a special case of a more general principle. Nevertheless, biological systems abound with unidirectional energy-driven cycles, although certain cycles may be completed only by the interaction of two or more cells or even different organisms. Indeed, the biosphere itself may be regarded as an elaborate unidirectional cyclical system driven by electromagnetic energy that is ultimately degraded to heat.

The concept of cycling has other important experimental and theoretical implications. Rates of flux through cyclical metabolic pathways will be underestimated by as much as several orders of magnitude when flux is determined by measurement of "end-product" accumulation. The measurement of isotopic exchange reactions would seem to give a much better indication of the true rate of metabolic turnover. In any steady-state system, determination of the concentrations of individual metabolites cannot give an indication of the overall flow rate. This is a well-recognized principle that obliges the use of isotopes for turnover studies [e.g., Reference (135)].

The concept of cyclical flows also calls into question the value of measurement of metabolite levels, even when attempts are made to separate the components of cell compartments.[136] It has been shown that periportal hepatic cells are primarily gluconeogenic, whereas in those lying close to the pericentral veins net flow is in favour of glycolysis.[137] This difference appears to be brought about by the gradient of O_2-tension from portal to central vein.[138] Clearly, under the circumstances, measurement of the metabolite content of whole liver would have serious limitations. In any cyclic system (occurring within one cell or over a multicellular array) an averaging of metabolite levels would be achieved by any sampling technique. The actual concentrations (activities) of the reactants for a particular step (including the redox reactions) cannot be accurately determined in this manner (cf. Figure 5).

3.5.2. Regulation of O_2-Uptake by Living Systems

Numerous studies using intact or broken cell preparations have attempted to identify the factors determining the rates of O_2-consumption by living systems. In general, conflicting conclusions have been reached. In early studies the availability of phosphate or ADP was considered the key controlling factor.[139–141] Later the "phosphorylation state" was regarded as crucial.[141–143] Others saw control being exerted at the level of the respiratory chain (e.g., cytochrome oxidase) or by adenine nucleotide translocation or individual dehydrogenases or a combination of these.[144–146]

The isolated liver cell preparation provides a useful tool for such studies. The characteristic feature of such preparations is the stimulation of O_2-consumption by added substrate, even though measurement of adenine nucleotide or inorganic phosphate concentrations fails to detect any changes resulting from substrate addition.[147] Moreover, the presence of an uncoupling agent which markedly lowers the phosphorylation state frequently fails to stimulate O_2-consumption to the extent brought about by substrate addition alone.†

Inspection of the scheme of Figure 5 provides a ready explanation for this. When substrate is added and the steady-state of the individual steps of the cycle perturbed, rates of flow through the system will necessarily increase as its homeostatic tendencies act to restore steady-state conditions. Since many of the individual steps are obligatorily coupled to proton flow, any increase in cycle flux will decrease the protonic potential, due to the increased flow of protons down their electrochemical gradient. The fall of protonic potential will bring about a corresponding fall in the electronic potential gradient since electron flow is obligatorily coupled to proton flow.[5,27] The resultant stimulation of O_2-uptake reflects the ultimate driving electrochemical reaction for aerobic metabolism

$$O_2 + 4e^- + 4H^+ \rightarrow 2H_2O$$

† Unpublished observations.

In effect we argue that the overall rate of metabolism is determined by the rate of consumption of proton current.[148] This may reflect resynthesis of ATP utilized in new bond synthesis, or the utilization of protonmotive force for ("electrogenic") metabolite translocation[33,34] or reversed electron transfer, associated with reductive syntheses.[111-118] In the basal resting state, it can be anticipated that, due to the very small free energy changes associated with the multiplicity of individual reactions which contribute to metabolic turnover, the overall fuel-cell reaction is poised close to electrochemical equilibrium, i.e., under thermodynamic rather than kinetic control. Only when the system is perturbed substantially—for example due to heavy current drain (cf. Section 2.3.)—do kinetic factors (such as enzyme activity or translocating capacity) become rate-limiting, giving rise to substantial departure from equilibrium ("over-potential").[149]

3.5.3. Regulation of Heat Production

In animals, chemical bond energy in the form of foodstuffs is utilized for the living processes of growth, repair, reproduction, movement, thought, and so forth, before being radiated from the body as heat. The amount of energy flow and corresponding heat loss is in part, a function of the degree of activity of the animal, and high activity requires increased food intake or alternatively accelerated breakdown of the body's energy stores. However, even at rest a substantial degree of energy flow occurs so that the heat loss of basal metabolism over a 24-hour period is about 50% of the total heat production for a moderately active man.[77]

The processes which go to make up the basal metabolism are not well delineated. Nevertheless, it is known that the basal metabolic rate is reasonably constant for a given individual. Although affected by age, sex, diet, training, etc., the major influence appears to be body size.[77] It has been established for many years that under similar conditions the basal metabolism of various mammals, as measured by the rate of heat loss in the basal state, is approximately proportional to the surface area, whereas heat loss is inversely related to body weight.[77] In consequence, small mammals such as the shrew with a large surface to volume ratio have a basal metabolic rate some sixty times greater than man and perhaps two hundred times that of an elephant.

Few biologists have addressed themselves to the question of how the tissues of a small mammal such as a mouse are capable of far greater metabolic flux than those of a man. Some twenty-five years ago Krebs[150] examined the problem by studying the respiration of a number of tissues from mammals of different sizes. In several cases no differences were found. This is rather surprisng considering the remarkable variation in metabolic rate between those same species *in vivo* based on the relationship to body weight. Current metabolic theory then is unable to answer the question of why the heat loss per gram of mouse is ten times that per gram of man since the discrepancies cannot be accounted for by differences in the proportions of inert tissues. It is not known what metabolic fluxes give rise to the extra heat in the smaller mammal. A question equally puzzling is how, for man and mouse alike, are the rates of flow in the metabolic pathways associated with heat production coordinated with the rates of heat loss. The same question arises when heat production and loss are considered for a single individual from birth to adulthood. At birth the surface area to body weight ratio is high and metabolism per unit mass must be much greater than in the adult. In mammals the mature foetus, which exists in protected surroundings, with presumably a major portion of its heat requirement derived from its mother, is on delivery suddenly exposed to a hostile cool environment and must rely on its own metabolic activity to maintain its body heat. The wonder is not that it achieves the transition rather precariously, but that it achieves it at all.

In the electrochemical interpretation of metabolism presented here, we discount the significance of chemical reactions occurring under conditions well removed from equilibrium as important sources of heat production. Rather, we identify those processes associated with electronic and protonic current flow as the critical elements in heat production. The relationship between body surface area and heat production becomes more readily understandable when it is appreciated that these heat-producing processes take place within cellular membranes (or the microtrabecular lattice). Thus, the finding that heat production is a function of body surface area rather than tissue mass may be a macroscopic reflection of the fact that cellular electrochemical reactions occur in two dimensions, in contrast with three-dimensional, scalar chemical reactions.

The existence of specialized mechanisms for heat production are now well recognized. Perhaps the best known example is brown fat tissue, which is of great importance in the hibernating animal.[7,151-152] Nevertheless, it is still necessary to account for the exact balance achieved between heat loss and heat production in living systems. Again, followng Mitchell,[5,27] we can assume that metabolism is directed towards the storage of separated charge across intracellular membranes and that this storage capacity is finite, being dependent on membrane area. Since membranes are unlikely to be perfect capacitors, leakage of charge is inevitable. It is only necessary to postulate that this charge dissipation is a function of the temperature gradient across the membrane, to explain how heat balance is achieved. (A current flow as a result of a temperature differential is a not uncommon circumstance in semiconductor systems and is used to advantage in the design of thermistors.) Heat loss reflects dissipation of protonic capacitance and this in turn promotes metabolic flow to restore the proton gradient. A steady-state of proton current generation and dissipation is achieved, corresponding to the basal metabolic rate.

On this basis it is easy to understand why a reptile at 37°C may have a far lower basal metabolic rate than a mammal. For in the poikilotherm the ambient temperature must also be close to 37°C, leading to a very small temperature gradient between animal and environment. In contrast the mammal will function at 37°C even (for example) when the ambient temperature is less than 20°C. It follows that much of the basal metabolism in the mammal is associated with maintenance of body temperature. These considerations in relation to heat production imply that living processes are probably far more efficient in terms of energy conservation than has previously been appreciated.

4. Electrochemical Processes and Disease

4.1. The Free Radical Theory of O_2 Toxicity

Despite the enormous amount of biochemical knowledge that has accumulated over the past one hundred years, we remain generally ignorant concerning the molecular mechanisms associated with the major diseases that still afflict mankind. None of them (with the possible exception of diabetes) are associated with obvious derangement of an important metabolic pathway. One cannot help but wonder if classical biochemical knowledge is deficient in crucial areas. The interpretation of metabolism on an electrochemical basis introduces a new dimension and raises the question of whether disease states may be related to disorders of fundamental electrochemical mechanisms, associated with the living state.

In fact, there is already in existence a theory of special relevance to cellular electrochemical processes—the "free radical theory of O_2 toxicity." This theory proposes that free radicals generated by redox reactions in living tissues interact with membranes to bring about lipid peroxidation.[153] The theory, first advanced by Gerschmann[154] in 1954, received a considerable boost 15 years later with the discovery of the superoxide dismutases[155] which were considered to represent defense mechanisms against the accumulation of toxic quantities of superoxide (O_2^-). It was further strengthened by the demonstration of the ubiquity of superoxide dismutases in aerobic organisms[156] and the discovery that superoxide is an important component of the mechanism utilized by phagocytic leukocytes to kill ingested microorganisms.[157]

Despite these advances, the free-radical theory of O_2 toxicity has not had a major impact on the elucidation of disease processes. The molecular mechanisms by which superoxide initiates cellular damage have not been determined; indeed, convincing arguments[158-160] have been advanced that superoxide is not, *per se*, a particularly reactive agent (since the unpaired electron is shared by both O_2 atoms[160]). A further conceptual difficulty arises from the assumption that superoxide production is generally a minor though unfortunate side event of aerobic metabolism, which induces damage due to the imperfection of the defences evolved to protect the cell. It is not easy to envisage why the selective pressures of evolution should fail in this instance to eliminate a side reaction that is not only of no apparent survival value but is harmful to life.

In this Section we address these difficulties by proposing that the formation of superoxide within energy-transducing membranes, by univalent reduction of O_2:

$$O_2 + e^- \rightarrow O_2^-$$

can be a fundamental step in the establishment of the protonic electrochemical potential.[5,27] It is further argued that superoxide may play an essential role in hormonal, neuromuscular, or trans-synaptic signal transmission. It follows that damage to membranes through lipid peroxidation would be expected to have deleterious effects on a multiplicity of cellular functions.

4.2. The Rate and Function of Superoxide Production

Although hydrogen peroxide (H_2O_2) production was thought, early in this century, to be an important pathway of O_2 metabolism, the subsequent discoveries of the cytochromes by Keilin and "atmungsferment" by Warburg (for review see Chance[161]) led to the conclusion that the tetravalent reduction of O_2 was the only physiological route. In the last ten years, however, there has been renewed interest in H_2O_2 production and it has been shown that the rate of formation of H_2O_2 in mitochondria,[162] which stoichiometrically reflects the rate of univalent reduction of O_2 to superoxide,

$$2O_2^- + 2H^+ \rightarrow 2HO_2 \rightarrow H_2O_2 + O_2$$

can in certain circumstances reach 15% of the total O_2 consumption.[163] When the rates of H_2O_2 and superoxide production by peroxisomes and the endoplasmic reticulum[164,165] are considered together, the rate of O_2 reduction in liver cells by these routes can equal that of the tetravalent pathway.[65] It seems an inescapable conclusion from the observation of rates of this magnitude that superoxide production must be involved in normal physiological functions.

Current concepts envisage the tetravalent reduction of O_2 to H_2O to be the only O_2-reducing process in living systems that can be coupled to energy transduction in animal cells. Accordingly, the inner membrane of the mitochondrion is assumed to be the only site of energy transduction, and ATP the primary end product. In fact, as discussed previously, we regard reversed electron transfer and ion and metabolite translocation as alternative mechanisms of free energy conservation not necessarily linked to the synthesis of ATP.[111,148] Thus, although only the inner mitochondrial membrane of the animal cell contains ATP synthetase, the possibility exists that energy-conserving processes may occur in other cellular membranes.

These membranes are devoid of cytochrome oxidase, and the alternative O_2-utilizing enzymes within them generate superoxide as a primary product. The question arises therefore as to whether superoxide formation could be linked to free energy conservation. A plausible mechanism can be suggested, dependent on the generation of superoxide in the aprotic environment of a lipid bilayer where it would not readily undergo further reduction.[160] If the superoxide formed enters the aqueous phase, it will immediately abstract a proton from water and undergo disproportionation, a reaction catalyzed by superoxide dismutase. This will result in an excess of positive charge within the membrane, and an excess of OH^- and negative charge in the aqueous compartment in which the superoxide has reacted (Figure 6). A membrane potential, negative inside, will be established simultaneously. It can be anticipated, as Mitchell[27] has postulated, that the combined pH gradient (ΔpH) and electric membrane potential ($\Delta\Psi$) provide a protonic potential (Δp) with a magnitude sufficient to drive redox reactions or transport processes against an unfavorable electrochemical potential gradient. Thus, potential energy is stored in both membranous and bulk aqueous phases. The scheme proposed here is very similar to that originally advanced by Mitchell[27] except that the membrane is not merely a passive barrier across which charge is separated. Rather, it is charge separation localized within the membrane itself which provides the primary driving force.[8,28]

It is tempting to speculate that the univalent reduction of O_2, which is the most favorable pathway for O_2 reduction, was the primitive pathway

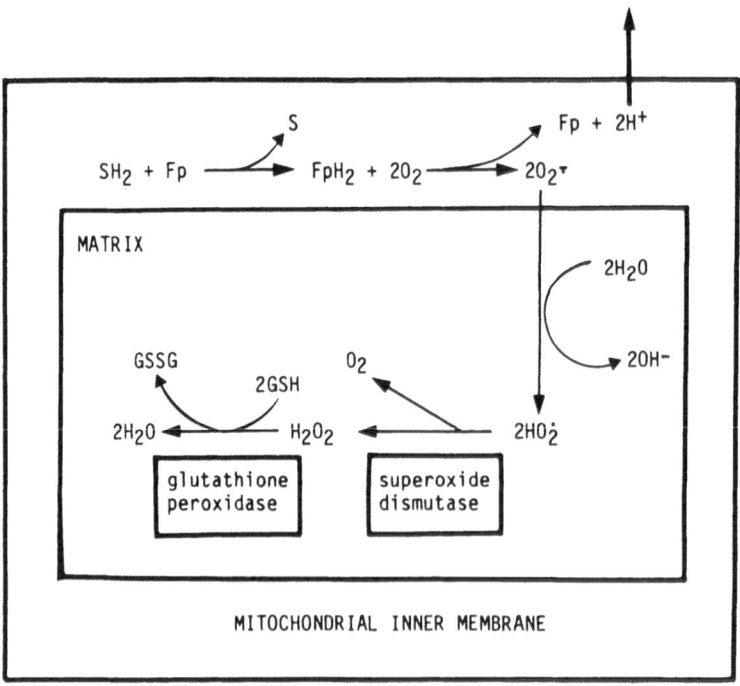

Figure 6. A possible mechanism for generation of a protonic electrochemical gradient across the inner mitochondrial membrane.

evolved by living systems when O_2 tension on earth was low. Later the tetravalent pathway to water formation, catalyzed by cytochrome oxidase, developed in association with the increase in atmospheric oxygen.[166] Under normal conditions the tetravalent reduction of O_2 will predominate since the corresponding fuel cell $(O_2 + 4e^- + 4H^+/2H_2O)$ has a potential of 1.3 V, whereas that of the fuel cell utilizing univalent O_2 reduction $(O_2 + 2e^- + 2H^+/H_2O_2)$ is only about 0.5 V (in aprotic solvents).[160] Hence under normal conditions of cellular O_2 tension, the cytochrome oxidase system will suppress the univalent O_2 reduction pathway. In fact, it will drive it backwards. In these circumstances O_2^- will act as a reductant rather than an oxidant. Indeed, it is possible that superoxide can act as a normal electron carrier between the relatively immobile macromolecular components of the mitochondrial respiratory chain. When cellular O_2 tension is so low that electron transfer to O_2 catalyzed by the cytochrome oxidase system is limited, the univalent pathway of O_2 reduction may become predominant. It can be anticipated that any increase in the activity of this pathway, with a corresponding increase in free-radical formation, may lead to membrane damage through lipid peroxidation. This should not be taken to imply that the univalent pathway is vestigial and only manifest in disease states. It seems likely that there are many important physiological functions which utilize superoxide for energy transduction, particularly when O_2 tensions are low; for example, developmental processes and growth.

Past emphasis on ATP synthesis as the exclusive means of energy transduction in animal cells has led to the assumption that redox reactions in membranes devoid of ATP synthetases must necessarily be energetically wasteful. However, modern views that charge separation is the primary energy-transducing mechanism encourage the suggestion that this process may be mediated by superoxide production within—and proton abstraction across—endoplasmic reticulum, and plasma membranes which lack ATP synthetase but are rich in elements that generate superoxide.[167] The role of the electric fields so generated in modulating enzymic reactions involving vectorial proton flux can be expected to be similar to that of fields established across the mitochondrial inner membrane.

4.3. Possible Roles of Superoxide in Hormonal and Neuromuscular Signal Transmission

The search for the "second messenger" for mediation of insulin action has up to now been as unsuccessful as the quest for "high-energy" intermediates of oxidative phosphorylation. A not unreasonable possibility, therefore, is that the second messenger does not exist as a separate chemical entity and that insulin action is mediated by electronic processes involving macromolecular semiconduction. Support for this view comes from the finding that the most rapid observed action of insulin is a hyperpolarization of the muscle cell plasma membrane (within 1 sec).[35] Insulin action is accompanied by H_2O_2 production.[168]

A plausible mechanism envisages insulin acting in the manner of a switch which allows a microvoltage to be established across the plasma membrane at the site of the insulin receptor, by means of superoxide production and proton abstraction. This signal would be amplified within the receptor complex and transmitted to sensitive intracellular sites by a process of solid-state semiconduction, possibly through the microtrabecular lattice. Transmembrane electric fields are thus established which promote membrane–protein association (e.g., hexokinase "binding" to mitochondria[97]) and the anabolic reactions of lipid and protein synthesis, known to be associated with insulin action. A likely target for activation would appear to be the mitochondrial energy-dependent transhydrogenase, responsible for the formation of NADPH from NADH.[133]

The possibility that the actions of other hormones or cellular surface-active agents (e.g., antibodies) are initiated by means of superoxide generation seems worthy of exploration. This may be true for the large group of hormones whose action is mediated by means of cyclic AMP formation.[169] For example, evidence has been obtained that glucagon initiates a hyperpolarization of the liver plasma membrane,[170] and lipid transformations within it,[171] while parathyroid hormone stimulates synthesis of glucose-6-phosphate dehydrogenase[172] which may be supplying NADPH for H_2O_2 removal.

It would also seem of value to explore further the molecular basis for the action of neurogenic amines capable of serving as transmitters at the neuromuscular or intersynaptic regions.[173] It is known that these amines are catabolized (with superoxide production) within the outer mitochondrial membrane.[174] The now well-recognized occurrence of flavin-linked superoxide-generating redox reactions within cellular plasma membranes[175] raises the possibility that amine-induced signal transmission may be dependent on superoxide-linked transmitter metabolism after the transmitter binds to specific receptor sites.

4.4. Possible Mechanisms by Which Superoxide Brings About Cell Damage and Promotes Lipid Peroxidation

A metal ion not infrequently implicated in superoxide-associated damage to cells is Fe^{3+}.[176] It is known that Fe^{3+} can form complexes with superoxide,[160] thereby leading to the formation of peroxide or even possibly OH·radicals, which are highly damaging.[176] It is also feasible that damage can result from the interaction of superoxide with other metal ions. Isolated mitochondria swell rapidly when incubated in the presence of Ca^{2+}.[177] Such cells can be protected for long periods by 10% dimethylsulfoxide,[178] an agent which readily destroys peroxide. Suspensions of liver cells tolerate ischemic conditions for far longer periods in the absence of added Ca^{2+} and especially when a chelating agent (e.g., EDTA) is present.†

It is known that superoxide forms complexes with Ca^{2+}, thereby permitting formation of calcium peroxide.[160] Since the majority of biological membranes translocate Ca^{2+} or other alkali metal ions, the opportunity for interaction with superoxide within the membrane may be considerable (Ca^{2+} and calmodulin are required for superoxide production by leukocytes[179]). Under pathological circumstances calcium peroxide could be formed and lipid peroxidation initiated. The breakdown of membrane capacitance as a consequence of this may be deleterious in a number of ways. For example, the loss of a critical electric field may remove the driving force for energy-dependent processes associated with charge transfer.[59] On the other hand, circumstances may be established which bring about unrestrained and unmodulated metabolic activity, as in cancer.[180]

Of course, superoxide itself may not be the radical species which initiates membrane damage. Its formation, however, signifies the operation of the univalent radical pathway of electron transport. It is tempting to speculate, for example, that the univalent reduction of O_2 catalyzed by alcohol dehydrogenase may contribute to the cellular damage induced by ethanol,[181–183] whereas the pathway of ethanol oxidation involving tetravalent O_2 reduction is relatively innocuous. Factors which alter the distribution of electron flow between the univalent and tetravalent pathway for O_2 reduction may be of major significance in determining the balance of health and disease.

4.5. Conclusion

The view that superoxide production is merely an occasional but toxic side-reaction of aerobic metabolism has, with a few exceptions, dominated thinking in this area for the last ten years.[184,185] This attitude would seem to devalue to some extent the free radical theory of O_2 toxicity. We argue here that superoxide production, far from being trivial, is a fundamental and universal feature of membrane function under aerobic conditions, and is intimately involved in energy transduction and signal transmission. Accordingly, it is not unexpected that gradual but cumulative membrane injury can occur over the lifetime of the cell and organism. It is easy to envisage how damage to the elaborate circuitry of the cell, brought about in this manner, might form the primary basis for a large variety of disease processes.

Thus, the electrochemical interpretation of metabolism adopted here allows not only a new approach to an understanding of metabolic regulation, but also provides possible insights into the mechanism of many diseases.

---

† G. Mannaerts, personal communication.

Acknowledgments

We thank Departmental colleagues, Drs. G. J. Barritt and A. M. Edwards for their constructive criticisms. Our thanks are also due to Drs. R. J. Havel, A. L. Lehninger, H. J. Morowitz, J. F. Nagle, H.-D. Soling, M. Stubbs, and particularly J. O'M. Bockris and G. R. Welch for helpful discussion.

We are grateful to the National Health and Medical Research Council of Australia, The National Heart Foundation of Australia, and the Alcohol and Drug Addicts Treatment Board of South Australia for support during the development of these concepts. We thank our colleagues Sara Norton, Stephanie Gay, Chris Farrington, Josephine Weekley, and Julie-Anne Burton for their dedicated assistance in preparing this Chapter.

References

1. A. Szent-Gyorgyi, *Nature* **148**, 157 (1941).
2. M. G. Del Duca and J. M. Fuscoe, *Int. Sci. Technol.* **3**, 56 (1965).
3. J. O'M. Bockris and S. Srinivasan, *Nature* **215**, 197 (1967).
4. J. O'M. Bockris, *Nature* **224**, 775 (1969).
5. P. Mitchell, *Nature* **191**, 144 (1961).
6. F. W. Cope, in *Bioelectrochemistry*, H. Keyzer and F. Gutmann, eds., Plenum Press, New York (1980), pp. 297–329.
7. D. G. Nicholls, *Biochim. Biophys. Acta* **549**, 1 (1979).
8. D. B. Kell, *Biochim. Biophys. Acta* **549**, 55 (1979).
9. J. O'M. Bockris, in *Bioelectrochemistry*, H. Keyzer and F. Gutmann, eds., Plenum Press, New York (1980), pp. 5–17.
10. J. S. Clegg, in *Cell-Associated Water*, W. Drost-Hansen and J. Clegg, eds., Academic Press, New York (1979), pp. 363–413.
11. E. Lazarides, *Nature* **283**, 249 (1980).
12. M. M. Fisher and M. J. Phillips, in *Progress in Liver Diseases*, Vol. 6, H. Popper and F. Schaffner, eds., Grune and Stratton, New York (1979), pp. 105–121.
13. J. J. Wolosewick and K. R. Porter, *J. Cell Biol.* **82**, 114 (1979).
14. K. R. Porter, J. B. Tucker, *Sci. Am.* **244**, 40 (1981).
15. M. N. Berry and F. O. Simpson, *J. Cell Biol.* **15**, 9 (1962).
16. M. N. Berry and D. S. Friend, *J. Cell Biol.* **43**, 506 (1969).
17. E. S. Kempner and J. H. Miller, *Exp. Cell Res.* **51**, 150 (1968).
18. M. Zalokar, *Exp. Cell Res.* **19**, 114 (1960).
19. R. Coleman, *Biochim. Biophys. Acta* **300**, 1 (1973).
20. C. J. Masters, *Curr. Top. Cell. Reg.* **12**, 75 (1977).
21. I. A. Rose and J. V. B. Warms, *J. Biol. Chem.* **242**, 1635 (1967).
22. H. S. Bachelard, *Biochem. J.* **104**, 286 (1967).
23. P. A. Craven and R. E. Basford, *Biochim. Biophys. Acta* **354**, 49 (1974).
24. T. Higashi, C. S. Richards, and K. Uyeda, *J. Biol. Chem.* **254**, 9542 (1979).
25. .A. V. Loud, *J. Cell Biol.* **37**, 27 (1968).
26. A. Blouin, R. P. Bolender, and E. R. Weibel, *J. Cell Biol.* **72**, 441 (1977).
27. P. Mitchell, *Biochem. Soc. Trans.* **4**, 399 (1976).
28. R. J. P. Williams, *FEBS Lett.* **85**, 9 (1978).
29. J. O'M. Bockris and M. S. Tunuli, *J. Electroanal. Chem.* **100**, 7 (1979).
30. E. Racker, *Fed. Proc. FASEB* **39**, 210 (1980).

31. J. Teissie, B. E. Knox, T. Y. Tsong, and J. Wehrle, *Proc. Nat. Acad. Sci. U.S.A.* **78**, 7473 (1981).
32. M. Avron, *Abst. 12th Int. Nat. Congr. Biochem.*, p. 26, (1982).
33. L. F. Jaffe, *Fed. Proc. FASEB* **40**, 125 (1981).
34. J. A. Williams, *Fed. Proc. FASEB* **40**, 128 (1981).
35. K. Zierler and E. M. Rogus, *Fed. Proc. FASEB* **40**, 121 (1981).
36. D. B. Wilson, *Ann. Rev. Biochem.* **47**, 933 (1978).
37. J. T. Wiskich, *Ann. Rev. Plant. Physiol.* **28**, 45 (1977).
38. R. I. Woodruff and W. H. Telfer, *Nature* **286**, 84 (1980).
39. K. R. Robinson and L. F. Jaffe, *Science* **187**, 70 (1975).
40. R. F. Stump, K. R. Robinson, R. L. Harold, and F. M. Harold, *Proc. Nat. Acad. Sci. U.S.A.* **77**, 6673 (1980).
41. J. Ladik, *Int. J. Quantum Chem. Quantum Biol. Symp.* **3**, 237 (1976).
42. J. Lewis, in *Submolecular Biology and Cancer*, G. E. W. Wolstenholme, D. W. Fitzsimons, and J. Whelan, eds., Ciba Foundation Symp. **67**, Elsevier Excerpta Medica, Amsterdam (1979), pp. 65–82.
43. R. Pethig, *Dielectric and Electronic Properties of Biological Materials*, John Wiley & Sons, Chichester (1979).
44. R. Pethig and A. Szent-Gyorgyi, in *Bioelectrochemistry*, H. Keyzer and F. Gutmann, eds., Plenum Press, New York (1980), pp. 227–265.
45. W. A. Pryor, in *Free Radicals in Biology*, Vol. 1, W. A. Pryor, ed., Academic Press, New York (1976), pp. 1–49.
46. H. Nohl and J. Hegner, *Eur. J. Biochem.* **82**, 563 (1978).
47. P. A. Loach and B. J. Hales, in *Free Radicals in Biology*, Vol. 1, W. A. Pryor, ed., Academic Press, New York (1976), pp. 199–237.
48. R. E. White and M. J. Coon, *Ann. Rev. Biochem.* **49**, 315 (1980).
49. T. Henriksen, T. B. Melo, and G. Saxebol, in *Free Radicals in Biology*, Vol. 2, W. A. Pryor, ed., Academic Press, New York (1976), pp. 213–256.
50. B. Hille, *Ann. Rev. Physiol.* **38**, 139 (1976).
51. D. E. Metzler, *Adv. Enzymol.* **50**, 1 (1979).
52. J. Kraut, *Ann. Rev. Biochem.* **46**, 331 (1977).
53. J. F. Nagle and H. J. Morowitz, *Proc. Nat. Acad. Sci. U.S.A.* **75**, 298 (1978).
54. J. F. Nagle, M. Mille, and H. J. Morowitz, *J. Chem. Phys.* **72**, 3959 (1980).
55. L. Onsager, *Science* **166**, 1359 (1969).
56. C. M. Anderson, F. H. Zucker, and T. A. Steitz, *Science* **204**, 375 (1979).
57. H. Gutfreund and D. R. Trentham, in *Energy Transformation in Biological Systems*, G. E. W. Wolstenholme and E. W. Fitzsimons, eds., Ciba Foundation Symp. **31**, Elsevier/Excerpta Medica, Amsterdam, New York (1975), pp. 69–86.
58. D. B. Kell and J. G. Morris, in *Vectorial Reactions in Electron and Ion Transport in Mitochondria and Bacteria*, F. Palmieri, F. Quagliariello, N. Siliprandi, and F. C. Slater, eds., Elsevier/North-Holland Biomedical Press (1981), pp. 339–347.
59. D. B. Kell, D. J. Clarke, and J. G. Morris, *FEMS Lett.* **11**, 1 (1981).
60. A. J. Hopfinger, *Intermolecular Interactions and Biological Organization*, J. Wiley & Sons, New York (1977).
61. V. P. Skulachev, *Can. J. Biochem.* **58**, 161 (1980).
62. D. E. Atkinson, *Biochemistry* **7**, 4030 (1968).
63. K. A. Schellenberg, in: *Pyridine Nucleotide-Dependent Dehydrogenases*, H. Sund, ed., Springer-Verlag, Berlin (1970), pp. 15–29.
64. M. N. Berry, *FEBS Lett.* **117** (Suppl.), K106 (1980).
65. M. N. Berry, A. R. Grivell, and P. G. Wallace, *Pharmac. Biochem. Behav.* **18** (Suppl.), 201 (1983).
66. H. H. Mollenhauer and D. J. Morré, *Subcell. Biochem.* **5**, 327 (1978).
67. J. P. Reeves and T. Reames, *J. Biol. Chem.* **256**, 6047 (1981).

68. A. Heinz, G. Sachs, and J. A. Schafer, *J. Mem. Biol.* **61**, 143 (1981).
69. R. N. Rosier, D. A. Tucker, S. Meerdink, I. Jain, and T. E. Gunter, *Arch. Biochem. Biophys.* **210**, 549 (1981).
70. G. R. Welch and J. A. DeMoss, in *Microenvironments and Metabolic Compartmentation*, P. A. Srere and R. W. Estabrook, eds., Academic Press, New York (1978), pp. 323–344.
71. G. R. Welch and T. Keleti, *J. Theor. Biol.* **93**, 701 (1981).
72. R. Lumry and R. Biltonen, *Struct. Stabil. Biol. Macromol.* **2**, 65 (1969).
73. B. E. C. Banks and C. A. Vernon, *Trends Biochem. Sci.* **3**, 156 (1978).
74. E. A. Newsholme and C. Start, *Regulation in Metabolism*, John Wiley & Sons, London (1973).
75. Th. Bucher and M. Klingenberg, *Angew. Chem.* **70**, 552 (1958).
76. B. Hess, *Symp. Soc. Exp. Biol.* **27**, 105 (1973).
77. M. Kleiber, *The Fire of Life*, Robert E. Krieger Publishing Company, New York (1975).
78. E. Schrodinger, *What is Life?*, Cambridge University Press, London (1944).
79. D. H. Williamson, P. Lund, and H. A. Krebs, *Biochem. J.* **103**, 514 (1967).
80. H. J. Hohorst, F. H. Kreutz, and Th. Bucher, *Biochem. Z.* **332**, 18 (1959).
81. Th. Bucher, B. Brauser, A. Conze, F. Klein, O. Langguth, and H. Sies, *Eur. J. Biochem.* **27**, 301 (1972).
82. M. E. Tischler, D. Friedrichs, K. Coll, and J. R. Williamson, *Arch. Biochem. Biophys.* **184**, 222 (1977).
83. H. A. Krebs, in *Advances in Enzyme Regulation*, Vol. 5, G. Weber, ed., Pergamon Press, Oxford (1967), pp. 409–432.
84. R. L. Veech, L. V. Eggleston, and H. A. Krebs, *Biochem. J.* **115**, 609 (1969).
85. H. A. Krebs, *Symp. Soc. Exp. Biol.* **27**, 299 (1973).
86. A. Sols and R. Marco, *Curr. Top. Cell. Reg.* **2**, 227 (1970).
87. M. G. Clark, D. P. Bloxham, P. C. Holland, and H. A. Lardy, *Biochem. J.* **134**, 589 (1973).
88. E. A. Newsholme and B. Crabtree, *Biochem. Soc. Symp.* **41**, 61 (1976).
89. J. Katz and R. Rognstad, *Curr. Top. Cell. Reg.* **10**, 237 (1976).
90. R. M. Cohn, in *Principles of Metabolic Control in Mammalian Systems*, R. H. Herman, R. M. Cohn, and P. C. McNamara, eds., Plenum Press, New York (1980), pp. 93–133.
91. K. Dalziel, *Symp. Soc. Exp. Biol.* **27**, 21 (1973).
92. R. K. Crane, *Rev. Physiol. Biochem. Pharmacol.* **78**, 99 (1977).
93. G. A. Kimmich, *Fed. Proc.* **40**, 2474 (1981).
94. E. Heinz, P. Geck, and C. Pietrzyk, *Ann. N.Y. Acad. Sci.* **264**, 428 (1975).
95. H. Low and F. L. Crane, *Biochim. Biophys. Acta* **515**, 141 (1978).
96. I. C. Gunsalus and S. G. Sligar, *Adv. Enzymol.* **47**, 1 (1978).
97. H. Rottenberg, *Methods Enzymol.* **55**, 547 (1979).
98. C. L. Bashford and J. C. Smith, *Methods Enzymol.* **55**, 569 (1979).
99. J. B. Hoek, D. G. Nicholls, and J. R. Williamson, *J. Biol. Chem.* **255**, 1458 (1980).
100. S. P. Bessman and P. J. Geiger, *Curr. Top. Cell. Reg.* **16**, 55 (1980).
101. R. H. Herman, in *Principles of Metabolic Control in Mammalian Systems*, R. H. Herman, R. M. Cohn, and P. D. McNamara, eds., Plenum Press, New York (1980), pp. 1–61.
102. A. Katchalsky and P. F. Curran, *Non-equilibrium Thermodynamics in Biophysics*, Harvard University Press, Cambridge, MA (1965).
103. G. F. Azzone, T. Pozzan, and S. Massari, *Biochim. Biophys Acta* **501**, 307 (1978).
104. P. C. Laris, D. P. Bahr, and R. R. J. Chaffee, *Biochim. Biophys Acta* **376**, 415 (1975).
105. K. E. O. Akerman and J. O. Jarvisalo, *Biochem. J.* **192**, 183 (1980).
106. B. D. Nageswara Rao, F. J. Kayne, and M. Cohn, *J. Biol. Chem.* **254**, 2689 (1979).
107. J. R. Knowles, *Ann. Rev. Biochem.* **49**, 877 (1980).
108. H. Sies, T. P. M. Akerboom, and J. M. Tager, *Eur. J. Biochem.* **72**, 301 (1977).
109. H. Sies, in *Alcohol and Aldehyde Metabolizing Systems* Vol. 3, R. G. Thurman, J. R. Williamson, H. Drott, and B. Chance, eds., Academic Press, London (1977), pp. 47–64.
110. M. N. Berry, A. R. Grivell, and P. G. Wallace, *FEBS Lett.* **119**, 317 (1980).
111. M. N. Berry, A. R. Grivell, and P. G. Wallace, *Eur. J. Biochem.* **131**, 205 (1983).

112. P. A. Loaoh, in *CRC Handbook of Biochemistry*, H. A. Sober, ed., The Chemical Rubber Co., Cleveland, OH (1968), pp. J2–J34.

113. M. N. Berry, *Anal. Biochem.* **6**, 352 (1963).

114. H. A. Krebs and H. L. Kornberg, *Ergebn. Physiol.* **49**, 212 (1957).

115. R. E. Davies and H. A. Krebs, *Biochem. Soc. Symp.* **8**, 77 (1952).

116. B. Chance and G. Hollunger, *J. Biol. Chem.* **236**, 1577 (1961).

117. M. Klingenberg and P. Schollmeyer, *Biochem. Z.* **335**, 243 (1961).

118. J. M. Tager, *Biochim. Biophys. Acta* **77**, 258 (1963).

119. L. Packer, *J. Biol. Chem.* **237**, 1327 (1962).

120. L. Danielson and L. Ernster, *Biochem Z.* **338**, 118 (1963).

121. A. M. Snoswell, *Biochim. Biophys. Acta* **60**, 143 (1962).

122. E. C. Slater, *Nature* **172**, 975 (1953).

123. D. F. Wilson, M. Stubbs, R. L. Veech, M. Erecinska, and H. A. Krebs, *Biochem. J.* **140**, 57 (1974).

124. M. Klingenberg, *Angew. Chem. Internat. Edit.* **3**, 54 (1964).

125. S. R. Earle and R. R. Fisher, *J. Biol. Chem.* **255**, 827 (1980).

126. L. L. Grinius, A. A. Jasaitis, J. P. Kadziauskas, E. A. Liberman, V. P. Skulachev, V. P. Topali, L. M. Tsofina, and M. A. Vladimirova, *Biochim. Biophys. Acta* **216**, 1 (1970).

127. J. Bremer and E. J. Davis, *Biochim. Biophys. Acta* **376**, 387 (1975).

128. B. Sacktor and A. Dick, *J. Biol. Chem.* **237**, 3259 (1962).

129. H. A. Krebs, T. Gascoyne, and B. M. Notton, *Biochem. J.* **102**, 275 (1967).

130. H. A. Lardy, V. Paetkau, and P. Walter, *Proc. Nat. Acad. Sci. U.S.A.* **53**, 1410 (1965).

131. H. Rottenberg and M. Gutman, *Biochemistry* **16**, 3220 (1977).

132. H. V. Westerhoff and K. Van Dam, *Curr. Top. Bioenerg.* **9**, 1 (1979).

133. M. Rydstrom, *Biochim. Biophys. Acta* **463**, 155 (1977).

134. H. J. Morowitz, *Foundations of Bioenergetics*, Academic Press, New York (1978).

135. E. O. Balasse and M. A. Neef, *Metabolism* **24**, 999 (1975).

136. P. F. Zuurendonk and J. M. Tager, *Biochim. Biophys. Acta* **333**, 393 (1974).

137. K. Jungermann, R. Heilbronn, N. Katz, and D. Sasse, *Eur. J. Biochem.* **123**, 429 (1982).

138. M. Nauck, D. Wolfle, N. Katz, and K. Jungermann, *Eur. J. Biochem.* **119**, 657 (1981).

139. E. Racker, *J. Cell Physiol.* **89**, 697 (1976).

140. B. Chance and G. R. Williams, *Adv. Enzymol.* **17**, 65 (1956).

141. M. Erecinska, M. Stubbs, Y. Miyata, C. M. Ditre, and D. F. Wilson, *Biochim. Biophys. Acta* **462**, 20 (1977).

142. R. van der Meer, H. V. Westerhoff, and K. van Dam, *Biochim. Biophys. Acta* **591**, 488 (1980).

143. J. W. Stucki, *Eur. J. Biochem.* **109**, 269 (1980).

144. M. Erecinska and D. F. Wilson, *J. Membrane Biol.* **70**, 1 (1982).

145. W. Kunz, R. Bohnensack, G. Bohme, U. Kuster, G. Letko, and P. Schonfeld, *Arch. Biochem. Biophys.* **209**, 219 (1981).

146. A. K. Groen, R. J. A. Wanders, H. V. Westerhoff, R. van der Meer, and J. M. Tager, *J. Biol. Chem.* **257**, 2754 (1982).

147. M. N. Berry, in *Regulation of Hepatic Metabolism*, Alfred Benzon Sympsium VI, Munksgaard, Copenhagen (1974), pp. 568–578.

148. A. L. Lehninger, *Adv. Exp. Med.* **111**, 1 (1979).

149. J. O'M. Bockris and A. K. N. Reddy, *Modern Electrochemistry*, Plenum Press, New York (1970).

150. H. A. Krebs, *Biochem. Biophys. Acta* **4**, 249 (1950).

151. J. Himms-Hagen, *Ann. Rev. Physiol.* **38**, 315 (1976).

152. J. Nedergaard and O. Lindberg, *Int. Rev. Cytol.* **74**, 188 (1982).

153. N. Haugaard, *Physiol. Rev.* **48**, 311 (1968).

154. R. Gerschmann, D. L. Gilbert, S. W. Nye, P. Dwyer, and W. O. Fenn, *Science* **119**, 623 (1954).

155. J. M. McCord and I. Fridovich, *J. Biol. Chem.* **244**, 6049 (1969).
156. K. Asada, S. Kanematzu, S. Okaka, and T. Hayakawa, in *Chemical and Biochemical Aspects of Superoxide and Superoxide Dismutases*, Vol. 62, J. V. Bannister and H. A. O. Hill, eds., FEBS Symp. (1980), pp. 136–153.
157. B. M. Babior, *New Engl. J. Med.* **98**, 659, 721 (1978).
158. J. A. Fee, A. C. Lees, P. L. Bloch, and F. C. Neidhardt, in *Oxygen and Oxy-radicals in Chemistry and Biology*, M. A. J. Rodgers and E. L. Powers, eds., Academic Press, New York (1981), pp. 205–239.
159. J. S. Valentine, in *Biochemical and Clinical Aspects of Oxygen*, W. S. Caughey, ed., Academic Press, New York (1979), pp. 659–677.
160. D. T. Sawyer and J. S. Valentine, *Acc. Chem. Res.* **14**, 393 (1981).
161. B. Chance, N. Oshino, T. Sugano, and D. Jamieson, in *Alcohol and Aldehyde Metabolising Systems*, Vol. 1, R. G. Thurman, T. Yonetani, J. R. Williamson, and B. Chance, eds., Academic Press, New York, (1974), pp. 169–182.
162. A. Boveris and J. F. Turrens, in *Chemical and Biochemical Aspects of Superoxide and Superoxide Dismutases*, Vol. 62, J. V. Bannister and H. A. O. Hill, eds., FEBS Symp. (1980). pp. 84–91.
163. J. F. Turrens and A. Boveris, *Biochem. J.* **191**, 421 (1980).
164. K. Takeshige and S. Minakami, *Biochem. J.* **180**, 129 (1979).
165. I. Fridovich, in *Oxygen and Living Processes. An Interdisciplinary Approach*, D. L. Gilbert, ed., Springer-Verlag, New York (1981), pp. 250–272.
166. D. L. Gilbert, in *Oxygen and Living Processes. An Interdisciplinary Approach*, D. L. Gilbert, ed., Springer-Verlag, New York (1981), pp. 73–101.
167. J. W. De Pierre and L. Ernster, *Ann. Rev. Biochem.* **46**, 201 (1977).
168. S. P. Mukherjee, R. H. Lane, and W. S. Lynn, *Biochem. Pharmacol.* **27**, 2589 (1978).
169. E. Sutherland, *Diabetes* **18**, 797 (1969).
170. N. Friedmann and G. Dambach, *Biochim. Biophys. Acta* **596**, 180 (1980).
171. J. G. Castano, S. Alemony, A. Nieto, and J. M. Mato, *J. Biol. Chem.* **255**, 9041 (1980).
172. D. Goltzman, B. Henderson, and N. Loveridge, *J. Clin. Invest.* **65**, 1309 (1980).
173. F. Franzen, and K. Eysell, *Biologically Active Amines Found in Man*, Pergamon Press, Oxford, (1969), pp. 2–62.
174. B. Chance, A. Boveris, and N. Oshino, in *Alcohol and Aldehyde Metabolizing Systems*, Vol. 2, R. G. Thurman, J. R. Williamson, H. R. Drott, and B. Chance, eds., Academic Press, New York (1977), pp. 261–274.
175. F. L. Crane, H. Goldenberg, D. J. Morre, and H. Low, *Subcell. Biochem.* **6**, 345 (1979).
176. K. L. Fong, P. B. McCay, J. L. Poyer, H. P. Misra, and B. B. Keele, *Chem. Biol. Interac.* **15**, 77 (1976).
177. L. Ernster and O. Lindberg, *Ann. Rev. Physiol.* **20**, 13 (1958).
178. M. G. Clark, B. J. Gannon, N. Bodkin, G. S. Patten, and M. N. Berry, *J. Molec. Cell. Cardiol.* **10**, 1101 (1978).
179. K. Takeshige and S. Minakani, *Biochem. Biophys. Res. Commun.* **99**, 484 (1981).
180. H. M. Swartz, in *Submolecular Biology and Cancer*, G. E. W. Wolstenholme, D. W. Fitzsimons, and J. Whelan, eds., Ciba Foundation Symp. **67**, Elsevier Excerpta Medica, Amsterdam (1979), pp. 107–141.
181. N. R. DiLuzio, *Fed. Proc.* **32**, 1875 (1973).
182. K. O. Lewis and A. Paton, *The Lancet* **ii**, 188 (1982).
183. H. Muller and H. Sies, *Biochem. J.* **206**, 153 (1982).
184. B. Halliwell, in *Age Pigments*, R. S. Sohal, ed., Elsevier/North Holland, New York (1981), pp. 1–62.
185. I. Fridovich, *Science* **201**, 875 (1978).

Amazing as the phenomena inherent in inorganic matter are, they can by no means be compared with those related with the activity of the nervous system and life processes. M. Faraday

7

Electrochemistry of the Nervous Impulse

YU. A. CHIZMADZHEV and V. F. PASTUSHENKO

1. Introduction

A broad range of very important processes in a living cell have electrochemical nature. Nowhere is the electrochemical nature more controlling than in the phenomena of nervous impulse generation and propagation. The study of the nervous impulse has had a rich and at times even dramatic history. It may be proper to recall here that electrochemistry as a science dates back to Galvani's (1798) experiments which eventually led to discovery of what he called "animal electricity." The last decade ushered in an impressive advance in this field which became possible owing to the use of new methods of investigation of cell membranes and new model systems and to the development of a quantitative theory of induced ionic transport. Today, it would not be an overstatement to say that there has emerged a new—bioelectrochemical— approach to problems concerning the functioning of biological systems. Its major characteristic is that attention is concentrated on the physical nature of the phenomena, while most details of a different nature, extremely important for physiology as they may be, are consciously ignored.

The present review follows this guideline precisely. It would be naive to try to incorporate in it the wealth of data amassed by now in this branch of biology. The reader in need of such information is referred to the fundamental

YU. A. CHIZMADZHEV and V. F. PASTUSHENKO • Institute of Electrochemistry, The Academy of Sciences of the U.S.S.R., 31 Leninsky Prospekt, Moscow V-71, U.S.S.R.

works given in the List of References at the end of this chapter. Our chief goal has been to epitomize the major achievements of investigators and describe the current state of the understanding of the processes taking place in excitable membranes which are responsible for the generation and propagation of the nervous impulse.

1.1. Historical Background

It is conventional to attribute the origination of the study of the nervous impulse to Galvani who observed contraction of a frog muscle in contact with various metals or under atmospheric electricity discharges, although bioelectric phenomena have been known to men from times immemorial. In ancient Rome discharges of electric fishes were prescribed by physicians against certain diseases. Such famous physicists as Faraday, Maxwell, Helmholtz, Hertz, and Planck, took an interest in bioelectric phenomena. The nervous impulse propagation velocity was measured in the last century (Helmholtz), and in 1902 Bernstein proposed the first electrochemical explanation of the electric potential across cell membranes. In 1939 Offner suggested an equivalent circuit for the nerve 'fiber membrane which attributed the nervous impulse propagation to a short-term increase in membrane conductance. In 1949 Cole worked out a voltage-clamp technique (fixed voltages across the nerve fiber membrane). His method was used in 1952 by Hodgkin and Huxley to carry out a detailed study of the ionic currents through a squid giant axon membrane. Their work, for which they received the Nobel Prize, initiated a multitude of investigations which had the common ultimate goal of determining the structure and activity mechanism of the molecular systems responsible for ionic transport and the related electric effects. The measurement by Armstrong and Bezanilla in 1973 of the so-called gating currents in a giant axon was an important milestone towards this goal. The currents are due to the movements of charged intramembrane particles which control the conductance of ionic channels and had in fact already been predicted on a theoretical basis by Hodgkin and Huxley. At about the same time or perhaps a little earlier, similar measurements were carried out by Schnider and Chandler on frog skeleton muscle. In the early seventies, another very important method of investigating ionic channels, based on measurements of current and voltage noise spectra in excitable membranes, was developed. Today, the nature of nerve excitation has been studied so thoroughly that we are in a position to formulate a certain conception of ionic channel structure. However, a final unambiguous decision about the chemical structure of ionic channels will be made only after it becomes possible to isolate individual channels from a biological membrane and study them by the methods of molecular biology.

1.2. The Nerve Cell

The nerve cell, or neuron, is the basic element of the nervous system of higher animals. A nerve cell is known to be composed of (cf. Figure 1): a body, dendrites—the short processes stemming off from the body, and an

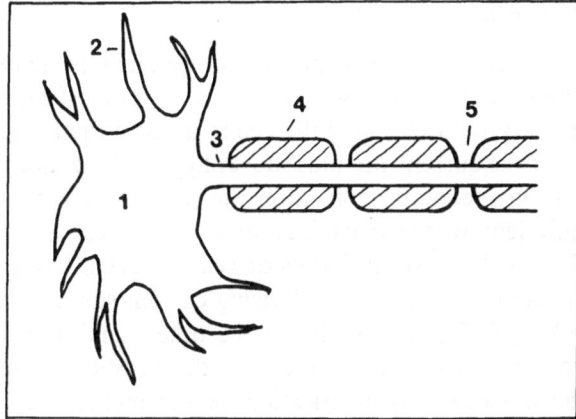

Figure 1. Nerve cell structure: 1. neuron body, 2. dendrite, 3. axon, 4. myelin sheath, 5. node of Ranvier.

axon, or nerve fiber—the longest process ramified at the end and contacting the dendrites and bodies of other cells by means of special formations, synapses. Many axons are coated with periodically interrupted fat or myelin sheaths. The gaps between adjacent myelinated fiber portions are called nodes of Ranvier. There are, in addition, smooth (nonmyelinated) axons, the squid giant axon (up to 1 mm in diameter) being one example. Within the cell there are membrane structures essential to the activity of the living cell; but in order to understand the manner in which a nerve cell functions it is first necessary to know the ionic composition of the axoplasm (the term used for the gel contained within a cell), and then the properties of the cell membrane. In a first approximation, the axoplasm may be regarded simply as an electrolytic solution in which K^+, Na^+, and Cl^- are the principal ions. Concentrations of these ions in the axoplasm are essentially different from ambient concentrations. Thus for the squid axon the inside and outside concentrations, c_i and c_o, (in mM) are approximately[b]: Na^+, $c_i = 50$, $c_o = 460$; K^+, $c_i = 400$, $c_o = 10$; Cl^-, $c_i = 40$–100, $c_o = 540$; isethionate, $c_i = 270$, $c_o = 0$; aspartate, $c_i = 75$, $c_o = 0$.

The nerve cell membrane, which is about 5 nm thick, consists primarily of lipids and proteins. When at rest it is permeable basically to potassium ions (although its resistance in this state is rather high, *ca.* 10^3 Ω cm^2), and therefore the electric potential difference between the inner and outer solutions (this difference is called the membrane potential or simply the potential) is negative at rest and amounts to a few tens of mV (about -60 mV on the giant axon). The membrane capacitance is of the order of 1 μF/cm^2. Thus a neuron membrane is already polarized when at rest. If some external action shifts the potential from its value at rest to more negative values (its absolute value increasing), the resultant situation is usually called hyperpolarization. Potential shift in the positive direction is called depolarization. If the potential reverses its sign and becomes positive the term is "overshoot."

1.3. Excitation Phenomenon

Normally, depolarization or hyperpolarization may be achieved with the aid of an electrode pair, one of which may be introduced into the cell. We will be interested in the response of the cell to a current pulse passed through the membrane. If the current stimulus causes hyperpolarization, the potential is observed to return monotonically, after the stimulus is discontinued, to the rest level, equivalent to the behaviour of a passive circuit consisting of a capacitor and a parallel rest-level resistor ($ca.$ $10^3 \, \Omega \, cm^2$) (Figure 2A). The behavior is the same with weak depolarizing current stimuli (potential shifts of up to 10 mV) (cf. Figure 2B). If, on the other hand, the depolarization exceeds a certain threshold value (20–30 mV), the membrane response changes drastically: the monotonic depolarization of the membrane goes on after termination of the stimulus, and then the potential becomes positive, attains a value of around +40 mV and, only after that, returns to the rest potential having passed through the hyperpolarized condition (Figure 2C). The observed potential variation with time is usually standard and is called action potential, spike, or nervous impulse. The above phenomenon has received the name of excitability, since, at least at the initial stage of the generation of action potential, LeChatelier's principle is inapplicable: the system response amplifies the external action. The nervous impulse spreads along the axon in the following manner: a zone neighboring on the active zone becomes charged to a superthreshold value by the local currents and the excitation is thus transmitted along the axon. A thorough study of the ionic currents in the process of nervous impulse generation led Hodgkin and Huxley[2] to the following membrane model (Figure 3): K^+ and Na^+ e.m.f.'s connected in parallel with the membrane capacitance and having different internal resistances (in the Hodgkin and Huxley scheme there was also a third e.m.f. due to other ions,

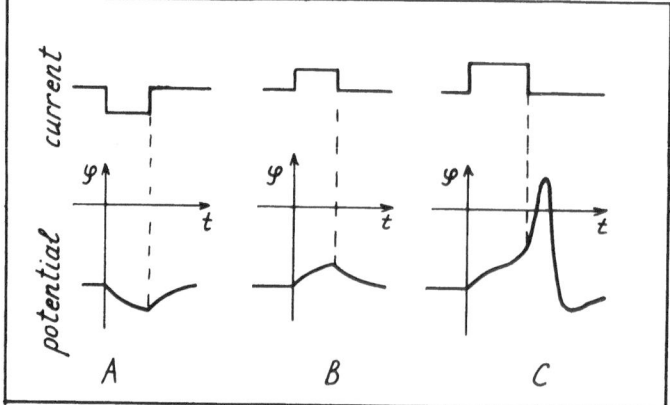

Figure 2. Nerve fiber responses to a hyperpolarizing (A) and a weakly depolarizing (B) current-pulse. (C) represents nervous impulse generation under superthreshold depolarization.

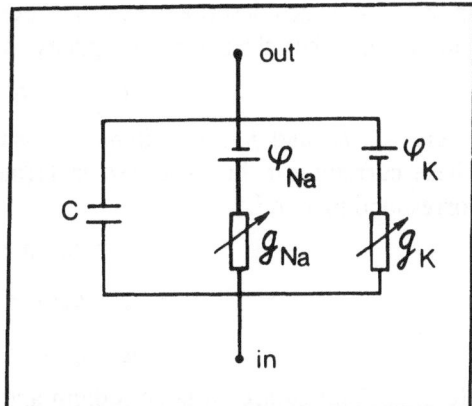

Figure 3. Equivalent circuit of a portion of nerve fiber membrane. Both φ_K and φ_{Na} are constant; the conductances g_K and g_{Na} are variable. The leakage conductance and e.m.f. are not shown.

called leakage e.m.f.; its role is secondary and it will be ignored in this review for simplicity). Under a superthreshold depolarization, the potassium permeability of the membrane (i.e., the conductance of the Na^+ battery) is sharply increased, whereby the membrane capacitance becomes charged to nearly a Na^+-e.m.f. value (overshoot). Thereafter the conductance of the Na^+-e.m.f. decreases and that of the K^+-e.m.f. increases, reestablishing the resting potential on the membrane. The intermediate hyperpolarized state occurs due to the fact that the potassium conductance of the membrane returns to the initial resting value with some delay.

1.4. Hodgkin–Huxley Equations

In a first approximation, the membrane capacitance is constant. Thus if the potential is the same over the entire membrane, the current through it may be written as:

$$I_e = C_m \frac{\partial \varphi}{\partial t} + I \qquad (1)$$

Here, I_e is the current in the external circuit, C_m is membrane capacitance, φ is potential, t is time, and I is ionic current through the membrane. In accordance with the definition of φ (the inside potential minus the outside potential), the current directed from the inside to the outside is considered positive. When the external circuit is open and there are no electrodes we have $I_e = 0$. The corresponding dynamic behavior of the potential under conditions of spatial homogeneity is described by:

$$C_0 \frac{\partial \varphi}{\partial t} = -i_0 \qquad (1')$$

where C_0 is capacitance per unit membrane area ($ca.$ 1 $\mu F/cm^2$) and i_0 is density of ionic current through the membrane. Hodgkin and Huxley[2] derived an equation which describes the properties of the active membrane of a squid giant axon rather closely (for simplicity the subscript "0" has been omitted):

$$i = i_{Na} + i_K + i_L \tag{2}$$

where i_{Na}, i_K, and i_L are sodium, potassium, and leakage current densities. These currents can be expressed in terms of potential deviations from the corresponding e.m.f.:

$$i_{Na} = g_{Na}(\varphi - \varphi_{Na}) \tag{3}$$

$$i_K = g_K(\varphi - \varphi_K) \tag{4}$$

$$i_L = g_L(\varphi - \varphi_L) \tag{5}$$

Here, g_{Na} and g_K are variable sodium and potassium conductances, g_L is the constant leakage conductance, and φ_{Na}, φ_K, and φ_L are the respective e.m.f.'s. The values g_{Na} and g_K can be expressed in terms of dynamic variables m, n, and h in the following manner:

$$g_{Na} = \bar{g}_{Na} m^3 h \tag{6}$$

$$g_K = \bar{g}_K n^4 \tag{7}$$

Here $\bar{g}_{Na}$ and $\bar{g}_K$ are coefficients corresponding to the maximum values of g_{Na} and g_K. The variables follow the first-order relaxation equations:

$$\frac{\partial u}{\partial t} = -\frac{u - u_\infty(\varphi)}{\tau_u(\varphi)} \tag{8}$$

Here, u is one of the variables m, n, or h; the functions $u_\infty(\varphi)$ and $\tau_u(\varphi)$ represent the steady-state value and characteristic relaxation time of the respective variable for a given φ. There are empirical formulas for these functions, which are shown in the general form in Figure 4. The other parameters used in the squid giant axon model are assumed to have the following values: $C_0 = 1\ \mu F/cm^2$, $\bar{g}_{Na} = 120\ mmho/cm^2$, $\bar{g}_K = 36\ mmho/cm^2$, $g_L = 0.3\ mmho/cm^2$, $\varphi_{Na} = 55\ mV$, $\varphi_K = -72\ mV$, $\varphi_L = -50\ mV$.

Note that in the case of a spatially nonuniform potential distribution neither eq. (1) nor (1') is valid. Thus in the case of a smooth axon of constant radius r (homogeneous fiber) the potential dynamics are given by:

$$C\frac{\partial \varphi}{\partial t} = \frac{1}{R}\frac{\partial^2 \varphi}{\partial x^2} - i \tag{9}$$

Here, $C = 2\pi r C_0$ and is capacitance per unit fiber length, $i = 2\pi r i_0$ is ionic current per unit fiber length, and R is the sum of external and internal resistances per unit fiber length. The Hodgkin–Huxley equations constitute a set of nonlinear equations which can only be solved numerically. These equations are adequate to describe many features of nerve excitation,[3] even though Hodgkin and Huxley themselves did not consider them as more than a first approximation to reality.

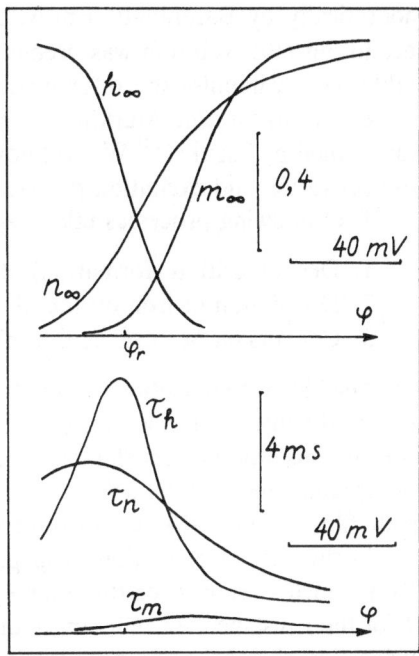

Figure 4. Effect of potential on the parametric functions u_∞ and τ_u in the Hodgkin–Huxley model ($u = m, n, h$).

2. Electrochemical Models of the Nerve Fiber

Physicochemical approaches to the excitation problem were developed alongside the pure biological investigations. A lot of trigger systems are known at present which can be regarded as some kind of excitation models. Two of them, currently known as the Lillie–Bonhoeffer model and Teorell's membrane system, have been found to be particularly suitable. Experimentation with these models has been particularly useful: the behavioral features common to all excitable objects have been established and the concrete electrochemical mechanisms responsible for the activity of these systems have been well-investigated. Here we will review only the major characteristics of these models.

2.1. The Lillie–Bonhoeffer Model

It was Ostwald who noticed the analogy between the behavior of an iron wire in concentrated nitric acid and nerve fiber excitation. The analogy is essentially this. In concentrated nitric acid, iron is covered with a passive oxide film. If a sufficiently large patch is freed from the film in some manner, one will observe an activation wave propagating along the wire, visible by gloss variation. Lillie[4–5] studied this effect thoroughly, and later Bonhoeffer[6–7] investigated and carried out calculations on the electrochemical mechanisms involved. The existence of a propagating activity wave was demonstrated theoretically by Barenblatt, Entov, and Salganik[8] and the existence of a second, unstable solution was predicted and an approximate formula for the stable activity impulse speed for small double layer capacitances was obtained by Levich, Mazur, and Markin.[9] Impulse propagation for high capacitances was studied by Suzuki.[10] We will now take a brief look at the mechanism of iron activation and excitation propagation.

The following processes take place on the iron surface:

1. Decomposition (formation) of the passivating film
2. Dissolution of iron on the film-free surface
3. Conversion of nitric acid to nitrous acid.

The film decomposition begins after the iron potential rises above a certain threshold value. This is accompanied by a gradual extension of the film-free surface. This process is accelerated as iron dissolution progresses. Subsequently the accumulated nitrous acid promotes the regeneration of the passivating oxide film and then the system returns to its original state.

In the case of excitation propagation, the potential is described by Eq. (9) in which C stands for the double layer capacitance per unit wire length, R for the resistance per unit length of solution, i for the current through unit surface length equal to $2\pi r i_0$ (r is wire radius, i_0 is current density). As noted above,

$$i = 2\pi r(i_1 + i_2 + i_3) \qquad (10)$$

Here, i_1, i_2, and i_3 are the respective current densities of the above processes. The currents themselves are expressed as follows:

$$i_1 = \begin{cases} (1 - \alpha)i_{10}(\varphi) & (\varphi > \varphi_* \rightarrow \text{decomposition}) \\ \alpha i_{10}(\varphi) & (\varphi < \varphi_* \rightarrow \text{formation}) \end{cases} \qquad (11)$$

Here, α is the film-free surface area fraction, $i_{10}(\varphi)$ is a certain function of the iron surface potential with the origin at equilibrium, and φ_* is the threshold potential. As follows from the definition of i_{10}, its sign must be reversed at the point $\varphi = \varphi_*$ (Figure 5, Curve 1). It is then obvious that $i_2 \sim \alpha$:

$$i_2 = \alpha i_{20}(\varphi) \qquad (12)$$

The function $i_{20}(\varphi)$ is shown as Curve 2 in Figure 5. Finally, there is one more relationship accounting for the self-catalyzed nature of the nitrous acid formation process:

$$i_3 = c_S i_{30}(\varphi) \qquad (13)$$

where c_S is the nitrous acid concentration at the surface. The $i_{30}(\varphi)$ function is shown in Figure 5, and c_S is found from the solution of the problem of nitrous acid diffusion in the electrolyte solution. The dynamic variable α

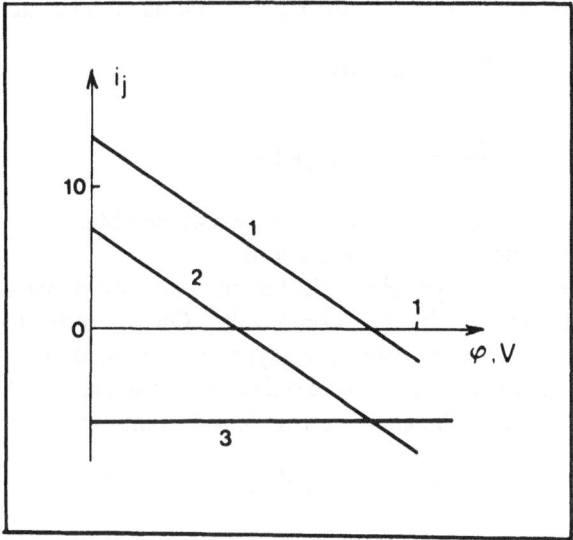

Figure 5. An approximation of the specific currents i_1 (curve 1), i_2 (curve 2), and i_3 (curve 3). The currents i_1 and i_2 are in A/cm^2, i_3 in A cm/mol.[11]

satisfies

$$\frac{\partial \alpha}{\partial t} = -\frac{1}{Q} i_1 \tag{14}$$

where Q is the charge passing through a unit surface area of wire upon complete dissolution of the passivating film.

The following approximate formulas were assumed in the analysis of the propagation velocity of the activation impulse:

$$i_{10} = A(\varphi_* - \varphi) \tag{15}$$

$$i_{20} = A(\varphi_1 - \varphi) \tag{16}$$

$$i_{30} = \begin{cases} 0 & \varphi = 0 \\ \text{const.} & \varphi > 0 \end{cases}$$

Here, A is a constant factor.

Consider the propagation process analyzed in Reference (9) in which the capacitance term in Eq. (9) may be neglected. As noted above, the current due to nitrous acid formation rises slowly, and it is this parameter which is actually responsible for the wire returning to the original passivated state. On the other hand, it is clear that the impulse propagation velocity is controlled by the processes in the activation zone where $\varphi > \varphi_*$. Therefore, in this zone the current due to nitrous acid formation may be ignored, and then, substituting the corresponding relationships into Eq. (9), we have in the activation zone:

$$\frac{\partial^2 \varphi}{\partial x^2} = 2\pi r A R (\varphi - \varphi_* - \alpha \varphi_1 + \alpha \varphi_*) \tag{17}$$

$$\frac{\partial \alpha}{\partial t} = \frac{A}{Q} (\varphi - \varphi_*)(1 - \alpha) \tag{18}$$

An approximate expression for the velocity may be obtained from the above equations from dimensional considerations.

Indeed, as follows from Eq. (17), the characteristic active zone length is determined unambiguously as $1/(2\pi r A R)^{1/2}$. On the other hand, the only parameter of the characteristic time type is contained in Eq. (18), i.e., $Q/[A(\varphi - \varphi_*)]$, in which φ can have only one characteristic value other than φ_*, i.e., φ_1. Therefore, the characteristic film decomposition time will be

$$\frac{Q}{A(\varphi_1 - \varphi_*)}$$

Clearly the impulse propagation velocity v is equal to the ratio of activation zone length to activation time, i.e.,

$$v \approx \frac{\varphi_1 - \varphi_*}{Q} \left(\frac{A}{2\pi r R} \right)^{1/2} \tag{19}$$

The more detailed analysis given in Reference (9) leads to the same formula. Below it will be shown that the above stimulus propagation behaviour bears a profound analogy with propagation of a nervous impulse in the sense that in both cases the impulse speed essentially depends on the characteristic activation time which in this model is the film decomposition time and, in the H–H model, the sodium channel activation time. The properties of the Lillie model have been described in more detail in Reference (11) where a long list of references can also be found.

2.2. Teorell's Electrokinetic Model

The system investigated by Teorell consists of two vessels containing electrolytes at different concentrations and in contact with each other via a sintered glass membrane (Figure 6). Normally, the liquid columns in the two vessels are different, so that there is a hydrostatic head drop across the membrane, p. In one of the vessels the surface area may be small, and then the passage of liquid from one vessel to another causes the pressure drop, p, to vary, as it does in the case of self-oscillations.[12,13] Otherwise p remains fixed, and it will be assumed to be so below in our discussion of the excitation impulse. The flow of liquid through the membrane is caused not only by the head drop p, but also by the electroosmosis that can be initiated by applying a potential difference to the membrane with the help of an electrode pair,

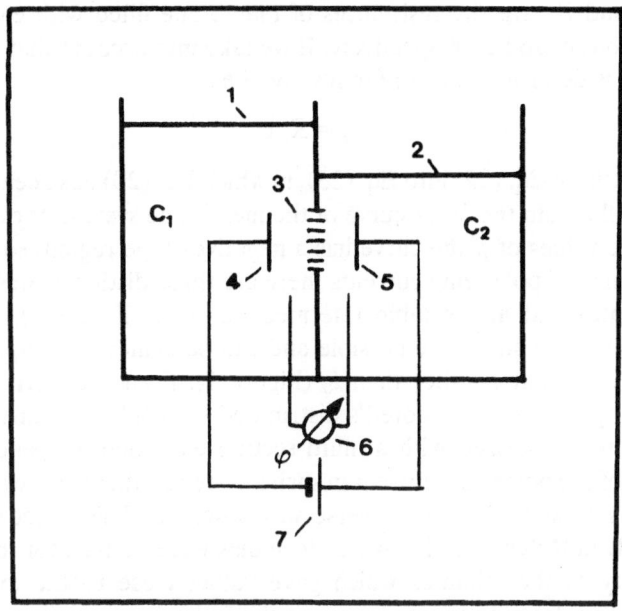

Figure 6. Schematic view of Teorell's apparatus. 1, 2: electrolyte levels in the left and right vessels, 3: porous membrane, 4, 5: polarizing electrodes, 6: voltage meter, 7: current source (the current stabilizer is not shown).

since there always are fixed charges on the glass pore walls. The usual way to do this is to pass a direct current i_0 through the membrane. Thus,

$$v = \alpha p - \gamma \varphi. \tag{20}$$

Here, v is the liquid flow rate, φ is the voltage on the membrane, and α and γ are constants (for more detail see Reference (11)). If v is constant, a constant electrolyte concentration profile becomes established within the membrane:

$$c(y) = \frac{c_1 - c_2 \exp(-V) + (c_2 - c_1) \exp\left[V\left(\frac{y}{\delta} - 1\right)\right]}{1 - \exp(-V)} \tag{21}$$

$$V \equiv \frac{v\delta}{D}$$

Here, c_1 and c_2 are electrolyte concentrations on opposite sides of the membrane, δ is the membrane thickness, y is the transverse coordinate, and D is the electrolyte diffusivity. From this it follows that under steady conditions the electric resistance R_m of the membrane depends on the liquid flow rate:

$$R_m = \frac{\exp(V) - 1}{\exp(V) - \frac{R_1}{R_2}} \cdot R_1 \left(1 + \frac{1}{V} \ln \frac{R_2}{R_1}\right) \tag{22}$$

where R_1 and R_2 are the resistances of membrane filled with electrolyte of concentrations c_1 and c_2, respectively. If we take into account that the polarizing current is determined from Ohm's law, i.e.,

$$i_0 = R_m^{-1}\varphi \tag{23}$$

then, substituting Eq. (22) into Eq. (23), in which Eq. (20) has been substituted for v, we will obtain the V–A curve of the membrane system. It can be shown that at large values of p the curve has a negative-slope region, so that within a certain range of polarizing currents there are three distinct states: the stable extreme states and an unstable intermediate state (Figure 7). Transitions between the stable states are possible and can be achieved by applying for a short time a current i' other than i_0 (Figure 7). Franck experimented with stimulus propagation in a Teorell's distributed system.[14] To this end a long membrane was polarized with a multi-section electrode. Depending on the magnitude of the polarizing current in such a system, either a wave of transition from state 1 to state 2 or the reverse was observed. This effect was treated theoretically in Reference (15) where formulas were derived for the propagation velocity of the stimulus which gave values close to the experimental ~ 1 mm/s. The description was based on the following equation:

$$\frac{1}{R}\frac{\partial^2 \varphi}{\partial x^2} + i_0 = \frac{\varphi}{R_m} \tag{24}$$

which is actually a version of Ohm's law for a distributed system. Here, R is resistance per unit length of electrolyte along the membrane, and i_0 and R_m are the constant polarizing current and resistance of unit membrane length. To define the dynamic variable R_m appearing in this equation, another equation

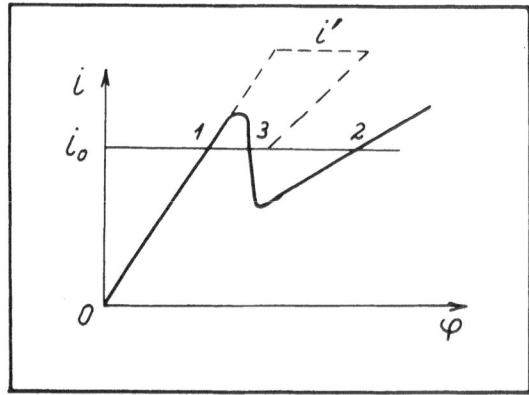

Figure 7. The V–A curve of Teorell's system under sufficiently large concentration and head differences. Under a polarizing current i_0 the states 1 and 2 are stable and state 3 is unstable. Transition from 1 to 2 can be effected by passing a current i', greater than the V–A peak current, for a sufficiently long time.

is necessary. Teorell used a first-order relaxation equation with a steady-state solution given by Eq. (22) for the purpose.

The case of predominantly convective electrolyte transfer through the membrane has been considered,[15] in which we have for R_m

$$\frac{\partial R_m}{\partial t} = -\frac{(R_2 - R_1)}{\delta} v \cdot \theta(R_m) \tag{25}$$

Here,

$$\theta(R_m) = \begin{cases} 1 & R_1 < R_m < R_2 \\ 0 & R_m = R_1, R_m = R_2 \end{cases} \tag{26}$$

The combination of Eqs. (25) and (24) may be solved in the analytical form if i_0 is close to i_*, where the critical current i_* was defined as

$$i_* = \frac{\varphi_*}{(R_1 R_2)^{1/2}} \tag{27}$$

and $\varphi_* = \alpha p / \gamma$ is the critical voltage. The result can be summarized as follows: the $1 \to 2$ transition wave is possible when $i_0 > i_*$, and, in addition, the transition propagation velocity v_{12} is given by:

$$v_{12} = \frac{\gamma \chi_1 \varphi_{*1}}{2\delta} \frac{(\chi_1 \varphi_{*1} - \chi_2 \varphi_{*2})^2}{[\chi_1 \chi_2 \varphi_{*1} - i_0 R + (R \varphi_* \omega / R_1)]^2} \tag{28}$$

Here,

$$\chi_k = \left(\frac{R}{R_k}\right)^{1/2}, \qquad k = 1,2$$

$$\varphi_{*1} = \varphi_* - i_0 R_1$$
$$\varphi_{*2} = i_0 R_2 - \varphi_* \tag{29}$$
$$\omega = \arctan\left[(R_2/R_1 - 1)/(R_2/R_1 + 1)\right]^{1/2}$$

For the $2 \to 1$ transition wave, which is possible when $i_0 < i_*$, the answer may be obtained from Eq. (28) by transposing the subscripts 1 and 2 and taking into account the identity

$$z = j \tan\left(\frac{1}{2j} \ln \frac{1+z}{1-z}\right), \qquad j^2 = -1 \tag{30}$$

whereby we obtain:

$$v_{21} = -\frac{\gamma \chi_2 \varphi_{*2}}{2\delta}$$

$$\times \frac{(\chi_1 \varphi_{*1} - \chi_2 \varphi_{*2})^2}{\left[\chi_1 \chi_2 \varphi_{*2} + R \cdot i_0 - \dfrac{R \varphi_*}{2R_2(1 - R_1/R_2)^{1/2}} \ln \dfrac{1 + (1 - R_1/R_2)^{1/2}}{1 - (1 - R_1/R_2)^{1/2}}\right]^2} \tag{31}$$

Figure 8 shows v_{12} and v_{21} as functions of the polarizing current. Near $i_0 = i_*$ the curves are semi-parabolas of different slopes, so that for a given $|i_0 - i_*|$ the $1 \rightarrow 2$ transition wave propagates faster.

To conclude this section we would note that, despite the fact that physically the two models are totally different, the stimulus propagation mechanisms in both are very similar: in both potential equations it is the second space derivative which is the highest, and if one compares specifically Eqs. (17) and (24) one will also notice the absence in both of the time derivative of potential. Further, in both cases there is another dynamic variable in the potential equation which determines the evolution of the process. In the next section we will show that to a good approximation the same situation occurs in a giant axon which is described by the Hodgkin–Huxley equations.

3. Excitation Propagation

In the last section we briefly discussed the mechanisms of the generation and propagation of the "excitation" impulse in the two physicochemical models of the nerve fiber. In a more general case one faces the problem of impulse propagation in a certain excitable biological medium, be it a nerve fiber, an electrically excitable syncytium, a neuron network, or some other object. As a rule, such a system may be characterized by the dynamic distribution of electric potential described by an equation of the type of Eq. (9):

$$C \frac{\partial \varphi}{\partial t} = \nabla \frac{1}{R} \nabla \varphi - i \tag{32}$$

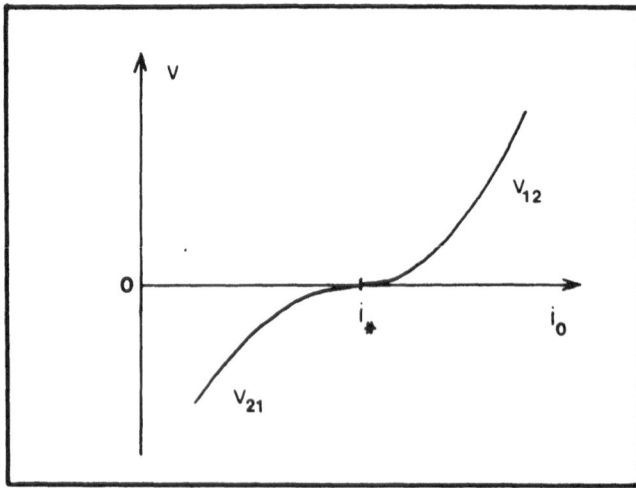

Figure 8. Relationship between excitation impulse speed in Teorell's model and polarizing current. The semi-parabola on the right is steeper than the one on the left.[11]

or some other field equation. Here, C and R are given functions of coordinates, i is a function of coordinates and time, and a functional of potential distribution. Actually, the investigation of electrochemical mechanisms of excitation generation and propagation has two aspects to it. The first consists in describing the excitation propagation for known functions C, R, and i. This problem is fundamental in the excitable medium theory[11] and will be at the focus in the present section.

The other aspect consists in determining the properties and molecular nature of the objects responsible for the specific features of the functions C, R, and i. Within the framework of the theory of excitable media one may ignore the molecular mechanisms of activity and introduce some phenomenological description of the local properties of the medium.

The excitable medium theory itself involves a variety of approaches and models. As examples we can cite axiomatic models,[16] computer simulation,[17,18] as well as various physical models.[19] Besides, men of different sciences pursue different aims: thus to a physiologist it is important that a model be as close to the real object as possible and be analyzed with utmost accuracy; this approach has engendered a series of works involving computer solutions of the Hodgkin–Huxley equations (as well as their analogues for other objects) (cf., e.g., Reference (20)). We will not, however, go into the details of the axiomatic models or discuss the results obtained by numerical methods. Instead we will step right over to the approximate methods which yield analytic solutions. The results have been reported in References (21—23). First we will consider the spread of nervous impulse in terms of the Hodgkin–Huxley model. One of the most important problems in the excitable medium theory is the calculation of the impulse speed. Therefore, we will center our attention on speed calculations, disregarding the waveform and other minor questions.

3.1. Reduced Hodgkin–Huxley Equations

Consider the problem of the passage of a single nervous impulse along a fiber; it is described by the Hodgkin–Huxley equations.[21-23] Our aim is not so much to obtain exact relationships between the speed and various other parameters (Huxley has already done this[24]), as to find out the most essential features of these equations. On the qualitative level an explanation of the nervous impulse looks very simple.

Figure 9 is a diagram of the local currents flowing along the nerve fiber during passage of an excitation impulse. To review the excitation propagation mechanism described in Section 1 it is convenient to consider Figure 9. The downstream-oriented axial current inside the fiber crosses the membrane as capacitive current, i.e., charges the membrane. When a certain threshold is exceeded, an inward ionic current appears apart from capacitive current. Further on, the outward current begins to dominate and returns the potential

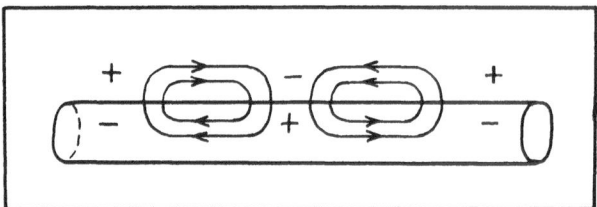

Figure 9. Diagram of local currents in the excited region accompanying nervous impulse propagation.

to the initial level. From this picture it must already be clear that the trailing edge can have only a little effect on the potential distribution in front of the impulse. For this reason the impulse speed is almost entirely dependent on the sodium current component (this condition will be discussed with more precision with a simpler model). Hereafter the potential level will be counted from its rest value. As noted in Section 1, the sodium current is given by the term:

$$i_{Na} = (\varphi - \varphi_{Na}) m^3 h \cdot \bar{g}_{Na} \tag{33}$$

Considering that h varies slower than m, we may assume in the excitable region $h = $ const. (~ 0.6). Similarly, the term $(\varphi - \varphi_{Na})$ can also be approximately considered constant in this region, since φ is small compared to φ_{Na}. Thus it is the factor m^3 which plays the most important role in determining the sodium current. As regards the variable m proper, we will assume for it a relaxation equation of the type of Eq. (8), wherein τ_m is constant and $m_\infty(\varphi)$ is a stepped function:

$$m_\infty(\varphi) = \eta(\varphi - \varphi_*) \tag{34}$$

where φ_* is the threshold potential corresponding to point $m_\infty(\varphi) = \frac{1}{2}$;

$$\eta(x) = \begin{cases} 0, & x < 0 \\ 1, & x > 0 \end{cases} \tag{35}$$

Introduce the characteristic sodium current $i_1 = \bar{g}_{Na} h(\varphi_{Na} - \varphi_*)$ and choose the following measurement units: time: $C\varphi_*/i_1$; speed: $(i_1/RC^2\varphi_*)^{1/2}$; coordinate: $(\varphi_*/R \cdot i_1)^{1/2}$; potential: φ_*. Introduce, further, the self-similar time:

$$\xi = (t - x/v) \frac{i_1}{C\varphi_*} \tag{36}$$

Using the above measurement units we can rewrite Eq. (9) in the following form:

$$\frac{d\psi}{d\xi} = \frac{1}{V^2} \frac{d^2\psi}{d\xi^2} + m^3 \tag{37}$$

$$\frac{dm}{d\xi} = -\alpha[m - \eta(\psi - 1)] \tag{38}$$

Here,

$$\psi = \frac{\varphi}{\varphi_*}$$

$$V = v\left(\frac{RC^2\varphi_*}{i_1}\right)^{1/2}$$

$$\alpha = \frac{C\varphi_*}{i_1\tau_m}$$

The additional conditions pertinent to Eqs. (37)–(38) are:

$$\psi(-\infty) = 0, \quad \psi(0) = 1, \tag{39}$$

$$m(-\infty) = 0 \tag{40}$$

The condition of Eq. (40) allows Eq. (38) to be solved, whereupon the conditions of Eq. (39) will enable solution of the potential equation [Eq. (37)]. It would seem that V would still remain unknown, but the condition for determining it will be found in the process of solving. Introduce the notation

$$W = V^2 \tag{41}$$

First, consider the set of equations (37)–(38) within the range $-\infty < \xi < 0$; Here, $m = 0$ and:

$$\psi = \exp(W\xi), \quad (-\infty < \xi < 0) \tag{42}$$

Hence,

$$\left.\frac{d\psi}{d\xi}\right|_{\xi=0} = W \tag{43}$$

Thus it is now possible to solve the set of equations (37)–(38) within the range $0 < \xi < \infty$, subject to the following conditions:

$$\psi(0) = 1, \quad \left.\frac{d\psi}{d\xi}\right|_{\xi=0} = W, \quad m(0) = 0 \tag{44}$$

In this range the potential increases monotonically, so that $\eta(\psi - 1) = 1$, and therefore, the variable m tends to relax to unity. Within this range Eq. (38) can be solved independently of Eq. (37) and the nonlinearity of Eq. (37) at $\xi > 0$ is, therefore, only apparent. We can, for this reason, use conventional methods for solving the set of equations (37)–(38) at $\xi > 0$. The most convenient way is to apply the Laplacian transformation:

$$\hat{\psi}(p) = \int_0^\infty \psi(\xi) \exp(-p\xi) \, d\xi \tag{45}$$

The image of the potential $\hat{\psi}$ may be found from Eqs. (37)–(38) in combination with Eq. (44):

$$\hat{\psi}(p) = \frac{-\dfrac{p}{W} + \dfrac{1}{p} - \dfrac{3}{p+\alpha} + \dfrac{3}{p+2\alpha} - \dfrac{1}{p+3\alpha}}{p\left(1 - \dfrac{p}{W}\right)} \tag{46}$$

The form of the denominator in this expression shows that the function possesses a singularity at point $p = W$. This means, generally speaking, that the equation for ψ contains a term of the type:

$$\text{res } \hat{\psi}(W) \cdot \exp(W \cdot \xi)$$

From physical considerations this is obviously impossible. Indeed, $m^3 \to 1$ and from Eq. (37) it follows that ψ cannot increase faster than linearly. It is clear, therefore, that the condition

$$\text{res } \hat{\psi}(W) = 0 \tag{47}$$

must be satisfied. This condition is in fact the formula from which W can be determined. Its particular form can easily be found by equating the numerator in Eq. (46) to zero and at the same time putting $p = W$. We thus come up with:

$$W(W + \alpha)(W + 2\alpha)(W + 3\alpha) = 6\alpha^3 \tag{48}$$

This equation has a unique positive root, whose dependence on α is plotted in Figure 10. The analytical form of this relationship is obtained through the parametric representation:

$$\alpha = \frac{6}{u(u + 1)(u + 2)(u + 3)} \tag{49}$$

$$W = \alpha \cdot u \tag{50}$$

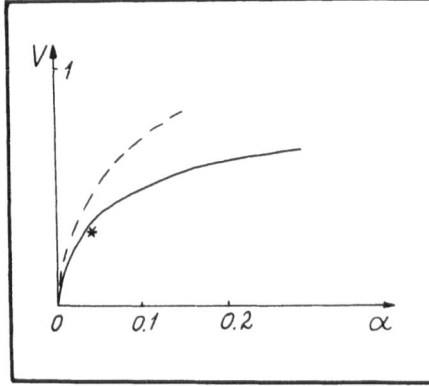

Figure 10. Relationship between nervous impulse propagation speed and rate of sodium current activation. The solid curve was calculated via Eqs. (49)–(50), the dotted curve via Eq. (52). The experimental value is marked with an asterisk.

wherein parameter u takes only positive values. The impulse speed versus α relationships are simple in two cases—for large and small α:

(a) $\alpha \ll 1$ (slow activation). From Eq. (49) it follows that in this case $u \gg 1$ and, in accordance with Eq. (50), $W \gg \alpha$. Then we have from Eq. (48):

$$W = (6\alpha^3)^{1/4} \quad \text{or} \quad V = (6\alpha^3)^{1/8} \tag{51}$$

Accordingly, in the dimensional variables

$$v = \frac{1.25}{R^{1/2}} \left(\frac{i_1}{\tau_m^3 C^5 \varphi_*} \right)^{1/8} \tag{52}$$

Remember that i_1 is the characteristic sodium current:

$$i_1 = \bar{g}_{Na}(\varphi_{Na} - \varphi_*)h \tag{53}$$

(b) $\alpha \gg 1$ (fast activation). From Eq. (49) it follows that in this case $u \ll 1$, and, therefore, $W \ll \alpha$. Hence we obtain from Eq. (48)

$$W = 1, \quad V = 1 \tag{54}$$

or in dimensional units

$$v = \left(\frac{i_1}{RC^2 \varphi_*} \right)^{1/2} \tag{55}$$

Let us now compare the theory as set forth above with experiment. Using the parameter values typical of the squid giant axon: $\bar{g}_{Na} = 0.12 \text{ mmho/cm}^2$, $\tau_m = 2$ ms, $C_0 = 1$ $\mu\text{F/cm}^2$, $\varphi_* = 30$ mV, at 18.5°C the parameter α is 0.0446. The experimental value $V = 0.274$ found for this α value is shown with an asterisk in Figure 10. The value of V calculated from Eq. (48) is 0.3 and that found from Eq. (51) is 0.39. If we compare these figures with $V = 1$ calculated for the fast relaxation of the variable m, it will become clear that the limiting case of slow relaxation is much closer to reality than rapid relaxation.

What we believe to be particularly important in the result [Eq. (52)] is that the impulse speed depends strongly on the sodium current activation rate. Thus by measuring the impulse speed we obtain information not only about passive electric characteristics of the nerve fiber but also about the dynamics of the molecular structures responsible for the fiber's activity. A more comprehensive comparison of the above theory with experiment, in particular with the computer-aided treatment of the H–H model carried out in Reference (24), is given elsewhere,[11] in which theory modifications that are more adequate to the H–H model are also analyzed. It should be noted, besides, that qualitatively similar results were obtained by Rinzel and Keller[25] who studied impulse propagation in a FitzHugh–Nagumo model (which takes into account the inertial nature of the variable in the same manner as it does potassium conductance).

3.2. Ionic Current Generator Model

By determining the most essential factors which control the nervous impulse propagation and introducing certain simplifications in the H–H model we were able in the last subsection to obtain simple formulas for the nervous impulse propagation speed. However, while in the simple case of a homogeneous fiber we were mainly concerned with determining as accurately as possible the functional dependences of the speed on the membrane characteristics, in problems of more complex nature other features of the excitable system come to the fore, viz., geometrical and functional inhomogeneity of single fibers or characteristics of the excitable syncytium. The ionic current generator model[26,27] has been found most useful for tackling these problems, for it allows qualitatively adequate (and sometimes even quantitatively satisfactory) conceptions of excitation impulse behavior in various systems to be obtained. The underlying principle of the model consists in specifying the current function of time which is counted down from the moment the potential has overshoot the threshold. In practice the simplest possible relationship was used in the form of two consecutive current steps of specified duration, nominally corresponding to the sodium and potassium currents, respectively:

$$
i = \begin{cases} -i_1, & 0 < t < t_1 \\ i_2, & t_1 < t < t_2 + t_1 \end{cases} \tag{56}
$$

As already pointed out, at $t = 0$, $\varphi = \varphi_*$. The current pulse durations, t_1 and t_2, are such that the total charge passing across the membrane equals zero:

$$
i_1 t_1 = i_2 t_2 \tag{57}
$$

We shall now consider the problem of impulse speed in a homogeneous fiber within the scope of this model.

3.2.1. Homogeneous Fiber

Introduce the notation

$$
I = i_2/i_1 \tag{58}
$$

and use the same units of measurement as in the preceding section. Then instead of Eq. (37) we will have

$$
\frac{d\psi}{d\xi} = \frac{1}{W} \frac{d^2\psi}{d\xi^2} - \frac{i}{i_1} \tag{59}
$$

By definition at point $\xi = 0$ we have $\psi = 1$. It is easy to see that at this point, in addition, $d\psi/\delta\xi = W$ [the second condition from Eq. (44)]. Substituting Eq. (56) into Eq. (59) and making the Laplace transformations identical to those made in the last section, we come up with the following equation for speed

$$
F_\infty(W) = 0 \tag{60}
$$

where

$$F_\infty(W) = \frac{1}{W}\{1 + I \exp[-W(\tau_1 + \tau_2)] - (1 + I) \exp(-W\tau_1)\} - 1 \quad (61)$$

where τ_1 and τ_2 are current step widths in new units:

$$\tau_1 = \frac{t_1 i_1}{C\varphi_*}, \quad \tau_2 = \frac{i_1 t_2}{C\varphi_*} \quad (62)$$

From the definition of τ_1 it is seen that the value is almost equal to the maximum/threshold potential ratio, i.e., to the safety factor. Normally, the maximum potential is several times higher than the threshold potential (it may be 4 to 5 times as great). Therefore, if W is not too small the exponentially small terms may be neglected to give:

$$F_\infty(W) \approx \frac{1}{W} - 1 \quad (63)$$

Hence, the solution of Eq. (60) is $W = 1$ which corresponds to the dimensional speed [Eq. (55)]. Thus the fact that the impulse propagation speed is independent of events sufficiently remote in time, which we mentioned in the preceding subsection when we calculated the speed, is actually due to the large enough safety factor ($\tau_1 \gg 1$) or, which is the same, to the sufficiently long duration of the inward current. To make sure that there is a second solution to Eq. (60) it is sufficient to consider the behavior of $F_\infty(W)$ at $W \ll 1$, where the first two terms of the expansion of $F_\infty(W)$ have the form:

$$F_\infty(W) \approx \frac{W\tau_1}{2}(\tau_1 + \tau_2) - 1 \quad (64)$$

Hence,

$$V^1 \approx \left[\frac{2}{\tau_1(\tau_1 + \tau_2)}\right]^{1/2} \approx \frac{1}{\tau_1}\left(\frac{2}{1 + 1/I}\right)^{1/2} \quad (65)$$

Considerations of a general nature suggest that the solution [Eq. (65)] is unstable. From the form of Eq. (65) it follows that the ratio of the stable speed to the unstable speed is close to the safety factor. It must be noted that the existence of unstable solutions has been demonstrated earlier by computer treatment of the H–H equations.[28] It may be added that the second unstable solution appears when the second current step is taken into account.

3.2.2. Impulse in a Closed Circuit

The practical interest in this problem lies primarily in the fact that the impulse behavior in a closed circuit is very similar to excitation circulation along the myocardium in the case of abnormal electric activity of the heart.[29]

Theoretically this problem has been studied with the FitzHugh–Nagumo[25] and current generator[11] models. It has been shown experimentally[30] that the shorter the ring, the smaller the propagation speed of excitation, and in a ring of a small diameter repetitive excitation becomes impossible, i.e., there exists a critical ring size.

Consider a closed circuit of unit length. We will solve Eq. (59) for this circuit under the conditions:

$$\psi(0) = \psi(T) = 1 \tag{66}$$

$$\frac{d\psi}{d\xi}(0) = \frac{d\psi}{d\xi}(T) \tag{67}$$

Here, T is the excitation period equal to the travel time of the impulse along the loop in the steady-state mode:

$$T = 1/v \tag{68}$$

In the same manner as Eqs. (47) and (60) were derived, we can obtain the following equations for the impulse speed $V = v(RC^2\varphi_*/i_1)^{1/2}$:

$$F_L(V) = 0 \tag{69}$$

the function $F_L(V)$ having the following form:

$$F_L(V) = F_\infty(V) + \frac{F_\infty(V) + 1}{\exp(LIV) - 1} - \frac{V\tau_1(\tau_1 + \tau_2)}{2L} \tag{70}$$

where L is the loop length in new units:

$$L = l \cdot (Ri_1/\varphi_*)^{1/2} \tag{71}$$

To determine $F_\infty(V)$ one has to substitute $W = V^2$ in Eq. (61). As before, τ_1 and τ_2 are current step widths [(Eq. (62)]. Figure 11 shows the V vs. $L/(\tau_1 + \tau_2)$ curve for $I = 1$ and $\tau_1 = 4$, 10, and 20 plotted numerically. The critical loop size is clearly apparent for every value of τ_1 in the figure and is seen to decrease as τ_1 increases (remember that at large τ_1 this value is close to the safety factor). It is also seen in the figure that for every loop length (for which propagation of excitation is possible) there exists a second—unstable—solution for the speed (the bottom branches of the curves), the confluence of the stable and unstable branches occurring at critical ring size.

3.2.3. Inhomogeneity Effects

The situations we have been discussing so far involved a homogeneous fiber, which allowed us to obtain analytical solutions using a simple ionic generator model. The excitation conditions are, however, much more complex in inhomogeneous media where inhomogeneity may be either geometrical[31] or functional.[32] The authors of Reference (31) obtained an analytical solution

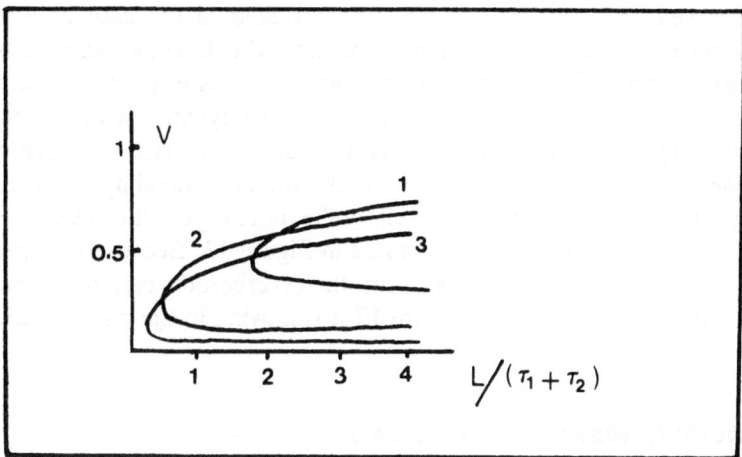

Figure 11. The impulse speed in the ring as a function of $L/(\tau_1 + \tau_2)$ at $I = 1$. The τ_1 values on the curves are: (1) 4, (2), 10, (3) 20.

for the speed of propagation of an excitation impulse along a geometrically inhomogeneous fiber composed of two homogeneous segments having different diameters. The propagation speed has been found to exhibit a hysteresis effect manifesting itself in its dependence upon the direction of impulse propagation (cf. Figure 12). Physically this effect is altogether clear—it is basically the same as that occurring in flame front propagation in an inhomogeneous system.

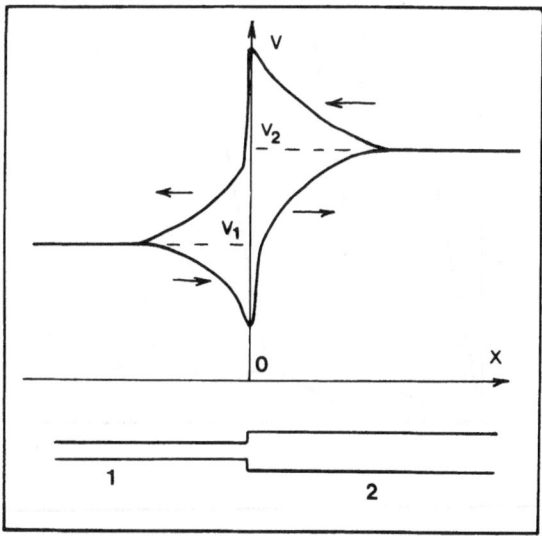

Figure 12. The hysteresis of excitation propagation along a nerve fiber whose outlines are shown beneath the figure. The impulse direction is marked with arrows beside the corresponding curves.

Complex excitation modes were also studied in the case of a micro-inhomogeneous medium, syncytium, consisting of uniformly mixed cells with different properties.[32] It was shown that in this case a reverberator-type behaviour was possible.[33] Here a reverberator is a dynamic structure periodically emitting excitation stimuli forward and backward relative to its travelling direction. The reverberator "center" is the impulse traveling along cells of one type in one direction and along cells of the other type in the other direction. A reverberator of this kind is illustrated in Figure 13. Because the speeds of the "partial" impulses may be different, the reverberator center will move in one direction or the other (in Figure 13 it is moving in the positive direction of the x-axis).

3.3. Activity Wave in a Neuron Net

We have already noted above that in analyzing excitable systems one has, more often than not, to deal with a parabolic equation with a nonlinear source. In this section we will concern ourselves with an excitable medium of a different type, where the signals are transmitted in the neuron network not by the local currents but by the nervous impulses traveling along the axons. The propagation speed of the activity wave will, if this transmission mode is possible at all, depend not only on the signal transmission speed but also on the other characteristics of nerve cells such as cell body capacitance, conductance, etc.

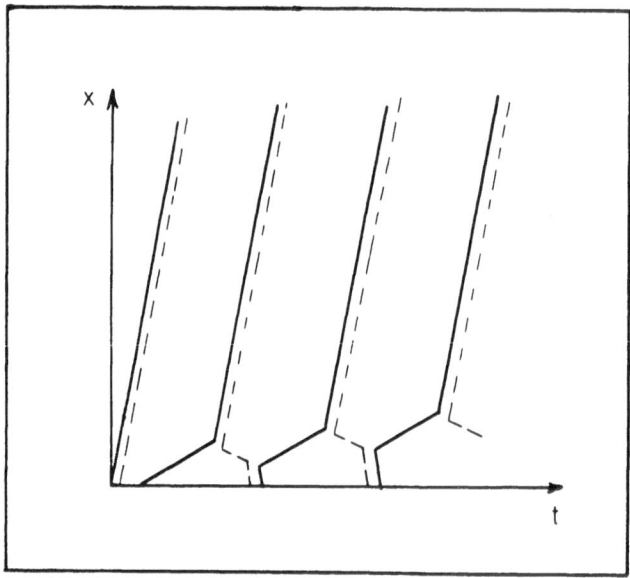

Figure 13. Reverberator in a two-component medium capable of conducting partial impulses. Single lines are for partial impulses, double lines are for collective impulses, x is excitation front coordinate, t is time.

To simplify the problem to the maximum possible extent, we will consider a homogeneous isotropic neuron net consisting of cells characterized by the same set of parameters.[34] The loss in generality will be lavishly compensated for by the considerable simplification of the analysis and the formal clarity in which the results can be obtained. Each neuron will be described as follows. Each unexcited ($\varphi < \varphi_*$) neuron is a summator of synaptic signals:

$$C\frac{\partial \varphi}{\partial t} = i_s - g\varphi \tag{72}$$

Here, C is neuron body capacitance, φ is the potential measured on the neuron body, t is time, i_s is total synaptic current of a given neuron, and g is passive conductance of the neuron membrane or "leakage" conductance (for simplicity the leakage e.m.f. is taken as equal to the resting potential). Assuming the net to be two-dimensional and considering the steady-state excitation mode in it, we have that the variables φ and i_s are functions of the self-similar coordinate:

$$\xi = x - vt \tag{73}$$

where v is the activity wave propagation speed and x is the coordinate perpendicular to the activity wavefront. During the active period, t_1, each excited neuron emits a constant signal (for example, a train of constant-frequency pulses) to all the neurons with which it is connected. The density of the interneuron connections is given by the connection density function $K(r)$, where r is the interneuron distance. The function $K(r)$ is normalized as follows:

$$\int_0^{2\pi} d\theta \int_0^\infty rK(r)\, dr = 1 \tag{74}$$

Let us place the point $\xi = 0$ on the wave front determined by the condition $\varphi = \varphi_*$. Then the synaptic current to the neuron with the coordinate ξ may be written as:

$$i_s(\xi) = i_0 \int_\Omega rK(r)\, dr\, d\theta \tag{75}$$

Here, i_0 is the maximum possible current and r, θ are polar coordinates with the pole at point ξ and the radius directed opposite the axis ξ. The symbol Ω represents the region from which the signal comes to a point with the coordinate $\xi = 0$ at a given moment of time (cf. Figure 14). The boundaries of this region are determined from the following considerations: the time it takes for the activity wave to travel from a given point within the region Ω to a point with the coordinate $\xi = 0$ must be greater than the time in which the signal can travel from the same point to a point with coordinate ξ but smaller than the

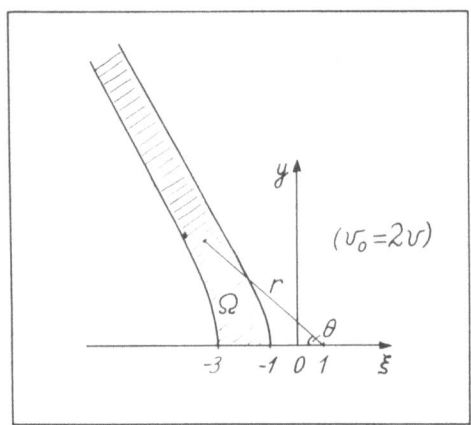

Figure 14. Selection of polar coordinates and the active region Ω (hatched) sending the signal to a point with coordinate ξ.

sum of this latter time and the neuron's activity time:

$$\frac{r}{v_0} \leq \frac{r \cos \theta - \xi}{v} \leq \frac{r}{v_0} + t_1 \tag{76}$$

where v_0 is signal speed.

Introducing the ratio

$$V = v/v_0 \tag{77}$$

we can define the region Ω in the following manner:

$$\frac{\xi}{\cos \theta - V} \leq r \leq \frac{\xi + vt_1}{\cos \theta - V}; \qquad \cos \theta > V \tag{78}$$

Equation (75) will then have the form:

$$i_s(\xi) = 2i_0 \int_0^{\arccos V} d\theta \int_{\xi/\cos \theta - V}^{\xi + vt_1/\cos \theta - V} rK(r)\, dr \tag{79}$$

Introducing the variable ξ of Eq. (73), we can rewrite Eq. (72) as follows:

$$-vC \frac{\partial \varphi}{\partial \xi} = i_s(\xi) - g\varphi, \qquad 0 < \xi \tag{80}$$

Making Laplacian transformations, we find

$$\hat{\varphi}(p) = \frac{\varphi_* - \dfrac{\hat{i}_s(p)}{vC}}{p - \dfrac{g}{vC}} \tag{81}$$

The function $\hat{i}_s(p)$ has no singular points within the range Re $p > 0$. Therefore,

$$p = g/vC \tag{82}$$

is the only singular point of $\hat{\varphi}(p)$.

In case of an arbitrary v, this point is the first-order pole. Therefore, as before, the residual of $\hat{\varphi}(p)$ at this point must be equal to zero—and this in fact yields the equation for v:

$$vC\varphi_* = \hat{i}_s\left(\frac{g}{vC}\right) \tag{83}$$

Consider the specific features of the solution of this equation; to this end we will study how the right-hand part of its depends on v:

$$\hat{i}_s\left(\frac{g}{vC}\right) = \int_0^\infty \exp\left(-\frac{g}{vC}\xi\right) i_s(\xi)\, d\xi \tag{84}$$

From Eq. (79) it follows directly that

$$\lim_{v \to 0} i_s(\xi) = 0, \quad \lim_{v \to v_0} i_s(\xi) = 0 \tag{85}$$

whence by virtue of Eq. (84) we obtain

$$\lim_{v \to 0} \hat{i}_s = 0 \quad \lim_{v \to v_o} \hat{i}_s = 0 \tag{86}$$

$$\lim_{v \to 0} \frac{\partial \hat{i}_s\left(\dfrac{g}{vC}\right)}{\partial v} = 0, \quad g > 0 \tag{87}$$

One function satisfying these conditions is shown in Figure 15A. This figure also shows points of intersection of the right- and left-hand portions of Eq. (83). The number of such points is either two (excluding the trivial point $v = 0$) or zero (the dashed line). This suggests that there is a critical threshold value above which signal propagation is impossible. Further, simple physical considerations show that the solution corresponding to the smallest speed is unstable. Therefore, the situation is very reminiscent of the nerve fiber although the potential equation is quite different. It is easy to see that, as in the case of the fiber, the unstable state is associated with the return of the potential to the initial value. Indeed, if we consider the no-return case, putting $g = 0$, we will find that the condition of Eq. (87) will not be satisfied and there will be no unstable solution (the situation shown in Figure 15B). Thus despite the considerable differences between a neuron net and the nerve fiber (the differential equations for the potential have different orders) there is a profound similarity between these media which manifests itself in specific features of the solution for excitation propagation speed.

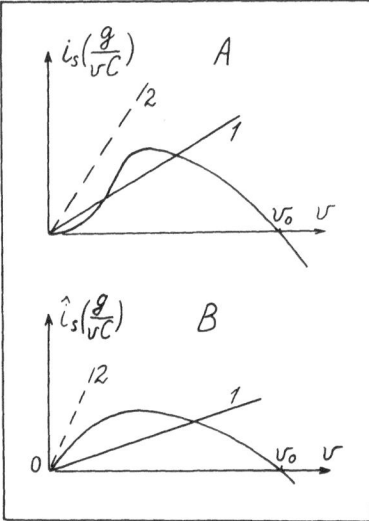

Figure 15. Graphic illustration of the solution of the equation for activity wave propagation speed. Straight lines correspond to left-hand side of Eq. (83), curve lines represent right-hand side of the Equation. A: finite-memory neuron net $(g > 0)$; B: infinite memory neuron net $(g = 0)$. The dashed lines represent the left-hand portion of Eq. (83) at sufficiently large φ_*.

4. Ionic Transport across Membranes

When we were discussing the excitation propagation processes in Section 3 we used phenomenological descriptions of the local properties of membranes, either in terms of the empirical Hodgkin–Huxley equations or with the help of ionic current generators with controlled characteristics. An indefinite parameter-free description of data transmission and processing must, obviously, be based on a molecular theory of the potential-dependent ionic transport across membranes. We will concern ourselves with these questions in the next two sections. A wealth of publications devoted to ionic transport across membranes have appeared,[35–40] manifesting a considerable evolution of the conceptual basis. A major distinguishing feature of this process is that, instead of describing a membrane as a homogeneous macroscopic phase to which classic electrodiffusion equations are applicable, one has to conceive ionic transport as a multistage vector reaction. Not claiming that the treatment which follows is comprehensive, we will start with discussion of the electrodiffusion theory to enable its comparison with the single-file transport theory that will be described in Section 5.

The considerable advance achieved in recent years in the discrete treatment of transport processes is primarily due to the development of the bilayer lipid membranes (BLM) modifiable with various ionophores. The most important results directly concerned with the functioning of excitable membranes are highlighted in Sections 4.2 and 4.3.

4.1. Electrodiffusion Equation

If a membrane is assumed to be a homogeneous medium in which ions can diffuse and migrate, the partial electric current of ions species l may be written as

$$I_l = z_l u_l RT \left(\beta z_l c_l E - \frac{\partial c_l}{\partial x} \right) \qquad (88)$$

where z_l is charge of ion species l (in proton units, e), u_l is the mobility, related to the diffusivity D_l by $D_l = u_l / \beta$ ($\beta = F/RT$) and E is electric field strength. The total electric current density is found as a sum for all ion species:

$$I = \sum_{l=1}^{n} I_l$$

Under steady-state conditions the partial currents persist, so that $I_l = \text{const.}$, and Eq. (88) is then a first-order nonlinear differential equation containing unknown functions c_l and E and unknown constant I_l. This fundamental electrodiffusion equation, conventionally referred to as the Nernst–Planck equation, may be obtained by direct differentiation of the expression for the electrochemical potential of ion species l present in the dilute solution

$$\bar{\mu}_l = \mu_l^0 + RT \ln c_l + F\varphi$$

If the system in question contains n ion species, we will have n equations (88) for $n + 1$ functions, including all c_l and E. To render the problem definite there must be one more equation. In the general case it will be the Poisson's equation

$$\frac{dc_l}{dx} - \beta z_l c_l E = \frac{I_l}{z_l u_l RT} \qquad (89)$$

$$\frac{dE}{dx} = \frac{F}{\varepsilon} \sum_{l=1}^{n} z_l c_l$$

Defining a total of $2n + 2$ boundary conditions, we will obtain a final formulation of the problem. Apart from this basic form of the electrodiffusion problem, it is sometimes convenient to use the integral form or an equation with excluded electric field. A comprehensive review of all questions pertinent to this problem can be found in Arndt and Roper's book[41]; the nonsteady-state case is also discussed in detail in this reference.

It is never possible to obtain an exact analytical solution of the electrodiffusion equation, and the standard practice is to seek approximate solutions under some sort of assumptions. In macroscopic membranes whose thickness is many times greater than the electric double layer thickness Planck's approximation, assuming that the system obeys the local electric neutrality condition, is valid. This means that instead of the Poisson equation in (89) one has to use $\sum z_i c_i = 0$. For a binary electrolyte A^+B^-, Planck's approximation yields a well-known equation for the diffusion potential

$$\varphi_d = \frac{RT}{F}\frac{u_A - u_B}{u_A + u_B}\ln\frac{c^o}{c^i} \tag{90}$$

where c^o, c^i are solution concentrations on both sides of the membrane. The criterion for validity of the approximation fails to be met by biological membranes where the screening radius is normally greater than their thickness. Goldman's constant field approximation[42] is found to be more adequate in this case; in this approximation the condition $E = $ const. is used instead of Poisson's eq. (A quantitative analysis of the scope of applicability of this approximation has been given in Reference (36).) When $E = $ const., the Nernst–Planck equation becomes readily integrable. The partial V–A characteristic of the ion species l becomes

$$I_l = \frac{z_l^2 RTu_l\psi\gamma_l}{\delta}\frac{c_l^o - c_l^i\exp{(-z_l\psi)}}{1 - \exp{(-z_l\psi)}} \tag{91}$$

where γ_l is a distribution factor relating ion concentrations in the solution and membrane, δ is membrane thickness, and $\psi = F\varphi/RT$. This formalism, based on ion diffusion and migration in the continuous phase, has been used until recently for describing the electric effects in excitable biomembranes. The rest potentials of nerve cells have so far been calculated via Eq. (91). As already mentioned, smooth fiber membranes are permeable to sodium, potassium, and chlorine ions. Therefore the resting state is a steady state rather than a thermodynamic equilibrium state. In an open circuit the stationarity condition reduces to mutual compensation of the partial currents, viz. $I = 0$. This condition actually defines an equation for the membrane potential whose solution has the form

$$\varphi = \frac{RT}{F}\ln\frac{P_K c_K^o + P_{Na}c_{Na}^o + P_{Cl}c_{Cl}^i}{P_K c_K^i + P_{Na}c_{Na}^i + P_{Cl}c_{Cl}^o} \tag{92}$$

where the $P_l = u_l/\gamma_l/\beta\delta$ are permeabilities. Equation (92) satisfactorily describes the experimental results provided the range of concentrations is not too wide. This can hardly be regarded as a serious argument in favor of applicability of the electrodiffusion theory for describing the electric characteristics of nerve cells in the resting state since the unknown parameters have to be determined substantially from the same experiments. Moreover, the available data suggest that the treatment of the membrane as a homogeneous phase, as currently used in the electrodiffusion theory, is inadequate. Therefore, the only argument that may justify the use of Eqs. (91) and (92) is that they are simple and that reliable information about the true mechanisms of ionic transport in the resting state is unavailable. Much more is known about ionic transport across membranes in the excited state, where the system conductance drastically increases. It has been firmly established that in this case the ionic transport occurs via the special, most probably lipoprotein, structures called ionic channels. We will discuss transport via the channels in Section 5, for it requires a special formalism, and at that point we will describe in brief one more corollary of the electrodiffusion approach which is very important for identification of the transport mechanism.

Apart from the electric characteristics we have discussed above, the labeling technique may be suitable for measuring ion fluxes. A formula for the latter may be derived from Eq. (91) by putting c_i^i and c_i^o alternately equal to zero and dividing I_l by z_iF:

$$\vec{j}_l = \frac{z_i u_l \psi \gamma_l c_l^i}{\beta\delta[1 - \exp(-z_l\psi)]}$$

$$\overleftarrow{j}_l = \frac{z_i u_l \psi \gamma_l \exp(-z_l\psi) c_l^o}{\beta\delta[1 - \exp(-z_l\psi)]} \tag{93}$$

Here, $\vec{j}_l$ is called efflux and $\overleftarrow{j}_l$ influx. The ionic current apparently is $I_l = z_kF(\vec{j}_l - \overleftarrow{j}_l)$. The efflux $\vec{j}_l$ linearly depends on the inside concentration but is independent of the outside concentration, while the influx depends only on the outside concentration. Thus, for instance, by varying the inside concentration we may affect the efflux while leaving the influx unchanged, provided of course the potential difference is kept constant. This means that in the Goldman approximation the effluxes and influxes are independent of each other. In the case of biological membranes it is essential to verify whether the independence principle is valid or not, for that would answer unambiguously whether the precepts of the theory are valid. Experiments have shown that there are cases where the independence principle does hold true. More than a few exceptions have also been found, however, to which the electrodiffusion theory has been found to be inapplicable. In Section 5 we will discuss this important issue in considerable detail.

Considering Eq. (93), the ratio of the unidirectional fluxes is:

$$\frac{\vec{j}_l}{\overleftarrow{j}_l} = \frac{c_l^i}{c_l^o}\exp(z_l\psi) \equiv \exp[z_l(\psi - \psi_l)] \tag{94}$$

This formula, called the Ussing equation,[43] is, furthermore, extremely important for the assessment of the fundamental principles of the electrodiffusion theory, although those unidirectional fluxes which satisfy the Ussing equation may, at the same time, not obey the independent principle.

The Ussing equation may be derived from the general electrodiffusion equations without the assumption of a constant electric field. To demonstrate this it may be convenient to represent the partial ionic flux in the form:

$$j_l = D_l \exp(-z_l\psi)\frac{d}{dx}(c_l\exp(z_l\psi)) \tag{95}$$

Consider the case where there is on the left of the membrane a solution containing penetrating ions A and on the right a solution containing penetrating ions B. Suppose that $D_A = D_B$ and $z_A = z_B$. We shall assume concentrations c_A^i and c_B^o are known, and $c_B^i = 0$ and $c_A^o = 0$. Substitute into the obvious identity $j_A j_B = j_B j_A$ Eq. (95) for ion fluxes:

$$j_A \frac{d}{dx}[c_B \exp(z\psi)] = j_B \frac{d}{dx}[c_A \exp(z\psi)]$$

and, using the flux constancy condition, include the fluxes inside the differentiation signs

$$\frac{d}{dx}[j_A c_B \exp(z\psi)] = \frac{d}{dx}[j_B c_A \exp(z\psi)] \tag{96}$$

By virtue of the assumed boundary conditions, integration of Eq. (96) gives:

$$j_A c_B(\delta) \exp[z\psi(\delta)] = -j_B c_A(0) \exp[z\psi(0)]$$

whence

$$\left|\frac{j_A}{j_B}\right| = \frac{c_A(0)}{c_B(\delta)} \exp(z\psi) \tag{97}$$

which is equivalent to Eq. (94). The same conclusion can be drawn for the case of three-dimensional diffusion as well.[44]

4.2. Bilayer Lipid Membranes

For several decades already the structure of cell membranes has been conceived in terms of the Danielli–Davson model[45] in which the membrane core is composed of a lipid bilayer and on the outside the "sandwich" is coated with a layer of proteins that may penetrate deep into the membrane giving rise to various functionally important structures such as polar pores. The overall thickness of such a structure is of the order of 100 Å. Presently, the fluid mosaic model of Singer and Nicolson[46] has become rather popular. According to this model the lipid matrix is inlaid with proteins of two types, surface and integral, the latter penetrating throughout the lipid matrix and apparently responsible for ionic transport as ion channels. This model itself prompts the idea of reconstitution of natural membranes, a first step in which should be the creation of an artificial lipid bilayer. Accordingly, as early as the 1930s the first attempts were undertaken to obtain thin stable lipid or proteolipid structures. These attempts did not meet with success until 1961 when Müller *et al.*[47] discovered that a suspension of an extract of oxidized phospholipids of bovine brain in aqueous solution tends to produce spontaneously a bimolecular black-coloured film. The method was found to be amazingly sinple and soon became widely popular.[48] A certain amount of phospholipid is dissolved in a liquid hydrocarbon such as *n*-decane. A Teflon baffle having a small orifice (*ca.* 1 mm²) is arranged in a vessel containing an electrolyte solution. A drop of phospholipid solution is applied to the orifice; gradually spreading, it first becomes a thick film with iridescent patterns, and then gets thinner and blacker. Capacitance measurements give a thickness of around 50 Å for the black membrane if the dielectric constant of the lipid is 2.5. This corresponds to double the hydrocarbon chain length in a phospholipid molecule. Optical measurements give a thickness of *ca.* 70 Å, which apparently includes the polar heads of phospholipids. Organic solvent molecules are present in each monolayer interspersed between the hydrophobic tails of phospholipid molecules. Investigations of phosphatidylcholine BLM[49] have shown that the membrane has an area of about 75 Å² per molecule, whereas the expected maximum is only 58 Å². This led to the conclusion that the system contains the lipid and solvent molecules in a 1:1 ratio.

Thus a BLM is a thin hydrocarbon film stabilized in the aqueous phase by lipid molecules which make up a considerable portion of its volume. From a thermodynamic analysis of such membranes[50] it follows that the molecules capable of forming such structures must have a high adsorption energy both on oil and water. Lipids which contain long hydrocarbon chains and short highly polar groups naturally satisfy this condition. Another requirement which lipids also meet is that the cross-sectional areas of their heads must not differ strongly from that of the hydrocarbon chains. What is absolutely unique to BLM's is that in one direction they have molecular dimensions and in the other macroscopic dimensions. The respective dimensions normally differ by a factor of 10^6. This gives rise to a host of questions which presently cannot be answered in a satisfactory manner. They primarily concern structure, stability, and phase transitions. Table 1 shows structural formulas of the basic phospholipids—phosphatidylcholine and phosphatidylserine—which are found in many cell membranes and which are conventionally used as BLM components. The polar heads of both charged and neutral phospholipids are characterized by a high dipole moment whose major component is tangential. Thus the potential distribution in a system consisting of a symmmetric membrane composed of charged lipids and bathed on both sides with electrolyte solutions is generally as shown in Figure 16. The transmembrane potential jump φ is made up of two surface jumps φ_{s1}, φ_{s2}, two dipole jumps φ_{d1}, φ_{d2}, and the intramembrane potential difference φ_{in}. The transmembrane potential jump φ can be measured by conventional electrochemical techniques. Special techniques had to be invented to measure the individual components of the total jump.[51-55] The surface potential jump is satisfactorily accounted for by the Gouy–Chapman–Stern theory. This holds equally true for natural membranes.[56] A circuit equivalent to the equilibrium system shown in Figure 16 consists of three series-connected capacitors whose capacitances obey the inequality: $C_m \ll C_{s1}, C_{s2}$, because $C_m/C_s \sim \varepsilon_m\lambda/\varepsilon_s\delta$, where λ is the screening distance in the electrolyte solution. Thus the applied external field is attenuated

Table 1

Structures of Basic Phospholipids

Phosphatidylcholine

```
R—CO—O—CH₂
              |
R₁—CO—O—CH        O⁻
              |      /
        H₂C—O—P=O                CH₃
                   \             /
                    O—CH₂—N⁺—CH₃
                             \
                              CH₃
```

Phosphatidylserine

```
R—CO—O—CH₂
              |
R₁—CO—O—CH        O⁻
              |      /
        H₂C—O—P=O                COO⁻
                   \             /
                    O—CH₂—CH
                             \
                              NH₂⁺
```

mostly by the membrane itself, whereas the ionic sheaths of the electric double layer remain substantially unperturbed.

The electric conductance of the BLM is extremely low and is within the range of 10^{-7} to 10^{-10} mho cm^{-2}.[48] For comparison, consider that electric conductance of a cell membrane in the resting state is *ca.* 10^{-3} mho cm^{-2} and that of a layer of an aqueous 0.01 *M* KCl solution of the same thickness as

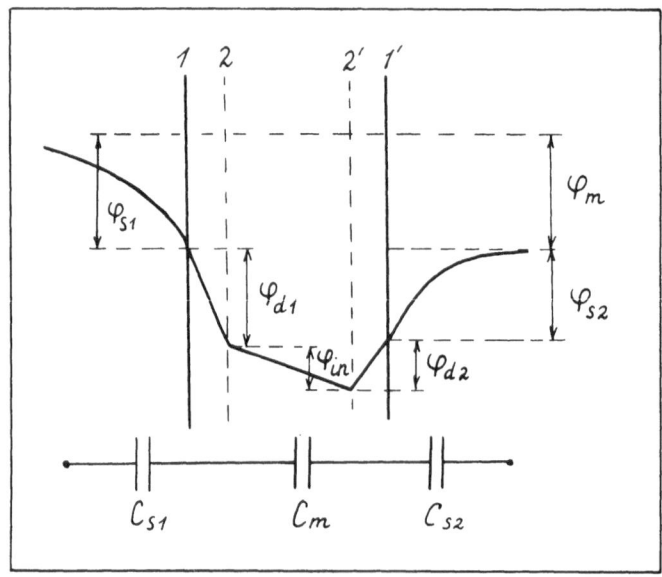

Figure 16. Potential distribution on membrane. See text for discussion.

the BLM is 10^4 mho cm^{-2}, a difference as great as 10^{12} times. The low conductance of the BLM is due to the very small dielectric constant of the lipid phase ($\varepsilon_m \sim 2.5$). The membrane is a high hydrophobic barrier preventing water-soluble ions from passing through it. The ion distribution between the lipid and water phases is given by

$$\gamma = \frac{c_m}{c_s} = \exp\left(-\frac{W}{kT}\right) \tag{98}$$

where W is the total energy of a particle in the lipid phase relative to that in the aqueous solution. Calculation of W by Born's formula, assuming that the particle has a radius a and a charge ze, yields the expression:

$$W = \frac{q_0}{a}\left(\frac{1}{\varepsilon_m} - \frac{1}{\varepsilon_s}\right) \qquad q_0 = \frac{z^2 e^2}{2kT} \tag{99}$$

At $T = 25°C$, $z = 1$, we have $q_0 = 282$ Å. If we put $a = 2$ Å, $\varepsilon_m = 3$, we have $\gamma = 10^{-20}$. In this case the conductance of the BLM would be much below the values given above. Apparently we have failed to take into account a number of factors reducing the hydrophobic barrier height, viz., small membrane thickness, ion pair generation, and ion hydration. Consider these effects one by one. The finite membrane thickness can be accounted for if we allow for repulsive forces acting on the ion near the membrane/water interface. At the center of the membrane calculation[57] shows the energy to go down by

$$\Delta W = \frac{e^2}{\varepsilon_m \delta} \ln \frac{2\varepsilon_s}{\varepsilon_s + \varepsilon_m} \tag{100}$$

The relative difference from the ion energy in an infinite medium is $(1.4a/\delta)$, i.e., does not exceed a few percent. Therefore, despite its small thickness the membrane is in fact a high barrier to the passage of dehydrated ions. Its height is several dozens of kcal/mol. The estimate of Eq. (99) is, of course, incorrect in the case of fat-soluble ions such as dipicrylamine, tetraphenyl borate, etc., since the contribution of the hydrophobic effects becomes dominant in these cases.

Generation of ion pairs also fails to produce a marked energy gain. The electrostatic energy of two particles having radii a_+ and a_- and separated by a distance d is:

$$W = \frac{e^2}{2\varepsilon_m a_+} + \frac{e^2}{2\varepsilon_m a_-} - \frac{e^2}{\varepsilon_m d} \tag{101}$$

It is seen hence that the maximum energy reduction will not be more than twofold. Only if the charged particles are covalently bonded to each other will the electric field be considerably attenuated in their vicinities, but that would mean a discharge of two associated particles.

Hydration causes a considerable reduction of the hydrophobic barrier height. A rigorous statistical treatment of this situation is fairly complicated.[58] Here, simple electrostatic evaluations[59] will suffice. Suppose a polar pore of radius b, has appeared in the membrane. For $b \ll \delta$, ion energy on the pore axis is:

$$W_p = \frac{e^2}{2\varepsilon_p a} + \frac{e^2}{\varepsilon_m b} P\left(\frac{\varepsilon_m}{\varepsilon_p}\right) \tag{102}$$

The second term in this equation is due to the image forces in the pore walls. It is inversely proportional to the pore radius. The function $P(x)$ has been numerically calculated.[59] Its maximum value does not exceed 0.25. If, e.g., $\varepsilon_p \sim \varepsilon_s$, then the barrier height for an ion crossing the membrane is defined simply by the second term. At $\varepsilon_m = 2$ it is

$$\frac{e^2}{2b} P(\tfrac{1}{40}) \simeq \frac{28}{b} \text{ kcal/mol} \tag{103}$$

in which b is in angstroms.

Consider, finally, the case of an ion surrounded in the membrane by a spherical polar "coat" ($\varepsilon = \varepsilon_c$) of radius b. Its energy has the form

$$W_c = \frac{e^2}{2\varepsilon_m b} + \frac{e^2}{2\varepsilon_c}\left(\frac{1}{a} - \frac{1}{b}\right) \tag{104}$$

If $\varepsilon_c \gg \varepsilon_m$ the first term is dominant, and for $b \sim 5–10$ Å and $T = 25°C$ we get $W_c = 16.5$ to 8.2 kcal/mol or 9.8 to 4.8 kT/ion, i.e., the hydrophobic barrier is sharply decreased.

It should be noted that an electric field promotes pore formation in the membrane. This has been clearly demonstrated in studies[60,61] of the electric breakdown of the BLM which involves formation and development of such pores.

The above estimates are important not only for the correct understanding of the background conductance mechanisms in lipid bilayers. They also underlie the physical interpretation of the induced conductivity of the BLM modified with various ionophores.

4.3. Induced Ionic Transport

The interest aroused by the BLM has been mostly due to the fact that these systems form the basis for reconstitution of complex transport biomembrane systems, such as ionic channels, ATPases, the acetylcholine receptor, bacteriorhodopsins, etc. A first step in this direction was the discovery of a class of compounds that were found capable of radically affecting the electric properties of membranes (see the review in Reference (35)). Such compounds are conventionally referred to as ionophores. In their presence the conductance of the BLM increases by many orders of magnitude. Very small quantities of ionophores are generally required, for they only serve to carry ions across the membrane. The conductance is in this case selective. Ionophores include fat-soluble acids—2,4-dinitrophenol, dicumarol, tetrachlorotrifluoromethyl-benzimidazole (TTFB), etc.—and polypeptides—valinomycin, the actin group, gramicidin A, B, C, etc. Using alameticin in combination with a surface-active protein—protamine—Müller and Rudin were able to simulate electric excitability in bilayer lipid membranes.[62] Later some authors reproduced a whole range of biomembrane phenomena using various compounds and cell fractions.

The sudden increase in BLM conductance in the presence of ionophores can be explained with the help of simple electrostatic considerations which we have already set forth above. For example, valinomycin has a polar cavity accommodating the potassium ion, while the periphery of the complex is hydrophobic, and this accounts for its good solubility in the hydrocarbon phase. As regards gramicidin A, its two monomers combine into a dimer spanning precisely the membrane thickness. Inside this dimer molecule there is a polar channel, 8 Å in diameter, permeable to water and hydrated ions. When Hladky and Haydon[63] studied BLM conductance at very low gramicidin concentrations, they observed discrete current steps, each corresponding to the appearance (and then disappearance) of a single channel. An example is given in Figure 17.

Unlike gramicidin, valinomycin and nonactin act as mobile carriers. It was most apparent in experiments[64] where the conductance of modified membranes was measured at different temperatures, both above and below the crystal–liquid transition point. In the presence of valinomycin and nonactin, the conductance was seen to decrease sharply on freezing of the membrane, while the conductance of the gramicidin channels remained unchanged. It only remains to add that fat-soluble weak acids such as TTFB assure a high conductance in the pH range where the charged form is dominant, simply via direct passage of the anions. At low pH values these compounds act as mobile proton carriers.

The Nernst–Planck equation [Eq. (88)] is suitable to describe the direct anion passage quantitatively in the constant field approximation. The mobile carrier mechanism is treated in the same way. The relevant diagram is shown

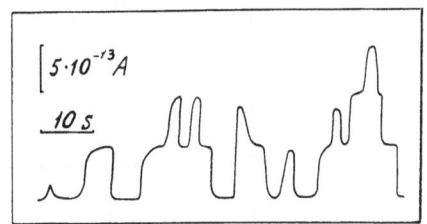

Figure 17. Current steps passed through the BLM in the presence of gramicidin A in an extremely low concentration (10^{-2} *M* KCl) (E. M. Egorova, unpublished data).

in Figure 18. We will now consider the problem in brief in the special case of the fat-soluble weak acid HT. In the membrane there are both HT and T⁻ particles and both are mobile. Free protons cannot penetrate into the membrane for energy reasons. They can cross the membrane only in the form of HT. On the left-hand boundary the protons combine with the anions T⁻ and the resultant molecules diffuse toward the right-hand boundary where the proton is released into the solution. The remaining anions migrate from right to left under the effect of the external field, and the cycle repeats. Thus the following reactions can take place at the membrane–solution interfaces:[65]

$$(T) \underset{k_2}{\overset{k_1}{\rightleftharpoons}} [T]$$

$$(H)[T] \underset{k_4}{\overset{k_3}{\rightleftharpoons}} [HT] \tag{105}$$

$$(HT) \underset{k_6}{\overset{k_5}{\rightleftharpoons}} [HT]$$

The external field induces fluxes of the T⁻ anions and HT molecules in the membrane:

$$j_{\mathrm{T}} = -D\frac{d[T]}{dx} - uE[T]$$

$$j_{\mathrm{HT}} = -D_1\frac{d[HT]}{dx} \tag{106}$$

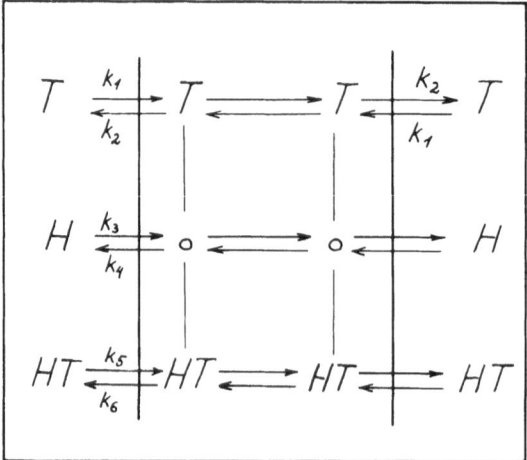

Figure 18. Diagram illustrating the function of mobile carriers.

The boundary conditions on the left wall of the membrane are:

$$k_1(T) - k_2[T]_1 + k_5(HT) - k_6[HT]_1 = j_T + j_{HT}$$

$$k_3(H)[T]_1 - k_4[HT]_1 + k_5(HT) - k_6[HT]_1 = j_{HT}$$

(107)

Those on the right wall are similar. The solution of this problem presents no difficulty. In the limiting case of the fast reaction of protons with T^- on the boundaries we obtain:

$$I = \frac{uFE\left(\dfrac{k_1}{k_2}\right)k_3(H)b\zeta}{[\zeta + (H)][2D/\delta + Dk_4/D_1 + k_3(H)]}$$

(108)

where $b = (HT) + (T)$, $\zeta = k_2 k_4 k_5 / k_1 k_3 k_6$. It is easy to see that this function has a maximum at a specific (H) and tends to zero for large and small (H). Physically this is quite apparent, since protons and vacant carriers of T^- are necessary for current to flow. It must be stressed that when alternating current is used the I vs. (H) relationship is already monotonic.[36] The reason is that at small (H) the alternating current is carried in the membrane by the T^- species whereas at large (H) the concentration $[T^-]$ becomes small and the current drops as described by Eq. (108).

We have given perhaps undue attention to the mobile carrier mechanism because at one time it was assumed that the Na and K transport in excitable cell membranes occurred precisely via this mechanism. In 1965, Chandler and Meves[66] undertook an experiment to assess the aforementioned specifics of the high-frequency conductance. A nerve fiber was placed in a solution containing no Na or K ions. This precluded direct current through the membrane. However, if there had been any mobile charged carriers in the membrane, the authors would have detected current on application of a variable field. The authors did not observe a detectable current under these conditions, from which it could be deduced that the transport systems of excitable membrane are structured as ion channels whose conductance is controlled by electric field.

It is time to discuss the theory of ionic transport by channels. What distinguishes this problem is that ion transport is single-file one, so that the interparticle interactions can be ignored only in the case of low filling ratios. The treatment should conveniently be carried out in discrete language[36] involving characterization of a channel by a potential energy profile with a certain number of potential pits and barriers. It will be taken into account that an ion can leap from the ith pit to a $(i+1)$th pit only if the latter is vacant. For clarity we will consider the simple case of transfer when there are only two pits (Figure 19). This will allow us to analyze the physics of the phenomenon and will yield results that are qualitatively similar to a more general case.

The channel state will be described by four binary functions each giving the probability of one of the possible events: $P(1, 1)$ is the probability of both pits being occupied with ions A; $P(0, 0)$, both pits are vacant; $P(0, 1)$ and $P(1, 0)$, one of the pits is occupied and the other vacant. Ions may pass to and from the solution. The rate constants of these heterogeneous processes will be denoted by k_1 and k_2. Leaps between the pits in a channel occur with the speed ν. Suppose the external potential φ applied to the membrane breaks up into φ_1, φ_2, φ_3 as shown in Figure 19. The external field sets ions A into directional motion. To calculate the rate of this process it is necessary to derive an equation for channel state probability.

For each of the binary functions P, one can write a "continuity" equation accounting for all the allowed transitions from a given state to a state symbolized by other binary functions and back. The probabilities of transition from one state to another are, in fact, probabilities of the corresponding leaps of the penetrating ion A. We thus have a set of simultaneous equations for the binary functions:

$$\frac{dP(1, 1)}{dt} = P(1, 0) k_1(A) \exp\left(-\frac{\psi_3}{2}\right) + P(0, 1) k_1(A) \exp\left(\frac{\psi_1}{2}\right)$$

$$-P(1, 1) k_2 \left[\exp\left(-\frac{\psi_1}{2}\right) + \exp\left(\frac{\psi_3}{2}\right)\right]$$

$$\frac{dP(0, 0)}{dt} = P(1, 0) k_2 \exp\left(-\frac{\psi_1}{2}\right) + P(0, 1) k_2 \exp\left(\frac{\psi_3}{2}\right)$$

$$-P(0, 0) k_1(A) \left[\exp\left(\frac{\psi_1}{2}\right) + \exp\left(-\frac{\psi_3}{2}\right)\right]$$

$$\frac{dP(1, 0)}{dt} = -P(1, 0) \left[k_2 \exp\left(-\frac{\psi_1}{2}\right) + \nu \exp\left(\frac{\psi_2}{2}\right)\right. \qquad (109)$$

$$\left. + k_1(A) \exp\left(-\frac{\psi_3}{2}\right)\right] + P(0, 1) \nu \exp\left(-\frac{\psi_2}{2}\right)$$

$$+ P(0, 0) k_1(A) \exp\left(\frac{\psi_1}{2}\right) + P(1, 1) k_2 \exp\left(\frac{\psi_3}{2}\right)$$

$$\frac{dP(0, 1)}{dt} = P(1, 0) \nu \exp\left(\frac{\psi_2}{2}\right) \sim P(0, 1) \left[k_2 \exp\left(\frac{\psi_3}{2}\right)\right.$$

$$\left. + \nu \exp\left(-\frac{\psi_2}{2}\right) + k_1(A) \exp\left(\frac{\psi_1}{2}\right)\right]$$

$$+ P(0, 0) k_1(A) \exp\left(-\frac{\psi_1}{2}\right) + P(1, 1) k_2 \exp\left(-\frac{\psi_1}{2}\right)$$

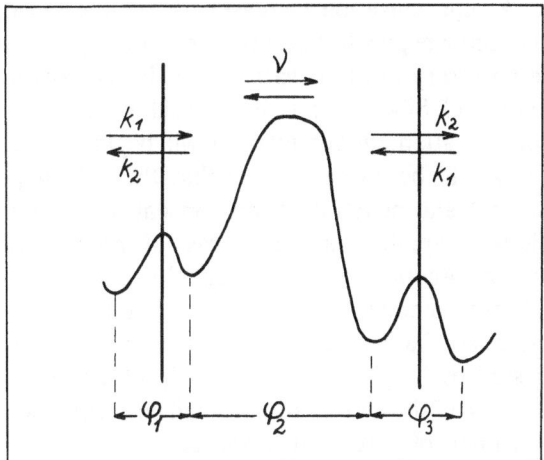

Figure 19. A simple ion energy profile in a channel with two potential pits.

Equations (109) are not independent, each one depending on the others. However, the probability normalization condition:

$$P(0, 0) + P(0, 1) + P(1, 0) + P(1, 1) = 1 \qquad (110)$$

completes the set of equations. The rate constants of ion leaps $k_1(A)$, k_2 and ν refer to one channel in this model, their dimension being, therefore, s^{-1}. Let a constant potential difference be applied to the membrane. It is quite easy to solve Eqs. (109) and (110), but the result is a rather cumbersome formula for P. We will not give it here, to save space.

The current across the membrane is obviously:

$$I = zFn\nu \left[P(1, 0) \exp\left(\frac{\psi_2}{2} \right) - P(0, 1) \exp\left(-\frac{\psi_2}{2} \right) \right]$$

where n is the number of channels per unit membrane surface. For small potentials the formulas for P become simpler, and we may finally obtain the following simple equation for membrane conductance

$$g = \frac{z^2 \beta Fn\nu k_2 k_1(A)}{[k_1(A) + k_2][k_1(A) + k_2 + 2\nu]} \qquad (111)$$

The electric conductance of a system of channels is identical with that calculated earlier for the carrier model, provided the carriers cannot penetrate through the membrane surface. Therefore, in the channel model under consideration the conductance is also a nonmonotonic function of the concentration of current-generating ions A in solution with the maximum at $A_{max} = [k_2(k_2 + 2\nu)]^{1/2}/k_1$. The reason for the nonmonotonic relationship between conductance and ion concentration is as follows. When the concentration of A is small the conductance is also small due to the shortage of the current-generating ions. As concentration increases, the conductance increases too, but soon another factor begins to tell. For generation of current not only the carried ions are necessary but also vacant sites in the conducting chains by which ions would travel. Since the number of such vacant sites decreases with increasing ion concentration, the conductance will begin to decrease at a certain concentration. It would be an easy guess that the bell-shaped relationship between conductance and penetrating ion concentration would remain even if a variable field were applied (unlike the case of mobile carriers).

In more general situations, when a channel has to be characterized by a chain of three of four potential pits the number of differential equations such as Eq. (109) increases sharply. They are solved either through the graphic procedure suggested by Hill[67] or with the help of a computer. In the next section the application of the above rationale will be exemplified for the sodium and potassium channels of natural membranes.

5. Channels in Biomembranes

5.1. Facts and Hypotheses

The phenomenological description of the excitability phenomenon given in Section 1.3 cannot claim to contain a final solution to the problem of the nature of transport systems of biological membranes responsible for nervous impuse generation. Where we stand, we can only conclude that the membrane as a whole is a nonlinear ion conductor whose properties are largely dependent upon the electrice field. For all that, the fact that the use of certain specific blocking compounds—tetrodotoxin and tetraethylammonium—allows the sodium and potassium ionic currents to be separated is alone sufficient to support the conception of selective transport systems located in the lipid matrix of a membrane. The entire body of data obtained for man-made bilayers modified with various ionophores (cf. Section 4.3) seems to verify the view that it is not the lipid phase but certain special structures that are responsible for the selective ionic transport. This conclusion is consistent with the Singer–Nicolson model which conceives of a membrane as a heterogeneous mosiac rather than a homogeneous phase, the integral proteins playing, most probably, the major role in ionic transport.

The number of transport systems per unit surface area for Na and K ions has been estimated according to the bound amounts of the specific blocking agents—tetrodotoxin and saxytoxin—as well as by measuring the ionic current fluctuations.[68,74] For example, a muscle cell membrane contains around 400 sodium transport systems per 1 micron^2. Conductance of each transport system is about $4 \times 10^{-12} \, \Omega^{-1}$ (for sodium) and $12 \times 10^{-12} \, \Omega^{-1}$ (for potassium).[75] By now there is plenty of evidence in favor of the fact that the conducting structures of membranes behave like channels rather than mobile carriers.

Note in this connection the previously mentioned experiment of Chandler and Meves.[66] As regards the chemical structure, ionic channels still remain a vague notion. It has not been possible thus far to identify the membrane proteins responsible for the electric field-controlled ionic transport. Therefore, the conceptions of the nature and structure of ionic channels advanced in current professional literature have been extremely varied. By analogy with the artificial channels in the BLM where the electric field influences the statistical process of channel buildup from subunits, some authors tend to believe that in biomembranes there are no preexisting channels at all and the observed increase of conductivity under depolarizing conditions is caused by their buildup. More investigators, however, uphold a different view according to which a channel is a sufficiently rigid macromolecular structure capable of minor conformational restructuring. There is a range of arguments to support the latter approach. The fact that tetrodotoxin binds with a closed sodium channel is a convincing one to support the presence of preexisting channels. The conception of an ionic channel as a lipoprotein complex characterized by a set of allowed conformational states permits it to be regarded as a "vector" enzyme which catalyzes the ion transfer process. A distinction of such an enzyme is that its activity can be controlled by an electric field.

The function of an ionic channel is threefold: it renders the membrane conductive, selective, and electric field-controllable. A natural question presents itself: is there a correlation between these particular functions of ionic channels and particular molecular groups? In a more modest (and more realistic) formulation the question is, may one assert that the different functions are performed by different portions or subunits of the channel? It is impossible so far to give a flat answer to this question, although it is held to be more probable that the selective–conductive and controlling functions are realized by different groups in the channel. Below we will give a more detailed analysis of these channel functions; meanwhile we will discuss what are currently the most widely adopted conceptions of channel structure which may be conveniently used as a working hypothesis. We wish to stress once again that the picture is not based on direct chemical structure data but is a result of functional reconstitution of the system.

An ion channel (Figure 20) comprises a rigid transport system (1) assuring a specific conductance and selectivity, a mobile group (3) called a gate, which can close or block up the channel, and finally, a membrane voltage sensor (4) which controls the gate function. The transport system (1) has a hollow space lined with polar groups. A hydrated ion first enters the wide mouth of the channel and, as it moves on, undergoes gradual dehydration, the process of substitution of the hydrate water with the polar groups present in the channel walls not requiring much activation energy. The narrowest portion (2) of the channel, called the selective filter, effectively selects ions according to their crystallographic radii. The selective filter length is very small, of the order of a few angstroms. In fact, it takes an ion *ca.* 200 ns to travel through the

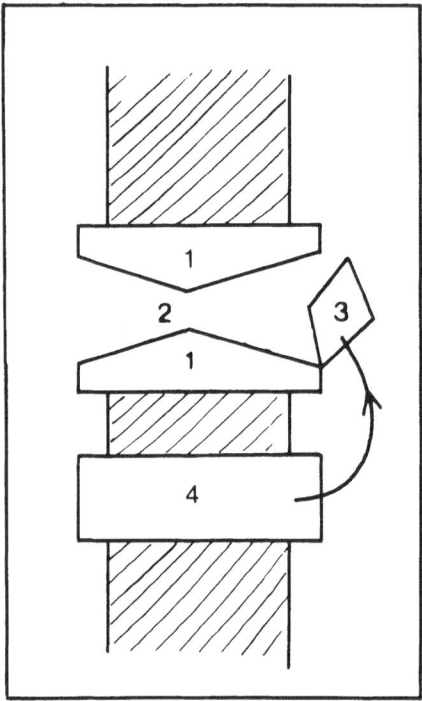

Figure 20. Ion channel structure. Discussion in text.

channel, which is about equal to the time it takes an ion to cross a layer of a standard aqueous electrolyte to equivalent thickness.

The reason for using a special kind of "subunits" in this particular model is that, for example, such enzymes as pronase can eliminate sodium channel inactivation without affecting its conductance. Generally speaking, the gate itself might be performing the function of field sensor, but the portion of potential drop for so small an element as a subunit is very small, and it would be hard to find a reasonable explanation for the rather steep shape of the conductance versus total membrane potential relationship. One therefore has to assume an independent electric field sensor capable of responding to nothing but the total membrane potential. The response actually amounts to a conformational restructuring of the sensor which manifests itself as the experimentally detectable "gating currents" (cf. Section 5.2.2.). An alternative to the model shown in Figure 20 might be a model in which all functions would be performed by one and the same conformationally mobile macromolecular system. There are no grounds to reject this alternative model offhand, but there is one more argument to support the first model we would like to adduce. It is of principal importance to decide whether the conductance of a single channel varies continuously in response to membrane depolarization or each channel operates according to the "all-or-none" principle. Ahead of the story, we mention here that fluctuation measurements seem to support discrete change of conductance states: the channel is *either* open *or* closed. This behavior can be more reasonably accounted for within the framework of the first model described.

5.2. Conductance Control by Electric Field

5.2.1. An Interpretation of the Hodgkin–Huxley Equations

As already pointed out, Hodgkin and Huxley considered their equations as nothing more than a first approximation to reality. Nevertheless, the unexpectedly good agreement with experiment stimulated attempts at their physical reinterpretation. Hodgkin and Huxley spoke about the m-, n-, and h-particles that can each take one of two fixed states in the respective channel: in one of the states the particle will block and in the other open the channel. In such an interpretation, the particles m, n, and h represent the probability for a particle to be in one of the two possible states. The particles themselves must be electrically charged and mutually independent. The dynamics of sodium channel conductance under depolarization ensuing from this approach are schematically illustrated in Figure 21: the resting state (A), in which three m-particles block the channel and the h-gate is open, is replaced by the open state (B), in whch all three m-particles, under the effect of the applied field,

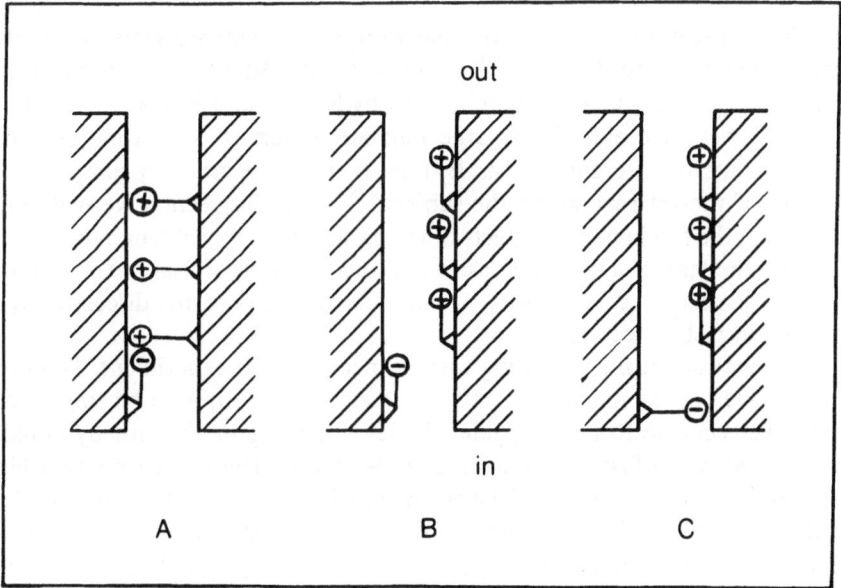

Figure 21. Schematic view of the different states the Na channel assumes under superthreshold depolarization: (A) closed state, (B) open state, (C) inactivated state. It is assumed that m- and h-particles may take only one of the shown states. (Strictly speaking, this diagram must be supplemented with the necessary elastic elements. We have not done this because the purpose of the diagram is purely illustrative.)

have been displaced and opened the channel and the h-particle has not yet had time to be displaced, and finally, by an inactivated state (C) in which the m-gate is open and the h-particle has been displaced, reclosing the channel. The potassium channel behavior is analogous, except that there are four n-particles instead of three m-particles and there is no inactivating particle of the h type.

While perhaps overly schematic, this approach proved useful for two reasons: first, because it provided terminology in which one can discuss the specific structural features of ionic channels, and second, because it allowed certain qualitative predictions to be made. Thus according to the scheme it is altogether clear that there must exist an electric transmembrane current due to displacement of the charged intramembrane particles in response to a field variation. This current has received the name of "gating current," the origin of which is obvious in the model in question.

It is quite clear that any information about the motion of the gating particles would be valuable toward understanding the functioning of ion channels and the first attempts to measure gating currents were made long ago,[66] although the insufficiency of the measurement techniques at that time rendered these unsuccessful.

5.2.2. Gating Currents

The first scientists to measure what were by all evidence gating currents were Armstrong and Bezanilla[76]; they used the squid axon. Very soon afterwards their results were corroborated by Keynes and Rojas[77] and their publiation was succeeded by a large number of reports on gating current measurements for different systems. By now, there are several very comprehensive state-if-the-art reviews on the subject.[78-81] In this section we will only touch upon the fundamental aspects of the gating current problem, not setting ourselves the task of a comprehensive exposition and appraisal of the role of this or that school or laboratory—some debatable issues are discussed, for example, in Reference (81).

Consider in outline the gating current measurement experiment. To have a clearer idea of the manner in which such an experiment is to be staged we must have recourse to the Hodgkin–Huxley model again. As already noted, the m-variable is a fast one compared to n- and h-. The $m_\infty(\varphi)$ relationship is such that within the H–H model it is attributed to the fact that the m-particle, which is characterized by a potential difference between its two possible states equal to the membrane voltage φ, has two elementary charges. The parameters of the n- and h-particles are about the same. For this reason, the gating current must be predominantly a flux of m-particles. In the resting state practically all the m-particles are displaced by the field (the resting potential is -60 to $-70\,mV$) in one direction (toward the outer side of the membrane in Figure 21a). Therefore, the m-particles can be expected to move under depolarization rather than hyperpolarization of the fiber. This, in fact, forced the investigators to fulfill the experiment as algebraic summation of the currents generated in response to voltage stimuli having equal amplitudes but different signs and applied to a resting membrane upon suppression of the transmembrane ionic transport. A typical experiment of this kind is illustrated in Figure 22. The experiment conducted in this manner allows one to isolate a signal which is small compared to the usual charging current of a constant capacitor. This small signal has been called asymmetric capacitive current. Obviously, the asymmetric current can be regarded as the result of a certain relationship between membrane capacitance and voltage. The following are the features which indicate a kinship between asymmetric current and gating current:

1. The current is carried by the intramembrane particles. This is indicated by the equality of charges carried by the asymmetric current with the voltage stimulus on and off.
2. The asymmetric current precedes the development of sodium channel conductance—a kinetic relationship which alone can exist between m and m^3. This feature is illustrated in Figure 23, presenting the results of asymmetric current measurements in the presence of a small quantity of Na^+ in the external medium.
3. The asymmetric current has practically exponential kinetics, the time constant depending on the voltage at the depolarizing stage almost in the same manner as $\tau_m(\varphi)$.

However, there are also a few discrepancies between the expected and observed properties of the gating current. Thus the displaced charge (the

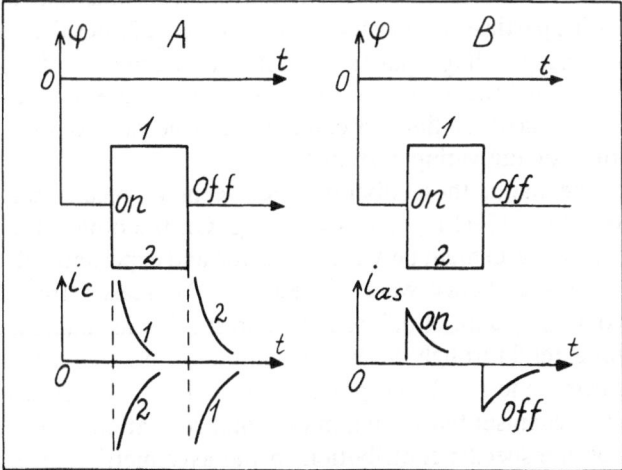

Figure 22. The gating current measurement experiment. (A) The capacitive currents (i_c) arising after switching on of the depolarizing (1) and hyperpolarizing (2) voltage pulses of equal amplitude are summed up algebraically. The resultant current is shown in (B): $i_{as} = i_{c1} + i_{c2}$.

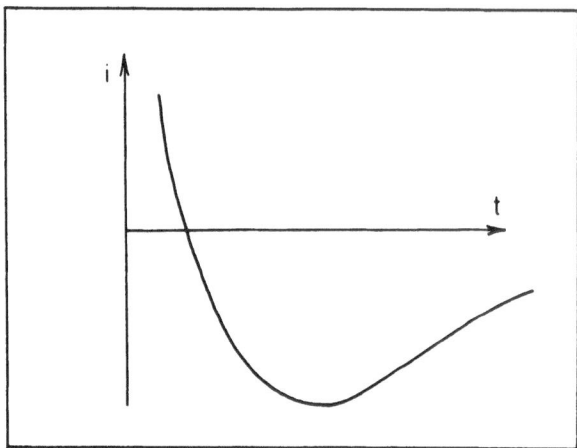

Figure 23. Asymmetric current through the giant axon membrane in the presence of a minor amount of Na ions in the external medium. The initial outward current coincides with the asymmetric current provided ionic transport has been completely eliminated; the inward current is due to the appearance of a detectable conductance of Na-channels.[78]

integral of the asymmetric current) is described by the following empirical dependence[77] on depolarising potential:

$$Q = Q_{max} \frac{1}{1 + \exp\left(\dfrac{1.3e(\varphi - \varphi')}{kT}\right)} \qquad (112)$$

where φ' is the potential corresponding to half the transferred charge, and Q_{max} is maximum transferred charge. It follows from this that the effective charge of a gating particle is 1.3 instead of the expected 2. Besides, on switching off the depolarizing voltage pulses the relaxation times of the asymmetric current and sodium conductance are practically the same, whereas the H–H model predicts a relaxation time for conductance (the value of m^3) three times as small as that for the gating current (m).

On the one hand, these discrepancies seem to mean that the above interpretation of the H–H equations is wrong. On the other, one can hardly expect an absolute agreement between predicted and experimental asymmetric current, even if the theory were perfect, for the reason that in a native membrane containing a lot of different protein species, including integral ones (such proteins extend through the entire membrane and, therefore, must react to the transmembrane field), the presence of mobile polar groups unrelated to the gating mechanism but having millisecond relaxation times and making for this reason nonspecific contribution to the asymmetric current is all too probable. Further investigation into the matter seemed to support this point of view, although new evidence emerged that at least one-half of the asymmetric current is due to the gating current. Thus it was shown[82] that a brief (a few ms) previous depolarization reduces the gating currents, and the reduction is concurrent with Na^+ conductance inactivation. The restitution of the gating currents after removal of the depolarizing voltage also occurs simultaneously with de-inactivation of the Na^+ channels. This parallelism leaves no doubt concerning the existence of a relation between the asymmetric current and motion of the gating particles, even though the H–H model fails to give an explanation of it. The relationship is verified by using pharmacological agents such as the proteolytic enzyme pronase which, while bringing Na^+-channels out of the inactivated state, also suppresses the reduction of asymmetric current which occurred when a depolarizing voltage pulse was previously applied.[83]

Therefore, the existence of such a thing as a gating current may be regarded as an established fact now, although its precise identification is a matter of future studies. Furthermore modern techniques of measurement have demonstrated that a simplistic interpretation of the Hodgkin–Huxley equations is imperfect.

To conclude this section, it should be noted that some of the conceptions suggested in previous attempts at reinterpretation of the Hodgkin–Huxley equations have become too conventional among electrophysiologists, considering that there are no reliable grounds on which to substantiate them. In particular, it is conventional in a theoretical treatment of gating particle motion to use the idea of a leap-like motion of a charged particle through the entire membrane thickness,[77] whereas the probability of such a process is near to zero. It is still traditional to imply that the field in the ionic channel region is constant, whereas the "field lumping" effects may be rather strong.[84] Finally, there is little chance that the gating particles, which are parts of the specific transport-promoting protein, are entirely independent. Moreover, by accounting for interparticle interactions it is possible to obviate some of the difficulties mentioned above and even give reasonable explanations for the experimentally observed effects for which no alternative explanations are available at present. For instance, a dipole cooperative model of the gating mechanism was proposed,[85] treating it as a system of charged interacting particles capable of traveling short distances. Under the assumption of independent particles, the experimentally observed large slope of the displaced charge versus potential curve would have to be attributed to unrealistically large particle charges. By introducing even the simplest form of interactions between particles one easily obtain as steep a curve as required. For example, a curve of the Frumkin isotherm type has been used[85]:

$$\frac{\theta}{1 - \theta} \exp\left(-W\theta\right) = K \exp\left(\frac{e\Delta\varphi}{kT}\right) \tag{113}$$

Here, W is a cooperation parameter, $kT/e \approx 25\,mV$, $\Delta\varphi$ is the fraction of membrane voltage measured between adjacent positions of the gating particle, $K = \text{const.}$; θ is analogous to the variable m and gives the fraction of displaced gating particles. Figure 24 shows a $\theta(e\varphi/kT)$ curve calculated for $W = 3.5$ and $\Delta = 0.2$. In principle, by choosing an appropriate $W = W(\theta)$ one can achieve perfect coincidence of the theoretical and experimental curves, but it would not matter much. What is more important is that in this model the relaxation time versus potential relationships are qualitatively close to the experimental ones: the relaxation time of the asymmetric current on application of a depolarizing pulse, τ_{on}, depends on the membrane voltage, φ, in the same manner as $\tau_m(\varphi)$, and the current relaxation time after switching off the depolarizing pulse, τ_{off}, has a sigmoid-shaped relationship to φ which is qualitatively similar to the experimental curve.[86] Note that the H–H model predicts no relationship between τ_{off} and φ at all. In a cooperative model the question of the change of ionic conductance remains open: principally it admits of both the conformational transformations of the channel, resulting in change of conductance state without intramembrane charge redistribution, and the control of conductance by charged group reorientation, whereby the electrostatic component of ion energy in the channel and, therefore, channel conductance are altered. Both variants admit of both a gradual change of conductance and a jump-like resetting in the "all-or-none" manner, as we assumed in the above interpretation of the Hodgkin–Huxley equations. It should be noted, however, that even the H–H equations themselves can be interpreted in a different way if one discards the assumption of a discrete conductance state alteration. Thus instead of a probability for the open state of the channel, n^4, one may operate with $G = n^4$ (i.e., the smoothly varying conductance normalized with respect to the maximum value), which is given by

$$\frac{1}{4}\frac{dG}{dt} = \alpha_n(\varphi)(G^{3/4} - G) - \beta_n(\varphi)G \tag{114}$$

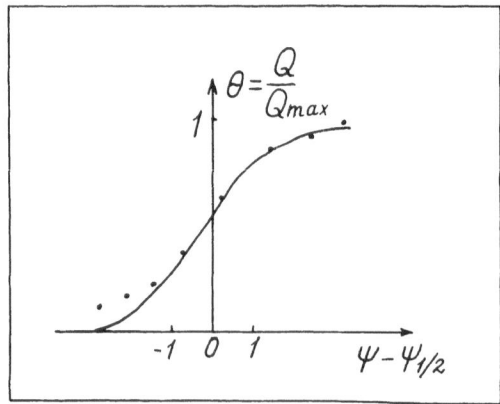

Figure 24. Displaced charge measured in Q_{max} units as a function of the potential in the cooperative model[85] (solid line) and from experiment[77] (points).

where $\alpha_n(\varphi)$ and $\beta_n(\varphi)$ are the usual coefficients of the Hodgkin–Huxley model. Which of the two mechanisms—the discrete or all continuous one—really takes place will apparently become clear only after it becomes possible to observe a single channel isolated from a biological membrane. The technique of measuring the current and voltage noise spectra, which has advanced considerably in the last decade, may be very helpful in this respect.

5.2.3. Electric Fluctuations

Measurement of the amplitude distributions of current or voltage fluctuations is a method of obtaining information about ionic channels which is supplementary to the standard relaxation measurement techniques. Generally, the fluctuation spectrum measurements allow one to determine parameters which it is generally impossible to determine through the conventional mean-value measurement techniques. This, for instance, holds true for single-channel conductance: in all the mean-value results it appears as a product with another unknown—channel density. It can be shown as well that by measuring the fluctuation spectrum of some physical parameter it is possible to distinguish between models predicting equal man-value results. In particular, it may be used for distinguishing between the two interpretations of the Hodgkin–Huxley equations. A number of reviews have already been published that are concerned with the theory and practice of noise spectrum measurements in biological membranes. [87–89]

The voltage noise spectrum measured under current-clamp conditions (fixed current passing through a small patch of membrane) is the parameter used most often in experiments. According to their origin the following kinds of noise are distinguished: 1) a thermal or Nyquist noise, 2) noise due to the "opening or closing of gates" or channel conductance fluctuations, and 3) a flicker noise, another name for which is transport noise, characterized by inverse proportionality between spectral density and fluctuation frequency. Spectra of the $1/f$ type have been observed on a wide variety of systems [89]: microelectrodes, small holes separating two electrolyte solutions, porous membranes, and artificial phospholipid membranes. There is no exact theory to explain a $1/f$-type spectrum and, most probably, wenever this type of spectrum is observed experimentally it is not a rigorous but rather an approximate $1/f$ relationship within a certain frequency range. One of the ways to account for the $1/f$-type noise has been to select the elementary noise sources with different fundamental frequencies and intensities so as to approximate the $1/f$ relationship.

Flicker noise has long been a nuisance in measurements of the spectral density of conductance fluctuations in experiments with biological membranes. In 1973, Fishman [90] measured the differential voltage noise spectra of a native membrane and a membrane whose potassium conductance has been blocked with tetraethylammonium. The differential spectrum he obtained fitted well into the following formula:

$$G_\varphi(f) = \frac{A(\varphi)}{f^2 + f_c^2(\varphi)} \tag{115}$$

where f_c is the so-called cutoff frequency and $A(\varphi)$'is the integral fluctuation power. A spectrum of the type represented by Eq. (115) is called a Lorentzian spectrum. As is known from the thermodynamical theory of fluctuations,[91] spectrum of this kind is typical of two-level systems or, in other words, systems described by a single thermodynamical variable. The $f_c(\varphi)$ relationship Fishman measured was close to $1/\tau_n(\varphi)$.

Note that at fixed current values the voltage fluctuations are substantially equivalent to conductance fluctuations. For potassium channels similar measurements[75,92,93] showed that the potassium conductance spectrum can be represented as a sum of the flicker and Lorentzian conponents. However, the $f_c(\varphi)$ relationships were quite different in all three studies undertaken,[90,92,93] perhaps due to the different systems and techniques involved.

Obviously, the treatment of the experimentally observed noise component as Lorentizian is also nothing more than an approximation, since in the case of several kinetic states, which we assumed in the interpretation of the H–H equations discussed above, there must be several Lorentzian terms in the integral spectrum.[88,87,94] It is hence clear that the measurement precision available at present does not allow us to distinguish between these variants of ionic channel models. Nevertheless, it has already been possible to estimate the conductance of single K^+ and Na^+ channels (several pmho)[92,95] using noise measurement techniques, although the results reported by different authors are largely at variance.

5.3. Open Channels: Selectivity and Conductance

Ionic channels are highly selective. Nonetheless, by replacing in the external solution sodium ions with other cations and blocking the potassium channel with tetraethylammonium it is possible to create an artificial situation in which the other cations are carried via the sodium channel. As a result, Hille succeeded in determining the series of conductances P_i for a broad range of cations.[96] Increasing the cation radius and varying the chemical compositions of the constituent groups, Hille was able to "probe" the size of the selective filter cavity and gain some understanding of the chemical nature of this channel region. He found, for example, that $P_K/P_{Na} = 0.09$, $P_{Tl}/P_{Na} = 0.33$, and $P_{aminoguanidine}/P_{Na} = 0.06$, and that methylammonium cannot permeate the channel. Addition of the CH_3 group to permeating organic cations rendered them practically nonpermeating. In this manner the cross-sectional area of the channel in the region of the selective sodium filter was estimated to be $3 \times 5 \text{ Å}^2$.

Similar experiments were carried out for the potassium channel.[97] It has been shown, for instance, that $P_{Na}/P_K = 0.010$. The investigation of a number of cations with different crystallographic radii made it possible to estimate the cross-sectional area of the selective filter region of the potassium channel to be *ca.* 3 Å in diameter.

Earlier we noted that sodium channels have a very high conductance. When open, the flux through such a channel is *ca.* 10^7 ions per second. One may therefore ask, is it not ion diffusion in the vicinity of membrane that limits the process? Estimates show that the maximum diffusion flux is about one order of magnitude higher than the measured current, i.e., the intramembrane transport is the dominant process. This is all the more true for potassium channels whose conductance is *ca.* 10^6 ions per second.

Turning now to the physical nature of selectivity, we note that the theory which has been most popular until very recently has been a thermodynamic one and involved calculation of the equilibrium ion distribution between the solution and the membrane. This means that if $P_i \sim u_i\gamma_i$, where u_i is mobility and γ_i is affinity of the ith ion for the channel, the selectivity is controlled solely by the thermodynamic factor γ_i, whereas the possibility of different kinetic factors u_i is disregarded altogether. Calculation of the difference in solvation energies, if conducted according to the simple Born's formula, would yield a selectivity series monotonical varying in accordance with the ion radius, whereas for real channels the relationship is obviously more complex. It was for this reason that Eisenmann[98] included consideration of the extra factor of ion interaction with the fixed charge localized in the membrane. Physically it is obvious that ions of small radius are then in a favorable position, since their coulombic interaction with a fixed charge of opposite sign will enhance their affinity for the channel. The selectivity series observed in a potassium channel prompts the assumption that there is an anion group in the vicinity of the selective filter. A nonmonotonic dependence of conductance on ion radius holds for the potassium channel as well. The discrimination of large size ions is due to steric factors. The question is, however, why is the potassium channel permeability to sodium, which has a radius smaller than potassium, so low? This can be explained as follows. If the channel cavity is rigid, the relatively small soium ion, which cannot be completely solvated by the polar groups of the selective filter, is in a state which is energetically unfavorable compared with its state in the solution where it is completely hydrated. Potassium, around which the polar groups of the channel form a dense solvate sheath, is in a much more favorable situation in this respect. Latest data on the relative permeabilities of sodium and potassium channels convincingly demonstrate that the thermodynamic theory of selectivity is incorrect. Ion transfer is a kinetic process and therefore kinetic parameters also affect the selectivity. We will return to this problem once again, after we have discussed characteristics of transport through ionic channels.

The experimental data suggest that under physiological conditions where sodium ions travel through the sodium channel the inward and outward fluxes are independent. If, however, the sodium ion concentration is unduly high or sodium has been substituted with other cations in the solution, the sodium channel ceases to obey the independence principle (cf. Section 4.1).[99] The transport of potassium ions through potassium channels fails to follow this principle even under physiological conditions.[100] For example, if potassium concentration in the medium is increased by a factor of 10, the inward current increases 30-fold whereas the outward current decreases 3- to 4-fold. The ratio of the inward and outward potassium currents does not satisfy Ussing's formula [Eq. (94)] and is instead described by the following empirical formula:

$$\frac{\vec{j}}{\overset{\leftarrow}{j}} = \exp\left[n(\psi - \psi_K)\right] \tag{116}$$

where $n \cong 2.5$. The corollary is that all attempts to describe the ionic transport in channels in terms of the electrodiffusion equation [Eq. (88)], which implies independent fluxes, whould be in vain. It can by only natural to resort to the approach set forth in Section 4.3, based on the discrete version of the single-file transport theory. This requires that the energy profile of the channel be given. The steric parameters of channels are such that simultaneous passage of two ions is impossible. It is difficult to tell how many potential barriers an ion has to surmount in its passage through the membrane. In any case, the number is not large. Neither the structure nor the chemical composition of the lipoprotein complex which forms the ionic channel is known. An authentic energy distribution chart for an ion in channel is, therefore, unavailable. The difficulty is aggravated by the fact that the local fields of the charged groups and dipoles may play an important role. For this reason one has to consider different energy distribution profiles in the channel. The qualitative effects which are common to all models thereby become particularly interesting.

In accordance with the model shown in Figure 20, the ion potential energy profile may be approximated by the curve given in Figure 25. The regions (1) and (3) correspond to channel inlets, and region (2) to the selective filter. As long as the number of potential pits in each region, pit depth, and barrier heights are not specified, the distribution will be general enough. For a more detailed description it is essential that an ion have a long enough residence time in each potential pit, the leaps between the pits being caused by thermal fluctuations and the probabilities of the leaps becoming asymmetrical on application of an electric field. It is natural to suppose that out of the complete set of barriers shown in Figure 25 several major or highest ones may be selected. Then only these barriers have to be treated as kinetic factors. The case of only three high barriers (Figure 26)[101] is found to be general enough. More complex four-barrier energy profiles have been considered[99,102] for potassium and calcium channels.

The next problem is to calculate the basic equilibrium and process rate data for a membrane containing channels of a specific type. All these para-

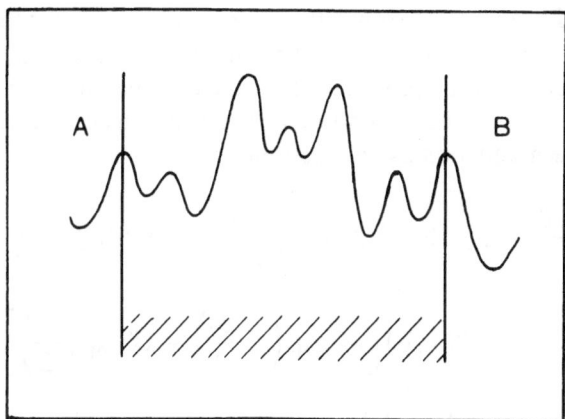

Figure 25. Ion energy profile in a selective ionic channel.

meters will depend primarily on the ion's potential energy profile in the channel used as the initial data in the calculations. The objective is, therefore, to determine the energy profile of the channel of a particular type from comparison of calculation with measurements. Strictly speaking, the problem is ambiguous. What one may hope to find are only certain qualitative features of the energy profiles of different channels when their measured characteristics exhibit qualitative differences. For example, ion fluxes through a sodium channel obey the Ussing formula, whereas those through a potassium channel follow Eq. (116). It will be shown below that this result suggests that there are some quite definite distinctions between the energy profiles of the sodium and potassium channels.

Consider a membrane to the left of which there is a solution containing single-charge cations A and to the right a solution containing single-charge cations B (Figure 26). The ions A and B can leap over the extreme barriers with rate constants k_1^A and k_1^B (inwards) and k_2^A, k_2^B (outwards). The ions can furthermore leap between the pits with rate constants ν_A and ν_B. They can jump only into a vacant pit. As shown in Section 4.3, the major characteristic of a channel state is the binary function $P(X, Y)$, where X is for the left-hand pit and Y for the right-hand one. In our example X and Y can both take values of A, B, 0, though not independently. The fluxes are expressed as:

$$j_A = n\nu_A \left[P(A, 0) \exp\left(\frac{\psi_2}{2}\right) - P(0, A) \exp\left(\frac{-\psi_2}{2}\right) \right]$$

$$j_B = n\nu_B \left[P(0, B) \exp\left(\frac{-\psi_2}{2}\right) - P(B, 0) \exp\left(\frac{\psi_2}{2}\right) \right]$$

(117)

and the current is

$$I = F \cdot (j_A - j_B)$$

where j_A and I are directed from left to right, and j_B in the opposite direction, n is the number of channels per unit membrane surface area, and ψ_2 is the voltage drop between the pits. A kinetic equation may be written for each of the binary functions, with regard to all the allowed transitions in the system.[101] Here is one such equation as an illustration:

$$\frac{dP(A,B)}{dt} = P(0,B)k_1^A(A)\exp\left(\frac{\psi_1}{2}\right) + P(A,0)k_1^B\exp\left(-\frac{\psi_3}{2}\right)$$

$$-P(A,B)\left[k_2^A\exp\left(-\frac{\psi_1}{2}\right) + k_2^B\exp\left(\frac{\psi_3}{2}\right)\right] \qquad (118)$$

where ψ_1 and ψ_3 are potential drops on the extreme barriers and (A) and (B) are concentrations in the respective solutions. The kinetic equations such as Eq. (118) are supplemented with normalization conditions $\sum_{i,j} P(X_i, Y_j) = 1$ and the obvious equality $P(B,A) = 0$, which shows that the state (B,A) is impossible since an ion A can get into the right-hand pit only from the left-hand one, but the latter is already occupied with ion B. Solutions of equations of the form of Eq. (118) have a rather complex form, so we will limit our treatment the analysis of the limiting cases.

Consider a three-barrier profile in which the middle barrier is high compared with the extreme ones, so that both the inlet pits are in equilibrium with the ambient solutions (Figure 26a). As we will show below, this simple model can form the basis for a quantitative description of the parameters of sodium channels. In this particular case the rate constants are subject to the following condition:

$$k_1^A(A), k_1^B(B), k_2^A, k_2^B \gg \nu_A, \nu_B \qquad (119)$$

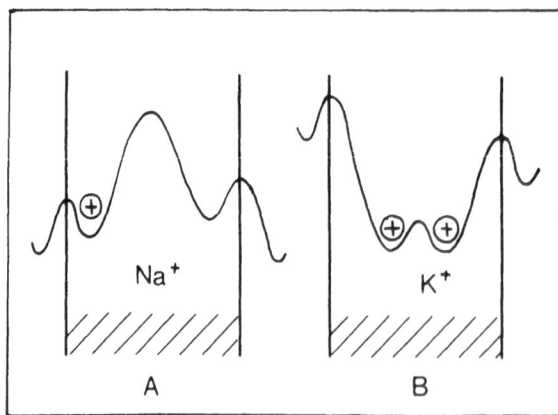

Figure 26. Ion energy profile: (A) sodium in sodium channel and (B) potassium in potassium channel.

Under these assumptions the set of equations (118) has a solution of the form[103]:

$$P(A, 0) = P(0, 0)(A)\gamma_A$$
$$P(A, B) = P(0, 0)(A)(B)\gamma_A\gamma_B$$
$$P(0, B) = P(0, 0)(B)\gamma_B \tag{120}$$
$$P(A, A) = P(B, B) = P(0, A) = P(B, 0) = 0$$

where

$$P(0, 0) = \frac{1}{[1 + \gamma_A(A)][1 + \gamma_B(B)]} \tag{121}$$

and the γ_i are distribution coefficients:

$$\gamma_A = \frac{k_1^A}{k_2^A} \exp(\psi_1) \quad \gamma_B = \frac{k_1^B}{k_2^B} \exp(-\psi_3) \tag{122}$$

The structure of these expressions suggests that in the case given by Eq. (119) the populations of the potential pits are not correlated. For example, $P(A, B) = \theta_1^A \theta_2^B$, where θ_1^A, θ_2^B are the equilibrium populations of ions A and B in the left and right-hand pits, respectively. Using the solutions (120), (121) and the definitions of the fluxes [Eq. (117)], we get:

$$j_A = \frac{n\gamma_A(A)\nu_A \exp\left(\frac{\psi_2}{2}\right)}{[1 + \gamma_A(A)][1 + \gamma_B(B)]}$$

$$j_B = \frac{n\gamma_B(B)\nu_B \exp\left(-\frac{\psi_2}{2}\right)}{[1 + \gamma_A(A)][1 + \gamma_B(B)]} \tag{123}$$

$$I = \frac{Fn \exp[0.5(\psi_1 - \psi_3)]}{[1 + \gamma_A(A)][1 + \gamma_B(B)]}$$

$$\times \left[\gamma_A^0(A)\nu_A \exp\left(\frac{\psi}{2}\right) - \gamma_B^0(B)\nu_B \exp\left(-\frac{\psi}{2}\right) \right]$$

where $\gamma_i^0 = k_1^i / k_2^i$, $i = A, B$. If we consider the equation for j_A, for example, we shall see that the set of factors $\gamma_A(A)/[1 + \gamma_A(A)]$ describe the probability of occupation of the inlet pit, the factor $[1 + \gamma_A(A)]^{-1}$ gives the probability that the second pit is vacant, and the product $\nu \exp(\psi_2/2)$ gives the rate of transition over the central barrier. If $\gamma_A(A) \ll 1$, $\gamma_B(B) \ll 1$, i.e., ion concentration is small in the solution or ion affinity for the inlet regions of the channels is small, the blocking factor $[1 + \gamma_A(A)]^{-1}[1 + \gamma_B(B)]^{-1}$ reduces to unity and Eq. (123) reduces to the standard formula of the Eyring theory, the physics behind which is the same as that behind the electrodiffusion approach.

Equating the electric current to zero, we find from Eq. (123) the membrane potential:

$$\psi_0 = \ln \frac{\nu_B \gamma_B^0(B)}{\nu_A \gamma_A^0(A)} \tag{124}$$

which is the same as the Goldman poential [Eq. (92)] and usually termed the reversion potential. The conductance under the reversion potential is:

$$g_0 = \frac{nF[\nu_A \nu_B \gamma_A \gamma_B(A)(B)]^{1/2}}{[1 + \gamma_A(A)][1 + \gamma_B(B)]} \tag{125}$$

Of special interest may be the ratio of unidirectional fluxes:

$$\frac{j_A}{j_B} = \frac{\nu_A \gamma_A^0(A)}{\nu_B \gamma_B^0(B)} = \exp(\psi - \psi_0) \tag{126}$$

Some important conclusions can be drawn from these results. First of all, note that the unidirectional fluxes across the blocking factor depend nonlinearly on ion concentrations on both sides of the membrane, i.e., the independence principle is not valid here. At the same time, the ratio of unidirectional fluxes is described by the Ussing formula as it was in the electrodiffusion theory. However, the conductance is already quite different. Only in the case of small populations are the unidirectional fluxes independent and, precisely as in the conventional theory, are determined only by the product of parameters $\nu\gamma$. What this relationship means is that the higher the concentration of carriers in the channels and the faster they cross the membrane the greater will be the corresponding flux. The physical limitations of such a conclusion are obvious. Just consider that if a particle is firmly adsorbed in the inlet region of the channel, it will block it, of course, and prevent the passage of a next particle. In the above transport scheme, the dependence of flux on ν is still linear, but the dependence on γ has a saturation point and if the solutions on both sides of the membrane contain ions of the same species, the flux has a peak. This means that to provide a maximum flux in the case of the same ions in solutions the distribution coefficient must not be too large. Another corollary of the theory must be stressed. The term "permeability" is widely used in application to membrane transport. This parameter is defined as the proportionality factor between unidirectional flux and the corresponding concentration. However, as follows from Eq. (123), the unidirectional fluxes depend in this case nonlinearly on concentrations on both sides of the membrane.

We will now illustrate the above rationale in application to the sodium channels of biomembranes. Since a sodium channel under physiological conditions obeys the independence principle, one has to conclude that $\gamma_{Na}(Na) \ll 1$. Now, consider the case where all the sodium ions in the solution have been replaced with another type of cations. Without going into details of the treatment[104] undertaken on the basis of the experimental data reported in Reference (96) and Eqs. (123)–(125), we will only cite the distribution coefficients and relative rate constants thus obtained (Table 2).

Table 2
Binding and Kinetic Data for Cations in Sodium Channel

Cation	γ_i, M^{-1}	ν_i / ν_{Na}
Sodium	2.6	1.0
Thallium	21.7	0.042
Hydroxyguanidinium	18.0	0.037
Aminoguanidinium	9.3	0.038
Methylguanidinium	16.3	0.039

The results in Table 2 were obtained under the assumption that $\psi_1 = \psi_3 = 0$, i.e., that all the membrane voltage drops on the central barrier. One is compelled, in fact, to make this assumption because the real potential profile in the channel is unknown. From the rate constant data it follows that the ions substituted for sodium are bound more strongly with the inlet portion of the channel, but their rate constants for passage through the selective filter are much smaller than that of sodium. Therefore, contrary to the thermo-dynamic theory, channel selectivity is determined to a greater extent by kinetic factors. The higher binding coefficients of other ions compared with sodium are responsible for the experimentally observed deviations from the independence principle.

A number of investigators[101-104] carried out calculation of the V–A characteristics of sodium channels, trying to take into account coulombic effects in ion–ion interactions. We will not, however, go into these questions in detail but will instead consider the case of the potassium channel in brief.

As we have already pointed out, the Ussing formula does not apply to potassium channels (the Ussing formula ensues from the electrodiffusion theory and holds true in the single-file transport theory when the energy profile is like that shown in Figure 26a). It follows that the energy profile in the channel must be modified so as to take into account the deviations from the Ussing formula. In fact, already the mechanical model of Huxley[100] and Heckmann's investigations[105] have shown how the solution of the problem may be found. The modification necessary to account for the conditions peculiar to the potassium channel must consist in adding to strongly correlated deep pits with a low barrier between them (Figure 26B). Particle transfer into the solution through the extreme barriers will now be the limiting process, i.e., the rate constants will satisfy the following inequalities:

$$\nu_A, \nu_B \gg k_1^A(A) \quad k_1^B(B) \gg k_2^A, k_2^B \tag{127}$$

The calculation procedure[107] is the same as described above, so we will only quote the result for the ratio of unidirectional fluxes:

$$\frac{j_A}{j_B} = \left[\frac{k_1^A(A)}{k_1^B(B)} \right]^2 \frac{k_2^B}{k_2^A} \exp(2\psi) \equiv \exp[2(\psi - \psi_0)] \tag{128}$$

since

$$\psi_0 \equiv \ln \left[\frac{k_1^B(B)}{k_1^A(A)} \left(\frac{k_2^A}{k_2^B} \right)^{1/2} \right] \tag{129}$$

The unidirectional flux formula which we are not quoting here (one can find it, e.g., in Reference (106)) satisfactorily describes their experimentally observed dependences on concentrations. The same is valid for Eq. (128), although one might achieve a better agreement with experiment by increasing the number of potential pits per channel while reducing at the same time their depth so that the mean population is *ca.* 2.5. There are grounds, therefore, to assert that the energy profile of the type shown in Figure 26B is a correct qualitative representation of the real properties of potassium channels.

Despite obvious successes, the single-file transport theory has a drawback in that the final formulas contain a considerable number of arbitrary parameters. Apparently, the molecular dynamics method, which is free of unknown parameters, is the most promising method available at present for solving the problems of ionic transport across open channels. So far we have little or no information about the nature and exact orientation of the groups that form the lining of the hydrophilic pore of an ionic channel; the method should perhaps be applied primarily to transport through the well-studied model channels made up of, say, gramicidin. One may hope that the computer-aided molecular dynamics method will also permit at some time in the future the problem of ionic channel selectivity to be solved unambiguously.

6. Conclusions

The development of the science of excitable biomembranes has been extremely rapid. The incubational period of data acquistion has given way to a qualitatively new status characterized by quite a different level of understanding of membrane processes. Thus ionic channels have ceased to be a mere nickname for a transport system but have gained functional flesh. Conductances of some single channels have already been measured, their number in different membranes determined, and their relaxation parameters found. However, as regards chemical structure data, the ionic channel still remains a "black box."

The cell-scale experiment was and remains a major source of information about the mechanisms of nervous impulse generation. The development of the new techniques of measuring gating currents and of fluctuation analysis has allowed information to be obtained about the properties of isolated channels. The dialysis of mollusc giant neurons carried out by Kostyuk *et al.*[108] opened up new possibilities and extended the class of analyzable systems. A characteristic tendency of the latest biophysical investigations in this field has been a desire to record to some way the functioning of a single channel directly in the real cell system.

Still, the problem of isolation, identification of the chemical structure, and reconstitution of channels on the basis of lipid model membranes remains the most urgent one. Therefore, investigations of the electrochemical properties of lipid bilayers and of their interaction and fusion with lipid vesicles gain particular importance, especially since there are no apparent alternative ways of introduction of hydrophobic proteins into flat membranes.

References

1. B. Katz, *Nerve, Muscle and Synapse*, McGraw-Hill, New York (1966).
2. A. L. Hodgkin and A. F. Huxley, *J. Physiol. (London)* **117**, 500 (1952).
3. A. L. Hodgkin, *The Conduction of the Nervous Impulse*, Liverpool University Press, Liverpool (1964).
4. R. S. Lillie, *J. Gen. Physiol.* **7**, 473 (1925).
5. R. S. Lillie, *Biol. Rev.* **11**, 181 (1936).
6. K. F. Bonhoeffer, *J. Gen. Physiol.* **32**, 69 (1948).
7. K. F. Bonhoeffer and K. J. Vetter., *Z. Phys. Chim.* **196**, 127 (1950).
8. G. I. Barenblatt, V. M. Entov, and R. L. Salganik, *Prikl. Matem. i Mekh.* **29**, 977 (1965).
9. V. G. Levich, N. G. Mazur, and V. S. Markin, *Dokl. Akad. Nauk. SSSR* **198**, 947 (1971).
10. R. Suzuki, *IEEE Trans. Bio-Med. Eng.*, **14**, 114 (1967).
11. V. S. Markin, V. F. Pastushenko, and Yu. A. Chizmadzhev, *Theory of Excitable Media*, Nauka Press, Moscow (1981) (In Russian).
12. T. Teorell, *J. Gen. Physiol.* **42**, 831 (1959).
13. V. F. Pastushenko, Yu. A. Chizmadzhev, and V. S. Markin, *Biofizika* **25**, 523 (1980).
14. U. F. Franck, *Ber. Bunsenges. Phys. Chem.* **71**, 789 (1967).
15. V. S. Markin, V. F. Pastushenko, and Yu. A. Chizmadzhev, *Elektrokhimiya* **7**, 337 (1971).
16. I. M. Gelfand and M. L. Tseltlin, *Dokl. Akad. Nauk. SSSR* **131**, 1242 (1960).
17. B. G. Farley and W. A. Clark, in *Information Theory*, Colin Cherry, ed., Butterworths, London (1961), pp. 242–251.
18. G. K. Moe, W. C. Pheinboldt, and T. A. Abildskov, *Amer. Heart J.* **67**, 200 (1964).
19. A. C. Scott, *Neurophysics*, Wiley-Interscience, New York (1977).
20. B. I. Khodorov, *Excitability Problems*, Meditsina Press, Leningrad (1969) (In Russian).
21. A. N. Kolmogorov, I. R. Petrovsky, and N. S. Piskunov, *Bull. MGU, Mathematics and Mechanics*, No. 6, 3 (1937).
22. P. J. Hunter, P. A. McNaughton, and D. Noble, *Progr. Biophys. Molec. Biol.* **30**, 99 (1975).
23. V. F. Pastushenko, Yu. A. Chizmadzhev, and V. S. Markin, *Biofizika* **20**, 894 (1975).
24. A. F. Huxley, *Ann. N.Y. Acad. Sci.* **81**, 221 (1959).
25. J. Pinzel and J. B. Keller, *Biophys. J.* **13**, 1313 (1973).
26. A. S. Kompaneets and V. Ts. Gurovich, *Biofizika* **11**, 913 (1966).
27. V. S. Markin and Yu. A. Chizmadzhev, *Biofizika* **12**, 900 (1967).
28. T. W. Cooley and F. A. Dodge, *Biophys. J.* **6**, 583 (1966).
29. V. I. Krinsky, Fibrillation in excitable media, in *Problems of Cybernetics*, Issue No. 20, Nauka Press, Moscow (1968), pp. 59–80 (In Russian).
30. Yu. I. Arshavsky, M. B. Merkinblit, and V. L. Dunin-Barkovsky, *Biofizika* **10**, 1048 (1965).
31. V. S. Markin and V. F. Pastushenko, *Biofizika* **14**, 316 (1969).
32. V. S. Markin and Yu. A. Chizmadzhev, *J. Theor. Biol.* **36**, 61 (1972).
33. G. R. Ivanitsky, V. I. Krinsky, and E. E. Selkov, *Mathematical Biophysics of Cell*, Nauka Press, Moscow (1978) (In Russian).
34. V. F. Pastushenko, *Biofizika* **20**, 494 (1975).

35. Yu. A. Ovchinnikov, V. T. Ivanov, and A. M. Shkrob, *Membrane-Active Complexons*, Elsevier, New York, London, Amsterdam (1974).
36. V. S. Markin and Yu. A. Chizmadzhev, *Induced Ionic Transport*, Nauka Press, Moscow (1974) (In Russian).
37. A. Kotyk and K. Janaček, *Membrane Transport*, Plenum Press New York, London (1977).
38. A. A. Lev, *Ionic Selectivity of Cell Membranes*, Nauka Press, Moscow (1975) (In Russian).
39. G. Eisenman, G. Szabo, S. Ciani, S. McLaughlin, and S. Krasne, in *Progress in Surface and Membrane Science*, Vol. 6, J. Danielly, ed., Academic Press, New York, London (1973), pp. 139–175.
40. B. Neumcke and E. Bamberg, in *Membranes*, Vol. 3, G. Eisenman, ed., Marcel Dekker, New York (1975), pp. 215–302.
41. R. Arndt and L. Roper, *Simple Membrane Electrodiffusion Theory, Physical Biological Sciences Misc.*, Blacksburg, Virginia (1972).
42. D. Goldman, *J. Gen. Physiol.* **27**, 37 (1943).
43. H. Ussing, *Physiol. Rev.* **29**, 127 (1949).
44. T. Schwartz, in *Biophysics and Physiology of Excitable Membranes*, Vol. 4, W. J. Adelman, ed., Van Nostrand Reinhold Co., (1971), pp. 47–92.
45. M. Davson and J. E. Danielli, *The Permeability of Natural Membranes*, Cambridge Univ. Press, Cambridge (1952).
46. S. J. Singer and G. L. Nicolson, *Science* **175**, 720 (1972).
47. P. Müller, D. O. Rudin, H. T. Tien, and W. C. Wescott, *Nature* **194**, 979 (1962).
48. H. T. Tien, *Bilayer Lipid Membranes (GLM). Theory and Practice*, Marcel Dekker, Inc., New York (1974).
49. T. Hanai, D. A. Haydon, and J. L. Taylor, *Proc. Roy. Soc., Ser. A* **281**, 377 (1964).
50. D. A. Haydon, in *Membranes and Ion Transport*, E. E. Bittar, ed., Wiley-Interscience, London (1970), pp. 64–92.
51. S. McLaughlin, G. Szabo, G. Eisenman, and S. Ciani, *Proc. Nat. Acad. Sci. U.S.A.* **67**, 1268 (1970).
52. O. Alvarez and R. Latorre, *Biophys. J.* **21**, 1 (1978).
53. D. F. Sargent, *J. Membrane Biol.* **23**, 227 (1975).
54. Yu. A. Chizmadzhev and I. Abidor, *Bioelectrochem. Bioenerg.* **7**, 83 (1980).
55. V. Cherny, V. Sokolov, and I. Abidor, *Bioelectrochem. Bioenerg.* **7**, 413 (1980).
56. K. Cole, *Ions, Potentials and Nerve Impulse*, Wiley, New York (1955).
57. B. Neumcke and P. Länger, *Biophys. J.* **9**, 1160 (1969).
58. Yu. A. Chizmadzhev and V. F. Pastushenko, *Biofizika* **26**, 829 (1981).
59. A. Parsegian, *Ann. N.Y. Acad. Sci.* **264**, 161 (1975).
60. U. Zimmermann, G. Pilwat, and F. Riemann, *Biophys. J.* **14**, 881 (1974).
61. I. G. Abidor, V. B. Arkelyan, V. F. Pastushenko, L. V. Chernomordik, and Yu. A. Chizmadzhev, *Bioelectrochem. Bioenerg.* **6**, 37 (1979).
62. P. Müller and D. Rudin, *Nature* **217**, 713 (1968).
63. S. B. Hladky and D. A. Haydon, *Biochim. Biophys. Acta* **274**, 294 (1972).
64. S. Krasne, G. Eisenman, and G. Szabo, *Science* **174**, 412 (1971).
65. V. Markin, L. Krishtalik, E. Liberman, and V. Topaly, *Biofizika* **14**, 256 (1969).
66. W. K. Chandler and H. Meves, *J. Physiol. (London)* **180**, 788 (1965).
67. T. L. Hill, Studies in irreversible thermodynamics.—IV. Diagrammatic representation of steady-state fluxes for unimolecular systems, *J. Theor. Biol.* **10**, 442 (1966).
68. W. Almers and R. Levinson, Tetrodotoxin binding to normal and depolarized frog muscle and the conductance of a single sodium channel, *J. Physiol. (London)* **247**, 483 (1975).
69. T. Begenisich and C. F. Stevens, How many conductance states do potassium channels have?, *Biophys. J.* **15**, 843 (1975).
70. F. Conti and E. Wanke, Channel noise in nerve membranes and lipid bilayers, *Quart. Rev. Biophys.* **10**, 1 (1977).

71. F. Conti, B. Hille, B. Neumcke, W. Nonner, and R. Stämpfli, Measurement of the conductance of the sodium channel from current fluctuations at the node of Ranvier, *J. Physiol. (London)* **262**, 699 (1976).

72. J. M. Ritchie and R. B. Rogart, The binding of saxitoxin and tetrodotoxin to excitable tissue, *Rev. Physiol. Biochem. Pharmacol.* **79**, 1 (1977).

73. J. M. Ritchie, R. B. Rogart, and G. R. Stichartz, A new method for labelling saxitoxin and its binding to non-myelinated fibres of the rabbit vagus, lobster walking leg and garfish olfactory nerves, *J. Physiol. (London)* **261**, 477 (1976).

74. F. J. Sigworth, Sodium channels in nerve apparently have two conductance states, *Nature* **270**, 265 (1977).

75. F. Conti, L. J. De Felice, and E. J. Wanke, *J. Physiol. (London)* **248**, 45 (1975).

76. C. M. Armstrong and F. Benzanilla, *Nature* **242**, 459 (1973).

77. R. D. Keynes and E. Rojas, *J. Physiol. (London)* **239**, 393 (1974).

78. C. M. Armstrong, *Quart. Rev. Biophys.* **7**, 179 (1975).

79. L. Goldman, *Quart. Rev. Biophys.* **9**, 491 (1976).

80. W. Ulbricht, *Ann. Rev. Biophys. Bioeng.* **6**, 7–31 (1977).

81. W. Almers, *Rev. Physiol. Biochem. Pharmacol.* **82**, 96 (1978).

82. F. Bezanilla and C. M. Armstrong, *Science* **183**, 753 (1974).

83. C. M. Armstrong and F. Bezanilla, *J. Gen. Physiol.* **70**, 567 (1977).

84. Yu. A. Chizmadzhev and V. F. Pastushenko, in *Topics in Bioelectrochemistry and Bioenergetics*, Vol. 2, G. Milazzo, ed., J. Wiley and Sons, (1978).

85. V. Pastushenko, Yu. Chizmadzhev and I. Kalandadze, *Biofizika* **23**, 74 (1978).

86. H. Meves, *J. Physiol. (London)* **243**, 847 (1974).

87. Yi-der Chen, in *Advances in Chemical Physics*, Vol. **37**, I. Progigine and S. Rice, eds., J. Wiley and Sons, New York (1978), pp. 67–97.

88. E. Neher and C. F. Stevens, *Ann. Rev. Biophys. Bioeng.* **6**, 345 (1977).

89. A. A. Verveen and L. J. DeFelice, *Progr. Biophys. Molec. Biol.* **28**, 189 (1974).

90. H. M. Fishman, *Proc. Nat. Acad. Sci. U.S.A.* **70**, 876 (1973).

91. L. D. Landau and E. M. Lifshits, *Statistical Physics*, Nauka Press, Moscow (1964) (In Russian).

92. H. M. Fishman, L. E. Moore, and D. J. M. Poussart, *J. Membrane Biol.* **24**, 305 (1975).

93. E. Siebenga, W. A. Meyer, and A. A. Verveen, *Pflügers Arch.* **341**, 87 (1973).

94. W. F. Pastuszenko and Ju. A. Chizmadžew, *Acta Univ. Lodz, Ser.* II, **19**, 419 (1976).

95. R. J. Van den Berg, J. De Goede, and A. A. Verveen, *Pflügers Arch.* **360**, 17 (1975).

96. B. Hille, *J. Gen. Physiol.* **58**, 599 (1971).

97. B. Hille, *J. Gen. Physiol.* **61**, 669 (1973).

98. G. Eisenmann, *Biophys. J.* **2** (Suppl.), 259 (1962).

99. B. Hille, in *Membranes*, Vol. 5, G. Eisenmann, Marcel Dekker, New York (1975), pp. 255–323.

100. A. Hodgkin and R. Keynes, *J. Physiol. (London)* **128**, 61 (1955).

101. Yu. A. Chizmadzhev, B. I. Khodorov, and S. Kh. Aityan, *Bioelectrochem. Bioenerg.* **1**, 301 (1974).

102. P. G. Kostyuk, S. L. Mironov, and P. A. Doroshenko, *Dokl. Akad. Nauk. SSSR* **253**, 978 (1980).

103. Yu. A. Chizmadzhev and S. Kh. Aityan, *J. Theor. Biol.* **64**, 429 (1977).

104. Yu. A. Chizmadzhev and S. Kh. Aityan, *Dokl. Akad. Nauk. SSSR* **220**, 1203 (1975).

105. K. Heckmann, *Z. Phys. Chem. (N.F.)* **44**, 184 (1965).

106. S. Kh. Aityan and I. L. Kalandadze, Yu. A. Chizmadzhev, *Bioelectrochem. Bioenerg.* **4**, 30 (1977).

107. S. Kh. Aityan and I. L. Kalandadze, *Bioelectrochem. Bioenerg.* **4**, 45 (1977).

108. P. G. Kostyuk and O. A. Krishtal, *J. Physiol.* **270**, 545 (1977).

8

An Electrochemical Approach for the Solution of Cardiovascular Problems

SUPRAMANIAM SRINIVASAN

Abstract

Thrombosis, an *in vivo* reaction at the blood vessel wall (or prosthetic material)/blood interface, has been the subject of investigations for over 150 years. It involves a series of reactions of blood coagulation factors and of blood cells, particularly platelets. The conventional approach in blood coagulation studies is the hematological one which, however, provides insights on the possibilities of electrochemical pathways for the thrombosis cascade. For a little over 30 years, detailed studies on blood vessel walls, blood cells, prosthetic materials, and blood coagulation factors using electrokinetic and electrode kinetic techniques have been carried out to unravel the intermediate steps in the thrombosis cascade. Most of the studies have been carried out *in vitro* but have necessarily also involved *in vivo* experiments coupled with clinical investigations. To provide a fundamental bioelectrochemical viewpoint to the reader, this chapter is presented in somewhat of a reverse order of the historical one, i.e., adsorption and charge transfer reactions of blood coagulation factors → electrokinetic characterization of the blood vessel walls and of blood cells → interactions of anti- or procoagulant drugs with the vascular components → selection of conducting and insular materials for vascular and heart valve prostheses.

SUPRAMANIAM SRINIVASAN • Los Alamos National Laboratory, P.O. Box 1663, MS D429, Los Alamos, New Mexico 87545, USA.

1. Introduction

1.1. Definition of Terms—Thrombosis, Thromboembolic Disease, Atherosclerosis, and Blood Clotting

The terms heart attack or myocardial infarction are more commonly used than thrombosis. The infarct-muscle destruction is simply the end result and thrombosis is the real cause of the heart attack. *Thrombosis* may be defined as the process of formation of a coalescent or agglutinated solid mass of blood components in the blood stream. Thrombi formed in either arteries or veins often cause occlusion in the vascular system and prevent blood flow. Obstruction to the blood vessel usually occurs at the site where the thrombi deposit. Furthermore, thrombi may break loose, travel through the circulating blood stream, and cause obstruction at some distal point of narrowing elsewhere. The mass or thrombus that moves is referred to as an "embolus." The two phenomena are lumped together under the term *thromboembolic disease*. Thrombosis that reduces blood supply to the heart is the primary factor in heart attacks.

The term *atherosclerosis* refers to the focal thickening of the intima of arteries.[1] The thickened intima consists of muscle cells, connective tissue such as collagen and elastin, mucopolysaccharides, and both intracellular and extracellular lipids. As it progresses to the more advanced condition, degenerating cells and cholesterol crystals are found in the lesions. Plasma proteins, lipoproteins, and the formed elements of blood (platelets, leukocytes, and erythrocytes) are also involved in the development of lesions. Progressive atherosclerosis leads to poor blood flow through blood vessels due to occlusive complications, which also promote the accumulation of formed elements.

The term *blood coagulation* may be defined as the spontaneous transformation of fluid blood to a solid clot.[2] Vroman in his book *Blood*,[3] has presented a sequential illustration of the clotting phenomena (Figure 1). In one cubic millimeter of blood, there are about five million erythrocytes (red cells), three hundred thousand platelets, and seven thousand leukocytes (white cells). The total volume of these cells is less than half of that of blood. The balance is occupied by the plasma fluids. As stated in the book *Human Blood Coagulation and Its Disorders*, by Biggs and Macfarlane,[2] "The blood contains within itself not only the clotting factors that will initiate, accelerate, and limit coagulation as required, but also a series of safety devices to prevent coagulation occurring within the vessels and for dissolving fibrin which is no longer useful."

1.2. Evidence for Electrochemical Mechanisms in Cardiovascular Phenomena

Thrombosis is an interfacial reaction which occurs at the blood vessel wall/blood or prothesis (vascular or heart valve)/blood interface. The total

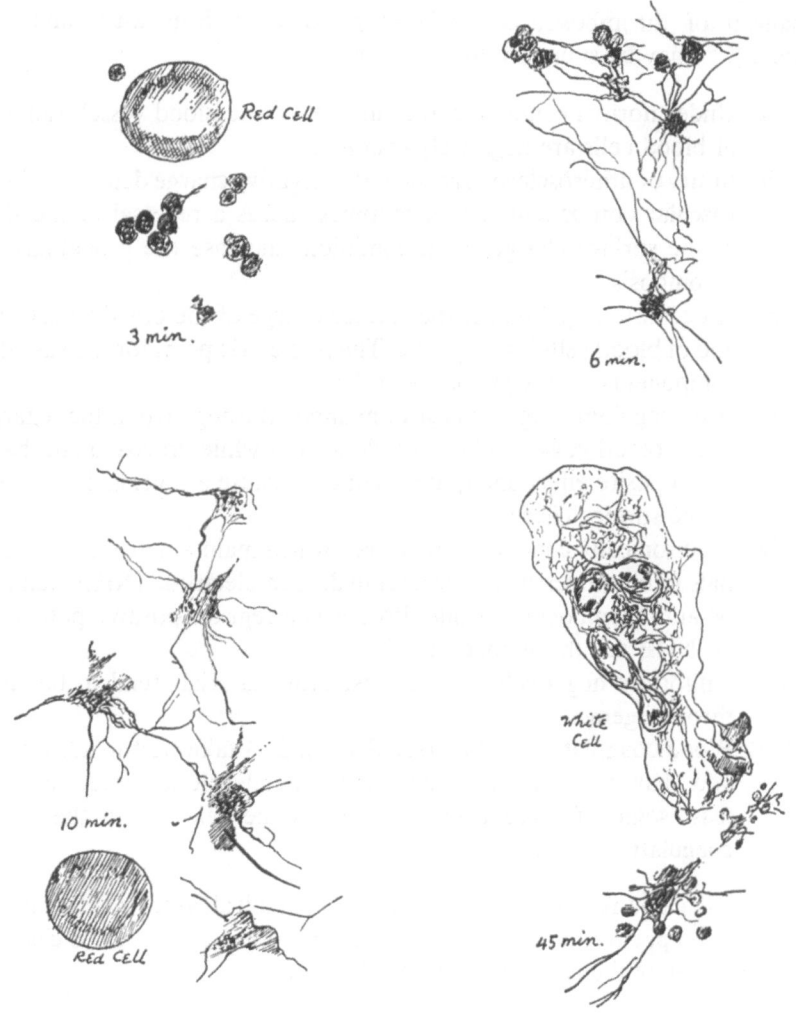

Figure 1. A sequential illustration of the clotting of blood by Vroman.[3]

molar concentration of the ionic constituents in blood is about 0.16 M. Apart from the cellular and ionic constituents, blood also contains several enzymes and proteins. The adsorption of blood proteins on the blood vessel wall or on a cardiovascular prothesis, their charge transfer reactions across the respective interfaces, the adhesion of blood cells on these surfaces, and the aggregation of blood cells either directly or via bridges (fibrin strands) are intermediate steps in the overall thrombosis or clotting reactions. Since the intermediate enzymatic reactions which lead to thrombosis are interfacial in nature and occur in electrolyte environments, one may expect many of them to be electrochemical in origin. The substantial evidence for an electrochemical mechanism of thrombosis, accumulated by Sawyer, Srinivasan, and co-workers,[4–10] may be summarized as follows:

(i) Under normal conditions, the surfaces of the blood vessel wall and of blood cells are negatively charged.

(ii) Injury or atherosclerosis reduces the negative charge density of these vascular components and sometimes causes a reversal of the sign of the surface charge. Such conditions increase the probability of thrombosis.

(iii) A decrease in pH makes the surface charge of the blood vessel wall and of blood cells less negative. The isoelectric point for the vascular components is at a pH of about 4.5.

(iv) Anticoagulant drugs increase or maintain the negative surface charge of the blood vessel wall and of blood cells while procoagulants have the opposite effect and quite often even cause a reversal in the sign of the surface charge.

(v) Electronically conducting materials which maintain negative potentials in blood versus the normal hydrogen electrode (NHE) tend to be antithrombogenic, while those which register positive potentials vs. NHE are thrombogenic.

(vi) Uniformly negatively charged insulator materials tend to be anti-thrombogenic.

(vii) Blood coagulation factors (see Section 2.3) take part in adsorption and charge transfer reactions across metal–solution interfaces. At least some of these reactions seem to be relevant to the blood coagulation sequence (see Section 2.1).

Several original and review articles, describing in detail all aspects of the above findings, have appeared in the literature. An attempt is made in this chapter to present a comprehensive review of all this work.

2. Hematological and Electrochemical Aspects of Blood Coagulation Mechanisms

2.1. The Hematologists' View of the Blood Coagulation Sequence

According to hematologists, several reactions between soluble proteins in blood or plasma lead to the formation of fibrin in the so-called "blood coagulation sequence." The blood coagulation factors, designated by Roman numerals according to an International Nomenclature Committee, are tabulated along with their synonyms in Table 1. The suffix "a" indicates factors

Table 1
Roman Numeral Designation and Most Common Synonyms for Blood
Coagulation Factors

Roman numeral designation	Most common synonym
Factor I	Fibrinogen
Factor II	Prothrombin
Factor III	Tissue thromboplastin
Factor IV	Calcium ions
Factor V	Proaccelerin
Factor VI	Not assigned
Factor VII	Proconvertin
Factor VIII	Antihemophilic factor
Factor IX	Christmas factor
Factor X	Stuart factor
Factor XI	Plasma thromboplastin antecedent
Factor XII	Hageman factor
Factor XIII	Fibrin stabilizing factor

which have been converted to an enzymatically activated form. A phospholipid, released by platelets, is also involved in blood coagulation but is not represented by a Roman numeral; it is designated as platelet Factor 3.

The "cascade system" or "waterfall sequence" for blood coagulation, first proposed by Macfarlane[11] and by Davie and Ratnoff,[12] and subsequently modified by Hemker and Kahn,[13] is presented in Figure 2. There is a significant difference between the triggering mechanism of the *in vitro* and *in vivo* coagulation processes. The *in vitro* reaction sequence, referred to as the "intrinsic pathway," involves a series of intermediate steps starting from contact activation of Factor XII by a foreign surface. The activated Factor XII serves as a proteolytic enzyme in activating Factor XI, which also appears to be a surface reaction. Calcium ions are needed for the enzymatic activation of Factor IX by Factor XIa. A complex is then formed by Factor IXa in which calcium ions, Factor VIII, and a phospholipid from platelet Factor 3 are involved. This complex (A) activates Factor X which in turn with Ca^{2+}, Factor V (or Factor Va), and a phospholipid yields another complex (prothrombinase or complex B). The enzyme prothrombinase converts prothrombin (Factor II) to thrombin. Thrombin catalyzes the polymerization of fibrinogen to fibrin.

The *in vivo* coagulation sequence, termed the "extrinsic pathway," starts fby the action of a tissue factor which directly activates Factor X. The remaining sequence of reactions is practically the same as in the "intrinsic pathway."

Even while the clot is being formed, reactions leading to eventual dissolution of the clot and repair of the injured site are triggered. Plasmin causes clot lysis which is not present in circulating blood. The precursor plasminogen in circulating blood is activated to form plasmin. The phenomenon of clot lysis

Intrinsic*

$XII \xrightarrow{\text{contact}} XII_a$

$XI \xrightarrow{XII_a} XI_a$

$IX \xrightarrow{XI_a} IX_a$

$IX_a + VIII + \text{phospholipid} + IV \longrightarrow \text{Complex A}$

$X \xrightarrow{\text{Complex A}} X_a$

$X_a + V + \text{phospholipid} + IV \xrightarrow{\text{Complex B}} \text{Complex B}$

$\text{Prothrombin} \xrightarrow{\text{Complex B}} \text{Thrombin}$

$\text{Fibrinogen} \xrightarrow{\text{Thrombin}} \text{Fibrin}$

Extrinsic†

$II + VII + IV \longrightarrow \text{Active product 1}$

$\text{Active product 1} + X \longrightarrow X_a$

$X_a + \text{phospholipid} + IV \longrightarrow \text{Active product 2}$

$\text{Active product 2} + V + IV \longrightarrow \text{Prothrombin activator}$

$\text{Prothrombin activator} + \text{Prothrombin} + IV \longrightarrow \text{Thrombin}$

$\text{Fibrinogen} \xrightarrow{\text{Thrombin}} \text{Fibrin}$

Figure 2. Intrinsic and extrinsic pathways in blood coagulation.[13]

is called fibrinolysis which is also an essential reaction in restoring homeostasis of blood.

2.2. The Role of Blood Cells, Particularly Platelets, in Blood Coagulation

Platelet aggregation is an important step in the formation of a thrombus deposit. The first step in arterial thrombosis appears to be the adhesion of platelets to the damaged vessel wall. It is followed by the aggregation of platelets. If the aggregation is reversible, the thrombus embolizes. Conversely, the irreversibility of the platelet aggregation reaction results in blockage of the blood vessels. Platelet survival time generally correlates with the incidence of thromboembolism. About 90% of patients with thromboembolism have a shortened platelet survival time. The platelet survival time may be of value in detecting thrombosis-prone patients with valvular heart disease and also to predict the thrombogenicity of newly developed prosthetic materials.

The blood clot formed in a test tube is strikingly different from a thrombus deposit formed *in vivo* in the vascular system. A clot in a test tube consists of a fibrin network in which red cells, white cells, and relatively small numbers of platelets are found. On the other hand, an *in vivo* thrombus deposit consists of large amorphous masses of platelets, surrounded by white cells with a few red cells. However, it is possible to form deposits closer to thrombi rather than to blood clots *in vitro* by using circulating plastic loops in which the blood is in motion during coagulation.[14]

Several hematologists are of the view that the primary step in hemostasis is the adhesion of platelets to exposed collagen fibers.[15–16] It has been proposed[17] that the adhesion is due to the formation of an enzyme–acceptor complex between the incomplete carbohydrate chains (galactosyl residues) of collagen and an enzyme of the platelet membrane (glycosyl transferase). Platelets in contact with collagen fibers contract and release adenosine diphosphate (ADP) as well as other platelet constituents. ADP causes more platelets to adhere and the autocatalytic reaction proceeds towards the formation of a hemostatic plug.

Since collagen is not present on foreign surfaces, another mechanism must be involved in the adhesion of platelets to such surfaces. According to some researchers,[17–20] this phenomenon is due to protein adsorption. However, until the present time the protein layers adsorbed from blood on foreign materials have not been chemically identified. With this brief background on the hematologic aspects of blood coagulation, the remainder of this chapter will be devoted to the electrochemical aspects of the blood coagulation mechanism, characterization of blood vessel wall and of blood cells, correlations between surface charge and antithrombogenic characteristics of drugs, and selection and evaluation of conducting and insulator materials for vascular and heart valve protheses.

2.3. Electrosorption and Electron Transfer Reactions of Some Blood Coagulation Factors at Metal–Electrolyte Interfaces

2.3.1. Fibrinogen

The adsorption and charge transfer reactions of fibrinogen have been investigated using electrochemical (electrocapillary, differential capacity, potentiodynamic, potentiostatic, and galvanostatic), ellipsometric, electron microscopic, and hematologic methods. A number of these studies also provided insight on the role of these reactions in blood coagulation. Fibrinogen plays a crucial role in blood clotting and thrombosis. Fibrinogen is present at a concentration of 3 mg/ml in blood, while some of the other blood coagulation factors are present only in trace amounts.[21] As illustrated in Figure 2, the polymerization of fibrinogen to fibrin is the last of the series of intermediate steps in thrombosis and in blood clotting. The terminal conversion of fibrinogen to fibrin is one of the most fundamental steps in thrombosis. Fibrinogen is a well-characterized globular type protein and has a molecular weight of 300,000. The molecule contains about 18 amino acids which include glycine, lysine, tyrosine, tryptophan, and arginine. It is believed that fibrinogen exists as a dimer with about 28 disulfide linkages in the molecule.

Fibrinogen is strongly adsorbed on mercury and on platinum over a wide range of potentials and even at relatively low concentrations. The adsorption behavior of fibrinogen on mercury[22] as a function of potential and of its concentration in the electrolyte, as ascertained from capacitance measurements, is shown in Figure 3. Studies on platinum, also using differential capacitance techniques, revealed a similar potential dependence and strong adsorption of fibrinogen[23] (Figure 4). Another technique which provided more direct information on the potential dependence of adsorption of fibrinogen on platinum was ellipsometry.[24] There was some disagreement between the results obtained using the two techniques—the capacitance method showed an adsorption maximum in the coverage potential plot, while

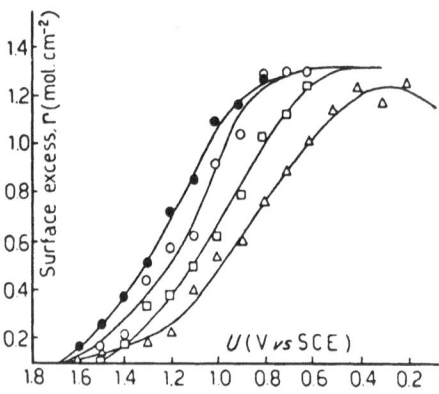

Figure 3. Potential-dependent adsorption of fibrinogen on mercury at 25°C from normal saline, as determined using the electrocapillary method. Concentrations of fibrinogen in electrolyte: 0.018 ($\triangle$); 0.06 ($\square$); 12 ($\bigcirc$); and 36 ($\bullet$) mg/cm^3.[22]

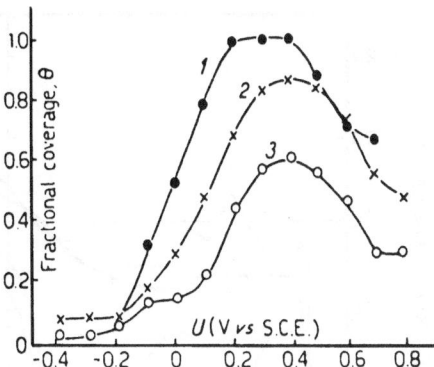

Figure 4. Potential-dependent adsorption of fibrinogen on platinum at 25°C from NaCl containing (1) 10^{-3} (2) 10^{-2} (3) 10^{-1} times physiologic concentration of fibrinogen (3 mg/cm^3).[23]

ellipsometry did not. Systematic studies of the adsorption behavior of amino acids, peptides, and proteins by Matthews[22] and by Stoner[23] revealed that the electrosorption behavior of these molecules can be identified with their π electron–metal interactions.

The potentiodynamic technique also revealed the strong adsorption of fibrinogen on platinum.[24] In addition, this technique was also used to demonstrate that fibrinogen undergoes charge transfer reactions at electrode-electrolyte interfaces. At a low concentration of fibrinogen in normal saline (0.154 M NaCl), the hydrogen peaks in the cyclic voltammogram on platinum are completely eliminated but the platinum oxide reduction peak remains, though reduced in peak height (Figure 5). Surprisingly, at a higher concentration, new peaks appear in the hydrogen region and the oxide reduction peak is greatly reduced (Figure 6). With continuous cycling, there is a slow but complete elimination of the oxide reduction peak and its replacement by a new peak at more anodic potentials. The results of the potentiodynamic studies for three concentrations of fibrinogen (0.007, 0.07, and 3.0 mg/ml) show that it takes part in at least three charge transfer reactions. These results were confirmed using potentiostatic techniques. Fibrinogen undergoes charge transfer reactions at anodic and cathodic potentials on a platinum electrode (Figure 7). In the intermediate potential range it is strongly adsorbed on the platinum surface. A linear relation in the potential–log current density plot for the electrooxidation of fibrinogen from saline containing this protein (concentration 3 mg/cm^2) was observed in the potential range -250 to 0 mV/SCE, by two groups of workers.[24,25] The product of the anodic reaction, examined electron microscopically, is similar to fibrin formed by thrombin addition to fibrinogen, i.e., an electropolymerization reaction of fibrinogen to a fibrin-like product occurs at anodic potentials.[25] Fibrinogen, subjected to anodic potentials, shortened coagulation times (thrombin time), while the reverse occurred with cathodized fibrinogen.[26]

Adsorption and conformational studies[27] showed a marked increase in adsorption at positive potentials, but the onset potential was highly positive.

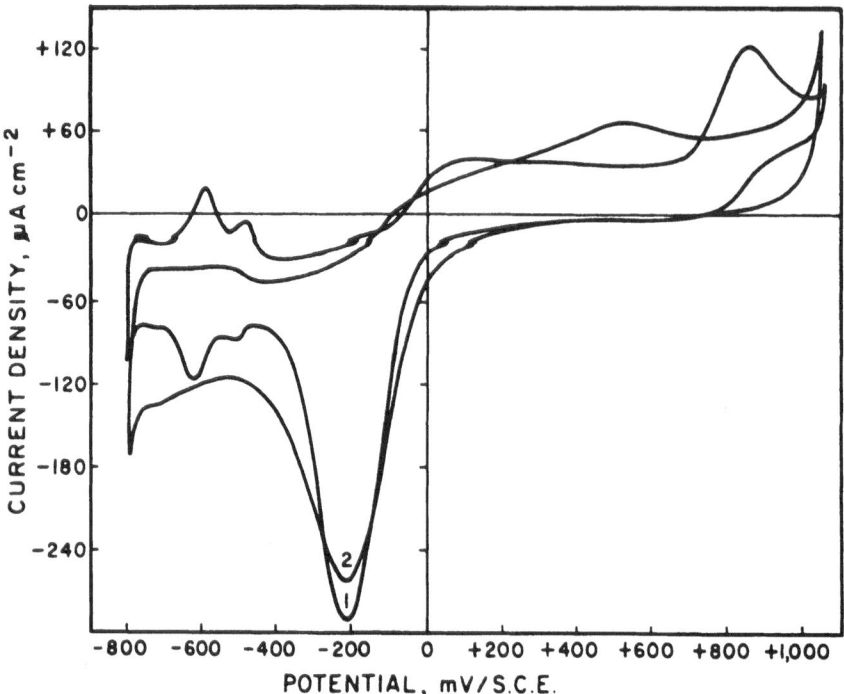

Figure 5. Cyclic voltammograms on platinum in saline at 25°C (1) before and (2) after addition of fibrinogen (0.007 mg/cm³). Scan rate 50 mV/sec.[24]

A conformational change of the adsorbed protein was observed at −200 mV/SCE, but it is not possible to correlate it with polymerization of fibrinogen to fibrin.

2.3.2. Prothrombin and Thrombin

The fibrinogen–fibrin conversion reaction is catalyzed by the enzyme thrombin. Thrombin, in turn, is formed from the blood coagulation factor prothrombin. There are two distinctive mechanisms for the prothrombin-thrombin conversion. In the intrinsic pathway, only the constituents of blood are involved. The extrinsic pathway occurs outside the vascular tree and requires tissue factors (see Figure 2). The concentration of prothrombin in bovine plasma is 0.1–0.15 mg/ml. Human and bovine prothrombin are similar in many respects. The molecular weight of purified prothrombin is 68,000–68,500. The main contaminants in prothrombin preparations are Factors VII, IX, and X. These contaminants can be removed by chromatography on DEAE cellulose. The molecule is ellipsoidal of length 119Å and width 34Å, and has 526 amino acid residues. The amino acid compositions of prothrombin and thrombin are shown in Table 2. The molecular weight of the amino acid fraction in prothrombin is 58,800. The other major biochemical constituents

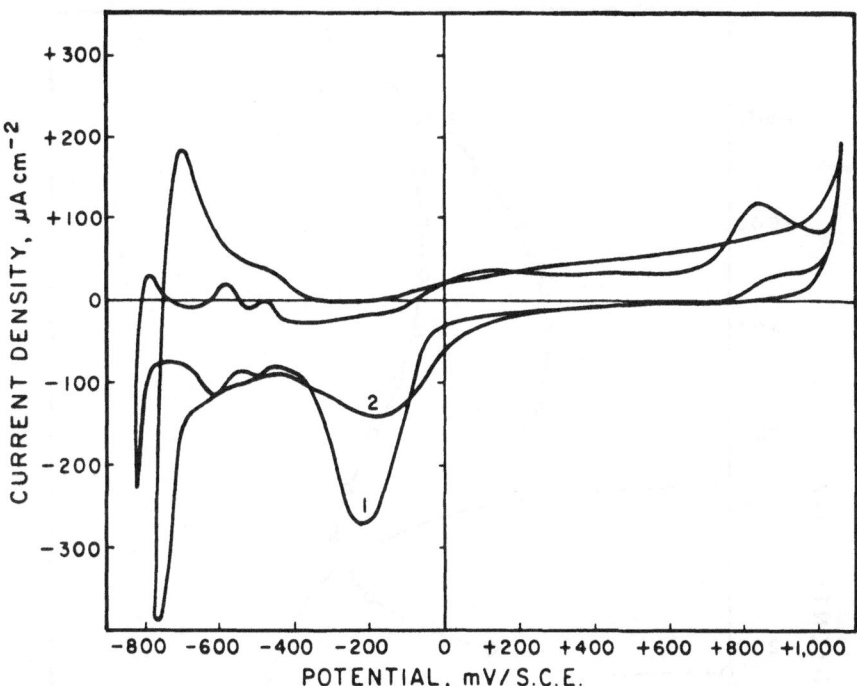

Figure 6. Cyclic voltammograms on platinum in saline at 25°C (1) before and (2) after addition of fibrinogen (0.07 mg/cm³). Scan rate 50 mV/sec.[24]

(carbohydrate) have a molecular weight of 8000. The carbohydrate fraction (11.2% by weight) consists of galactose (3.3%), mannose (1.53%), fucose (0.09%), hexosamine (2.3%), and sialic acid (4.2%).

Thrombin can be obtained from prothrombin in different ways. The molecular weight of thrombin is 33,700. The length of the molecule is 84 Å and its width 34 Å. Its amino acid composition is also found in Table 2. Bovine thrombin contains 258 amino acid residues, corresponding to a molecular weight for this fraction of 28,400. There are 2.15 moles of disulfide per mole of thrombin. It has two N-terminal amino acids, isoleucine, and threonine. The total carbohydrate content of thrombin, 9.68%, is made up of galactose (2.34%), mannose (1.17%), fucose (0.07%), hexosamine (2.2%), and sialic acid (3.9%). Physicochemical evidence indicates that thrombin has half the molecular weight of prothrombin. It has been hypothesized that prothrombin is composed of thrombin and autoprothrombin III (inactive factor X). More recent evidence suggests that prothrombin might be a dimer of two thrombin molecules.

Cyclic voltammetric studies were conducted on a platinum electrode in normal saline solution containing different concentrations of prothrombin or thrombin.[28] With addition of small amounts of prothrombin to 0.154 M NaCl, the hydrogen peaks are suppressed, the currents in the platinum oxide

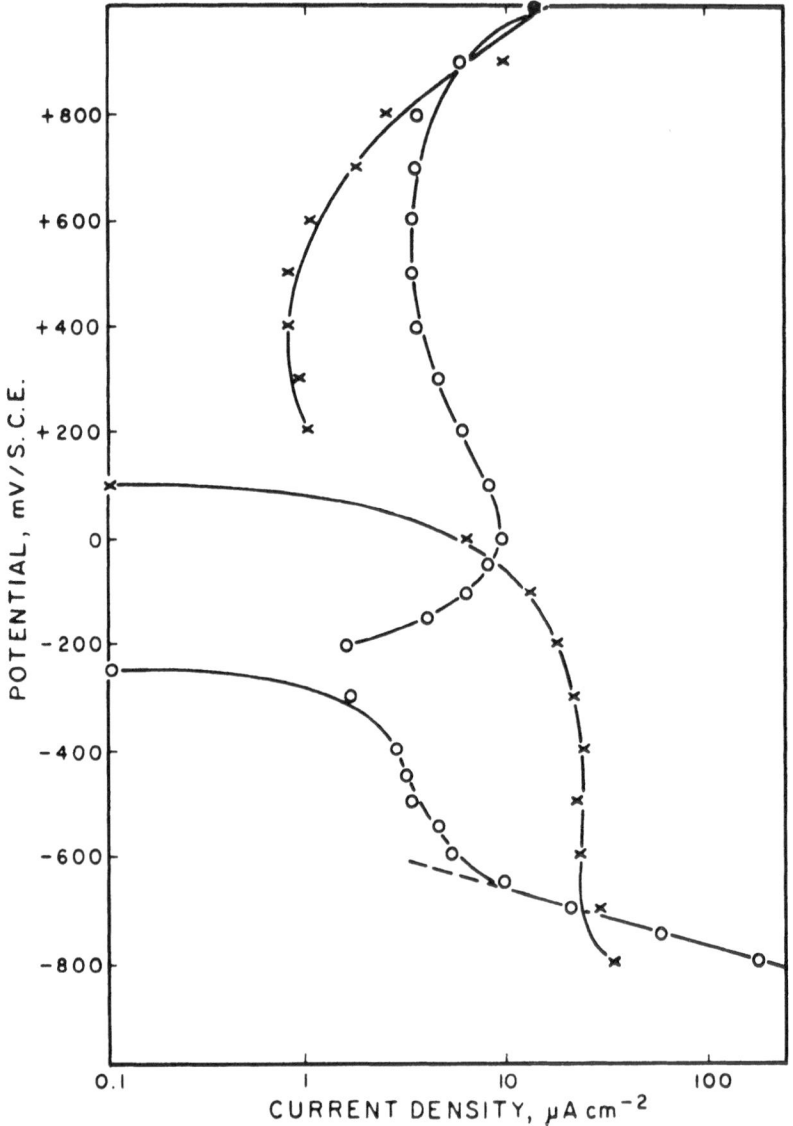

Figure 7. Potentiostatic potential—current density plots on platinum at 25°C in aerated saline (×) and in aerated saline containing $0.07\,mg/cm^3$ of fibrinogen (○).[24]

formation region is lowered, and the area representing platinum oxide reduction is decreased. At higher concentrations of prothrombin in the electrolyte, a significant time variation of the cyclic voltammograms (Figure 8) indicates the occurrence of some faradaic process. By limiting the potential scan to the negative potential region, the peak currents were reduced (Figure 9). It thus appears that the species participating in the electrode reaction at negative potentials is formed at high positive potentials. By carrying out the cyclic

Table 2
Amino Acid Composition of Bovine Prothrombin
and Thrombin

Amino acid	Ratio of appearance using methionine as reference	
	Prothrombin	Thrombin
Aspartic acid	9.0	7.0
Threonine	5.0	3.2
Serine	6.2	3.5
Glutamic acid	12.8	7.2
Proline	5.8	3.5
Glycine	8.3	5.2
Alanine	6.0	3.0
Cystine/2	2.0	1.5
Valine	6.2	4.0
Methionine	1.0	1.0
Isoleucine	3.3	2.7
Leucine	7.6	6.0
Tyrosine	4.2	2.5
Phenylalanine	3.8	2.5
Lysine	5.8	4.5
Histidine	1.6	1.3
Arginine	8.2	4.5
Tryptophan	1.8	1.5

voltammetric experiments at different sweep rates, it was shown that both the anodic and cathodic peaks, due to the electrochemical reaction of prothrombin, shift by 30 mV for a tenfold change of sweep rate (Figure 10). Further, the ratio of the anodic to cathodic peak currents increases from about 0.45 to 0.85 for a corresponding increase in sweep rate from 10^{-2} to 2 V/sec. This observation indicates that chemical reactions interfere with the charge transfer reactions.

At a first glance, the cyclic voltammograms on Pt with thrombin in 0.154 M NaCl may appear to be similar to those with prothrombin. However, with thrombin, there is no time dependence of the cyclic voltammograms. Also, the peak potential in the cathodic region is more negative with thrombin than with prothrombin (Figure 11). As with prothrombin, scanning only in the negative region of potential reduces the peak currents due to the electrochemical reactions of thrombin. The plots of peak potential versus the logarithm of scan rate are linear both for the oxidation and reduction reactions and have a slope of 60 mV/decade. With prothrombin, the slopes were only half this value. A possible explanation is that twice the number of electrons are involved in the electrochemical reactions of prothrombin as compared with the ones for thrombin.

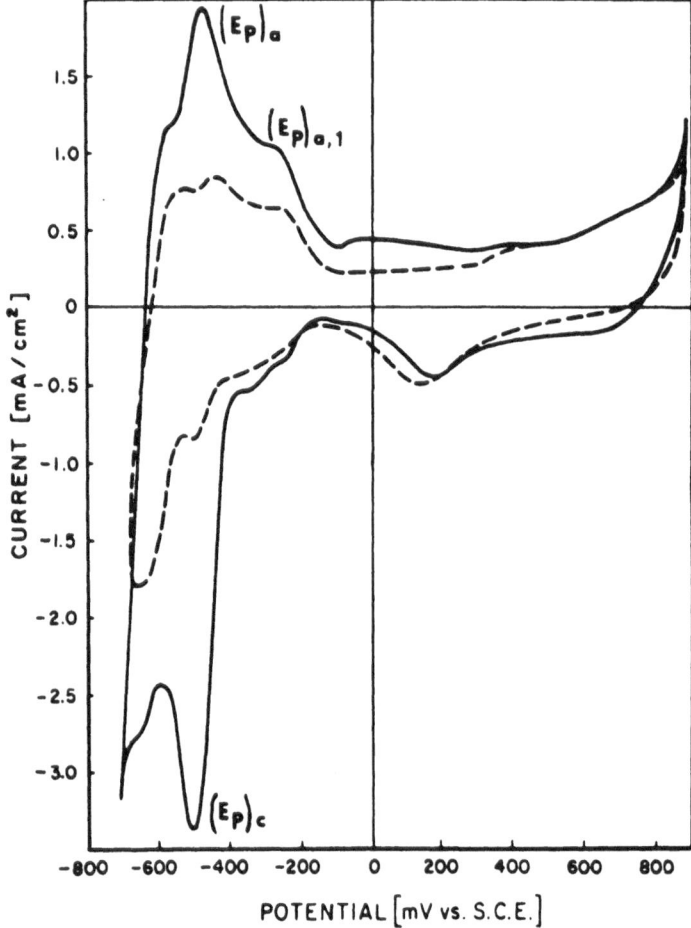

Figure 8. Cyclic voltammograms on platinum in saline containing prothrombin (1 unit/cm³) at 25°C after cycling 10 (– – – –) and 60 (——) min at a scan rate of 50 mV/sec.[28]

A relevant question is "Is there an electrochemical conversion of pro-thrombin to thrombin?" One of the simplest methods of determining whether electrochemical activation causes a prothrombin–thrombin conversion is to check whether there are any changes in hematological activity after electrochemical treatment. Thus, experiments were carried out to test the effects of electrochemically activated prothrombin on the clotting time of blood. The tests showed a decrease in clotting time but not a measurable increase in thrombin activity. The latter may be due to the lack of sensitivity of the analytical method for small amounts of conversion of prothrombin to thrombin. In an alternative experiment, solutions containing prothrombin were exposed to a platinum electrode maintained at constant anodic potentials. The test on the clotting time of fibrinogen was carried out with the samples of

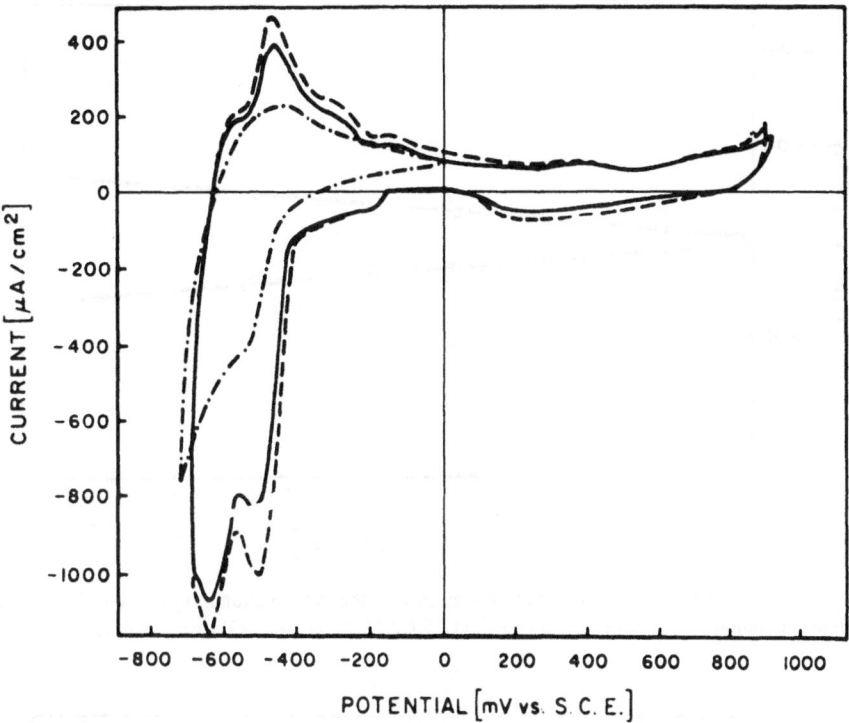

Figure 9. Effect of limiting potential scan to only positive and to only negative range on cyclic voltammogram on platinum in saline containing prothrombin (1 unit/cm^3) at 25°C: scanning in entire region (——), after scanning only in anodic region (– – – –), and after scanning only in cathodic region (· – · – ·) for 30 min at 50 mV/sec.[28]

prothrombin that were taken after 15 minutes of maintaining the electrode at a constant potential. The results showed that the clotting times decreased with increase of anodic potentials of the platinum electrodes to which the prothrombin samples were exposed. The results cannot be attributed to the concomitant pH change since a lower pH inhibits the clotting activity of thrombin. Thus it appears that during electrolysis of prothrombin a product with thrombin-like activity is formed.

2.3.3. Proaccelerin (Factor V) and Antihemophilic Factor (Factor VIII)

The electrochemical reactions of Factors V and VIII at the platinum/normal saline interface have been investigated using cyclic voltammetric techniques.[29] Factor V is necessary for the conversion of prothrombin to thrombin by either the extrinsic or intrinsic pathway (Figure 2). The protein is present in human plasma at a concentration of 0.01 mg/ml. The molecular weight of Factor V of bovine origin is 290,000.[30] This protein is highly sensitive to pH and temperature and is consumed during clotting. Factor VIII is essential for

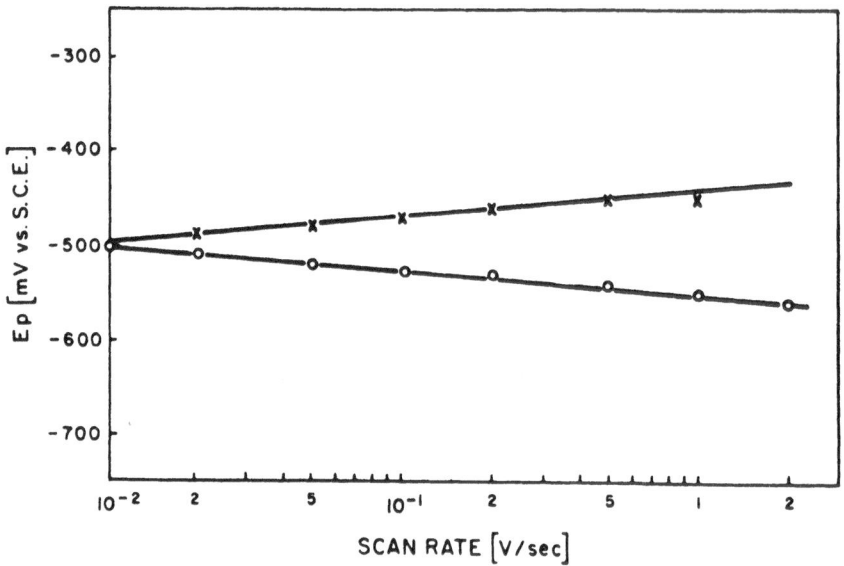

Figure 10. Peak potential as a function of scan rate of the cyclic voltammograms on platinum in saline containing prothrombin (1 unit/cm^3) at 25°C.[28] (×) anodic, (○) cathodic peak potentials.

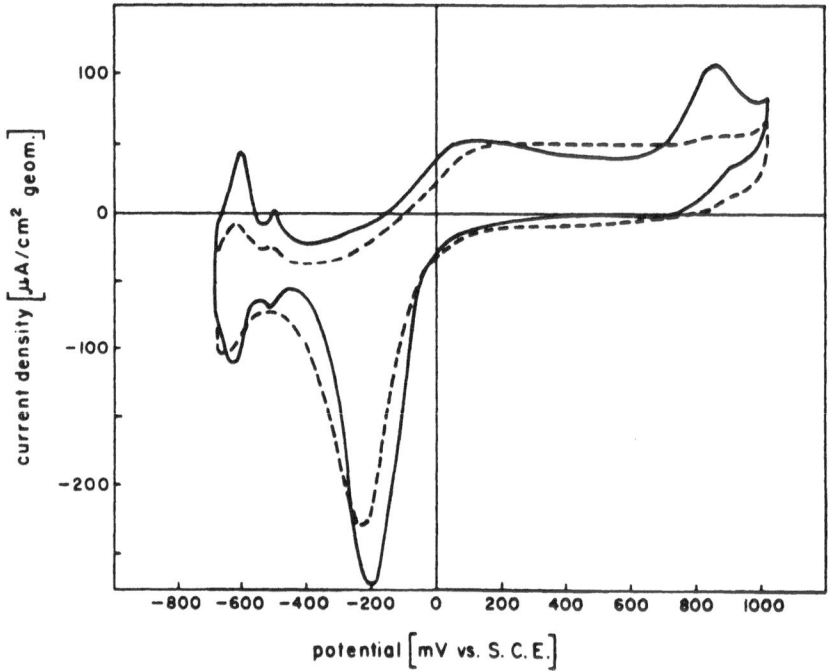

Figure 11. Cyclic voltammograms on platinum in saline at 25°C before (——) and immediately after (----) addition of thrombin (1 unit/cm^3).[28]

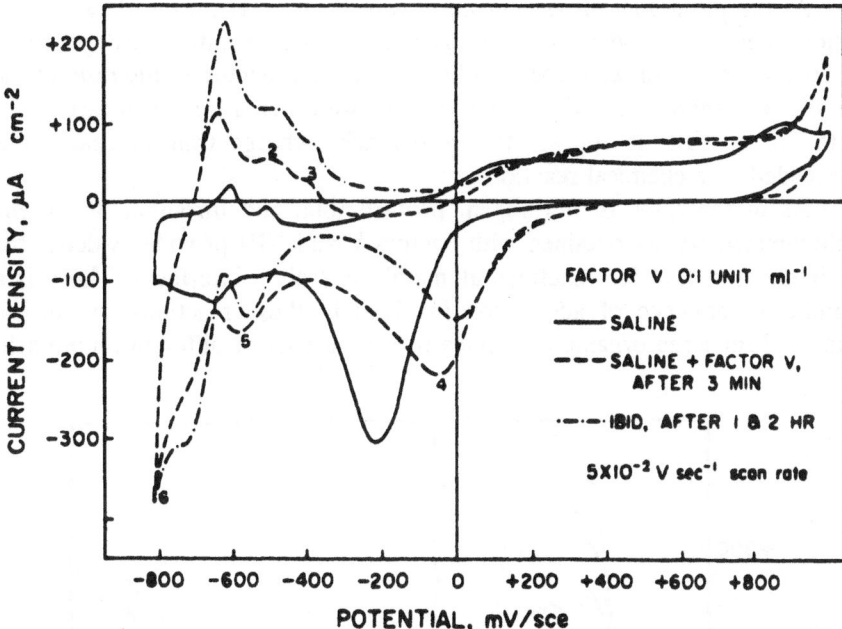

Figure 12. Effect of cycling time on cyclic voltammogram at platinum in saline containing Factor V at 25°C (0.1 unit/cm³).[29]

the coagulation of blood in the intrinsic system. Absence of this factor in plasma causes hemophilia, one of the oldest and most frequently occurring congenital defects. The protein is present in human plasma at a level of 0.02–0.05 mg/ml. It is a highly labile protein with a molecular weight of about 180,000 and while separation is easy, purification is difficult.[31] Scanty structural information is available on both proteins, which are β-globulins, as revealed by electrophoretic studies.

For the cyclic voltammetric studies on smooth platinum, the proteins were dissolved in 0.15 M NaCl (saline of physiological concentration). Two concentrations of Factor V (0.1 and 1 unit†/ml) and three of Factor VIII (0.1, 1, and 10 units/ml) were used for the study. At both concentrations of Factor V, the anodic and cathodic peak potentials (peaks 1, 2, 3, and 5 in Figure 12) showed a shift of 30 mV/decade of scan rate. The ratio of anodic to cathodic peak currents (peaks 2 and 5) steadily increases from 0.6 at 30 mV/sec to 0.9 at 300 mV/sec. Further electrochemical evidence suggests that these peaks represent a reversible charge transfer step preceded by a chemical one. The other anodic (peak 1) and cathodic (peak 3) peaks indicate catalytic influence on the charge transfer reaction (probably hydrogen discharge/ionization).

---

† The concentrations of blood coagulation factors are generally expressed in units/ml. One unit is defined as the amount of protein present in 1 ml of normal plasma.

At the physiological concentration with Factor VIII, a set of two peaks (anodic peak 2, cathodic peak 5 in Figure 13) exhibits a shift of peak potential of 30 mV/decade of scan rate. Unlike in the case of Factor V, the ratio of the anodic to cathodic peak currents decreases with increase of scan rate. Thus with Factor VIII, it appears that a reversible charge transfer reaction is succeeded by a chemical reaction.

As in the cases of fibrinogen, prothrombin, and thrombin, the cyclic voltammetric results obtained with Factors V and VIII provide evidence for their electron transfer reactions at metal–electrolyte interfaces. There is a significant influence of adsorption involved in these reactions, as can be expected for large organic molecules with a number of different amino acid

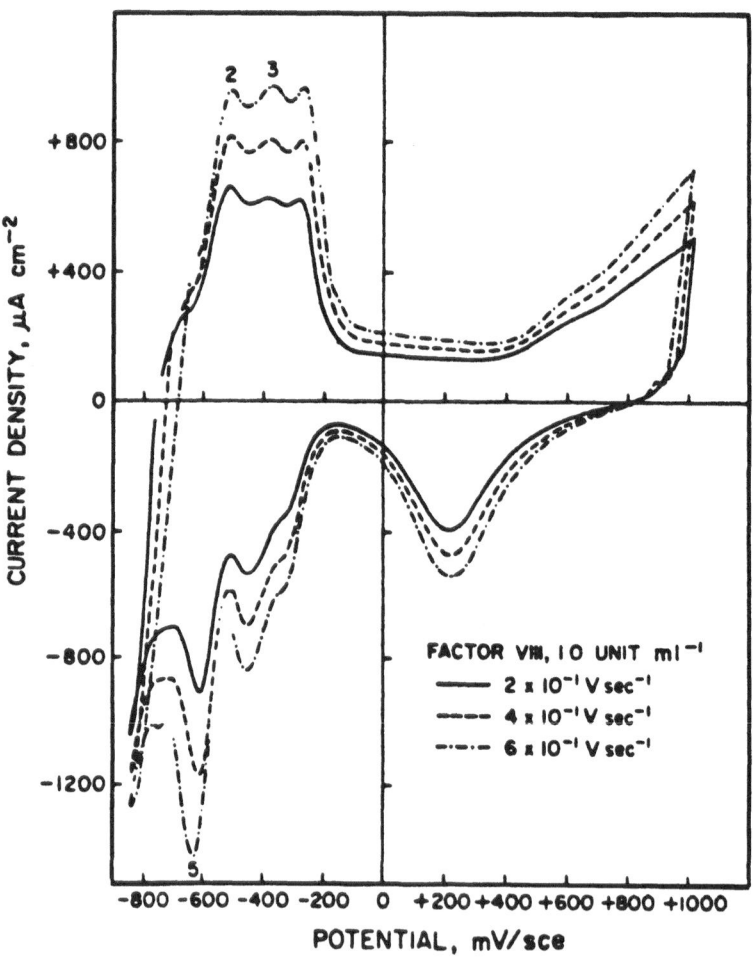

Figure 13. Effect of scan rate on cyclic voltammogram at platinum in saline containing Factor VIII (1 unit/cm³) at 25°C.[29]

groups. The observations that the voltammetric peak currents and potentials obey some of the diagnostic criteria developed for relatively simple electrochemical reactions indicate that some specific sites of these complex molecules are conformationally more prone to electrochemical reactions.

The demonstration of the hematological and cardiovascular relevance of such basic bioelectrochemical studies is necessary and the former has to some extent been exemplified in this and the earlier subsections. In the cardiovascular system, especially when metallic implants are used, it is likely that such types of electrochemical reactions are triggered and provide an alternate pathway for the blood coagulation cascade. This aspect will be dealt with in Section 4.1.

2.4. Adsorption Reactions of Some Blood Coagulation Factors at Insulator–Solution Interfaces

The adsorption of proteins is the first interaction which occurs when a foreign surface comes in contact with blood. The processes which then lead to hemostasis are attachment of cellular elements to the surface, platelet adhesion and release, and triggering of the blood coagulation cascade. Investigations of interactions of proteins at the interface, which are the primary steps in blood clotting and determine whether it will occur, are hence of fundamental importance.

The interactions of blood proteins with surfaces have been extensively studied using biochemical, hemotological, and physicochemical methods. A rapid method for the study of adsorption of proteins on insulator materials was developed by Vroman and Adams.[32] It is based on the hydrophobicity of protein layers and interactions of antisera with their matching proteins. Using this method, it was shown that fibrinogen deposits within 5 seconds onto a glass plate immersed in plasma. The reactions appear to be far more complex with preadsorbed protein films which are treated with human plasma followed by specific antisera. The ellipsometric technique was also used by Vroman[33] to show that fibrinogen films on oxidized silicon crystal surfaces lose thickness on exposure to all proteins, antisera, and plasma. The adsorption of fibrinogen on mica is inhibited by the prior adsorption of heparin, as revealed in an electron microscopic study by Stoner, Srinivasan, and Gileadi.[34] It was proposed in this investigation that the highly negatively charged groups in heparin prevent the adsorption of fibrinogen.

The concentrations of the globulins are about ten times higher than that of fibrinogen in human plasma. Irrespective of whether the materials are hydrophilic or hydrophobic, fibrinogen and globulins are adsorbed on their surfaces.[35] Adsorption of albumin reduces the subsequent adsorption of fibrinogen. According to the investigations of Morrissey and Stromberg,[36] there are no conformational changes with time or with concentration when the proteins albumin, prothrombin, and fibrinogen are adsorbed on silica.

3. An Electrokinetic Approach for the Characterization of Blood Vessel Walls and of Blood Cells and for the Selection of Anticoagulant Drugs

3.1. Surface Charge Characteristics of Blood Vessel Walls Using Streaming Potential and Electroosmosis Techniques

An advantage of the streaming potential method for the surface charge characterization of blood vessel walls is that both *in vitro* and *in vivo* measurements can be made. The apparatus, which was used by Srinivasan *et al.*[37] for the *in vitro* measurements, is shown in Figure 14. Canine blood vessels (thoracic aorta or carotid artery) were connected to the two T-joints which were in turn connected to the electrolyte reservoirs and reference electrodes. Nitrogen pressures (measured with a manometer) were used to attain the desired flow rates in either direction. Graduations were made on the bottles to measure volume changes for fixed times. One advantage of the symmetrical arrangement is that streaming potentials can be measured in both directions under pratically identical conditions and any artifacts can be eliminated by taking the average value of the streaming potentials for flow in both directions. The electrolytes used in these studies were varying concentrations of physiologic solutions—the normal physiologic solution (Krebs solution) contains 64.3 g NaCl, 29.0 g NaHCO$_3$, 2.47 g MgSO$_4$, 1.26 g KH$_2$PO$_4$, 2.24 g KCl, and

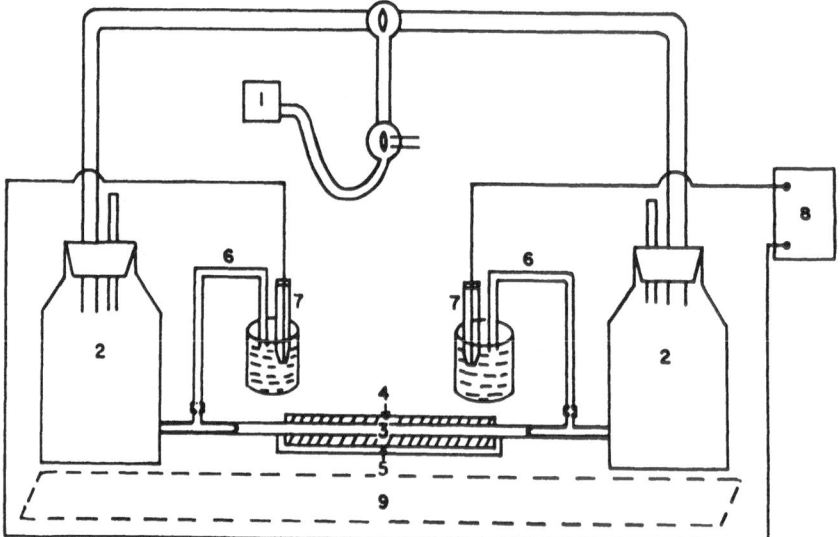

Figure 14. Experimental setup for measurements of *in vitro* streaming potentials during electrolyte flow through canine aorta or carotid artery: (1) nitrogen tank; (2) graduated aluminum foil-covered bottles; (3) aorta or carotid artery; (4) sponge soaked with electrolyte; (5) insulation; (6) agar bridge; (7) calomel electrode; (8) Keithley electrometer; and (9) grounded aluminum plate.[37]

3.68 $CaCl_2 \cdot 2H_2O$ per liter. The pH of the solution was adjusted to 7.1–7.3 by bubbling O_2–CO_2 mixtures. The sponge on the outside of the blood vessel was soaked with the electrolyte to maintain the biological condition. Linear relations were obtained between the streaming potential (E) and pressure (P). The ζ potentials were calculated from the slopes of the streaming potential–pressure relations according to the equation:

$$\zeta = \frac{4\pi\eta\kappa}{\varepsilon}\frac{dE}{dP} \qquad (1)$$

where η, κ, and ε are the viscosity, specific conductance, and dielectric constant of the electrolyte. A positive slope corresponds to a negative ζ potential. It was shown that there is a marked decrease in the slope of the streaming-pressure (or flow rate) line on an aged blood vessel as compared with that on a fresh one (Figure 15). Quite often, even negative streaming potentials were observed with aged blood vessels, indicating a reversal in their surface charge from normal behavior.

Streaming potentials have also been measured *in vivo*.[38] Healthy mongrel dogs, with an average weight of 20 kg, were used. General anesthesia was induced by injecting sodium pentobarbital, 30 mg/kg, intravenously. Poly-ethylene tubes were inserted through side branches into the aorta and femoral arteries at known distances apart and connected to beakers containing calomel electrodes. The backflow of blood through the polyethylene tubes, which then clotted, served as electrolyte bridges. The calomel electrodes were connected to a battery-operated Keithley electrometer—thus both measuring electrodes

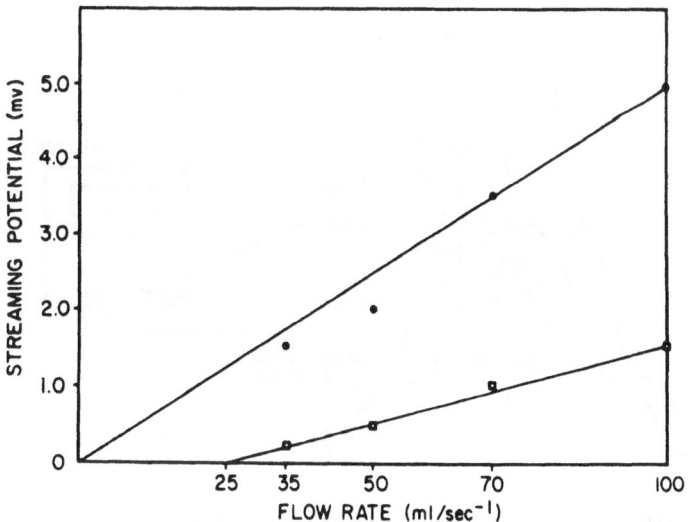

Figure 15. *In vitro* streaming potential versus flow rate plot for aorta. Times (h) after removal of the blood vessel from dog are 4 (●) and 24 (□).[37]

were effectively off ground. As a substitute for this type of reference electrodes, silver–silver chloride wire electrodes have also been directly inserted into the T-joints. The experimental arrangement is schematically represented in Figure 16. Measurements were made at the normal flow rate of blood. Streaming potentials under normal conditions averaged 0.2 mV across the aorta and 0.5 mV across the carotid artery. Injury to the blood vessels reduced the magnitudes of these streaming potentials.

In the electroosmotic technique, an electric current is passed across the porous blood vessel and the rate of resulting transport of fluid is measured across it. A simple apparatus, which has been used for such measurements, is illustrated in Figure 17. The membrane was clamped between the two compartments and the initial readings of the positions of the electrolyte (Krebs solution) on the horizontal manometers noted. Current was passed between the two silver–silver chloride electrodes, and the displacement of the liquids in the

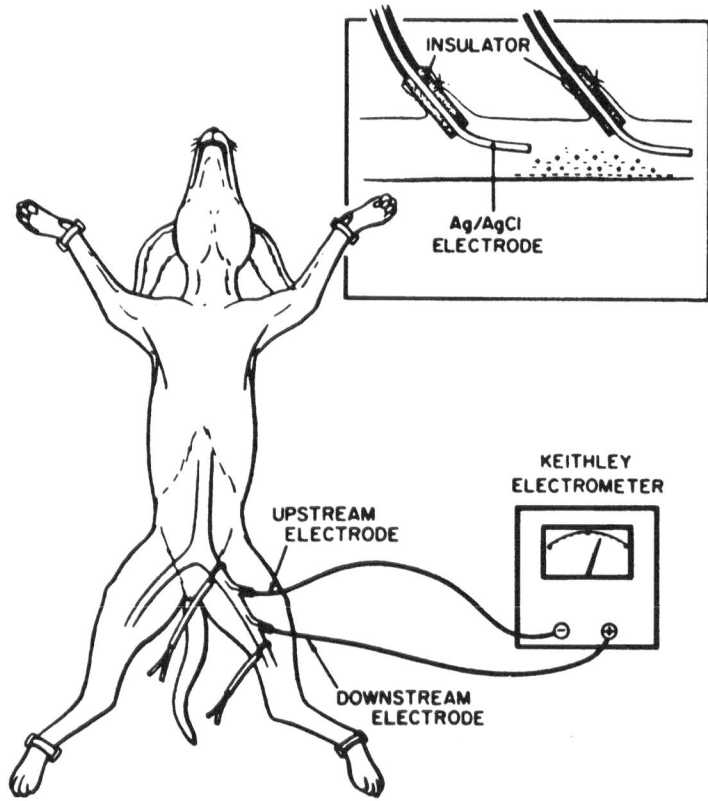

Figure 16. Schematic representation of experimental arrangement for measurement of *in vivo* streaming potential in femoral artery. Insert shows insertion of electrodes through side branches.[38]

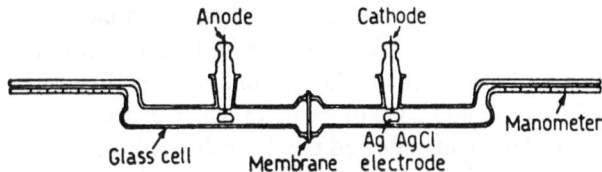

Figure 17. All glass cell for electroosmosis measurements across blood vessels.

manometers after a known period of time noted. The direction of the current was reversed and the measurements repeated. The zeta potential of the membrane–electrolyte interface was calculated from the relation:

$$\zeta = \frac{4\pi\kappa\eta\nu}{\varepsilon I} \tag{2}$$

where ν is the rate of electroosmotic flow in milliliters and I is the current in electrostatic units. Systematic electroosmotic studies were carried out on normal and atherosclerotic blood vessels.[39] The experiments revealed that the electroosmotic behavior was nearly identical on normal and progressively increasing atherosclerotic vessels but showed a sudden change for vessels with maximum degree of atherosclerosis. For the latter case, the ζ potential was lower in magnitude, which corresponds to a lower negative surface charge density. The experimental findings and the conclusions drawn from these results are consistent with the observations that the blood vessel wall is significantly resistant to thrombosis even in progressive atherosclerotic aortae until an advanced stage of atherosclerosis is produced. This study lends further support to an electrochemical mechanism of thrombosis.

3.2. Surface Charge Characteristics of Blood Cells Using Mainly Electrophoresis and to a Limited Extent Sedimentation Potential Techniques

Electrophoretic techniques have found wide applicability in the biomedical areas.[40] Electrophoresis is the migration of colloidal particles in an electrolyte under the influence of an electric field. The zeta potential across the colloidal particle/electrolyte interface may be obtained from the electrophoretic mobility (ν) according to the equation:

$$\zeta = \frac{4\pi\eta\nu}{\varepsilon E} \tag{3}$$

where E is the electric field in the electrolyte. The Seaman microelectrophoresis apparatus[41] has been extensively used for electrophoretic measurements on blood cells. Recently, an automated instrument, which was developed for measurements of electrophoretic mobilities of nonbiological colloidal particles, has been successfully used for obtaining highly reproducible results on the electrophoretic mobilities of platelets. This instrument—the Laser Zee System 3000 (Pen Kem Inc., NY)—employs a helium–neon laser and optics to provide illuminations of cells in the stationary layer (Figure 18). The beam is split to provide a second image of particles which is projected on the grating of a rotating disc. The light reflected from the cell will be alternately blocked and transmitted as the disc turns. The output voltage from the photomultiplier tube has a single component at a frequency proportional to the speed of the disc. If the cell moves in one direction, the image will cross fewer line pairs per unit time and hence the single component will be at a slightly lower frequency. If the cell moves in the opposite direction, the signal will be at a higher frequency. This frequency shift is proportional to the electrophoretic mobility. Within five minutes, approximately 700 cells can be counted and the result with the mean standard error of each measurement recorded. For measurements on blood cells, less than one ml of cell suspension is required in the electrophoretic chamber (diameter of 1 mm and length of 20 mm).

Like the blood vessel wall, blood cells (red cells, white cells, platelets) also exhibit negative charge densities on their surfaces. A wide variety of studies have been conducted which shows the effects of chemicals on the electrophoretic mobilities of blood cells (see Section 3.3). It is interesting to state in this section that Yazamaki and co-workers[48] showed using the Laser Zee meter that the electrophoretic mobilities of platelets are 1.441 ±

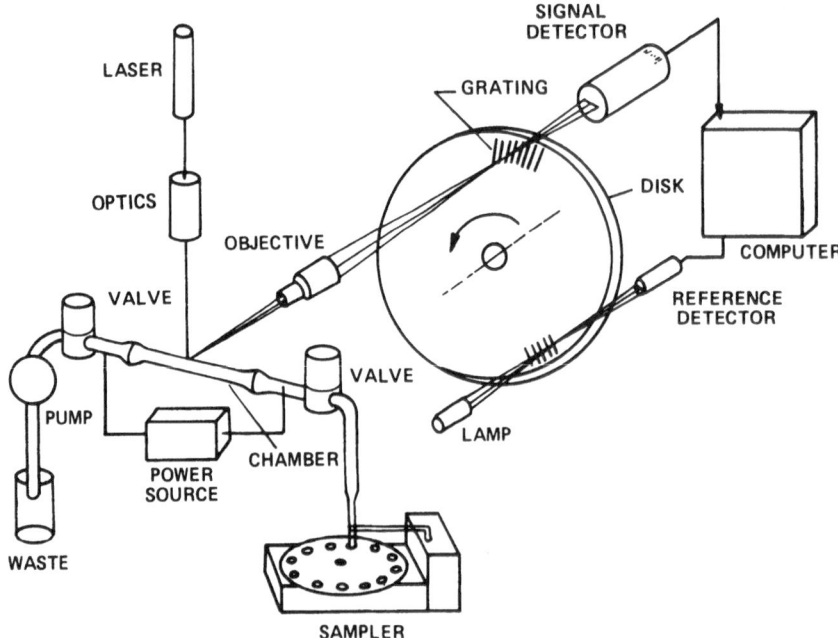

Figure 18. The Pen Kem Laser Zee Meter for electrophoresis measurements on blood cells.[48]

0.015 μm/sec/V for healthy males and 1.494 ± 0.012 μm/sec/V for healthy females. No significant correlation was found between mobility and age of the subjects.

The sedimentation potential method has been used only to a limited extent to determine the surface charge characteristics of blood cells. In this method, the colloidal particles (e.g., blood cells) are allowed to fall through a vertical column and the potential difference between two electrodes, vertically separated, is measured. An apparatus, used for measurements on erythrocytes,[42] is shown in Figure 19. The zeta potential across the colloidal particle–solution interface is calculated from the sedimentation potential (E) according to the equation:

$$E = \frac{\varepsilon \gamma^3 (\rho - \rho') n g \zeta l}{3 \eta \kappa} \tag{4}$$

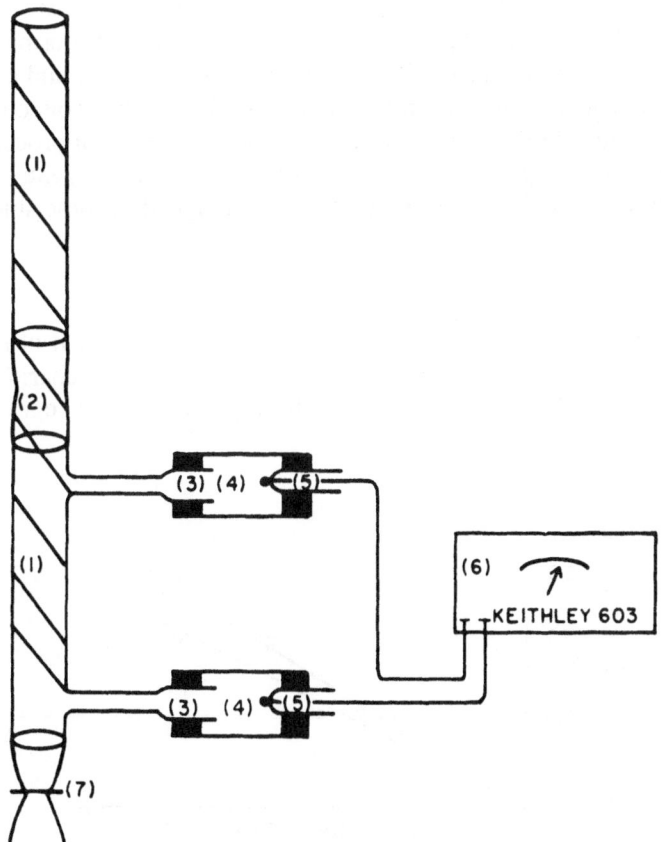

Figure 19. Apparatus for determinations of sedimentation potentials of blood cells: (1) glass tube; (2) Tygon tube; (3) agar bridge; (4) KCl chamber; (5) Ag–AgCl electrode; (6) Keithley electrometer; and (7) clamp.[42]

where γ is the radius of the sedimenting particles, ρ and ρ' are the densities of the particle and the solution, n is the number of such particles in a unit volume, g is the acceleration due to gravity, and l is the distance between the electrodes. The sedimentation potential is linearly related to the concentration of particles in solution, as was observed for erythrocytes in Krebs solution and in plasma (Figure 20).

3.3. Correlations between Effects of Drugs on the Surface Charge Characteristics of the Vascular System and Their Pro- or Antithrombogenic Properties

The knowledge that a decrease in the negative surface charge density of the blood vessel wall and of blood cells is conducive to thrombosis lends credence to the concept that thrombosis is an electrochemical phenomenon. Thus, the measurement of the alteration in surface charge characteristics of the vascular components, produced by chemicals either *in vivo* or *in vitro*, using electrokinetic techniques (streaming potential, electroosmosis, electrophoresis) should be a promising approach for the selection and evaluation of antithrombogenic drugs. Extensive studies have been carried out on a variety of chemicals, including the presently used and potential anti- or procoagulants, oral contraceptives, platelet aggregating or disaggregating agents, etc. It is worthwhile summarizing the significant results obtained in these studies.

3.3.1. Heparin

The most widely used antithrombogenic drug is heparin. It originates in connective tissue mast cells. Heparin is an acid mucopolysaccharide (Figure 21) of high molecular weight (about 17,000). Hydrolysis yields repeating units

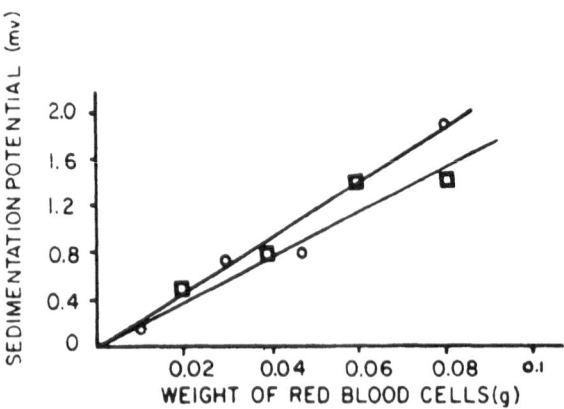

Figure 20. Sedimentation potential of erythrocytes in Krebs solution (○) and in plasma (□) as a function of weight of blood cells in 50-cm solution.[42]

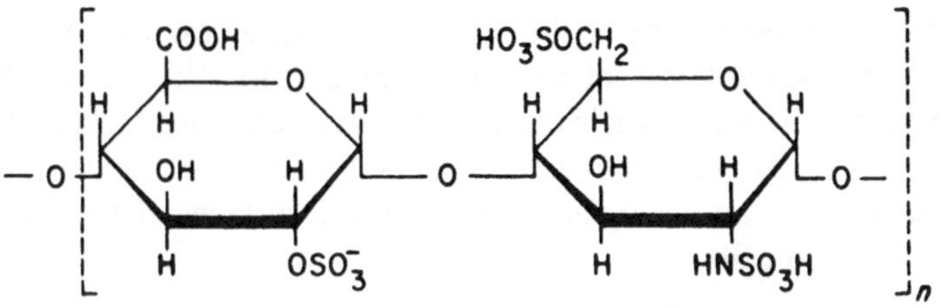

Figure 21. The structure of the tetrasaccharide unit of heparin.

of D-glucoronic acid with an O-sulfate group at C-2 and D-glucosamine N-sulfate with an additional O-sulfate at C-6. Heparin differs from the other mucopolysaccharides in that there are sulfate groups bound to amino groups to form sulfamic linkages.[43] This type of linkage is unique in nature. The content of esterified sulfuric acid is approximately 40%. This structure appears to make heparin the strongest organic acid ocurring in mammalia.

In vitro and *in vivo* streaming-potential measurements in canine blood vessels[38] (carotid arteries) showed that there is a significant increase in the negative surface charge density with infusion of heparin (Figure 22). Incubation of platelets with heparin causes an increase in electrophoretic mobility and therefore of the zeta potentials at the platelet–electrolyte interface.[44] The relation between sulfonic acid content and antithrombogenic activity has been

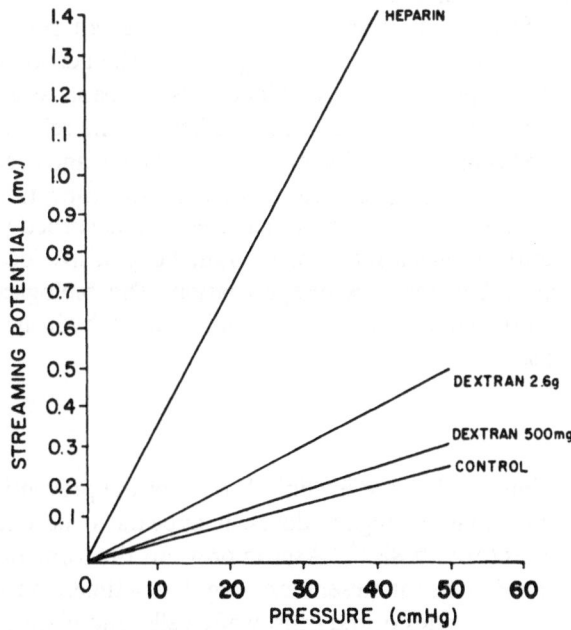

Figure 22. Effect of heparin and of dextran on *in vitro* streaming potential–pressure relations in canine carotid arteries. Control electrolyte—Krebs solutions.[38]

demonstrated by hydrolysis of the ester linkage as a result of desulfation of heparin with resultant loss of anticoagulant activity.[45] The highly potent anticoagulant activity of heparin can be given an electrochemical interpretation in terms of its high content of acid groups (sulfonic and carboxylic acid), high molecular weight, and strong adsorption on the surfaces of the blood vessel wall and of blood cells.

3.3.2. Protamine

Protamine is a strongly basic molecule and it decreases the streaming potentials in blood vessels[38] and the electrophoretic mobilities of blood cells.[44] Protamine has a low molecular weight and contains a large amount of arginine. It has a powerful antagonistic action on heparin. Protamine attacks the acidic groups in heparin, forming a stable salt with loss of anticoagulant activity.

3.3.3. Dextran

Dextran is also widely used in anticoagulant therapy. A low-molecular weight dextran (MW of 40,000) was also found to increase the slope of the streaming potential–pressure plot in canine blood vessels[38] (Figure 22). Dextran is a branched polysaccharide composed of glucose units, connected primarily by glucosidic linkages.

Brooks and Seaman have shown that in simple salt solutions dextran adsorption is related to polymer concentrations.[46] The adsorption results in an increase in the zeta potentials of red blood cells, probably as a consequence of the expansion of the double layer. Figure 23 shows the relative increase in zeta potential of red cells as a function of molecular weight and of concentration of dextran. Polymer bridging, the result of dextran adsorption, can cause aggregation but electrostatic repulsion increases with adsorbed polymer concentration. At a critical concentration of dextran, the repulsive forces overcome the attractive ones. The result is disaggregation. The biological activity of dextran as an antithrombotic agent is possibly due to hydrolysis producing free acidic groups.

3.3.4. Aspirin

Aspirin continues to be used in clinical trials for the prevention of thromboembolic complications. In higher doses, it appears to be effective in the treatment of venous thrombosis.[47] Aspirin prevents the formation of platelet aggregates. It enhances the rat mesenteric coagulation times and also increases the negative charge densities of red cells, white cells, and platelets. Clinically, it has been shown to prolong bleeding time. There is a progressive increase in the electrophoretic mobility of platelets to which increasing concentrations

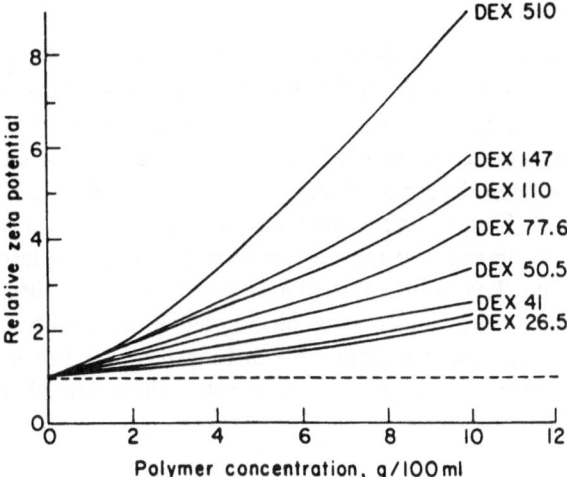

Figure 23. Relative zeta potentials as a function of polymer concentration for human erythrocytes suspended in saline containing different molecular weight dextrans.[46]

of aspirin are added (Figure 24).[48] It appears from this figure that the plateau starts at 10 mM aspirin. It appears that there is a direct interaction (acetylation) between this drug and platelets resulting in surface charge modification. There are some side effects caused by the clinical use of aspirin—surgical bleeding, gastrointestinal bleeding, aggravation of peptic ulcers. However, some of the advantages of the drug are that it can be administered orally and used in combination with other drugs (see Section 3.3.6).

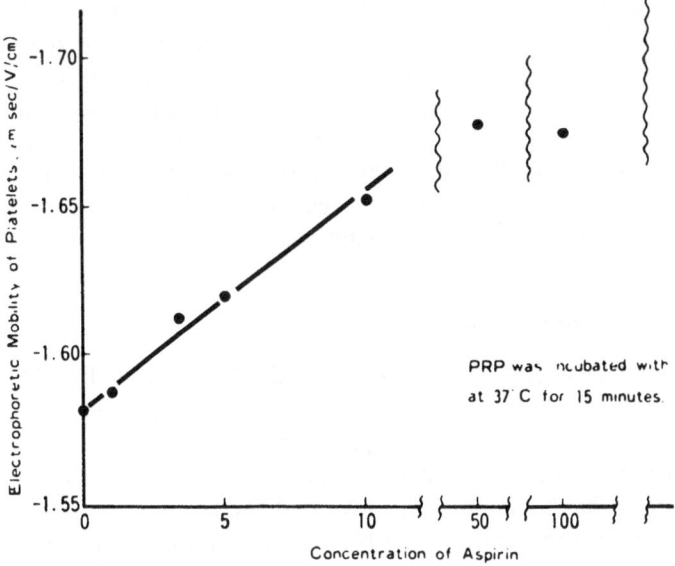

PRP was incubated with
at 37 C for 15 minutes.

Figure 24. Effect of aspirin on electrophoretic mobility of human platelets in plasma.[10]

3.3.5. Flavonoids

A mixture of flavonoids, derivatives of vitamin P, have been used for the treatment of venous disorders in Europe. The effects of a mixture of mono-, di- and trihydroxyethyl rutoside—catechin A and two of its soluble derivatives (Zyma)—on the surface charge characteristics of the blood vessel wall and of blood cells have been determined and correlated with their anthrombogenic characteristics.[49] Electrokinetic techniques for the former and current-induced occlusion times in rat mesenteric techniques for the latter were used in this study. The flavonoids, particularly in the pure condition, made the surfaces of blood vessel walls and of blood cells more negative. Flavonoids, like aspirin, are oral anticoagulants. From long-term studies it appears that the breakdown products of these compounds, which are phenolic, have a stronger anticoagulant action.

3.3.6. Oral Contraceptives

The relatively high incidence of thromboembolic phenomena in women on the "pill" prompted an examination on the effects of the estrogenic and/or progesteronic components in the "pill" on the surface charge characteristics of the vascular system.[50,51] Electrophoretic mobilities of erythrocytes and platelets and blood coagulation times were determined with samples of blood drawn from 33 women on the "pill" and 30 control subjects. The electrophoretic mobilities were reduced; thrombin recalcification times were not affected but thrombin times and partial thromboplastin times were shortened.

In vivo investigations in dogs showed more significant electrokinetic effects of the oral contraceptives on blood vessel walls than on blood cells. Further, the negative surface charge densities were reduced more for the veins than for the arteries. This result is an agreement with the finding that there is a higher incidence of venous rather than arterial thrombosis in women taking oral contraceptives. Further, the estrogenic component is more damaging than the progesteronic one. To reduce these side effects of the oral contraceptive, it may be worthwhile introducing a negatively charged group in the steroid ring or it may be necessary to add an anticoagulant (e.g., aspirin) to the "pill."

3.3.7. Noradrenaline and ADP

Aggregating agents noradrenaline and adenosine diphosphate (ADP) exhibit a biphasic effect on platelet electrophoretic mobility, i.e., increase at low concentration and decrease at higher concentration.[52-54] Figure 25 shows that the concentration for maximum electrophoretic mobility is 0.05 mg/l. Stoltz *et al.*[52] have qualitatively illustrated the regions of optimal stability, reversible aggregation, and irreversible aggregation by plotting the percentage change in electrophoretic mobility of platelets as a function of concentration

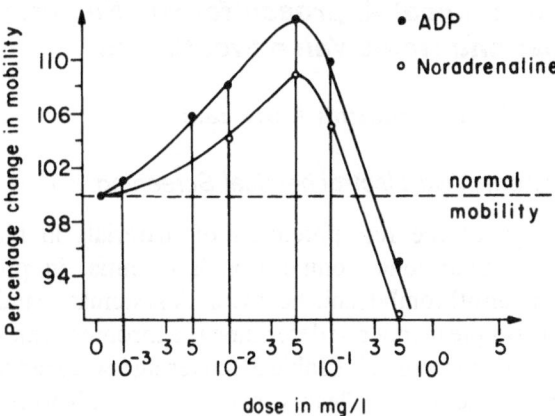

Figure 25. Percentage change in electrophoretic mobility of platelets from control values as a function of concentration of (i) ADP and (ii) noradrenaline.[52]

of aggregating agents (Figure 26). Confirmation of this model, by observation of the effect on rat mesenteric circulation, verifies the basic theory although the phase for rapid irreversible aggregation is encountered at a lower adrenaline concentration than would be expected.

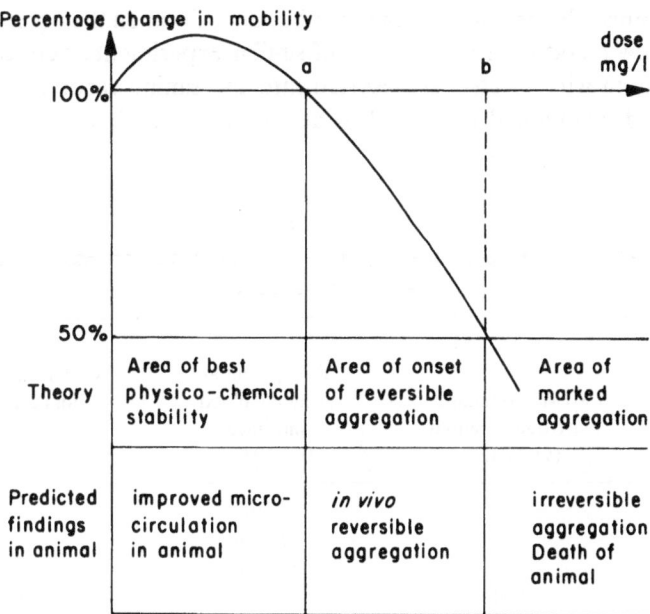

Figure 26. Qualitative representation of percentage change in electrophoretic mobility from control value as a function of concentration of chemical, showing regions of stability or aggregation of cells.[52]

4. An Electrochemical Approach for the Selection of Vascular and Heart Valve Prostheses

4.1. Electronically Conducting Materials

4.1.1. In Vitro and In Vivo Electrochemical Screening

A knowledge of the rest potentials of materials in blood or other electrolytes with similar ionic composition is essential in predicting their thrombogenic and antithrombogenic behavior. A systematic study was carried out in a relatively simple manner to determine the thrombogenic or antithrombogenic properties of some metals (Table 3), covering a wide range of potentials in the electromotive series.[55] Clean wires of the metals were inserted into canine carotid and femoral arteries through side branches and left in position for 30–40 minutes. The potentials of these electrodes in contact with flowing blood were measured with respect to the standard calomel electrode. At termination, the vessels were clamped above and below the electrodes. After injection of formalin into the vessels for fixation of the thrombus deposits, the dogs were sacrificed. The vessels were slit open and the electrodes examined. The results proved to be quite interesting (Table 3). Electrodes of metals establishing negative potentials in blood vs. the normal hydrogen electrode (NHE) were antithrombogenic while those exhibiting positive potentials were thrombogenic.

To confirm the dependence of thrombus deposition on the potential across an electrode–blood interface, a series of similar experiments were carried out with platinum wire electrodes, inserted into the canine carotid and femoral arteries and maintained at several values of potentials between -1 and $+1$

Table 3

Dependence of Thrombus Deposition at Metal Electrodes on Position of Metal in Electromotive Series

Metal	M/M^{n+} standard electrode potential (vs. NHE)	Rest potential at metal–blood interface (vs. NHE)	Occurrence ($\sqrt{}$) or nonoccurrence ($\times$) of thrombus deposition
Mg	-2.375	-1.360	$\times$
Al	-1.670	-0.750	$\times$
Cd	-0.402	-0.050	$\times$
Cu	$+0.346$	$+0.025$	$\sqrt{}$
Ni	-0.230	$+0.029$	$\sqrt{}$
Au	$+1.420$	$+0.120$	$\sqrt{}$
Pt	$+1.200$	$+0.125$	$\sqrt{}$

volts/NHE.[56] The remaining procedure was the same as in the experiments with the metal wires implanted at their rest potentials. These experiments clearly demonstrated that there was an increasing degree of thrombus deposition with increase of anodic potentials above +200 mV/NHE. The extent of thrombus deposition below this potential was quite small.

Prosthetic materials, implanted in the body, can interfere with the platelet function. Platelets tend to adhere to the surfaces of foreign materials and of injured blood vessels. A platelet plug is subsequently formed by aggregation of more platelets at the site of adhesion. Adenosine diphosphate (ADP) is known to induce platelet aggregation.[57] Platelets are also the source of ADP. When platelets aggregate, the intraplatelet constituents are released to the surrounding media. The influence of some metals on the platelet release reaction, *in vitro*, has been investigated.[58] For this purpose, washed platelets were treated with some metal powders and the extent of release of adenine nucleotides determined. The results showed that the release reaction is inhibited by aluminum, while noble metals including platinum and gold cause an increased release of adenosine mono-, di-, and triphosphates and glucose-6-phosphate (Figure 27). These preliminary results on the platelet release reaction indicate that metals which are potentially thrombogenic release more nucleotides from platelets. This process is likely to cause platelet aggregation *in vivo* with possible activation of the extrinsic pathway of blood coagulation.

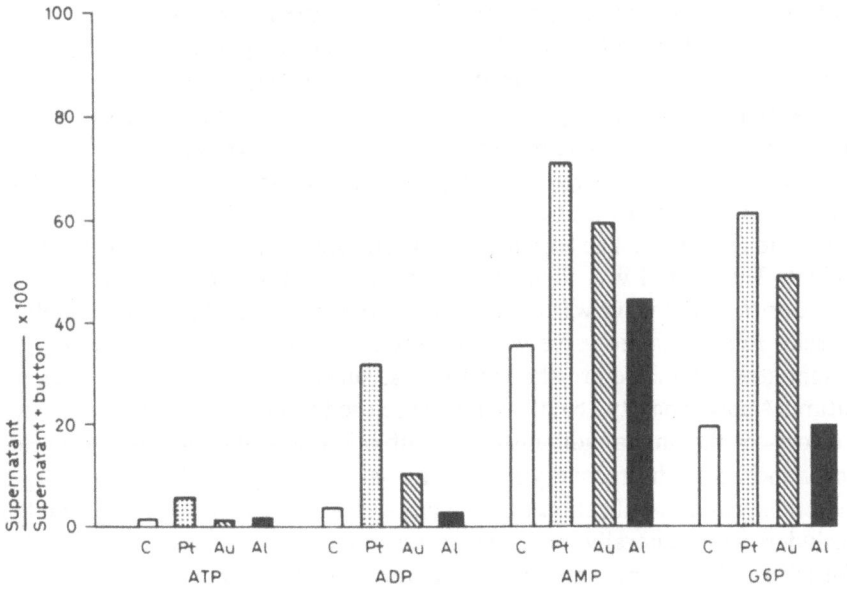

Figure 27. Percentage of adenine nucleotides and glucose-6-phosphate (G6P) in supernatant after centrifugation of platelet suspensions exposed to metal powders.[58]

4.1.2. In Vitro and In Vivo Evaluation of Vascular and Heart Valve Implants

The most convenient method for testing vascular implants is by use of materials in tubular form. Several studies have been conducted to evaluate the pro- or antithrombogenic properties of metallic materials, used in tubular form to replace segments of the descending thoracic aorta (DTA) or thoracic inferior vena cava (TIVC).[59] In a typical experiment, a dog with an approximate body weight of 20–25 kg was anesthesized with pentobarbitol (0.5 cm^3/kg body weight). A right posterio-lateral thoractomy was performed. The animal was intubated and positive pressure ventilated. In the case of the TIVC implant, the phrenic nerve was dissected out and the TIVC clamped proximally and distally. The vena cava was incised, the prosthetic tube slipped into position and tied in place bridging the gap between the distal and proximal ends of the TIVC. The chest was then closed. The entire surgical procedure was carried out in 30 minutes. Implantation in the thoracic aorta involved the opening of the left chest and substitution of a metal tube for a 2-cm segment of the aorta. Two sets of intercostal arteries were ligated to prepare the DTA for the prosthetic implant. In one series of experiments, the tubes were allowed to remain in position until death occurred. In Figure 28, a plot is made of the duration of patency of the tubes as a function of the standard electrode potential. As observed in the *in vitro* and *in vivo* electrochemical screening experiments (Section 4.1.1), there is an increasing antithrombogenic tendency with decrease in the standard electrode potential. The unfortunate aspect is that the most antithrombogenic metals are the corrodible ones. Use of alloys of the corrodible metals, instead of the pure metals, appears to be the better approach for the selection of antithrombogenic metals.

A calf with an average body weight of 75–80 kg has been the animal commonly used to evaluate materials used as heart valve cages.[60,61] Prior to anesthesia, the calf received a 200 mg dose of solumedrol intravenously. The calf was anesthesized with nembutol (0–5 cm^3/kg) and placed on a pentathol-fluothane respirator. The animal was heparinized at the level of 3 mg/kg body weight. The animal was then placed on total cardiopulmonary bypass. The aortic or tricuspid valve was then excised and its corda tendineae and papillary muscles removed. The valve was implanted through a right thoractomy and a right atriotomy. The prosthetic valve was sutured in place using 3–0 dacron suture. Approximately 18–20 sutures were used to insert the valve. The chest was closed, the animal administered antibiotic (penicillin and streptomycin), and removed to the recovery room. The valve design used for the testing of materials was a modification of the Starr–Edwards ball valve.[62] Silastic ball occluders were generally used. The aortic annulus is a high-flow region unlike the tricuspid or mitral annulus. The latter regions are more suitable for the rapid evaluation of blood compatibility. The results in Table 4 again show that there is a good correlation between the thrombogenic characteristics of conducting materials and their potentials in blood.

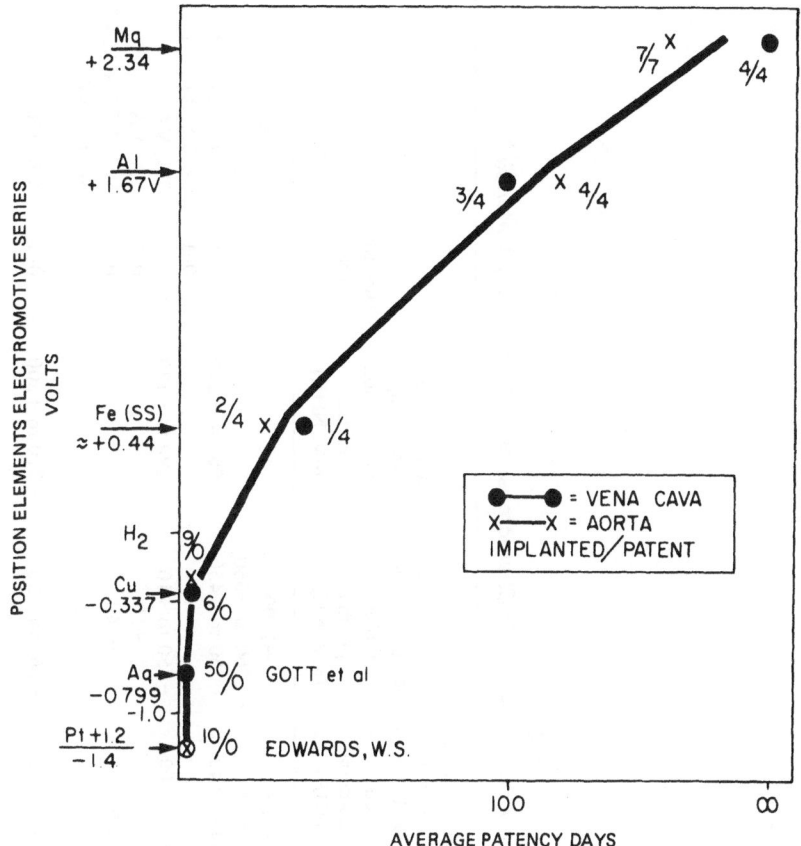

Figure 28. Duration of patency of metal tubes implanted in canine thoracic aorta or vena cava as a function of the standard electrode potential of the metal.[59]

The purity of the surface of a valve prior to implantation was found to be an essential criterion in determining its thrombogenic or antithrombogenic characteristics. Thus, in several instances, contaminated (possibly organic) aluminum valves which were implanted were coated with a thrombus deposit. However, if the aluminum valves were first cleaned in concentrated hydrochloric acid, washed with distilled water and implanted, their surfaces were antithrombogenic. An electropolished aluminum valve, implanted in a calf for 82 days, was found to be strikingly clean (Figure 29), while valves of the same material implanted without pretreatment occluded readily (Figure 30).

4.1.3. Antithrombogenic Characteristics of Cathodically Polarized Vascular Prostheses

Blood-compatible metals and alloys are essential for certain types of medical devices. The experiments, described in the previous section, demon-

Table 4

Correlations between Electrochemical and Antithrombogenic Characteristics of Conducting Materials[a]

Metal	Alloy	Corrosion electrode potential (mV, NHE)	Potential following surface reduction in 10 N-HCl—reduced surface potential (mV, NHE)	Rest potential in blood plasma (mV, NHE)	Potential change following implantation in calf (mV, NHE)	Thrombosis index[b] (0–5)	Test position[c]
Magnesium	Mg	−2370	−1610	−1500			W, TA, TV
Magnesium alloys	Mg, Al, Zn		−1500	−1360 to −660			VA
Aluminum	Al	−1850	−760 to −540	−750 to −450	−560 to −290	0–1	W, TA, TV, VA
Aluminum alloy	Al, Cu					0–1	W, TA, TV, VT
Titanium	Ti	−1630	−350 to −200	−250 to +10	−190 to −130		TA, VA
Cadmium	Cd	−400		+50		0–1	TA, VA
Cobalt	Co	−277	−240	−240		0–2	WA, WV, VA, VT
Nickel	Ni	−250	−300			4	W, TA, VA
Stellite 21, Sawyer–Arwood	Co, Fe, W, Cr		−285 to −160	−260 to −30	−210 to +200	0–1	VA, VT
Stellite 21, Starr–Edwards	Co, Fe, W, Cr		−285 to −160	−60 to +300	−20 to +130	0–1	VA, VT

Material	Composition		Not pretreated			Degree[b]	Location[c]
Starr–Edwards, Stellite 21 covered with Teflon fabric	Co, Fe, W, Cr			−180 to +100	−470 to +110	0–1	VA, VT
Haynes 25	Co, W, Cr		−180 to −80	−40 to +360	−100 to +190	1–3	TV, VA, VT
Inconel	Ni, Cr, Mn, Si		−140 to −120	−30 to +70	−10 to +220	3	TV, VA, VT
Hastalloy B	Ni, Cr, Co, Mo, Mn, Si, Fe		−180 to −160	−110 to +185	−185 to −140	4	TV, VA, VT
Stainless 304	Cr, Ni, Mn, Si, Fe		−160	+110 to +180	+30 to +440	3	VA, VT
Stainless 309	Cr, Ni, Mn, Si, W, Fe		−260	+90 to +110	+10 to +230	3–5	W, TA, TV, VA
N 155	Ni, Co		−310 to −130	+70 to +120	+30 to +200	5	W, TA, TV, VA
Copper	Cu	+350	−60 to +40	+90 to +260	−40 to +70	3	W, TA, TV, VA
Silver	Ag	+800	+280			2	W, TA, TV, VA
Platinum	Pt	+860				5	W, TA, TV, VA
Gold	Au	+1500	+330				

[a] Reference 59.
[b] 0, antithrombogenic; 1–5 increasing degree of thrombosis.
[c] TV: tube—vein; W: wire; VA: valve—aortic; VT: valve—tricuspid; TA: tube—aorta.

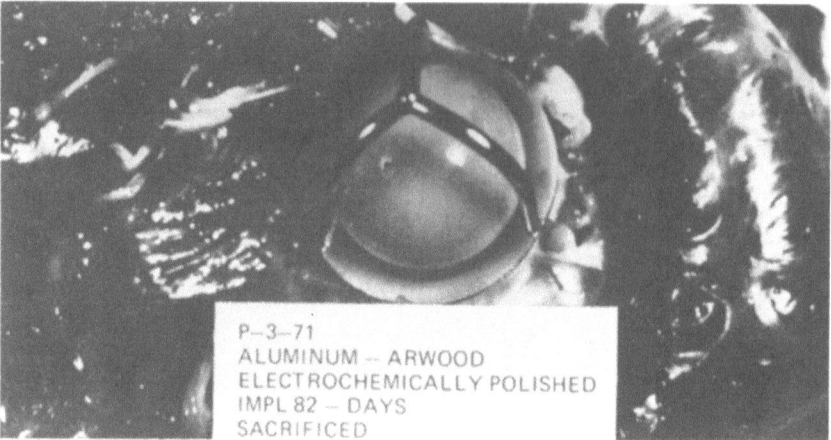

Figure 29. Photograph demonstrates the antithrombogenic character of an electropolished aluminum valve after being implanted in the tricuspid annulus of a calf for 82 days.[9]

strate that one of the criteria for the selection of blood-compatible conducting materials is that the potential across the material/blood interface must be negative with respect to the potential of NHE. One way of achieving negative potentials with the less electropositive (i.e., more noble) metals is by cathodic polarization. Experiments have thus been carried out to test the antithrombogenic characteristics of cathodically polarized vascular prostheses.[63] One

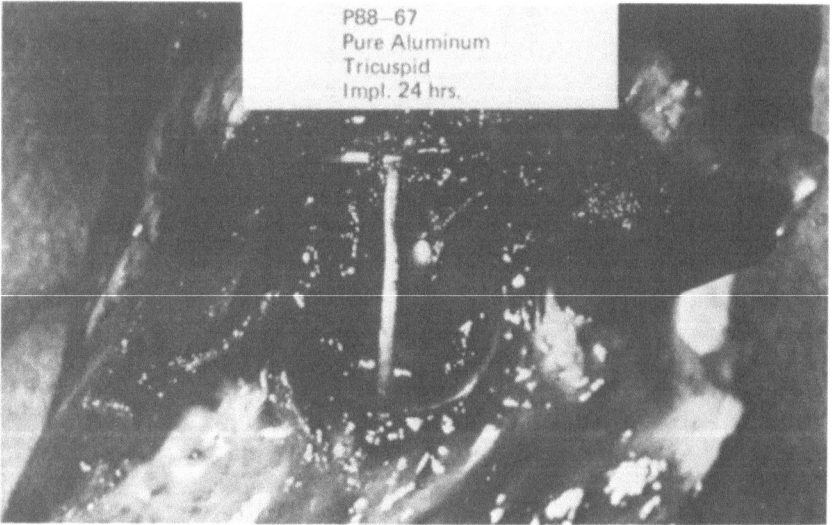

Figure 30. Photograph demonstrates the thrombogenic character of an "as received" aluminum valve after being implanted in the tricuspid annulus of a calf for only 24 h.[9]

of the best sites for such testing is the canine TIVC. A copper prosthesis (a tube 5-cm long, 1-cm internal diameter, and 1-mm wall thickness) was implanted in the canine TIVC. It was maintained at a cathodic potential using a suitable polarizing circuit (Figure 31). Three experiments were conducted in which the potentials of the copper tubes ranged from -160 to -60 mV/NHE. The tubes were left in position for 6, 8, and 14 days. All tubes were free of thrombosis deposits. In two control experiments with no polarizing circuit, the copper tubes registered positive potentials and were found to be occluded in 8 and 14 days. These experiments confirm that the potential across the metal/blood interface is a more basic parameter than the chemical nature of the surface in determining its blood compatibility characteristics.

4.2. Insulator Materials

4.2.1. In Vitro Electrokinetic Screening

The surface charge characteristics of insulator materials, potentially useful for vascular and heart valve prostheses, have been mostly determined using

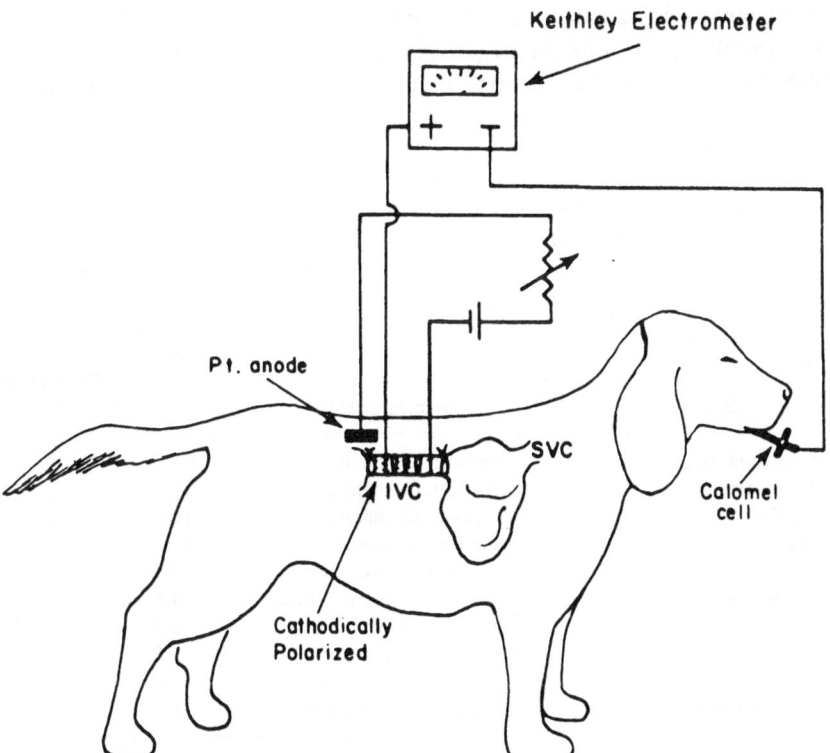

Figure 31. Experimental arrangement for maintaining a copper prosthesis, implanted in the canine thoracic inferior vena cava, at negative potentials.[63]

streaming-potential and, to a lesser extent, electroosmosis techniques in physiologic fluids (0.154 N NaCl or Krebs solution). Streaming-potential measurements have also been carried out *in vivo* on protheses fabricated using insulator materials and implanted in canine carotid and femoral arteries. The experimental technique used for this purpose is akin to that described in Section 3.1.

Several types of insulator materials—Teflon, polyethylene, polystyrene, silicon, dacron, ioplexes, ethylene–vinyl acetate and ethylene acrylic acid copolymers, glass, etc.—have been examined for their electrokinetic characteristics. In some of these cases, the surface charge characteristics have been modified by prior electrical or chemical treatment. The correlations which have been drawn between the electrokinetic and anti- or prothrombogenic characteristics of these materials are presented in the next section.

4.2.2. In Vitro and In Vivo Evaluation of Vascular and Heart Valve Implants

Implantation experiments with vascular and heart valve prostheses fabricated with insulator materials were similar to those with the conducting materials (Section 4.1.2). In general, the positively charged materials (i.e., charge characteristics indicated from the electrokinetic screening experiments) were thrombogenic while the negatively charged materials tended to be nonthrombogenic (Table 5). The reason that not all of the negatively charged

Table 5

Correlations between Electrochemical and Antithrombogenic
Characteristics of Some Insulator Materials[a]

Material	Source of material	Zeta potential in Krebs/1000[b] mV	Thrombosis index[c]
1. Untreated Teflon	American Cyanamid	−9.0	4
2. Teflon electret	American Cyanamid	−11.0	1
3. Carboxylated Teflon	American Cyanamid	−13.0	2
4. Sulfonated Teflon	American Cyanamid	−32.0	0
5. Untreated silicone rubber	Battelle Memorial Institute	−3.2	1
6. Aminated silicone rubber	Battelle Memorial Institute	+4.5	5
7. Heparinized silicone rubber	Battelle Memorial Institute	−6.3	0
8. Untreated epoxy	Monsanto Corporation	−5.3	2
9. Detergent-treated epoxy	Monsanto Corporation	−4.8	5
10. Heparinized epoxy	Monsanto Corporation	−5.8	1

[a] Reference 64.
[b] 1 ml of Krebs solution diluted to 1000 ml.
[c] Antithrombogenic; 1–5, increasing degree of thrombosis.

materials are antithrombogenic is probably the heterogeneity of the surface charge. A clear example is glass, which is highly negatively charged and is quite thrombogenic. However, glow discharge cleaning of glass[65] makes the material blood compatible. Untreated glass tubes, implanted in the canine TIVC, were completely occluded in a short time. Similarly implanted glow discharge-cleaned tubes remained patent for more than one year (Figure 32). The introduction of sulfonate or carboxylate groups on the surface of Teflon markedly increases the magnitude of the zeta potential (since Teflon is negatively charged, these groups make Teflon even more negative) and also improve the blood compatibility. Hydroxyl groups have a similar effect. Teflon electrets with a negative surface charge density tend to be antithrombogenic. Electrification does not increase the negative charge density but results in a more uniform distribution of charges.

Fluorosilicone, dimethylsilicone homopolymers, as well as their copolymers, are negatively charged and inhibit the blood clotting reaction. There are a number of polyelectrolytes which have negative zeta potentials and are antithrombogenic. The most promising is the ethylene–acrylic acid copolymer, neutralized to the extent of 60% with sodium ions. Another useful and related material is the vinyl acetate–crotonic acid copolymer.

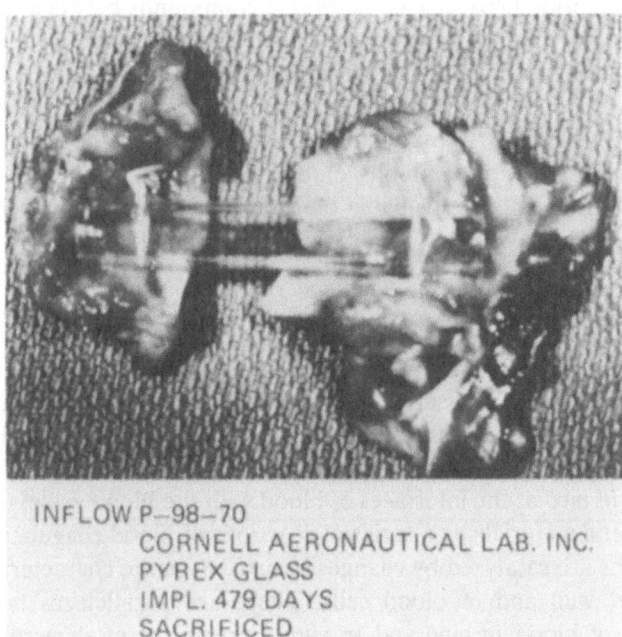

INFLOW P–98–70
 CORNELL AERONAUTICAL LAB. INC.
 PYREX GLASS
 IMPL. 479 DAYS
 SACRIFICED

Figure 32. Photograph demonstrates the antithrombogenic character of a glow discharge-cleaned glass tube after being implanted in the canine thoracic inferior vena cava for 479 days.[65]

The evaluation of insulator materials as heart valve prostheses has not been as extensive as with conducting materials.[64] The Starr–Edwards ball and cage design heart valve has been generally used for the *in vivo* evaluation of materials for fabrication of such valves. This type of valve is composed of three components: a cage, most often made with an alloy such as cobalt–chromium (Stellite 21), a silicon or Teflon ball, and a sewn cuff of woven Teflon fabric. This design has been used for the testing of numerous other metallic and polymeric materials.

4.2.3. Antithrombogenic Characteristics of Heparinized Vascular Prostheses

As stated in Section 3.3, heparin is one of the most potent anticoagulants. Several attempts have been made to bond heparin to plastic surfaces. In one type of study, heparin was attached ionically to a plastic via a bridge, e.g., a polymerizable quarternary ammonium salt such as the methyl iodide salt of dimethylaminoethyl methacrylate. Many of these surfaces after heparinization, initially antithrombogenic, lost their activity as a function of time due to elution of heparin in the electrolyte "blood." This was also found to be the case with ionically bonded heparin surfaces, prepared with hydroxy-3-methacryloyloxypropyltrimethyl ammonium chloride (GMAC) or with tridodecylmethyl ammonium chloride (TDMAC) as the bridge compounds between heparin and the substrate polymer.[66,67] Attempts have also been made to bond heparin covalently to surfaces. Though heparin in such surfaces is more stable than when ionically bonded, the former type had less heparin activity (in respect to antithrombogenicity). Cross-linking of ionically bonded heparin to surfaces, using gluteraldehyde, also inhibits elution of heparin from surfaces[68] but there is no long-term data available on their antithrombogenic behavior.

5. Conclusions

Cardiovascular disease is the number one killer in the U.S. It is due to the obstruction of blood vessels at the site where thrombic deposit has occurred or due to the disintegrated thrombi breaking loose and travelling through the blood stream and causing narrowing of blood vessels elsewhere. Thrombosis is triggered *in vivo* at the interfaces of blood with the blood vessel wall and/or with prosthetic materials by a series of reactions of blood coagulation factors. The reactions are catalyzed by changes in surface charge characteristics of the blood vessel wall and of blood cells. There are parallelisms between the coagulation of blood *in vivo* and *in vitro*. Foreign materials accelerate both processes. *In vitro* blood clots consist of fibrin networks in which are red cells, white cells, and a relatively small number of platelets. Conversely, *in vivo* thrombus deposits consist of large amorphous masses of platelets, surrounded by white cells with a few red cells. Blood coagulation factors takes part in adsorption and charge transfer reactions at the blood vessel wall, as well as at prosthetic material/blood interfaces. The present chapter describes the hematological and cardiovascular relevance of such reactions (Section 2.3). Such types of electrochemical reactions can especially be catalyzed when metallic implants are present in the cardiovascular system.

Evidence for an electrochemical mechanism of thrombosis is summarized in Section 1.2. Parallelisms are drawn between the hematological and electrochemical approaches in the investigations of blood coagulation. The electrochemical approach involves the use of electrode kinetic and/or electrokinetic techniques in (i) studies of reactions of the blood coagulation factors; (ii) investigations of the surface charge characteristics of the blood vessel wall and of blood cells; (iii) examination of the interactions of anti- or procoagulant drugs with components of the vascular system; and (iv) the selection and evaluation of vascular and heart valve prostheses. The complex nature of the blood coagulation process has mystified scientific and clinical investigators for over one hundred and fifty years. The hematological and electrochemical approaches are essential for shedding further light on the intricate bioelectrochemical reactions involved in cardiovascular reactions, and in the selection of anticoagulant drugs and nonthrombogenic prosthetic materials.

Acknowledgments

Most grateful appreciation is expressed to Professor Philip N. Sawyer of the State University of New York, Downstate Medical Center, Brooklyn, New York, for introducing the author to this vitally important bioelectrochemical field. The collaboration with Professor Sawyer and our other associates at Downstate Medical Center (please see references) was most illuminating and productive.

References

1. J. F. Mustard, in *Thromboembolism—A New Approach to Therapy*, J. R. A. Mitchell and J. G. Domenet, eds., Academic Press, New York (1977), Chapter 1.
2. R. Biggs and R. G. Macfarlane, *Human Blood Coagulation*, F. A. David Company, Philadelphia (1962).
3. L. Vroman, *Blood*, The Natural History Press, Garden City, New York (1967).
4. P. N. Sawyer, ed., *Biophysical Mechanisms in Vascular Homeostasis and Intravascular Thrombosis*, Appleton-Century Crofts, New York (1965).
5. S. Srinivasan and P. N. Sawyer, *J. Colloid Interface Sci.* **32**, 456 (1970).
6. S. Srinivasan, L. Duic, N. Ramasamy, P. N. Sawyer, and G. E. Stoner, *Ber. Bunsenges. Phys. Chem.* **77**, 798 (1973).
7. S. Srinivasan and B. R. Weiss, in *Colloidal Dispersions and Micellar Behavior*, K. L. Mittal, ed., ACS Symposium Series 9, Washington, D.C. (1975), Chapter 24.

8. S. Srinivasan, G. L. Cahen, and G. E. Stoner, in *Electrochemistry the Past Thirty Years and the Next Thirty Years, A Volume in Honor of J. O'M. Bockris*, H. Bloom and F. Gutmann, eds., Plenum Press, New York (1977), Chapter 4.

9. S. Srinivasan, N. Ramasamy, B. Stanczewski, and P. N. Sawyer, *Advances in Cardiovascular Physics, Vol. 3*, D. N. Ghista, ed., S. Karger, Basel, Switzerland (1979), p. 133.

10. S. M. Jung, K. Kinoshita, K. Tanou, I. Isohisa, and H. Yazamaki, *Thrombosis and Hemostasis* **47**, 203 (1982).

11. R. G. Macfarlane, *Nature* **202**, 498 (1964).

12. E. W. Davie and O. D. Ratnoff, *Science* **145**, 1310 (1964).

13. A. C. Hemker and M. J. P. Kahn, *Nature* **215**, 1201 (1967).

14. E. Genton and P. Steele, Reference 1, p. 107.

15. J. C. F. Poole, *Quart. J. Exp. Physiol.* **44**, 377 (1959).

16. M. B. Zucker, in *Hematology*, J. Williams, E. Beutler, A. J. Erslev, and R. W. Rundles, eds., McGraw-Hill, New York (1972), pp. 1014–1022.

17. J. Feijen, T. Beugeling, A. Bantjes, and C. Th. Smit Sibinga, Reference 1, p. 100.

18. E. W. Salzman, E. W. Merrill, A. Binder, C. F. W. Wolf, T. P. Ashford, and W. G. Austen, *J. Biomedical Mater. Res.* **3**, 69 (1969).

19. M. B. Zucker and L. Vroman, *Proc. Soc. Exp. Biol. Med.* **131**, 318 (1969).

20. S. W. Kim, R. G. Lee, H. Oster, D. J. Lentz, D. L. Coleman, J. D. Andrade, and D. Olsen, *Trans. Am. Soc. Artificial Internal Organs* **20**, 449 (1974).

21. G. E. Stoner, *J. Biomedical Mater. Res.* **3**, 655, 1969.

22. D. B. Matthews, *J. Biomedical Mater. Res.* **3**, 475 (1969).

23. G. E. Stoner and S. Srinivasan, *J. Phys. Chem.* **74**, 1088 (1970).

24. N. Ramasamy, M. Ranganathan, L. Duic, S. Srinivasan, and P. N. Sawyer, *J. Electrochem. Soc.* **120**, 354 (1973).

25. G. E. Stoner and L. Walker, *J. Biomedical Mater. Res.* **3**, 645 (1969).

26. N. Ramasamy, J. S. Keates, S. Srinivasan, and P. N. Sawyer, *Bioelectrochem. Bioenerg.* **1**, 244 (1976).

27. B. W. Morrisey, L. E. Smith, R. R. Stromberg, and C. A. Fensternmaker, *J. Colloid Interface Sci.* **56**, 557 (1976).

28. L. Duic, S. Srinivasan, and P. N. Sawyer, *J. Electrochen. Soc.* **120**, 348 (1973).

29. N. Ramasamy, S. Srinivasan, and P. N. Sawyer, *Electrochim. Acta* **19**, 137 (1974).

30. M. P. Esnouf and F. Jobin, *Biochem. J.* **102**, 660 (1967).

31. E. F. Mamman, *Thrombosis and Bleeding Disorders*, N. B. Bang, ed., Academic Press, New York (1971), Chapter 4.

32. L. Vroman and A. L. Adams, *J. Biomedical Mater. Res.* **3**, 669 (1971).

33. L. Vroman, *Fed. Proc.* **30**, 5 (1971).

34. G. E. Stoner, S. Srinivasan, and E. Gileadi, *J. Phys. Chem.* **75**, 2107 (1971).

35. L. Vroman, A. L. Adams, M. Klings, and G. C. Fischer, *Ann. N.Y. Acad. Sci.* **283**, 65 (1977).

36. B. W. Morrissey and R. R. Stromberg, *J. Colloid Interface Sci.* **46**, 152 (1974).

37. S. Srinivasan, C. B. Burrowes, T. R. Lucas, S. B. Bauer, and P. N. Sawyer, *J. Biomedical Mater. Res.* **1**, 355 (1967).

38. S. Srinivasan, R. Aaron, P. S. Chopra, T. Lucas, and P. N. Sawyer, *Surgery* **64**, 827 (1968).

39. P. N. Sawyer, S. Seto, and S. Srinivasan, *Surgery* **63**, 822 (1968).

40. H. A. Abramson, L. S. Moyer, and M. H. Gorin, *Electrophoresis of Proteins and the Chemistry of Cell Surfaces*, Hafner Publishing Company, Inc., New York (1965).

41. G. V. F. Seaman, in *Cell Electrophoresis*, E. J. Ambrose, ed., Little Brown and Company, Boston (1965).

42. J. L. Cohen, S. Srinivasan, and P. N. Sawyer, *Arch. Biochem. Biophys.* **124**, 556 (1968).

43. G. J. Durant, H. R. Hendrickson, and R. Montgomery, *Arch. Biochem.* **99**, 418 (1962).

44. J. F. Stoltz, M. Stoltz, P. Alexander, F. Struff, and A. Larcan, *Bibl. Anatomica* **10**, 343 (1969).

45. C. Eika, *Thrombosis Research* **2**, 349 (1973).

46. D. E. Brooks and G. V. F. Seaman, *Bibl. Anatomica* **11**, 251 (1972).

47. R. McKenna, F. Bachmann, J. Galante, S. P. Kanshal, and P. Meredith, *Blood*, **48**, 977 (1976).
48. H. Yamazaki, R. Tsukui, T. Motomiya, S. M. Jung, M. Sonoda, C. Watanabe, M. Ogino, and N. Miyagawa, *Thrombosis Research* **22** (1981).
49. S. Srinivasan, T. Lucas, C. B. Burrowes, N. A. Wanderman, A. Redner, Bernstein, and P. N. Sawyer, VI Conference on Microcirculation, European Society for Microcirculation, Aalborg, Denmark (1970), p. 40.
50. S. Srinivasan, J. Solash *et al.*, *Contraception* **9**, 291 (1974).
51. A. Schwann *et al.*, *Microcirculation Vol. 1*, J. Grayson and W. Zingg, eds., Plenum Publishing Corporation, New York (1976), p. 218.
52. J. F. Stoltz, M. Stoltz, and A. Larcan, *Bibl. Anatomica* **10**, 474, Karger, Base, Switzerland (1969).
53. J. R. Hampton and R. A. Mitchell, *British Medical Journal* **1**, 1074 (1966).
54. I. Hawkins, *Nature, New Biol.* **233**, 92 (1971).
55. P. S. Chopra, S. Srinivasan, T. Lucas, and P. N. Sawyer, *Nature* **215**, 1494 (1967).
56. P. N. Sawyer and S. Srinivasan, *Am. J. Surgery* **114**, 42 (1967).
57. A. J. Marcus and M. B. Zucker, *The Physiology of Blood Platelets*, Grune and Stratton, New York (1965).
58. R. V. Zivkovic, S. Srinivasan, and P. N. Sawyer, *Abstracts of the International Society of Thrombosis and Hemostasis, IIIrd Congress*, Washington (1972), p. 266.
59. P. N. Sawyer and S. Srinivasan, in *Medical Engineering*, C. D. Ray, ed., Year Book Medical Publishers, Inc., Chicago, Illinois (1973), Chapter 82.
60. P. N. Sawyer, S. Srinivasan, S. A. Wesolowski, S. A. Berger, K. E. Campbell, A. A. Samma, S. J. Wood, and L. R. Sauvage, *Trans. Am. Soc. Artificial Internal Organs* **13**, 124 (1967).
61. P. N. Sawyer, S. Srinivasan, M. E. Lee, J. G. Martin, T. Murakami, and B. Stanczewski, in *Prosthetic Heart Valves*, L. A. Brewer, ed., Charles C. Thomas Publisher, Springfield, Illinois (1969), Chapter 13.
62. A. Starr, M. L. Edwards *et al.*, in *Prosthetic Heart Valves*, L. A. Brewer, ed., Charles C. Thomas Publisher, Springfield, Illinois (1969), Chapter 7.
63. E. Gileadi, B. Stanczewski, A. Parmeggiani, T. R. Lucas, M. Ranganathan, S. Srinivasan, and P. N. Sawyer, *J. Biomedical Medical Res.* **6**, 489 (1972).
64. R. K. Aaron, S. Srinivasan, and P. N. Sawyer, Reference 59, Chapter 83.
65. R. E. Baier, V. A. DePalma, A. Furuse, V. L. Gott, G. W. Kamloff, T. Lucas, P. N. Sawyer, S. Srinivasan, and B. Stanczewski, *J. Biomedical Mater. Res.* **9**, 547 (1975).
66. G. A. Grode, R. D. Falb, and S. J. Anderson, in *Artificial Heart Program Conference—Proceedings June 9-13, 1969, Washington, D.C.*, R. J. Hegyoli, ed., US Government Printing Office, Washington, D.C. (1969), Chapter 11.
67. R. I. Leininger, R. D. Falb, and G. A. Grode, *Materials in Biomedical Engineering*, S. N. Levine, ed., *Ann. N.Y. Acad. Sci.* **146**, 11 (1969).
68. H. R. Lagagren and J. C. Eriksson, *Trans. Am. Soc. Artificial Internal Organs* **17**, 10 (1971).

9

Electrochemical Techniques in the Biological Sciences

**EUGENE FINDL, ELAINE R. STROPE, and
JAMES C. CONTI**

Although the initial application of electricity to biological materials goes back to the first experiments of Luigi Galvani (*circa* 1786–1791), it is only recently that the coupling of electrochemistry and biology has resulted in a synergistic field of science, i.e., bioelectrochemistry. The field is growing rapidly and is beginning to have a significant impact on the practice of medicine, biology, pollution control, energy conversion, food technology, and other areas. Among the facets of bioelectrochemistry that are being investigated are mechanisms of electron transfer by enzymes and other macromolecules, electrochemically stimulated tissue and bone growth, electrical homeostasis, uses of biologically active molecules and organisms in combination with electrodes, and many more. Since there is such a diversity of topics that could be put under the aegis of bioelectrochemistry, it would be well nigh impossible to cover the field in depth in a single chapter. Therefore, emphasis will be placed on three aspects, electroanalytical techniques, electrophysiology, and electrobiology.

EUGENE FINDL, ELAINE R. STROPE, and JAMES C. CONTI • Bioresearch Inc., 315 Smith Street, Farmingdale, New York. Eugene Findl's current address is Brookhaven National Laboratory, Bldg. 801, Upton, New York 11973. Elaine R. Strope's current address is Personal Products Co. (Div. J&J), Van Liew Ave., Milltown, New Jersey 08850. James C. Conti's current address is Ethicon Inc. (Div. J&J), Rt. 22, Somerville, New Jersey 08876.

1. Electroanalytical Techniques

1.1. Bioelectrodes

There are two basic types of electrodes used as detectors of electrochemical potentials and/or currents in the biological sciences, i.e., selective and nonselective electrodes. Selective electrodes are those electrodes that exhibit a high degree of specificity to the chemical activity of one type of chemically reactive species in an electrochemical system. Conversely, nonselective electrodes are nonspecific and indicate the electrochemical activity of many different types of chemical constituents of an electrochemical system. The most common examples of these two types are the pH electrode (high H^{+5} selectivity) and the platinum metal (general purpose) electrode.

Electrodes are used as sensors in either a potentiometric mode or an amperometric mode. As the names imply, potentiometric electrodes measure electrochemical activity by relating it to a potential (voltage). Amperometric electrodes measure electrochemical activity by relating it to a quantity of current (amperes). Both modes have found wide application. Ion-selective electrodes generally operate in the potentiometric mode. Amperometric sensors, conversely, generally use nonselective electrodes which can be made selective by electrochemical and nonelectrochemical modification. Potentiometric electrodes operate via a number of presently ill-defined mechanisms. However, regardless of the mechanism, the measured potential is due to an interfacial chemical equilibrium that does not involve a bulk transfer of material. Amperometric electrodes, on the other hand, do involve the bulk transfer of material.

1.1.1. Microelectrodes

Microelectrodes, as the name implies, are miniaturized versions of conventional electrodes. The term "micro" usually denotes electrodes whose tip diameters are less than $10\ \mu$ while "ultramicro" refers to dip diameters less than $1\ \mu$. There are four basic types of microelectrodes commonly used in bioelectrochemistry: glass, ion-selective, metal, and carbon. This section deals with their fabrication and characteristics.

Glass microelectrodes are made from 1–2 mm O.D. conventional pyrex tubing. The tubing is pulled into micropipets either by hand over a gas burner or on a commercial pipet puller. This is a machine which heats up a coiled element positioned around the glass capillary. As the glass softens, the capillary ends are rapidly pulled apart using a controlled force, resulting in 2 nearly identical micropipets. The micropipet is then filled with a conducting solution, typically 0.5 to 3 M potassium chloride or acetate. Until recently this was not an easy step because bubbles commonly became trapped in the very fine tip. Workers had to resort to laborious methods such as boiling in ethanol under reduced pressure followed by storage of the tips in salt solutions, which gradually replaced the alcohol. The advent of microfilament capillary tubing changed all that and revolutionized the production of microelectrodes. Microfilament capillary tubing has a very thin glass filament attached to the inside of the capillary. When a pipet is pulled, capillary action along the microfilament allows for very rapid and easy filling. A chloridized silver wire inserted into the shank completes the electrode. Once filled, the electrodes are usually stored in a dilute salt solution. Concentrated solutions tend to etch the very thin glass at the tip. A glass microelectrode is diagrammed in Figure 1.

While these electrodes are merely miniaturizations of the familiar Ag/AgCl reference electrode, the very small tips impart a few unfamiliar problems. One of these is very high electrode resistance, with resistance

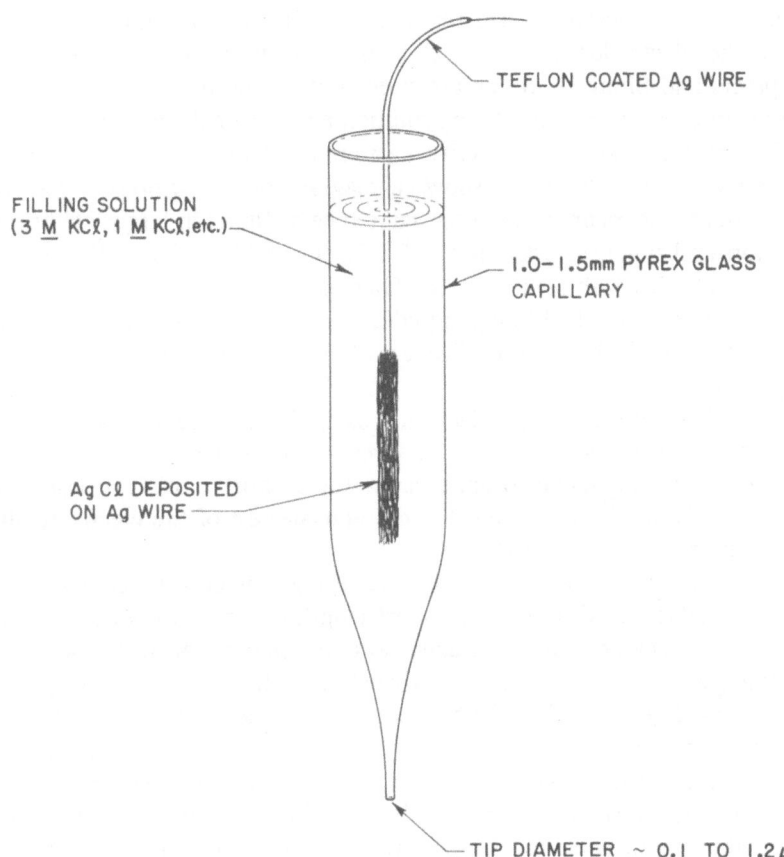

Figure 1. Glass microelectrode.

increasing as tip diameter decreases. For example, a 1 μ-tip diameter Ag/AgCl electrode filled with 3 M KCl commonly exhibits a resistance of 10–100 megohms. Capacitance can also be a problem, particularly in the measurement of rapidly changing signals such as action potentials.

Another unfamiliar problem is that of tip potential (TP). Tip potential is an anomalous potential, on the order of 1–30 mV, that arises with very small tip diameter electrodes. It is not a liquid junction potential because the tip potential disappears when the very tip of the electrode is broken off. Because the tip potential may change when the tip is immersed in different solutions, an uncontrollable error can be introduced into the measurement. Several hypotheses have been advanced to account for this potential, including an ion-exchange reaction of the thin glass near the tip and leaching of contaminants out of the glass, but no mechanism has been universally accepted. The standard method of dealing with the problem is to measure the tip potential by breaking the tip off and to assume that if the absolute magnitude is small, then changes in the TP will also be small.

Ion-selective electrodes are really a subgroup of glass microelectrodes because the electrodes are identical in many respects. Conventional glass micropipets and filling solutions are used, but in addition, an ion-selective barrier is interposed between filling solution and external solution.

The most common ion-selective barriers are liquid ion-exchangers and ion-selective glasses. Figure 2 shows a diagram of the various types. The liquid ion-exchangers are usually the same type as found in larger electrodes. Liquid ion-exchangers selective for Na^+, K^+, Cl^-, and Ca^{++} have all been used in microelectrodes with some success. Glasses sensitive to pH, Na^+, and K^+ are also commonly used, although liquid ion-exchangers have come to replace glasses in the cases of Na^+ and K^+ because of insulation problems and difficulty in fabrication.

In the fabrication of microelectrodes using liquid ion-exchangers, the tip of a conventional micropipet is rendered hydrophobic by, for example, exposure to dimethyldichlorosilane vapor for a short time. The liquid ion exchanger is then wicked up into the tip, and the rest of the electrode filled with the appropriate electrolyte.

Fabrication of ion-selective electrodes using ion-sensitive glass is somewhat more difficult. All but a very short length of ion-sensitive glass at the tip must be insulated. This is usually accomplished by an outer sheath of insulating glass sealed to the ion-sensitive glass, with the ion-sensitive portion either recessed or protruding. The recessed tip method has the advantage of a less tricky glass seal, but the disadvantage of a slower response time.

In general, ion-selective microelectrodes have very high resistances (10^9–10^{12} Ω) and they must be frequently calibrated. However, the valuable information obtained with these electrodes warrants the effort involved. Several excellent texts are available on the fabrication and use of ion-selective microelectrodes; the reader is referred to these for further study.[1,2]

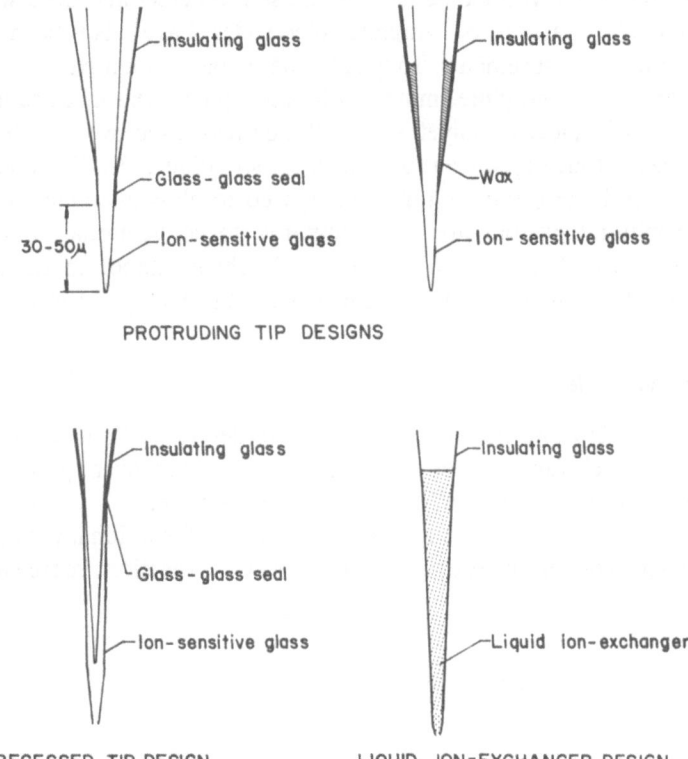

Figure 2. Various ion-selective microelectrode configurations.

Metal microelectrodes are primarily used in two ways: as potential-sensing electrodes and/or as voltammetric electrodes. Potential-sensing metal microelectrodes are used in electrophysiology only for extracellular measurements because it is difficult to fabricate wires with tips small enough to be used intracellularly. The tips are usually made by electropointing (electrolytically removing material) to produce a tapered tip with a diameter of $1-10 \mu$. All but the tip is then insulated with an epoxy or lacquer coating. Glass sheaths are sometimes used, but this is technically difficulty because the glass and metal must have the same coefficient of expansion or cracks will form in the glass. Platinum, tungsten, stainless steel, and aluminum are a few of the many metals used for these electrodes. They are characterized by minimal high-frequency impedance, making them better suited for the measurement of fast-changing signals than glass microelectrodes.

Voltammetric electrodes are much less commonly used. However, small Pt wires pretreated with iodide have been used to detect neurotransmitters in brain tissue.[3] Carbon or graphite electrodes have also been used in neurochemical measurements. They are usually made of a mixture of graphite and mineral oil or graphite, mineral oil, and epoxy. The mixture is packed into the end of a piece of polyethylene tubing press-fitted over a stainless steel wire or packed into a conventional micropipet (Figure 3). These electrodes are relatively large (20 μ to 1 mm) compared to glass microelectrodes used in electrophysiology, but are tiny compared to conventional voltammetric electrodes. Several types of voltammetric and chronoamperometric measurements have been made in intact brain tissue with this type of electrode.[4,5,6]

1.1.2. Enzyme Electrodes

The term enzyme electrode was coined to describe a "miniature chemical transducer which functions by combining an electrochemical procedure with immobilized enzyme activity." In simplistic terms all that is required is to coat the active surface of a specific ion or gas electrode with a thin layer of enzyme, through which the products of an enzyme–substrate reaction are free to diffuse.

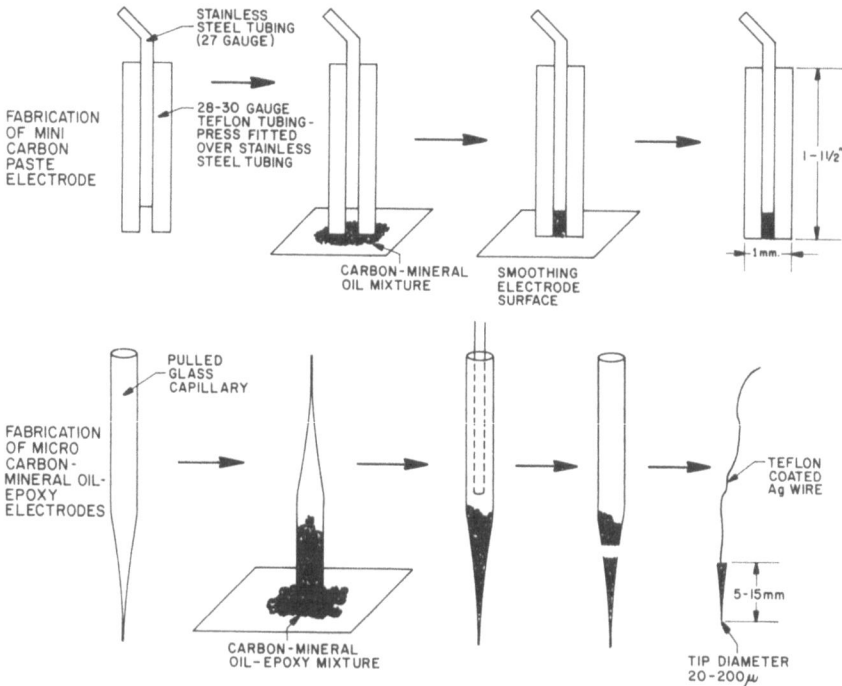

Figure 3. Carbon paste electrodes.

Enzymes, unlike conventional catalysts, are high-molecular weight proteins with very specific reaction mechanisms. As with other catalysts, enzymes do not alter the equilibrium constant of a reaction but do increase the rate at which equilibrium is attained.

Enzymes are found in all living matter. They are not however, living matter; they are simply complex organic molecules. One of the purposes of enzyme-catalyzed reactions in living matter is to either convert a "food" to energy or to convert such dietary material to the materials necessary to sustain and/or extend life. In general, the smaller the living organism, the greater its unit metabolic activity.

In order to utilize enzymes for analytical purposes, one must either have a concentration of enzyme far in excess of that of the substrate, or a known quantity of enzyme. The excess quantity of enzyme is useful and practical where small samples are to be analyzed. Such is the case for enzyme electrodes, where an excess of enzyme is still a very small quantity of material, i.e., a few milligrams. However, maintaining this excess at a fixed location, i.e., at the electrode–test solution interface, requires some method of immobilizing the enzyme.

For continuous analysis of large samples at low substrate concentrations, a known, fixed quantity of enzyme, whose rate of conversion of the substrate is constant, is the preferred approach. Maintaining a known, fixed quantity of enzyme is most easily accomplished, again, by immobilization.

A multitude of processes have evolved over the past two decades for enzyme immobilization. Figure 4 represents one scheme[7] for classifying the various processes that have been employed.

The microenvironment produced by immobilization can present very unique situations. For example, the pH of the environment in the vicinity of the enzyme can be controlled by the material used to immobilize it. Thus, the pH of the reaction medium can be made less important. Similarly, the immobilizing agent can be designed to improve stability via built-in electrostatic charge effects. Conversely, unless care is taken, the immobilizing agent can increase the rate of denaturation (unfolding) of the enzyme. Overall, it is of great importance to select a method of immobilizing enzymes that is (a) efficacious with regard to enzyme stability and (b) conducive to rapid substrate–enzyme interaction.

The technology of enzyme electrodes was apparently initiated by Updike and Hicks in 1966 with their development of a glucose detector incorporating glucose oxidase in an oxygen electrode.[8,9] Since 1966, the number of enzyme electrodes reported in the literature has rapidly increased. An oxygen electrode can be converted to an enzyme electrode by simply adding a second membrane and containing the enzyme between the outer membrane and the electrode membrane. Figure 5 illustrates some of the techniques that can be used. Methods for incorporating enzymes onto electrodes are limited only by the imagination of the electrode designer.

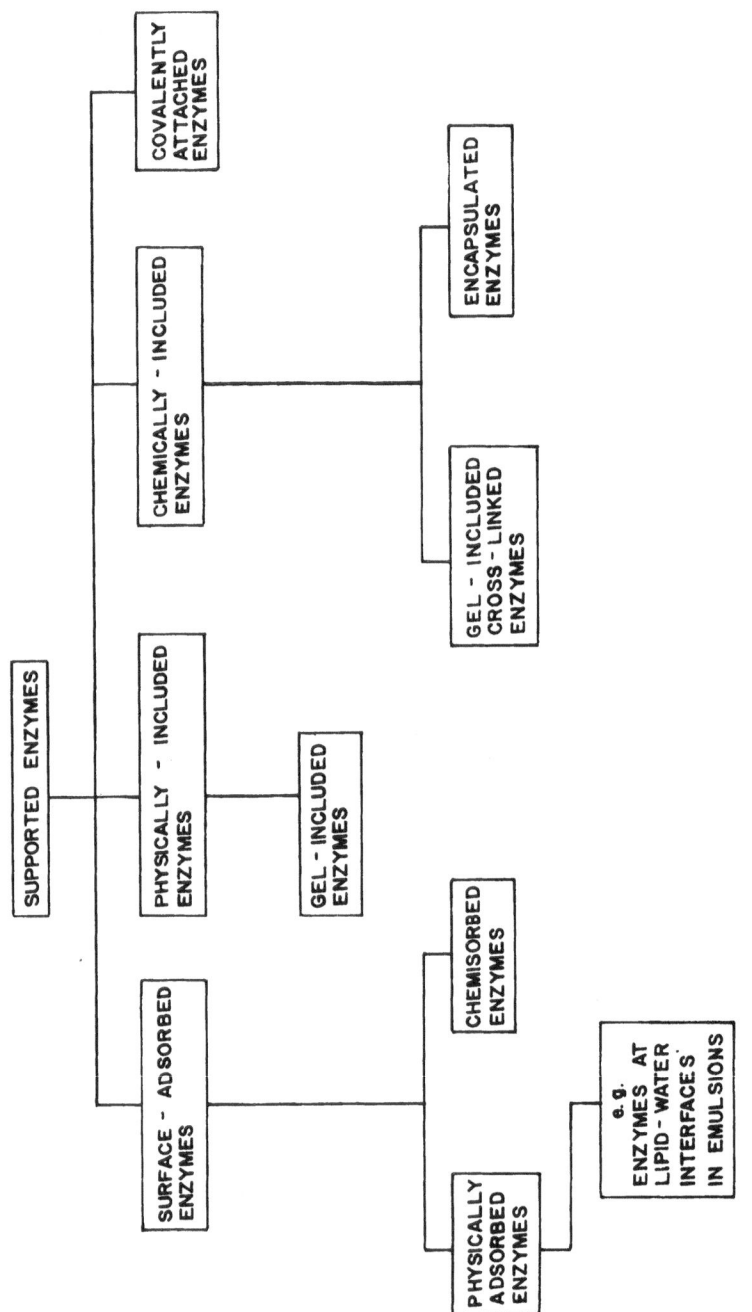

Figure 4. Classification of immobilization processes.

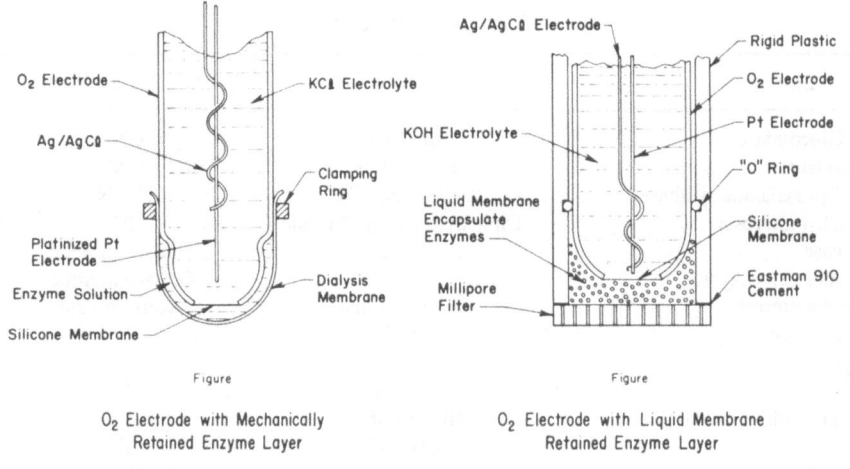

O$_2$ Electrode with Mechanically
Retained Enzyme Layer

O$_2$ Electrode with Liquid Membrane
Retained Enzyme Layer

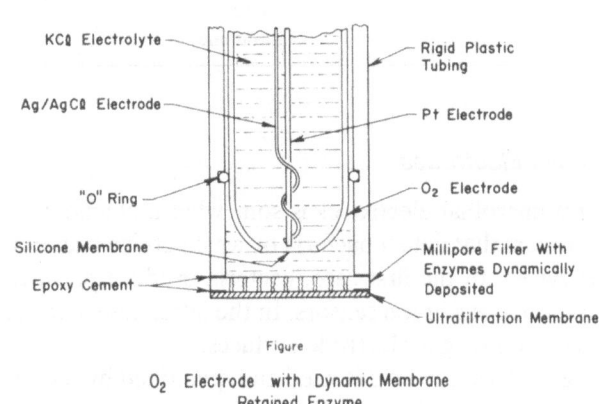

O$_2$ Electrode with Dynamic Membrane
Retained Enzyme

Figure 5. Enzyme electrode immobilization methods.

Table 1 lists a few of the enzyme electrodes reported.[10] Both potentiometric and amperometric devices have been employed with these electrodes.
Test results with enzyme electrodes have been reported as ranging from
excellent to poor. Problems reported deal principally with lack of stability of
the enzymes used, extended response times, and temperature sensitivity. There
does not appear to be any pattern to the problems reported. It is apparent
however, that (a) care must be taken in the choice of enzyme immobilization
method, (b) the immobilization technique should provide a minimal barrier
to reactant and product diffusivity, and (c) temperature stabilization is
desirable.

Table 1
Enzyme Electrodes

Enzyme	Substrate	Sensing electrode
β-Glucosidase	Amygdalin	CN^-
Rhodanese	$S_2O_3^{2-}/CN^-$	CN^-
β-Cyanoalanine synthase	L-Cysteine/CN^-	S^2/CN^-
α-Chymotrypsin	Diphenylcarbamyl fluoride	F^-
Urease	Urea	Glass/antibiotic
Deaminase enzymes		Glass/antibiotic
Cholinesterase	Acetylcholine	Acetylcholine
Peroxidase	H_2O_2	I^-
Catalase	H_2O_2	I^-
Glucose oxidase	β-D-Glucose	I^-
Hypoxanthine oxidase	Hypoxanthine	I^-
Uricase	Uric acid	I^-
Cytochrome b_2	Lactate	$Fe(CN)_6/Pt$
l-Amino acid oxidase	l-Amino acids	NH_4^+
Arginase/urease	l-Arginine	NH_4^+
Urease	Urea	CO_2
Glutaminase	Glutamine	Cation

1.1.3. Microbial Electrodes

The term microbial electrodes is somewhat ambiguous in that it is used to describe sensors that detect bacteria or which utilize bacteria as a renewable source of enzymes. In the first case, bacteria produce products which can be detected by specific gas or ion sensors. In the latter case, bacteria are immobilized on specific ion or gas electrode surfaces.

Growing bacteria can be detected and quantified by the reaction products they produce or the growth media they consume. There are several approaches that can be taken to measure the change in products or reagents. One method is to utilize specific ion or gas electrodes to measure O_2 consumption, H_2 evolution, CO_2 evolution, sulfide ion production, ammonium ion production, etc. Another technique is to measure changes in bulk electrolyte conductivity due to microbial biochemistry. A third method involves the detection of bacterial enzymes using specific substrates and ion-specific electrodes. A brief description of the three approaches follows.

In suitable growth media, all aerobic bacteria produce CO_2 and many produce H_2. There are various electrochemical approaches to detection of H_2 and/or CO_2. One approach[11,12] utilizes a platinum electrode in conjunction with a reference electrode. At a given pH, potential developed by this electrode couple is proportional to the log of the hydrogen activity (partial pressure) in the vicinity of the platinum electrode. (See Figure 6.) Another approach uses a coulometric hydrogen electrode in which hydrogen is quantitatively oxidized

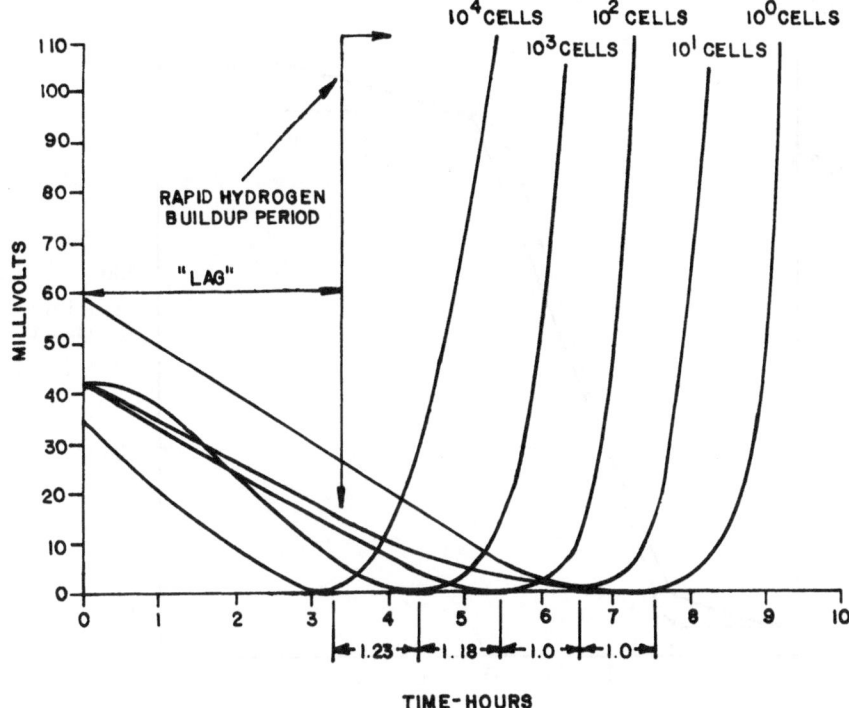

Figure 6. Response of platinum/calomel electrodes to *E. coli* metabolic hydrogen.

as a function of partial pressure. Carbon dioxide can be monitored by either a CO_2 gas electrode or by directly measuring the change of pH of the broth as a function of CO_2 concentration, assuming it is not buffered.

The conversion of nonionic growth media components to ionic materials is used as the basis for bacterial detection via measurement of electrical resistivity.[13] Conversion of lactose to CO_2 and thence to HCO_3^- is a typical example of such a conversion. Rather than measure the absolute value of resistivity as a function of bacterial growth, it is more convenient to measure the difference in resistivity between identical growth media, one sterile and one containing the growing bacteria. By connecting the resistivity-measuring cells in a bridge circuit, changes in resistivity as a function of the number of viable organisms are easily detected. Figure 7 illustrated how resistivity of a growth media for *E. coli* decreases as a function of bacterial concentration.

Enzyme electrodes can also be used to detect bacteria by detecting the presence of bacterial enzymes or their reaction products. For example, β-D-galactosidase is present in coliforms but not in other bacteria. Lactose is converted to galactose and glucose by β-D-galactosidase. Glucose, in the presence of glucose oxidase, is converted to hydrogen peroxide (H_2O_2) and gluconic acid. Hydrogen peroxide is readily detected by an enzyme electrode using catalase to convert H_2O_2 to O_2 and H_2O.

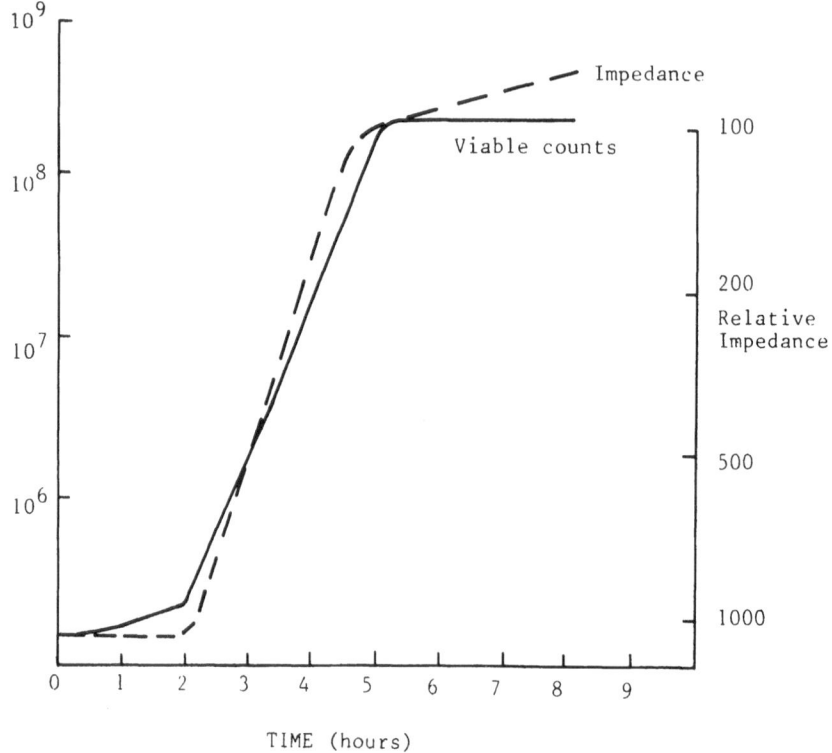

Figure 7. Impedance growth curve and viable counts curve of *Escherichia coli.*

Rechnitz and co-workers have pioneered the development of renewable enzyme electrodes. Their reasoning was based upon the thought that if a purified enzyme works in an electrode matrix, why not just use the organism that makes the enzyme and forget purification? There are two advantages to this approach. First, many enzymes require cofactors in order to function properly. These cofactors are present and are regenerated in living organisms. Second, many enzymes are unstable in environments typically found in analytical practice. Enzyme stability is not a severe problem in living organisms since they are readily regenerated *in vivo* and the environment is close to ideal inside a cell. However, care must be taken to ensure that unwanted side reactions, involving other enzymes present in the bacteria, do not complicate the analysis.

Specific ion and specific gas electrodes, as well as other electrode types, can be employed. Examples of the types of bacterial electrodes include: the use of *Streptococcus foecium* immobilized on an ammonia electrode to sense arginine,[14] the use of *Sarcina flava* and an ammonium electrode to sense L-glutamine,[15] and the use of *Lactobacillus fermenti* and a platinum, silver peroxide electrode to sense vitamin B_1.[16]

Techniques that are employed to immobilize bacteria on ion-specific electrodes are generally limited to (a) entrapment within membranes and gels, (b) adsorption onto polymeric and porous surfaces, and (c) covalent bonding. In general, entrapment techniques involve the inclusion of cells in a rigid network that prevents the bacteria from migrating, while permitting penetration of the surrounding medium. The most common entrapment materials are polyacrylamide gels, agar, collagen, and cellulosics. Table 2 lists examples of bacteria entrapped in various supports along with some of the reactions or products with certain substrates.[17]

Adsorption of bacteria onto porous surfaces is effective in certain circumstances. An advantage of adsorption over gel entrapment is that a monocellular layer of bacteria can be utilized, thus reducing the diffusion problems inherent with thicker layers required for gels and films. However, the quantity and type of bacteria that will be adsorbed by a surface is highly variable.

Covalent bonding of bacteria to surfaces is typically accomplished by the use of glutaraldehyde or similar cross-linking agents. It is difficult to predict the effect of bonding of bacteria onto various surfaces. However, cells of various types have been successfully bound to polyethylene, carboxymethyl cellulose, and amine glass beads.

1.2. Analytical Applications of Bioelectrodes

The preceding section discussed the tools of electroanalysis in the biological sciences. How these tools are utilized is described in this section.

1.2.1. Neurochemical Analysis

The brain is a complex web of interconnecting neurons, each individual cell directly communicating with as many as 10,000 of its neighbors. A primary mode of this communication occurs when the sending cell releases a neurotransmitter that diffuses across the synapse, or gap between the cells, to a receptor site on the recipient neuron. Its functional importance finished, this transmitter molecule will then be metabolized or reabsorbed into the nervous tissue. There is an endless list of stimuli capable of initiating this act of neurotransmission, including behavioral and pharmalogical perturbations. An additional point to note is that physically parallel or crossing neuronal pathways are kept from interfering with each other simply by utilizing different neurotransmitters and highly specific receptor sites.

For many years neuropharmacologists, neurochemists, and behavioral psychologists have been obtaining qualitative and quantitative localized chemical information from the brains of test animals in a rather unsatisfying manner. Following a controlled behavioral or chemical manipulation the animal is sacrificed, usually via cervical dislocation, decapitated, and its brain

Table 2
Reactions of Bacteria Entrapped in Supports

Bacterial type	Support	Substrate	Product of reaction
Saccharomyces cerevisiae	Brick, wood	Glucose	Ethanol
Saccharomyces cerevisiae	Various polymers	Glucose	Ethanol
Bacillus megaterium	DEAE-Cellulose	6-APA	Ampicillin
Achromobacter sp.	CM/DEAE/TEAE Cellulose	7-ADCA	Cephalexin
Mycobacterium globiforme	Cellulose	Steroids	Reduction
Escherichia coli	Polyacrylamide gel	Fumaric acid	Aspartic acid
Brevibacterium ammoniagenes	Polyacrylamide gel	Fumaric acid	L-Malic acid
Microbacterium ammoniaphilum	Polyacrylamide gel	Diaminopimelic acid	L-Lysine
Corynebacterium simplex	Polyacrylamide gel	Cortisol	Prednisolone
Curvularia lunata	Polyacrylamide gel	Reichstein's S	Cortisol
Achromobacter liquidum	Polyacrylamide gel	L–Histidine	Urocanic acid
Pseudomonas putida	Polyacrylamide gel	L-Arginine	L-Citrulline
Bacillus ammoniagenes	Polyacrylamide gel	Pantothenic acid	Coenzyme A
Corynebacterium glutamicum	Polyacrylamide gel	Glucose	Glutamic acid
Mycobacterium globiforme	Polyacrylamide gel	Reichstein's S	Hydrocortisone
Mycobacterium globiforme	Polyacrylamide gel	Steroids	Reduction
Clostridium butyricum	Polyacrylamide gel	Glucose	Hydrogen
Lactobacillus casei	Polyacrylamide gel	L-Malic acid	L-Malic acid
Achromobacter guttatus	Polyacrylamide gel	β-Aminocaproic acid cyclic dimer	Hydrolysis
Escherichia freundii	Polyacrylamide gel	Glucose + Phenylphosphate	Glucose-6(1)-phosphate
Escherichia coli	Polyacrylamide gel	Penicillin G	6-APA
Lactobacillus bulgaricus *Kluyveromyces lactis* }	Polyacrylamide gel	Lactose	Hydrolysis
Streptomyces phaechromogenes	Collagen membrane	Glucose	Isomerization
Streptomyces Venezuelae	Collagen membrane	Glucose	Isomerization
Corynebacterium glutamicum *Arthrobacter simplex* }	Collagen membrane	Glucose Steroid dienediol	Glutamic acid } Prednisolone }

Table 2 (continued)

Bacterial type	Support	Substrate	Product of reaction
Serratia marcescens	Collagen membrane	Glucose	2-Ketogluconic acid
Saccharomyces pastorianus	Agar gel	Sucrose	Hydrolysis
Escherichia coli	Agar gel	Lactose	Hydrolysis
Rhodospirillum rubrum	Agar gel	Malate	$H_2 + CO_2$
Candida tropicalis	Aluminum alginate	Phenol	$CO_2 + H_2O$
Saccharomyces cerevisiae ⎫	Calcium alginate	Glucose	Ethanol ⎫
Kluyveromyces marxianus ⎬	Calcium alginate	Inuline	Ethanol ⎭
Erwinia aroidea	Cellulose triacetate	Acylase	6-APA
Escherichia coli	Cellulose triacetate	Lactose	Hydrolysis
Aspergillus oryzae	Cellulose nitrate	Amino acid acylase	L-Tryptophan
Actinoplanes missouriensis	Cellulose fibres + glutaraldehyde	Glucose	Isomerization
Escherichia coli	Hollow fiber	Fumaric acid	L-Aspartic acid
Brevibacterium ammoniagenes	Cellophane tube	Glucose	Acetyl-CoA
Candida tropicalis	Polystyrene	Phenol	$CO_2 + H_2O$
Penicillium notatum	Polyurethane (sheets)	Glucose	Gluconic acid
Acetobacter	Metal hydroxides	Ethanol	Acetic acid
Micrococcus dentirificans	Liquid membrane	NO_2 and NO_3^-	N_2

excised into cold physiological buffer. This neuronal tissue is then dissected or chopped to localize the region of interest. A standard chemical analysis usually follows. Since the brain is an extremely dynamic organ responding to even the slightest manipulation, a scientist realizes that the process of analysis could easily override the effect of the controlled perturbation.

In vivo electroanalytical techniques offer the advantages of localized, relatively nondestructive procedures that can be repeated on the same animal several times. This last point allows any particular animal to be its own control and thus greatly reduces the number of animals sacrificed per experiment. The amount of damage done during an experiment will, of course, depend on the size of the electrode system utilized as well as the neurosurgical skills of the experimenter and the degree of postoperative recovery each animal experiences.

One final but important point that an investigator should realize is that animals which are chronically implanted with electrodes can be tested without interference from anesthetizing drugs or stressful restraint.

Qualitative information can be obtained via any of three voltammetric techniques. These include classical, differential pulse, and differential double pulse voltammetry. Lane[3,18] has shown that the differential pulse and double pulse procedures using a variety of graphite-based electrodes do provide, as would be expected, more peak resolution than normal voltammetry. This is important because the brain contains several species that are all oxidized in the same range (*ca.* +0.8 V vs. SCE at graphite/epoxy). Unfortunately these techniques do not answer all of the interference or resolution problems. There are, however, three ways in which one can increase "effective" resolution. Implanting the working electrodes in a region of the brain that has been shown to contain predominantly one neurotransmitter has been described.[19] Electrical stimulation of a fiber bundle that is mediated by a single neurotransmitter or the utilization of a drug that has been shown to release primarily a single transmitter have also been outlined.[4–6,19–21]

Absolute qualitative identification can be assured only if samples are removed and analyzed. Two examples of such a procedure have been reported.[19,22] The first was an attempt to determine if direct electrical stimulation of the caudate nucleus resulted in the release of dopamine as well as ascorbic acid from that tissue. Micro voltammetric and stimulating electrodes were micromanipulated into excised caudate tissue which was flushed with warmed, oxygenated buffer. Reference and auxiliary electrodes were nearby. Quantitative information was taken, stored, manipulated, and displayed by a minicomputer. Simultaneously a push–pull cannula device sampled the caudate and delivered the perfusate to an iced vial. Changes in the electrochemical signal that followed stimulation were correlated with changes in the dopamine and ascorbic acid content of the perfusate as determined via HPLC with electrochemical detection. It was found that little if any ascorbic acid was released as a result of electrical stimulation in these experiments. Although there is some question concerning the stability of ascorbate in an iced vial, the above example does illustrate this coincident analytical technique.

The second example[22] involved a similar procedure; however this time the electrochemical and chromatographic analyses were used to monitor cerebral spinal fluid (CSF) in the lateral ventricles of a living rat. Electrical stimulations and drug administrations were clearly followed via the electrochemical techniques while HPLC coupled with electrochemical detection identified the transmitters and metabolites that were involved.

Quantitative chemical information is somewhat easier to come by from an *in vivo* experiment. Mini- or microelectrodes should be calibrated with standard solutions of well-behaved electroactive species before and, if possible, after implantation and experimentation. *o*-Dianisidine (3,3'-dimethoxybenzidine) in 0.1 M H_2SO_4 was found very satisfactory. After implantation (and, in the chronic experiments, recovery), data can be obtained using chronoamperometry. Using the calibration data as a standard it is possible to monitor the concentration of *all* species that are oxidizable at the applied potential.

It is important that no purposeful stimulation be initiated until a fairly constant baseline is obtained. At that point behavioral, electrical, or pharmacological manipulations can be followed in the brain region of choice.

Electrochemical monitoring of discrete brain regions can now help unlock the mysterious door that leads to a better understanding of brain function. In addition to eliminating the substantial trauma associated with the death, decapitation, excision, and tissue-chopping procedures, it is now possible to monitor the levels of electroactive species at pinpoint locations from the brain of an unanesthetized, unrestrained, seemingly unperturbed animal.

1.2.2. Mutagenicity and Toxicity Detection

In recent years, attention has become focused on the problem of chemical reagents being released into our environment that are potential health hazards. The problem is complicated by the socioeconomic aspects of production and use of these chemicals and by our limited knowledge of the long-term effects of such reagents, with regard to their toxicity, carcinogenicity, and mutagenicity. We obviously cannot afford to wait 5–25 years after the introduction of a new reagent to see if it is an ecological or health hazard. Methods for evaluation and screening of potentially hazardous materials are therefore needed that provide an indication of the degree of hazard in relatively short time periods.

There have been a number of approaches to evaluation and/or screening of chemicals to evaluate their hazard potential. Since the problem is one of the effect of chemicals on biological entities, the approaches have centered on bioassay procedures. The range of biota chosen for controlled exposure to potentially hazardous chemicals has been enormous. At one end of the spectrum are single cells such as bacteria and yeasts. At the other end are multifunctional organisms, such as plants and animals. In general, the tendency has been towards screening with less biologically complex entities, followed by more definitive testing with complex biota of those chemicals found hazardous in the screening effort.

Among the generally accepted screening procedures for carcinogens and mutagens is the so-called Ames test developed by Dr. Bruce N. Ames and his associates.[23,24] The Ames test is based upon the assumption that carcinogens will cause the genetic reversion of certain mutant strains of *Salmonella typhimurium*. In other words, the mutant strains revert to their normal "wild" form in the presence of carcinogens and mutagens. Extensive testing with a wide variety of potential carcinogens has indicated the general validity of the basic assumption that mutagens cause bacterial reversion. (Since the majority of mutagens or their metabolic products are also carcinogens, the Ames test is also used to screen this class of chemicals.)

Basically, the mutant *Salmonella typhimurium* strains are selected because they lack the ability to produce histidine, an essential amino acid. These mutants are unable to multiply unless histidine is present in their growth media. In the presence of carcinogens and mutagens, the mutated strains revert to their "wild" form. Since the "wild" forms can manufacture histidine from other materials, it need not be present in the growth media in order for the reverted strains to multiply. Thus, by culturing mutant strains of *Salmonella typhimurium* in a medium that does not contain histidine but does contain a suspected chemical, one can evaluate the carcinogenicity or mutagenicity by determining the growth characteristics. It is not however, a go, no-go situation since a certain portion of the mutant bacteria reverts spontaneously, with no experimental stimulus. This "natural" reversion rate phenomena represents a "noise level" that limits the overall sensitivity of the test. It is not a critical factor with properly chosen bacterial strains, i.e., those with low spontaneous reversion rates.

As with most bacterial tests, plate-counting techniques are used to determine the number of revertant bacteria. For the Ames test, plate counting requires a 48-hour growth period, plus time to count both the test plates and control plates used to monitor spontaneous reversion.

There is an electrochemical method which greatly reduces the time required to evaluate the degree of spontaneous reversion. The technique involves the use of oxygen electrodes in conjunction with the test ingredients (bacteria, growth media, homogenates, etc.) specified by Dr. Ames. Figure 8 illustrates the approach schematically. As shown, two commercial oxygen electrodes of the Clark type are employed. One electrode, designated the control, measures the respiration rate (O_2 uptake) of a fixed volume and concentration of Ames test bacteria in a medium that does not contain either histidine or a chemical to be evaluated for carcinogenicity and/or mutagenicity. The second electrode, designated the test electrode, measures the respiration rate (O_2 uptake) of the same fixed volume of the medium with the same concentration of Ames test bacteria as the control electrode. In the second case, the medium also contains the chemical to be evaluated. If the chemical is a carcinogen or mutagen (CM) the O_2 uptake of the test electrode will increase at a much faster rate than that of the control. This occurs because the reversion rate, and thus the growth rate and O_2 consumption rate, will be higher in the test than in the control electrode. Therefore, by measuring the difference in O_2 uptake rates between a control and a test electrode, the carcinogenicity and/or mutagenicity of a chemical can be evaluated. The difference measurement can be made by any of the many electronic techniques used for such measurements or by simple manual subtraction. Figure 9 illustrates the effect of a mutagen on the uptake of oxygen as measured in the apparatus shown in Figure 8.

If the chemical to be analyzed is toxic, rather than simply carcinogenic, the O_2 uptake of the bacteria will decrease. This effect is easily determined and is discussed in greater depth later.

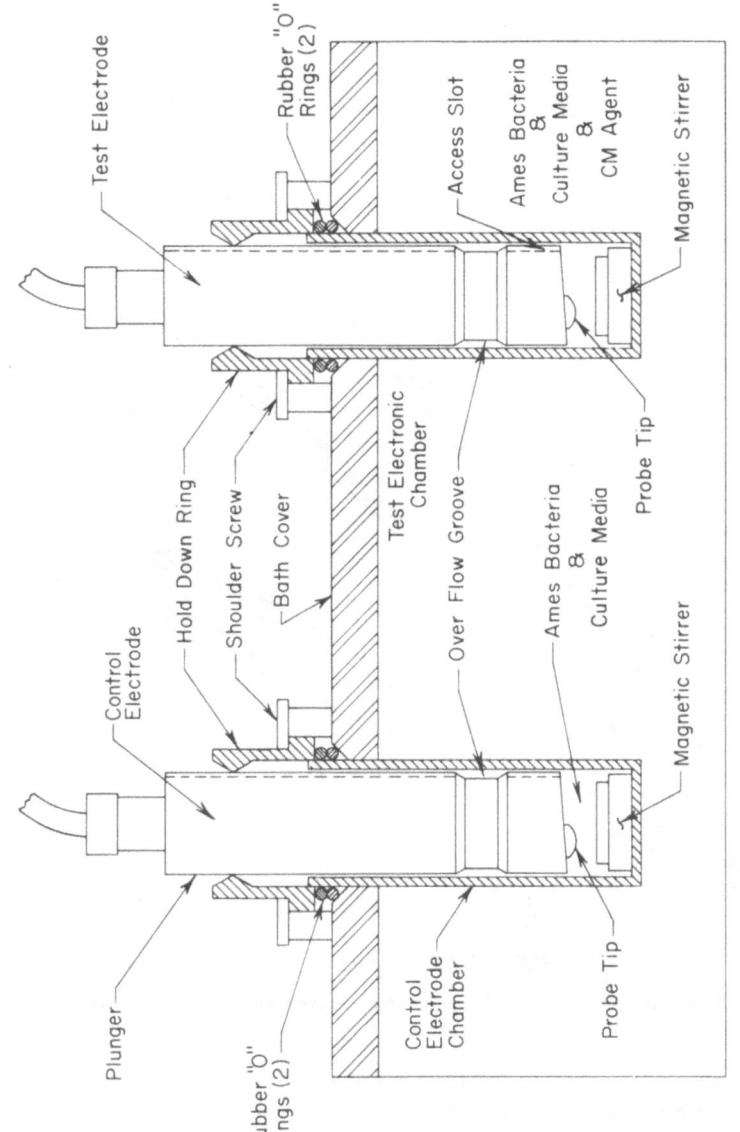

Figure 8. Ames test dual electrode test setup.

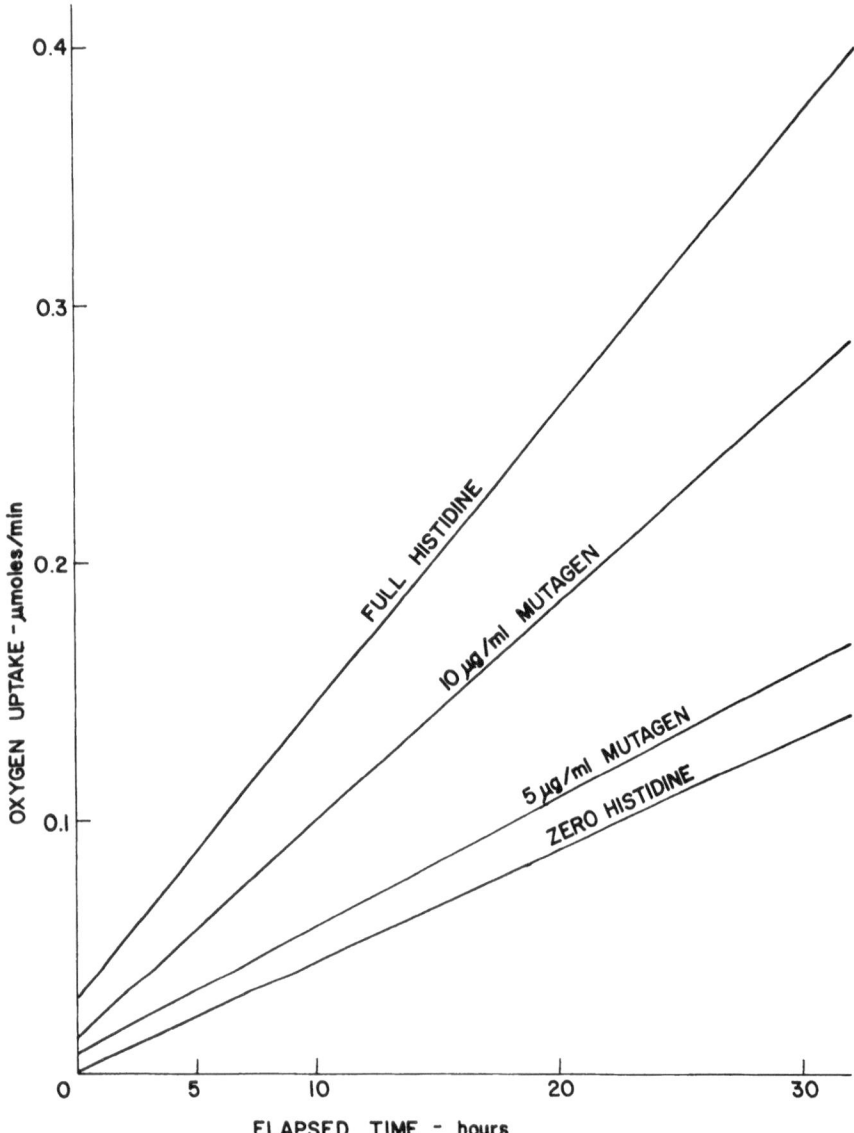

Figure 9. Effect of the mutagen 9-amino acridine on *S. typhimurium* TA1535 oxygen uptake.

Since O_2 uptake does not require the growth of colonies of bacteria needed for visual counting of plates, four hours or less of growth is sufficient for O_2 uptake difference measurements to be made. Thus, test time probably can be reduced from 48 hours to 4 hours (neglecting sample preparation time, which would be similar for both). Further, plate-counting time and cost is eliminated, because differences in reversions are measured directly by the difference current output of the electrodes. In principle, the greater the current difference, the greater the carcinogenic/mutagenic potential of the chemical, assuming identical test conditions for each animal.

When the effects of unknown substances are tested and one of the test criteria involves the measurement of O_2 uptake, there is always the possibility that the unknown substance is an uncoupler of oxidative phosphorylation. If this is the case, a *rapid* increase in oxygen uptake will be registered. This oxygen uptake, however, is independent of "real" respiration and no ATP will be produced. One of the best known uncouplers of oxidative phosphorylation is dinitrophenol (DNP). Naturally occurring substances, such as the thyroid hormones thyroxine and triiodothyronine, stimulate respiration and ATP production in low concentrations (10^{-9} M) and uncouple at higher concentrations. Thus, any substance that produces an increase in oxygen uptake, independent of increased numbers of organisms, should be tested for uncoupling activity.

The same general technique noted in the discussion on carcinogenicity and mutagenicity can be applied to toxic chemical detection.[25] In this case, bacteria, yeasts, or mammalian cells can be used. A prime criterion for use is that the chemical agent be toxic to the sensing "cell" at very low concentration levels. Additionally, it is desirable that the selected cell's die-off rate be proportional to the concentration of the toxic chemical. Both criteria can be met by a variety of cells.

One mammalian cell type which is suitable for detection of a variety of toxic reagents is the rabbit kidney cell. These cells are relatively easy to culture, are commercially available, and have high O_2 uptake rates. A typical rabbit kidney O_2 uptake rate curve, and the effect of a toxic chemical on same, is presented in Figure 10.

1.2.3. Analysis of Blood Ions and Gases

Analytical schemes concerned with the determination of blood ions and gases can be divided into two categories: analyses done *in vivo* and those done *in vitro*. By far the most common method of determining blood ions *in vitro* involves atomic spectroscopy. Atomic absorption and flame emission have both been used although the latter is the most popular. In the clinical lab nearly all of the remaining determinations (both *in vivo* and *in vitro*) are performed with ion-selective (for ions, NH_3 and CO_2) or amperometric electrodes (O_2 and H_2). Two important characteristics of ion-selective electrodes, sensitivity and selectivity, should be mentioned. The applicability of a specific electrode in any particular situation can be determined by considering, on one hand, the ionic constituents of the solution to be measured and, on the other hand, the sensitivity and specificity of the electrode in question. Proper consideration of these points will allow an investigator to determine the accuracy and validity of the measurement.

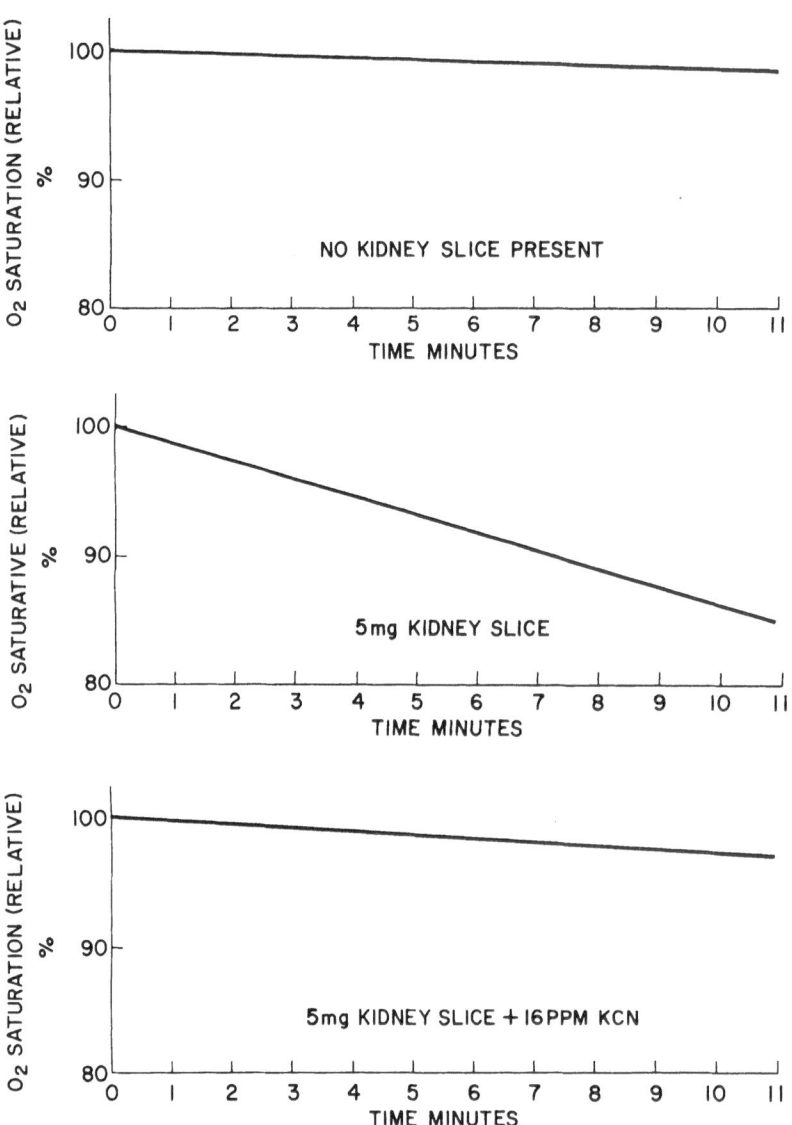

Figure 10. Effect of KCN on O_2 uptake of rabbit kidney cells.

There are eight ions in blood, not including H^+, that are commonly determined potentiometrically: Na^+, K^+, NH_4^+, Ca^{++}, F^-, Br^-, Cl^-, and I^-. In addition, CO_2 levels can be calculated if CO_3^- as well as pH is known.

Calcium analysis via ion-selective electrode has now become an inexpensive and dependable method for determining the free, ionized form of this element in blood and sera. A substantial portion of this technology has been due to an electrode which analyzes ionic calcium in blood (500 μL sample) in 3 minutes with a standard deviation of 2%.[26]

Fluoride is present in normal serum at about 0.01 ppm. Before the technology of fluoride ion-selective electrodes became available, procedures for the determination of this ion were so tedious and time-consuming that few were performed in the clinical lab. Presently, inorganic fluoride and, with an ashing step, total bound fluoride can be quickly and accurately assayed.

Chloride analyses in the clinical setting are primarily concerned with levels in the sweat for diagnosing cystic fibrosis, although *in vitro* blood analysis is easily accomplished.

Bromide determinations in the blood are performed primarily for the monitoring of patients treated with bromo-sedatives, anesthetized with halothane, or suffering from drug abuse.

Sodium and potassium ions are usually measured with an electrode that incorporates either a liquid ion exchanger or Na^+ sensitive glass. Fabrication of microelectrodes, however, is much easier when ion exchanger liquids are used.

Blood gas analysis is usually limited to the determination of O_2, CO_2, and NH_3. Of the three the most common analysis, that of oxygen, was one of the first determinations done with an electrode that was molecularly specific. Basically an amperometric determination, this type of device utilizes a semipermeable membrane coupled with the proper applied voltage to obtain its specificity. Ammonia and CO_2 determinations, on the other hand, are forms of ion-selective potentiometric analyses.

1.2.4. Blood Flow Analysis

Electrochemical monitoring of blood flow rate and direction has been used to successfully detect circulatory system defects as well as actual transport times from one location in the body to another. In 1956, Clark and Misrahy[27,28] showed that platinized platinum electrodes, inserted into the circulatory system of patients, could detect inhaled hydrogen. Following inhalation and circulation of the dissolved hydrogen gas, potentials developed at the platinum electrodes. By placing the electrodes at key locations throughout the circulatory system, the rate of blood flow from the lungs to each of these electrodes could be measured.

A variation of this type of methodology is also used to measure blood flow rates. In this case appropriate anodic potentials are applied to implanted platinum electrodes. A solution of ascorbic acid is then injected into the vein or artery at a location of choice. Blood flow rate between the location of the platinum wire and injection site is determined from the time of injection to the time of a marked increase in oxidation current.[29] Similar techniques using cathodic potentials can be used to monitor blood flow and oxygen levels during cardiac surgery.

One of the most efficient and definitive ways of determining a left-to-right cardiac shunt is to place a platinum electrode in the pulmonary artery and then have the patient inhale hydrogen gas or inject a solution of ascorbic acid into the pulmonary vein. The time delay and magnitude of the signal from the electrode gives information concerning the presence of a shunt.

2. Electrophysiology

Although electrochemistry provides the fundamental basis of electrophysiology, the "language" of electrophysiology bears little resemblance to that of electrochemistry. Accordingly, a brief description of some of the commonly held concepts in electrophysiology will be given in more familiar terms.

A living cell is a complex matrix of organelles and cytoplasm (or intracellular fluid) enclosed in a semipermeable, highly organized, biomolecular lipoprotein layer called the cell membrane. The intracellular fluid is different from interstitial (or extracellular) fluid in many ways but two differences are particularly important in electrophysiology:

(1) The concentration of ions are quite different. Na^+ and Cl^- concentrations are much higher in interstitial fluid than in intracellular fluid—just the reverse of K^+. Large, membrane-impermeable anions on the inside of the cell help to maintain osmotic balance.

(2) An electrical potential difference exists between the inside and the outside of the cell, with the inside negative with respect to the outside. This voltage difference means that electric charges are separated by the membrane. It typically ranges from -20˙ to -100 mV and is called the transmembrane potential or the resting membrane potential.

The concentration gradients between the inside and the outside of the cell would be expected to quickly disappear due to diffusion. However, the membrane is impermeable to the large mucopolysaccharide anions inside the cell and semipermeable to Na^+. Still, one would expect eventual dissipation of the concentration gradients, albeit slowly. The most widely accepted mechanism to account for maintenance of these concentration gradients is active transport or, in the case of Na^+ and K^+, the "Na^+–K^+ pump" (more simply, the "Na^+ pump"). Active transport is simply a term used to describe movement of a substance across a membrane *against* its concentration gradient with a corresponding consumption of metabolic energy. The energy is supplied by or "coupled to" the cell's energy production system. The "Na^+ pump" is a rather unfortunate term used to specifically denote the active transport of Na^+ and K^+. Its physical form is believed to be a specific protein assembly in the cell membrane called the Na^+–K^+ ATPase. This protein assembly cleaves ATP and uses the energy to transport Na^+ out of the cell and K^+ into the cell.

Active transport, concentration gradients, and differential ionic permeabilities are all involved in the maintenance of the transmembrane potential.

2.1. Measurement of Transmembrane Potential

The transmembrane potential is commonly measured with glass microelectrodes described in Section 1.1.1. One electrode, the measuring electrode, is carefully advanced through the cell membrane under micromanipulator control. The reference electrode is usually a broken-tip glass microelectrode or simply a Ag/AgCl wire immersed in the bathing solution. The voltage difference observed between the inside and the outside of the cell is the transmembrane potential. Because the resistance of a glass microelectrode is quite large, a high-input impedance measuring device must be used and great care must be taken to properly shield the apparatus and instrumentation and avoid ground loops.

The measurement of transmembrane potentials can be a tricky affair, particularly with small cells. In order to achieve stable, meaningful voltages, the cell membrane must seal around the glass tip and damage to the cell must be kept to a minimum; hence, small tip diameters and vibration-proofing of the apparatus are very desirable. The cells often require special, controlled environments including optimum temperature, proper air/CO_2 mixtures, and biological media. Fluctuation in these conditions can easily cause marked changes in the membrane potential.

Because the transmembrane potential is so sensitive to environmental perturbations as well as internal events, it is a useful test parameter for studies of cell function, particularly transport mechanisms. The other major area of importance is in the study of excitable cells.

2.2. Stimulation of Excitable Cells

A neuron, or nerve cell, is an example of an excitable cell. These cells have cell bodies (or soma) containing the nucleus and elongated processes called axons and dendrites. These cells form a complex web with many connections. Figure 11 shows a schematic diagram of a connection between two such nerve cells. The presynaptic neuron is separated from the postsynaptic neuron by a small gap known as the synapse, typically 2–800 Å wide. Presynaptic nerve endings contain small sacs or vesicles filled with one of several compounds called neurotransmitters. The postsynaptic neuron has receptor sites for specific neurotransmitters located on the cell membrane. When appropriately stimulated, and area of a presynaptic neuron membrane becomes depolarized (the transmembrane potential becomes more positive). The depolarized area propagates down the axon very rapidly. This wave-like movement of depolarization is called the action potential. When the depolarized areas reaches the nerve ending, the vesicles move to the cell wall, fuse with it, and dump their contents into the synaptic cleft—a process called exocytosis. (Exocytosis is accepted as the mechanism of neurotransmitter release in the peripheral nervous system, but it has not yet been demonstrated

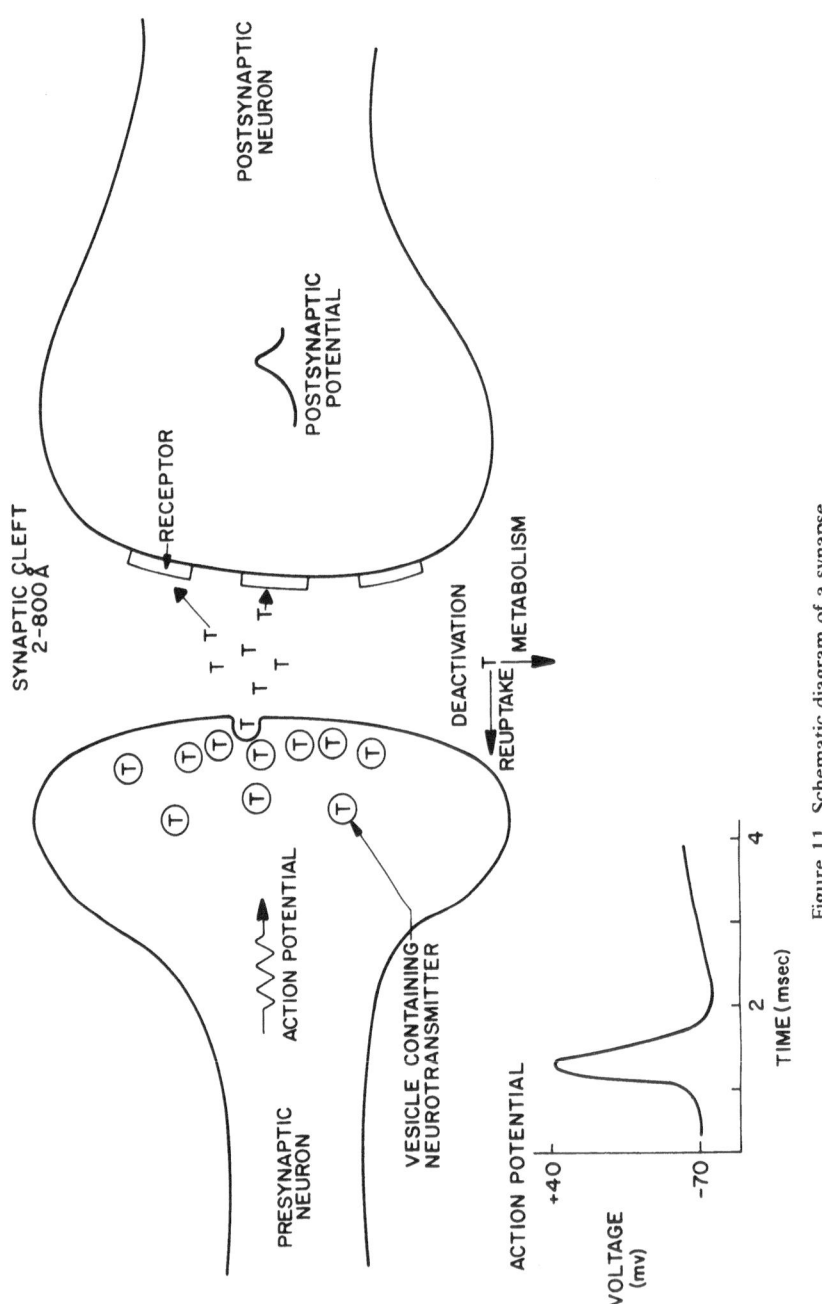

Figure 11. Schematic diagram of a synapse.

in the central nervous system.) The transmitter diffuses across the synapse to bind with the receptor, initiating a chain of events in the postsynaptic neuron.

The action potential, then, occurs in excitable cells and the depolarization is triggered when the membrane's permeability to Na^+ and K^+ increases suddenly, causing Na^+ to rush into the cell and K^+ ions to rush out. The resulting changes in transmembrane potential are also shown in Figure 11. The increased permeability lasts for a very short time (less than a millisecond), then rapidly returns to the initial state. Many events can trigger an action potential, including application of drugs, the action of excitatory neurotransmitters on receptor sites, and electrical stimulation.

In electrical stimulation, current is passed in the vicinity of the cell and causes a portion of the membrane to depolarize. If the depolarization is great enough, an action potential is initiated. The stimulation can be done with monopolar, dipolar, or tripolar electrodes. Various stimulation parameters (pulse width, magnitude, and frequency) can be used, with the exact conditions tailored to the particular application. Electrical stimulation is a valuable tool in the investigation of neuronal pathways of connections between one portion of the brain and another. It also plays a key role in correlating the activity of specific brain regions with behavioral or physiological changes in the animal.

2.3. Control of Membrane Potential (Voltage Clamping)

Another very useful electrochemical technique in electrophysiology is the "voltage clamp" technique, a means of controlling the membrane potential. With this technique the membrane potential can be held at any desired level, making it possible to maintain the resting potential (effectively eliminating action potentials), to hyperpolarize or depolarize the cell, etc. A voltage-measuring and a current-injecting electrode are inserted into the cell while another electrode is immersed in the bathing solution. The desired voltage is impressed between the voltage-measuring and the extracellular electrodes by means of a potentiostat and current is passed through the current-injecting electrode as necessary to maintain that voltage. The voltage clamp technique evolved in the early days of electrophysiology and has aided greatly in the understanding of such questions as how ionic permeability varies with time and membrane potential, and what component of the ionic current through the membrane is carried by a particular ion. It is still widely used today as evidenced by the following example of the work of Spray et al.[30] Two cells joined by a gap junction (a sort of channel from the inside of one cell to the inside of another) were investigated to see if the membrane potential affected the junctional conductance. Each cell was impaled with voltage clamp electrodes and the membrane potentials were varied independently. They found that while the transmembrane voltage did not affect the conductivity, the transjunctional voltage did. Thus, the transjunctional voltage provides a means for varying the extent to which a cell communicates with its neighbors.

2.4. Iontophoresis

Iontophoresis is the procedure whereby ions or charged molecules are transported from a reservoir to a target site by means of an electric current. Since the passage of current through an aqueous solution is accomplished by this movement of charged species, it is possible to accurately determine and even control the number of these species that are delivered by a consideration of the following charge/current relationships. 1 Faraday = 1 mole electrons = 6.02252×10^{23} electrons and the magnitude of an elementary charge is 1.6021×10^{-19} coulombs. Therefore, the charge of one Faraday (mole) of monovalently charged species equals (6.02252×10^{23} electrons) (1.60210×10^{-19} coulombs/electron) = 96,488 coulombs. An example of how the above relationships can be used to deliver controlled amounts of charged species to a specific location follows.

Let us suppose that we need to raise the K^+ concentration of a fibroblast cell to 200×10^{-3} M. The average fibroblast cell volume is 10^{-12} liters with an internal K^+ concentration of 150×10^{-3} M. Therefore, we need to add: (50×10^{-3} moles/liter) (10^{-12} liters) = 5×10^{-14} moles K^+. (This assumes, of course, no leakage from the cell during the process.) This quantity of K^+ corresponds to (5×10^{-14} moles) (96,488 coulombs/mole) = 4.82×10^{-9} coulombs of charge. Assuming a constant current d.c. power source is available to deliver 10^{-9} amps, current must be passed for 4.82 seconds to achieve the desired concentration: (4.82 seconds) (10^{-9} amps) = 4.82×10^{-19} coulombs.

The actual experiment requires the insertion of a micropipet, filled with KCl, into the cell. The positive lead of the power supply will be connected to a noble metal wire located in this micropipet. In most cases a second micropipet would be used to complete the circuit via the negative terminal of the power supply.

Charged, pharmacologically active compounds are, by far, the most common materials delivered via iontophoresis. With the recent advent of convenient micropipet pulling and filling technology, along with a new generation of micromanipulators, this technique has become very prevalent in single cell investigations.

Historically, iontophoretic delivery of medicines has been confined to rather superficial, epidermal applications. Ions of zinc and copper have been employed in the treatment of some skin infections and chronic infections of sinuses and cavities. Chloride ions can be utilized to loosen superficial scars. Many drugs used for general metabolic or specific hormonal effects can be driven into the circulating blood by iontophoresis, although currently most physicians use the more common forms of drug delivery.

2.5. EEG, EMG, ECG

The electroencephalogram (EEG), electromyogram (EMG), and electrocardiogram (ECG) are all measurements of electrical signals generated in the body. Individual signals from many sources are summed together and can be detected on or near the surface of the body. These measurements are accomplished in one of two ways: surface signals can be obtained by simply placing two silver/silver chloride wires, plates, or cups at positions chosen to optimize the qualitative and quantitative aspects of the signal. Depth measurements specifically restrict the area from which signals are taken and are accomplished using fine wire or micro capillary Ag/AgCl electrodes.

Electroencephalograms are electrical recordings taken from the skin of the head or the surface or interior of the brain. The summation of nerve cell activities in the brain gives rise to the complex patterns of the EEG. The intensities of the recorded voltages range from 0 to 300 μV, and their frequencies range from one every few seconds to more than 50 Hz. Much of the time no distinct pattern can be found in the EEG; however, several patterns associated with specific mental and physiological states do occur frequently in normal persons. Alpha waves occur at frequencies in the range of 8–13 Hz and are found in quiet, awake, and resting states. Beta waves occur at 14–50 Hz and are associated with the degree of activation of the central nervous system. Theta waves have a frequency of 4–7 Hz and occur mainly in children; however, they can also occur during emotional stress in some adults. Delta waves include all EEG waves below 3.5 Hz, and occur in deep sleep, in infancy, and in very serious organic brain disease.

The EEG is routinely used to diagnose different types of epilepsy and to localize the area in the brain causing the epilepsy. In addition, it is used to localize brain tumors and diagnose certain types of psychopathic disorders.

Like the EEG, the EMG is a summation of cellular activity; however, the EMG is primarily a result of the firing of muscle cells. If many muscle cells fire simultaneously (as happens when a muscle contracts), the electrical potentials generated near the surface of the skin will reflect that muscular activity. In practice, a very fine electrode is inserted into the muscle to stimulate a specific region. A second electrode or set of electrodes can then detect the resultant muscle activity. The EMG, then, is used primarily to detect abnormalities in muscle excitation.

Classically the electrocardiogram (ECG) was believed to be a graphic tracing of the electric current produced by the myoneural activity associated with heart muscle excitation. The normal ECG, it was believed, showed deflections resulting from atrial and ventricular activity. The first signal, P, is due to atrial excitation with the QRS deflections arising from ventricular activity. T waves are believed to be due to ventricular recovery (repolarization) while the U waves are seen in the normal ECG and are accentuated in hypokalaemia (low potassium levels in the blood).

More recently[31] reports have indicated that measureable electrokinetic potentials are present in rats and humans suggesting that the ECG are, at least in part, due to streaming potentials.

Careful analysis of the ECG can quite often give the physician or scientist a noninvasive handle on aberrant cardiac functioning. Right and left ventricular hypertrophy, myocardial infarction, as well as intraventricular conduction abnormalities are identifiable from their characteristically aberrant ECG. In addition, the diagnosis of the exact nature of an arrhythmia almost always is dependent upon careful consideration of the electrocardiogram.

3. Electrobiology

It is generally accepted that living cells generate and are modified by electrical currents that cross their boundary cellular membranes. These electrical currents, due to the movement of ions and the asymmetric nature of the cellular membranes, create an imbalance of electric charge across the membrane. This imbalance sets up a potential difference known by bioelectrochemists as the transmembrane potential or, more exactly, the intracellular potential, as referenced to the surrounding bulk medium. In addition to the transmembrane potential, an interfacial double layer potential of different absolute magnitude is established on either side of the membrane. This occurs at the membrane because of its nature as an interface between two fluids of differing ionic and macromolecular composition. Unfortunately, the absolute magnitude of these double layer potentials, although of major significance to ionic transport across the membrane, are not physically measurable. (However, changes in their magnitude can be measured.)

In recent years, another facet of the double layer potential has also become of interest. This aspect deals with the interfacial surface charge between living tissue and the bodily fluid in which it is bathed. Movement of the fluid relative to the tissue or vice versa sets up electrokinetic potentials and currents that play a significant role in the compatibility of biomaterials, medical diagnosis, and overall homeostasis.

A third example of electrochemistry in the biological sciences deals with the effects of electric fields on microorganisms. Such field effects range from lethality to simple control of mobility.

3.1. Stimulation of Nonexcitable Cells

It is generally acknowledged that electric fields play a dominant role in cellular chemical actions. Such electric fields arise due to the presence of membranes that separate internal cell electrolytes from external bathing solutions. Differing double layer potentials are established on both sides of the membrane; membrane asymmetries set up ionic charge separation potentials; membrane components set up concentration difference potentials; last, but not least, externally applied fields can be used to modify cellular potentials.

There appear to be several mechanisms by which electric fields modify cell actions. First, the transport rate of ions across cellular membranes appears to be a major factor affected by transmembrane potential (i.e., the potential difference measured between the cell interior and exterior).[32-34] Second, the magnitude of the transmembrane potential appears to be involved in cellular division, with low potentials < 20 mV (interior negative) favoring growth and > 40 mV halting growth. Third, electric fields cause electrophoretic migration of charged species in cell membranes, altering the charge distribution at the membrane.[35] Such alterations can cause local variations in ion transport across cellular membranes. These local variations have been noted to cause variations in cellular morphology, growth of "dendrites," pseudopods, and so on.

The specialty field of membrane ion transport studies is growing rapidly. It is, however, not the intent of this brief discussion to do more than touch on the subject of mechanisms by which electric fields control membrane transport. Rather, emphasis has been placed on describing the effects of electric fields on cells. At present, much of the research in this experimental field is Edisonian. However, the large body of experimental data that has accumulated strongly supports the hypothesis that ionic gradients are critically involved in the control of cell processes. That this is true can best be illustrated by the practical medical uses of electric fields to control or inhibit growth of tissue and bone.

A prominent example of the application of electrochemistry in medicine is as a technique to stimulate bone growth.[36-43] There is some controversy over the type of stimulation that is optimal for growth, i.e., direct application of current via implanted electrodes or induced currents via pulsating electromagnetic or electrostatic fields. Both approaches have been successfully applied.

When implantable electrodes are used for direct current stimulation of bone growth, current density levels of 0.1–0.5 mA/cm^2 of geometric electrode area appear to be effective. Above the 0.5 mA/cm^2 level, cell damage occurs. Below 0.1 mA/cm^2, stimulation is ineffective. Unfortunately, little if anything has been reported on the actual fields, at these current levels, that individual cells see or the actual portion of the current that passes through the cell compared to that which flows in the medium surrounding the cells.

It has been well-established, however, that growth is stimulated in the region of the cathode when d.c. currents are used. Necrosis (death) of cellular material occurs in the vicinity of the anode. Therefore, care must be taken to keep anodic current densities down to the microamp per square centimeter level in order to minimize damage. This is accomplished typically by using large-area electrodes placed on the epidermis. For long-term stimulation, the anode position is also shifted at intervals to permit healing of damaged tissue. In most cases, damage is due to anodic reaction products.

Induced currents, using electromagnetic or electrostatic fields, bypass the problem of having to implant electrodes and eliminate necrosis of tissue by anodic currents. However, inducing currents in tissue requires the use of pulsating fields.

Until recently, it was believed that symmetrical alternating fields were generally ineffective in stimulating cellular activity, presumably due to field reversal averaging out to zero cellular activity. Therefore, electronic manipulation of such fields was desired to achieve asymmetric fields in which the polarity, magnitude, and time of one-half of the cycle, e.g., the positive halves, would stimulate cellular activity, while the negative portion is essentially inactive cellularly. This can be accomplished by reducing the magnitude of the negative portion below the stimulating threshold level, while maintaining this subthreshold level for a period of time such that the overall energy levels of both polarities are equal. Thus, although both positive and negative halves of the cycle are energetically equal, the effect on cellular stimulation is such that only the positive half-cycle is effective.

Electronic manipulation, such as described, is not difficult. However, it has been found that pulsating fields act on cells in a more complex manner than direct current, i.e., steady state fields. Cellular interaction is dependent upon pulse width and frequency as well as amplitude. Thus, so-called "windows" have been described in which imposed pulsatile fields are effective for cellular stimulation, as well as inhibition. In general, pulsatile frequencies greater than approximately 500 Hz are not effective. For bone growth, frequencies of <100 Hz are used, with different investigators suggesting 5–20 Hz, 30–40 Hz, or 70–80 Hz as optimal for growth stimulation.

Other medical areas that have successfully utilized electric fields are wound healing[44-48] and anesthesiology.[49-52] For wound healing, fields similar to those described above in the bone healing discussion are utilized. Again, electrical currents appear to stimulate cell growth at a rate above the normal healing rate.

The application of electrochemistry to anesthesiology is one of the oldest applications of electricity, dating back to the mid-18th century. Unfortunately, practioners of this form of medicine tended to be either naive or charlatans. As a result, the use of electrochemical stimulation to control pain became a medical joke, promoted only by quacks. Recently this field is again receiving attention, partially due to the stimulus of research into the Oriental art of acupuncture.

There are a number of terms used to describe the application of electric currents to relieve pain, such as electroanalgesia, electroanesthesiology, electronarcosis, electrosleep, electroacupuncture, anodal electronics, etc. Electrical stimulation is accomplished both by transcutaneous insertion and topical application of electrodes. Both pulsatile and steady state currents are deemed useful.

The mode of action of electric currents in the relief of pain is not well-established. Two theories have been suggested to account for the results noted. One theory states that extended low-level stimulation of nerves tends to block their response to pain stimuli. The other theory states that low-level stimulation of nerves works by stimulating the release of opiates from the brain or hormonal system. Perhaps both theories are correct.

3.2. In Vivo Electrokinetic Potentials

There are two types of electrokinetic phenomena, namely those in which an electric potential is generated by the mechanical notion of a surface in a liquid and those in which a particle or liquid is caused to move by an electric potential. Classically there are four major effects, i.e., streaming potentials, sedimentation potentials, electrophoresis, and electroosmosis. (There are also several secondary effects that have been noted more recently such as acoustic potentials,[53] K effect potentials,[54] and U effect potentials.[55]) Each of the major effects has found a niche in physical chemistry, analytical chemistry, and chemical engineering. However, the application and understanding of electrokinetic phenomena in the biological sciences has been very spotty.

Surface charge of biomaterials apparently plays a large role in the clotting of blood and the formation of thrombi in blood vessels. It has been known for over 150 years that blood coagulates on a positive electrode but not on a negative electrode. More recent experimentation has shown that metallic materials appear to have antithrombogenic characteristics that follow an electromotive series.[56] Materials positive to the normal hydrogen electrode (NHE) cause the formation of blood clots. Conversely, materials negative to the NHE do not. (However, one must remember that highly negative materials may be chemically toxic even though not thrombus formers.)

Selection of biomaterials for circulatory assist devices is a task made difficult by the complexity of the interfacial interactions between blood and the solid surfaces of candidate materials. Among the major factors contributing to the complexity are surface charge, surface energy, surface roughness, chemical reactivity, adsorptivity, hydrophobicity, etc. To further aggravate the complexity, the factors noted interact with each other in ways that are not as yet understood or well-defined.

Of the factors listed, surface charge and surface roughness can be related to and measured by electrokinetic phenomena. This is not to say, however, that by measuring electrokinetic parameters one can predict the biocompatibility of materials. Rather, electrokinetics can provide some of the information needed to predict *in vivo* behavior of materials.

Streaming potential measurements are probably the most useful of the electrokinetic techniques in evaluating the surface charge of biomaterials. For such measurements, care must be taken to ensure that artifacts such as the moto-electric effect and pressure-induced potentials do not interfere with test results. The moto-electric effect is a little known but significant artifact caused by the disturbance of the double layer around metallic electrodes immersed in a moving electrolyte. In general, it is highly desirable to keep metallic electrodes used for sensing streaming potentials out of the moving electrolyte. This is sometimes accomplished by the use of salt bridges or porous plugs. Unfortunately such devices can lead to pressure-induced potentials. Such potentials are essentially streaming or piezoelectric potentials caused by a pressure differential. In the case of salt bridges or porous plugs, if a pressure differential exists across the bridge or plug, a flow of ions into or out of the bridge or plug generates a streaming potential that may be additive to or subtractive from the measurement desired. Piezoelectric effects are typically encountered when dealing with flexible polymers. Such materials tend to expand when pressurized, setting up a charge separation. This charge separation is detected by the measuring electrodes, introducing an error.

If care is taken, streaming potentials will provide a measure of the surface charge of the biomaterial of interest. One can of course relate streaming potentials to zeta potentials by the use of the appropriate streaming potential equation. However, care must be taken in doing this since there is a growing quantity of evidence that the zeta potential is a function of flow. That is, zeta potential changes with flow velocity. (This change in zeta potential with flow has been called the K effect potential.) If zeta potential does change with flow, other factors in the electrokinetic potential equations must also change, since it is well-established that streaming potential is a linear function of all of the factors that go into its defining equations. The factors most likely to change with flow are the dielectric constant and conductivity of the electrolyte in the double layer.

Another aspect of electrokinetics in the biological sciences deals with the generation of potentials due to the flow of blood *in vivo*. It was recently demonstrated[57] that the ECG is due, at least partially, to electrokinetic rather than myoneural potentials. It was also shown that electrokinetic potentials are

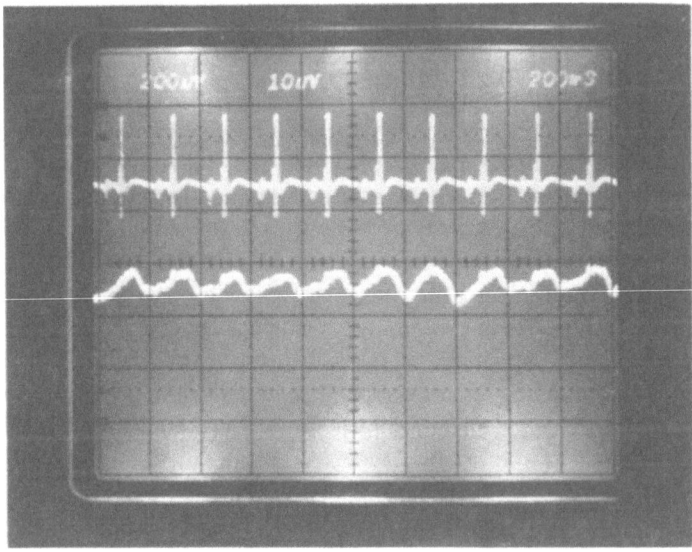

Figure 12. Electrokinetic arterial potentials. Upper trace: Lead I ECG; lower trace: arterial pulse wave measured on the skin of a rat over the right femoral artery.

generated in the arterial system and that these potentials may form a basis for the detection of atherosclerotic lesions. Figure 12 illustrates the type of electrokinetic arterial potentials generated *in vivo*.

3.3. Microbial Electrochemistry

Microorganisms, particularly bacteria (prokaryotes), provide a unique "laboratory" for conducting experiments designed to uncover electro-chemical/biological interactions in living entities. Although bacteria are by no means simple experimental models, they are generally less complex than mammalian or plant cells (eukaryotes), which contain a nucleus, mitochondria, and other organelles. The relative abundance of bacteria and their ease of handling lend them readily to experimentation. One major drawback in electrochemical experimentation with bacteria, however, is their small size (diameters typically less than 5 μm). This makes it extremely difficult to use even the smallest of microelectrodes for potential measurements. There are, however, indirect techniques for the measurement of potentials and ionic concentrations, such as the use of dyes sensitive to ion and oxidation–reduction potential.[58] These indicators have been shown to be useful in evaluating the ionic behavior of cells. One of the major research uses of bacteria deals with the transport of ions and molecules across their membranes. Other areas of microbial electrochemistry include electrochemical techniques for causing their demise in water purification systems and their use in fuel cells.

Much of the prevailing theory of how cells make adenosine triphosphate (ATP), the energy storehouse of living cells, was developed from experimental observations of bacteria. This theory, called the chemiosmotic hypothesis by its chief proponent, P. Mitchell,[59] emphasizes the cause and effect of proton transport across cell membranes. The chemiosmotic hypothesis is presented in Figure 13, and may be summarized as follows[60]:

> Electrons and protons are donated to flavin adenine dinucleotide (FADH$_2$) which exports the 2 protons; the electrons are returned to the inner surface through iron–sulphur proteins (FeS). The electrons, together with 2 protons from the internal medium then reduce a single molecule of ubiquinone to hydroquinone (QH$_2$), which diffuses across the membrane and releases protons outside. Finally, the electrons proceed through cytochromes *b* and *o* to water.

Also shown in Figure 13 are pathways through the bacterial bilipid layer by which protons assist or are transported in conjunction with amino acids, a carbon source (lactose), and other ions such as Na^+, Mg^{+2}, Ca^{+2}, etc. In the case of lactose transport, it has been shown that one proton is transferred per molecule of lactose. Interestingly, this 1:1 ratio remains constant over a pH range of 6–8 in the external medium. Apparently, the bacterium compensates for pH changes by changing its transmembrane potential and/or Na^+ concentration.

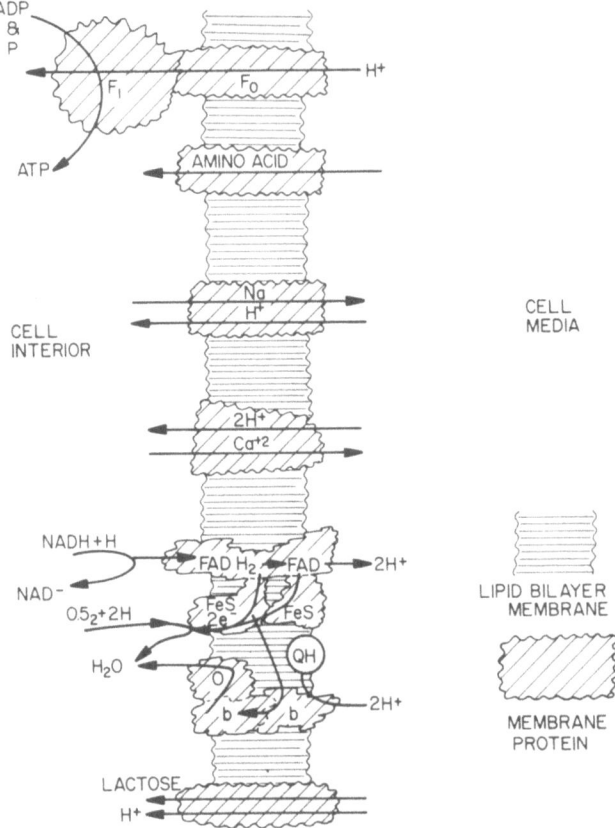

Figure 13. Model of bioenergetic mechanism of *Escherichia coli.*

Unlike eukaryotes, which maintain low Na^+ concentrations within their intercellular media by ATP-driven Na^+ "pumps," prokaryotes possess no such Na^+ extrusion mechanism. The mechanism by which bacteria control sodium, according to the chemiosmotic theory, is a sodium–proton exchange. This has been verified by a number of investigators. Why bacteria require low internal Na^+ concentrations is not readily apparent, however. It is presumed that the Na^+ gradient represents an energy storage system as well as a means of compensating for environmental pH changes.

The interplay of H^+, K^+, Na^+, and the transmembrane potential $\Delta\Psi$ and their effect on membrane transport is a complex phenomenon. Which factors are the independent variables in a transport process and which are the dependent variables is difficult to determine. However, it is well-established that transport of many amino acids, sugars, and other carbon and nitrogen sources requires the co-transport (symport) of an ion along with it and probably the reverse transport (antiport) of a similarly charged ion.

The application of electrochemical technology to control microbiological growth has been described in the patent literature for many decades. Although it is well-established[61,62] that passing of electrical currents milliamp/cm^2 level between two electrodes immersed in a liquid containing microorganisms will result in the death of those organisms, the question arises as to the mechanism(s) of the biocidal action of this current passage, and how to optimize the effect at minimal cost.

There are a number of mechanisms by which cellular death might occur, depending upon the level of current, type of current (a.c., d.c., r.f., pulsatile, etc.), electrolyte type, and concentration and type of electrodes (inert or electroactive). Among the mechanisms purported to occur are (a) generation of reactive chemical species at the electrodes, such as Cl_2, $HOCl$, I_2, Ag^+, etc., (b) oxidation/reduction of the microorganisms at the anode or cathode surface, (c) bipolar plate effect, with high ionic current flow through the microorganism, and (d) thermal heating at high current or radio freqency levels.

Bacteria and certain other biological agents can be utilized in three basic types of fuel cells: a product cell, depolarized cell, and regenerative cell.[63] In the first type, bacteria are used to produce an electroactive species that can be oxidized or reduced at an appropriate electrode. In the second type, bacteria are utilized to consume one or more of the electrode products, thus acting as a depolarizer. Lastly, bacteria are utilized to regenerate an electroactive species, by combining the first and second types.

Examples of bacterial fuel cells are presented in Table 3. Of the fuel cells noted, the production of hydrogen as an end product of bacterial metabolism appears to be the most useful biological mechanism for use in electrochemical energy conversion.

Table 3
Bacterial Fuel Cells

Bacterial type	Type of cell	Comments
Escherichia coli	Regenerative	Ferricyanide reduced to ferrocyanide as bacteria oxidize glucose[a]
Aerobacter rancens	Regenerative	Same as above as bacteria oxidize ethanol[a]
Micrococcus cerificans	Product	Products oxidized are bacterial metabolites of n-hexadecane[b]
Escherichia coli	Product	Bacteria oxidize glucose[c]
Clostridia butyricum	Product	Hydrogen, produced from fermentation of glucose, is oxidized[d]
Rhodospirillum rubrium	Product	H_2, produced from fermentation of glucose, is oxidized[d]
Clostridia butyricum	Product	H_2, produced from various waste products, is oxidized[d]

[a] Reference (63).
[b] Reference (64).
[c] Reference (65).
[d] Reference (66).

References

1. R. C. Thomas, *Ion-Sensitive Intracellular Microelectrodes*, Academic Press, New York (1978).
2. L. A. Geddes, *Electrodes and the Measurement of Bioelectric Events*, Wiley-Interscience, New York (1972).
3. R. F. Lane and A. T. Hubbard, *Anal. chem.* **48**, 1287 (1976).
4. R. M. Wightman, E. R. Strope, P. Plotsky, and R. N. Adams, *Brain Research* **159**, 55 (1978).
5. J. C. Conti, E. R. Strope, R. N. Adams, and C. A. Marsden, *Life Sci.* **23**, 2705 (1978).
6. R. F. Lane, A. T. Hubbard, and C. D. Blaha, *J. Electroanal. Chem.* **95**, 117 (1979).
7. K. J. Laidler and P. S. Bunting, *The Chemical Kinetics of Enzyme Action*, Clarendon Press, Oxford (1973), Chapter 12.
8. S. J. Updike and G. P. Hicks, *Nature* **214**, 986 (1967).
9. G. P. Hicks and S. J. Updike, U.S. Patent 3, 542, 662, Nov. 24, 1967.
10. G. A. Rechnitz, *Research and Development*, August, 1973.
11. R. E. Taylor, W. P. Dill, and E. L. Jeffers, *Method and Automated Apparatus for Detecting Coliform Organisms*, NTIS Document N78-22588.
12. J. R. Wilken, G. F. Stoner, E. H. Boykin, *Med. Microbiology* **27**, 949 (1974).
13. A. Ur and D. F. K. Brown, *Biomed. Eng.*, Jan., 1974, p. 18.
14. G. A. Rechnitz and R. K. Kobos, *Chem. Eng. News*, Oct. 25, 1976, p. 23.
15. G. A. Rechnitz, T. L. Reichel, R. K. Kobos, and M. E. Meyerhoff, *Science* **199**, 440 (1978).
16. S. Suzuki and I. Karube, *Annals N.Y. Acad. Sci.* **326**, 255 (1979).
17. G. Durand and J. M. Navarro, *Process Biochem.* Sept., 1978, p. 14.
18. R. F. Lane, A. T. Hubbard, and C. D. Blaha, *Bioelectrochem. Bioenerg.* **5**, 506 (1978).
19. J. C. Conti, *The Development and Application of Micro Analytical Techniques and Devices*, Ph.D. Thesis, Univ. of Kansas (1978).
20. R. M. Wightman, E. R. Strope, P. M. Plotsky, and R. N. Adams, *Nature* **262**, 145 (1976).
21. R. N. Adams, J. C. Conti, C. A. Marsden, and E. R. Strope, *Br. J. Pharmacol.* **64**, 470 (1978).
22. E. R. Strope, "Electroanalytical Techniques in the Study of Neurotransmitters and Neuronal Pathways in Small Animal Central Nervous Systems." Ph.D. Thesis, Univ. of Kansas (1978).
23. B. N. Ames, F. D. Lee, and W. E. Durston, *Proc. Nat. Acad Sci. U.S.A.* **70**, 782 (1973).
24. B. N. Ames, J. McCann, and E. Yamasaki, *Mutation Research* **31**, 347 (1975).
25. D. J. Boyles, *Biotechnol. Bioeng.* **20**, 1101 (1978).
26. Orion Model 99–20 and Model 55–20 electrodes.
27. L. C. Clark, Jr. and G. A. Misrahy, XXth International Physiological Congress, Belgium, Abstract No. 177 (1956).
28. G. A. Misrahy and L. C. Clark, Jr., XXth International Physiological Congress, Belgium, Abstract No. 650 (1956).
29. P. L. Frommer, W. W. Pfaff, and E. Braunwald, *Circulation* **24**, 1227 (1961).
30. D. C. Spray, A. L. Harris, and M. V. L. Bennett, *Science* **204**, 432 (1979).
31. R. J. Kurtz, E. Findl, P. O. Milch, and C. B. Moore, *Abstracts J. Am. Assoc. Med. Instr.*, in preparation, 14th Annual Meeting, p. 155, Las Vegas, May 1979.
32. C. D. Cone, Jr., *J. Theor. Biol.* **30**, 183 (1971).
33. J. L. Howland, *Cell Physiology*, p. 277, Macmillan Co., New York (1973), p. 277.
34. P. C. Hinkles and R. E. McCarty, *Sci. Am.* **238**, 104 (1978).
35. L. F. Jaffe, *Nature* **265**, 600 (1977).
36. F. Burney, E. Herbst, and M. Hinsenkamp, *Electric Stimulation of Bone Growth and Repair*, Springer-Verlag, New York (1978).
37. R. O. Becker and A. A. Pilla, in *Modern Aspects of Electrochemistry*, J. O'M. Bockris, ed., Vol. 10, Plenum Press, New York (1975), p. 289.
38. Z. B. Friedenberg, E. T. Andrews, B. I. Smolenski, B. W. Pearl, and C. T. Brighton, *Surgery Gyn. and Obst.* **11**, 894 (1970).
39. L. S. Lavine and I. Lustren, *Science* **175**, 1118 (1972).

40. G. V. B. Cochran, *Ann. N.Y. Acad. Sci.* **48**, 899 (1972).
41. D. P. Levy and B. Rubin, *Clin. Orthopaedics* **88**, 218 (1971).
42. J. P. Jacobs and L. A. Norton, *J. Periodintol.* June, 311 (1976).
43. J. A. Spadaro, *Bioelectrochem. Bioenerg.* **5**, 232 (1978).
44. R. Dueland, R. E. Hoffer, W. A. Seleen, and R. O. Becker, *Cörnell Vet.* **68**, 51 (1978).
45. B. A. Rowley, J. M. McKenna, and L. E. Wolcott, *Biomed. Sci. Instr.* **10**, 111 (1974).
46. D. H. Wilson, *Brit. Med. J.* **2**, 269 (1972).
47. L. C. Carey and D. Lepley, Jr., *Surgical Forum* **13**, 33 (1962).
48. D. Assimacopoulos, *Am. J. Surgery* **115**, 683 (1968).
49. C. J. Woolf *et. al.*, *S.A. Med. J.* **53**, 179 (1978).
50. C. R. Chapman and C. Benedette, *Life Sci.* **21**, 1645 (1977).
51. R. H. Smith, T. Kano, G. S. M. Cowan, and R. E. Barber, *Anesth. Analg.* **56**, 678 (1977).
52. R. S. Hertert and D. E. Cutright, *Milit. Med.* **142**, 929 (1977).
53. E. Yeager and F. Hovorka, *J. Acoustical Soc. America* **25**, 445 (1953).
54. R. J. Kurtz, E. Findl, A. Kurtz, and L. Stormo, *J. Coll. Interface Sci.* **57**, 28 (1976).
55. S. Ueda, A. Watanabe, and F. Tsuji, *J. Electrochem. Soç. Japan* **19**, 142 (1951).
56. P. N. Sawyer and S. Srinivasan, *Bull. N.Y. Acad. Medicine* **48**, 235 (1972).
57. E. Findl and R. J. Kurtz, in *Electrochemical Studies of Biological Systems*, D. T. Sawyer, ed., ACS Symposium Series (1977), p. 180.
58. F. Hartline, *Science* **203**, 992 (1979).
59. P. Mitchell, *Biol. Rev.* **41**, 445 (1966).
60. P. C. Hinkle and R. E. McCarty, *Sci. Am.* **238**, 104 (1978).
61. G. E. Stoner, G. L. Cahen, and J. Parcells, *J. Electrochem. Soc.* **122**, 106C, 109C (1975).
62. G. E. Stoner, U.S. Pat. 3,725,226.
63. T. G. Young, in *Biological Aspects of Electrochemistry*, G. Milazzo, ed., Birkhauser Verlag, Basel (1971), p. 675.
64. H. A. Videla, in *Biological Aspects of Electrochemistry*, G. Milazzo, ed., Berkhauser Verlag, Basel (1971), p. 667.
65. J. B. Davy and H. F. Yarbrough, Jr., *Science* **137**, 615 (1962).
66. S. Suzuke, I. Karube, and T. Matsunaga, *Biotech. Bioeng. Symp.* # 8, J. Wiley & Sons, New York (1978), p. 501.

Index

531

Errata

In Chapter 3 of Volume 5, errors were inadvertently introduced into several equations. The corrected equations appear below:

$$\sigma_2(a) = 2E'_2\left(Q_h^{\text{rel}}(b) + Q(b) - (k_h + k_e)\ln\frac{2\kappa q}{c^{1/2}} \right) \tag{192}$$

$$\alpha = 0.8204 \times 10^6 z^3 (DT)^{3/2} \tag{194}$$

$$\beta = 82.501 \times z^2/\eta (DT)^{1/2} \tag{195}$$

$$E'_1 = 2.9422 \times 10^{12} \times z^6/(DT)^3 \tag{196}$$

$$E'_2 = 0.43329 \times 10^8 \times z^5/\eta (DT)^2 \tag{197}$$

$$b = 16.708 \times 10^{-4} \times z^2/aDT \tag{198}$$

$$\frac{\kappa q}{c^{1/2}} = 4.20155 \times 10^6 \times z/(DT)^{3/2} \tag{199}$$

$$\Lambda = \gamma(\Lambda_0 - Sc^{1/2}\gamma^{1/2} + Ec\gamma\log c\gamma + J(R)c\gamma + J_{3/2}(R)c^{3/2}\gamma^{3/2}) \tag{203}$$

$$Q_{e1}(b) = 2\left(E_i(-b) - \frac{e^{-b}}{b^2} - \frac{e^{-b}}{b} \right) + 2\left(E_i(b) - \frac{e^b}{b^2} - \frac{e^b}{b} \right) \tag{270}$$

$$\nabla \cdot \mathbf{v} = 0 \tag{317}$$

$$z(r) = -r^4 \Big/ \left[\frac{|ze|^2}{16\pi\eta}\left(\frac{\varepsilon_0 - \varepsilon_\infty}{\varepsilon_0^2} \right)\tau_s \right] \tag{318}$$

$$\frac{1}{\omega_i} = 6\pi\eta R_i + \frac{17}{280}\frac{|ze|^2}{R_i^3}\frac{\varepsilon_0 - \varepsilon_\infty}{\varepsilon_0^2}\tau_s + O(z_i^{-2}) \tag{320}$$

$$\frac{1}{\omega_i} = 4\pi\eta R_i + \frac{1}{15}\frac{|ze|^2}{R_i^3}\frac{\varepsilon_0 - \varepsilon_\infty}{\varepsilon_0^2}\tau_s + O(z_i^{-2}) \tag{321}$$

$$\frac{1}{\omega}(\text{point}) = 15.624\eta\left[\frac{|ze|^2}{16\pi\eta}\frac{\varepsilon - \varepsilon_\infty}{\varepsilon_0^2}\tau_s \right]^{1/4} \tag{322}$$

$$\nu_{ji} = \frac{1}{Z_i}\frac{\lambda_i/Z_i - \lambda_j/Z_j}{\lambda_i^0/Z_i^2 + \lambda_j^0/Z_j^2} \tag{323}$$